Michael J. Pelczar
1976

INDUSTRIAL MICROBIOLOGY

INDUSTRIAL MICROBIOLOGY

Brinton M. Miller, Ph.D.
Director, Animal Infections
Merck Institute for Therapeutic Research

Warren Litsky, Ph.D.
Commonwealth Professor of Environmental Science
University of Massachusetts

McGraw-Hill Book Company

New York St. Louis San Francisco Auckland Düsseldorf Johannesburg
Kuala Lumpur London Mexico Montreal New Delhi Panama
Paris São Paulo Singapore Sydney Tokyo Toronto

INDUSTRIAL MICROBIOLOGY

Copyright © 1976 by McGraw-Hill, Inc. All rights reserved.
Printed in the United States of America. No part of this publication
may be reproduced, stored in a retrieval system, or transmitted, in any
form or by any means, electronic, mechanical, photocopying, recording, or
otherwise, without the prior written permission of the publisher.

1 2 3 4 5 6 7 8 9 0 KPKP 7 8 3 2 1 0 9 8 7 6

This book was set in Primer by Black Dot, Inc.
The editors were William J. Willey and Anne T. Vinnicombe;
the designer was Jo Jones;
the production supervisor was Robert C. Pedersen.
The drawings were done by J & R Services, Inc.
Kingsport Press, Inc., was printer and binder.

Library of Congress Cataloging in Publication Data
Main entry under title:

Industrial microbiology.

    1. Industrial microbiology.   I.   Miller, Brinton
Marshall, date   II.   Litsky, Warren, date
QR53.I52   660'.62   76-10207
ISBN 0-07-042142-0

CONTENTS

List of Contributors   vii

Preface   xi

1
EXPERIMENTAL DESIGN IN APPLIED MICROBIOLOGY
H. Boyd Woodruff   1

2
ANALYTICAL MICROBIOLOGY   Frederick Kavanagh   13

3
VITAMINS   Helen R. Skeggs   47

4
ANTIBIOTICS   John N. Porter   60

5
MICROBIOLOGY OF STEROIDS   Herbert C. Murray   79

6
MICROBIAL SYNTHESIS OF AMINO ACIDS   Donald R. Daoust   106

7
MICROBIAL ENZYMES   Leland A. Underkofler   128

8
ALCOHOLIC BEVERAGES AND FERMENTED FOODS
Gerhard J. Haas   165

9
THE ROLE OF MICROBIAL GENETICS IN INDUSTRIAL MICROBIOLOGY
Richard P. Elander and Marlin A. Espenshade   192

10
FOOD MICROBIOLOGY   John H. Litchfield   257

11
MICROBIOLOGICAL DETERIORATION OF PULPWOOD, PAPER, AND
PAINT   Richard T. Ross and C. George Hollis   309

12
POLLUTANTS AND AQUATIC ECOSYSTEMS: BIOLOGICAL ASPECTS OF
WATER QUALITY PROBLEMS   Robert A. Coler and Warren Litsky   355

13
PETROLEUM MICROBIOLOGY Robert D. Schwartz and
William W. Leathen   384

14
ELEMENTS OF HEAT AND GASEOUS STERILIZATION
Charles R. Stumbo   412

Indexes   451
Species Index
Subject Index

LIST OF CONTRIBUTORS

Robert A. Coler, Ph.D.
University of Massachusetts, Amherst, Mass.

Dr. Coler's primary research contribution is in riverine ecosystems and the assessment of the benthic macroinvertebrate community response to stress. His publications range from pesticide studies to urban runoff and heavy-metal toxicity. Presently, he is investigating chronic and subacute effects of ozone applications on selected fish and benthic species.

Donald R. Daoust, Ph.D.
Armour Pharmaceutical Co., Kankakee, Ill.

Dr. Daoust has been an active investigator of the microbial synthesis of amino acids. He also developed assay techniques used to screen and evaluate potential anthelmintic products which led to the first known discovery of the microbial synthesis of physostigmine.

Richard P. Elander, Ph.D.
Bristol-Myers Co., Syracuse, N.Y.

Dr. Elander's research has been concerned with microbial genetics and industrial applications of it. He has directed such projects as (1) induced mutations and improved productivity mutants in the penicillin, cephalosporin, erythromycin, tylosin, and pyrrolinitrin fermentations and (2) mutants requiring penicillin and cephalosporin precursor amino acids analogs.

Marlin A. Espenshade, Ph.D.
W. R. Grace Company, Clarksville, Md.

Dr. Espenshade's major efforts have been in screening programs for antitumor compounds, antifungal agents, and uses for microbial metabolites in the chemical industry. He is responsible for maintaining a culture collection of industrially important organisms.

Gerhard J. Haas, Ph.D.
General Foods, Tarrytown, N.Y.

Dr. Haas has been engaged in investigations relating to the development of resistance to polymyxin by gram-negative organisms, effect of antibiotics on organisms important in brewing, application of enzymes in foods, and microbial control in breweries.

He holds numerous patents including riboflavin derivatives and a process of preparing them, aqueous solutions of lipoid-soluble vitamins, process of improving the properties of fermented malt beverages, and process of preventing freezing precipitation in malt beverages and products.

C. George Hollis, Ph.D.
Department of Biology, Memphis State University, Memphis, Tenn.

Dr. Hollis, in his previous position at Buckman Laboratories, Inc., was engaged in the prevention of biodeterioration in the pulp and paper industry.

Frederick Kavanagh, Ph.D.
Formerly of Eli Lilly and Co., Inc., Indianapolis, Ind.

Dr. Kavanagh was editor of "Analytical Microbiology," vol. I, 1963, and vol. II, 1972 (Academic Press, Inc., New York). He has also published extensively on topics including vitamin requirements of fungi, antibiotics from basidiomycetes and actinomycetes, use of fluorescence as an analytical tool, and automation of analytical microbiological procedures.

William W. Leathen, M.S.
Gulf Oil Corporation, Pittsburgh, Pa.

Mr. Leathen has published extensively in the field of hydrocarbon microbiology, specializing early in the effect of induced sulfuric acid production from the ferrous ion of bituminous sludge by *Thiobacillus ferrooxidans* (originally isolated and characterized by Leathen as *Ferrobacillus ferrooxidans*). Later, he reported on microbial degradation of petroleum products, especially jet fuels, and the production of single-cell protein from hydrocarbons; he holds five patents in these areas.

John H. Litchfield, Ph.D.
Battell Memorial Institute, Columbus, Ohio

Dr. Litchfield is the coauthor of "Food Plant Sanitation" (Reinhold Publishing Corp., New York, 1962), and a contributor to "Microbial Technology" (Reinhold Publishing Corp., New York, 1967) and "Single Cell Protein" (The M.I.T. Press, Cambridge, Mass., 1968). He is an active investigator in the fields of food science and technology and of applied microbiology.

Warren Litsky, Ph.D.
University of Massachusetts, Amherst, Mass.

Dr. Litsky has published nearly 200 papers, primarily in the areas of sterilization, disinfection, control of microbes in various environments, and the isolation and enumeration of indicator organisms of pollution. Recently he reported a fluorescent antibody technique for the rapid identification of fecal streptococci. His laboratory was the first to demonstrate the feasibility of "come-up time" pasteurization for milk and other dairy products and holds a patent on the use of this method for the production of virus vaccines. He is on the editorial board of *Applied Microbiology, Journal of Environmental Health,* and *Health Laboratory Science.*

Brinton M. Miller, Ph.D.
Basic Animal Science Research, Merck Institute for Therapeutic Research, Rahway, N.J.

Dr. Miller holds patents on the discoveries of (1) thermomycin, a broad-spectrum antibiotic produced by a thermophilic streptomycete, (2) bottromycin as an antimycoplasmal agent, and (3) several coccidiostats; he also participated in the discovery of MK436, an anti-*Trypanosoma cruzi* (Chagas' disease) compound. He is past president of the Society for Industrial Microbiology, present treasurer of the American Society for Microbiology, and cochairman of Biotech Advisory Council, AIBS/NSF.

Herbert C. Murray, Ph.D.
Upjohn Company, Kalamazoo, Mich.

Dr. Murray has been an active investigator in the area of selection and use of microorganisms in corticosteroid syntheses. He holds several patents concerning the use of microorganisms for organic synthesis.

John N. Porter, Ph.D.
Formerly of Lederle Laboratories, Division of American Cyanamid Co., Pearl River, N.Y.

Dr. Porter has participated in antibiotic research since 1943. He discovered puromycin, a biochemically useful antibiotic. He is a specialist in the various species and genera of Actinomycetes. His most significant recent publications are related to the prevalence and distribution of antibiotic-producing actinomycetes and cultural conditions for antibiotic-producing microorganisms.

Richard T. Ross, Ph.D.
Buckman Laboratories, Inc., Memphis, Tenn.

For the past 20 years Dr. Ross has been involved in the study of biological deterioration of industrial products. Specifically, he has been engaged in research on the control of microflora that disfigures and deteriorates paints, plastics, and leather and in the development of test methods to evaluate chemical agents used to inhibit these microorganisms. He holds two patents in this area.

Robert D. Schwartz, Ph.D.
Exxon Research and Engineering Company, Linden, N.J.

Dr. Schwartz has specialized in hydrocarbon microbiology, especially as it pertains to the mechanism of incorporation of molecular O_2 by oxygenases into hydrocarbons and the factors responsible for microbial enzyme specificity. Recently he participated in the research resulting in the isolation of highly active hydrocarbon-oxidizing microorganisms that enzymatically epoxidate α-ω-dienes with a high degree of stereospecificity.

Helen R. Skeggs, B.A.
Merck Institute of Therapeutic Research, West Point, Pa.

Mrs. Skeggs has distinguished herself in the areas of microbial physiology, vitamin synthesis by bacteria, and animal nutrition. She has published extensively on methods for assaying vitamins and other compounds such as biotin, pantothenic acid, biocytin, orotic acid, B_{12}, strepogenin, and thiamine. Her name is synonymous with vitamin assay.

Charles R. Stumbo, Ph.D.
Formerly of University of Massachusetts, Amherst, Mass.

Prior to going to the University of Massachusetts in 1963, Dr. Stumbo held administrative research positions at Owens Illinois Glass Co., H.J. Heinz Co., and Producers Creamery Co. His primary interest is in the field of microbiological research relating to mathematical and engineering applications in the evaluation of preservation processes with regard to sterilization and enzyme degradation. He has served as associate editor of *Food Research* and *Food Technology*. He is also author of the book "Thermobacteriology in Food Processing" (Academic Press, Inc., New York, 1965) and hold two patents.

Leland A. Underkofler, Ph.D.
Formerly of Miles Laboratories, Inc., Elkhart, Ind.

Dr. Underkofler is author of a textbook "Introduction to Organic Chemistry" (D. Van Nostrand Company, New York, 1954) and coeditor of "Industrial Microbiology," two volumes (Chemical Publishing Company, New York, 1954).

He has published extensively on fermentation, microbiological, and enzyme chemistry. In 1953 he was awarded an honorary Doctor of Science degree from Nebraska Wesleyan University. He holds seven patents on fermentation processes and enzyme applications. He is a past president of the Society for Industrial Microbiology.

H. Boyd Woodruff, Ph.D.
Merck, Sharp & Dohme, Ltd. (Japan)

Dr. Woodruff has published extensively in the areas of antibiotic production. As early as 1940, he published the descriptions of actinomycin A and B. He holds patents for production processes of actinomycin, penicillin, streptomycin, streptothricin, subtilin, vitamin B_{12}, coenzyme Q, and dextranase. He was editor-in-chief of *Applied Microbiology* (a journal of the American Society for Microbiology) 1953–1962. He was also editor of "Scientific Contributions of Selman A. Waksman, Selected Articles Published in Honor of His 80th Birthday, 22 July, 1968" (Rutgers University Press, New Brunswick, N.J.).

PREFACE

The diversity and number of microorganisms which man has sought to control for economic gain or for prevention of economic loss—that is, use of microorganisms to make something we want or prevention of them from making something we do not want—has grown tremendously over the past thirty-five years. This statement formally and informally expresses our idea of what modern industrial microbiology is all about.

As will be noted when reading the chapters of this text, industrial microbiologists strive to learn how to manipulate microorganisms to do their bidding. The spectrum of areas in which they make these attempts is constantly increasing, apparently bound only by their abilities to explain the mysteries of life in the smallest of living cells. The utilization of microorganisms can be directly related to man's economic livelihood whether it be his person, his environment, his food, or even his pleasure.

Until the advent of antibiotics in the 1940s, industrial aspects of microbiological research were fairly limited. Even though fermented food and drink date to prerecorded time, it was not until Schwann's report on alcohol fermentation in 1837 and Pasteur's monumental explanation of lactic acid fermentation in 1857 that the science of microbiology was initiated. Most would agree that microbiology, influenced in large measure by these two scientists, developed into a science in the nineteenth century. However, the emphasis in that century and even into the twentieth century was primarily medical or veterinary. Until the 1940s industrial microbiology was essentially amplification of early alcoholic fermentation research and efforts to stabilize the production of fermented foods by the use of in vitro cultured microorganisms. Also, the first third of this century saw some research into the prevention of microbial deterioration, more often carried out by mycologists than bacteriologists.

The development of submerged fermentation methods for large-scale production of antibiotics and other microbial products aided greatly in establishing industrial microbiology as a permanent subdivision of the science of microbiology. The concomitant development of analytical microbiology was an absolute necessity for the evaluation of microbiological products' presence and activity. During these years of development two important features of applied research became evident. First, the scientific method, which had for generations been the basis of academic research, was found to be applicable and necessary for finding solutions to industrial research problems. Second, basic research (work on a problem where no work was previously done) not just developmental research (work starting from previously discovered facts) was done by industry. An example of basic research in industry was the discovery and production of vitamin B_{12}.

Industrial microbiologists have had to concentrate on how to frame a proper question, how to design their research to answer the question efficiently, and what the probable answers will direct them to next, all at or as near to the initiation of the research program as possible; we call this method of organization research management by objective. We asked Dr. H. Boyd Woodruff, one of the earliest industrial microbiologists in the field of antibiotics, in the first chapter to set the tone for this new type of text. The reader will note he emphasizes the scientific method as adapted for applied microbiology and the need to be able to solve both basic and developmental problems.

The next seven chapters, by Kavanagh, Skeggs, Porter, Murray, Daoust, Underkofler and Haas, describe in some detail selected areas of microbiology concerned with microbial processes which man controls for economic gain. The chapter on microbial genetics by Elander and Espenshade deals with application of the principles and practices of modern genetics to microbial processes. If these practices were not used, our utilization of microorganisms would be relegated to the status existing during the first third of this century. The last five chapters by Litchfield, Ross and Hollis, Litsky and Coler, Schwartz and Leathen, and Stumbo deal primarily with selected areas involving efforts to prevent microorganisms from causing economic losses. Paradoxically, studies in petroleum microbiology which were designed originally to find methods of controlling or preventing microbial growth have lead in recent years to a potentially important new high-quality source of food for man or his livestock; we mean production of single-cell protein in the form of microorganisms grown on petroleum by-products. However, if the student will return to Chapter 1 after reading about biosynthesis of single-cell protein, the illogical becomes logical because the petroleum microbiologists have followed an axiom Woodruff discusses—observation should be followed by application.

Obviously this book could have included many other areas of microbiological study which might be considered industrial. For example, microbial corrosion of metals and plastics, the action of microbicides other than heat and gases, and the production of nucleic acids would all have been extensions of the examples presented. Still other subdisciplines concerned with microorganisms, such as phytopathology and clinical microbiology, were not included as they are separate and distinct and not ordinarily considered industrial even though their principles and practices are comparable in many ways to those of industrial microbiology. The purpose of this text is to bring to the advanced undergraduate or graduate student descriptions of selected microbial processes which are generally carried out on a large-scale industrial basis for the gainful use of man.

The authors of these chapters have spent a considerable proportion of their professional careers working in those special fields of microbiology about which they have written. All the authors have been involved at the bench level of microbiological research. In most instances, the examples of research which they describe are taken from personal experiences in their recent past. In fact, the use of practical examples of the microbiological principles discussed is one of the important features of this text. Throughout the text, the symbol ⧺ will be used to alert the reader to the beginning of such an example.

The editors have made a conscious effort to revise each chapter no more than necessary so that the reader might feel he was in the laboratory of the author. By interfering as little as possible with each author's contribution, the editors hope every reader will be able to follow the author's investigation and understand his elation with success or despair with failure.

ACKNOWLEDGMENTS

All of us, the authors and editors alike, wish to acknowledge and thank those who also served, our spouses.

We are grateful to the following persons for their helpful comments upon reading the manuscript: J. E. Zajic, Walter J. Mallman, Milo Don Appleman, Charles L. Goldman, Richard N. Kinsley, Jr., and W. W. Umbreit.

Brinton M. Miller

Warren Litsky

1
EXPERIMENTAL DESIGN IN APPLIED MICROBIOLOGY
H. Boyd Woodruff

"Every man and woman, even every boy and girl, can be a scientist. This is so because science is based on *common sense*." Thus begins an encyclopedic definition of "science."[1] The statement is verified by the writings of T. G. Huxley, who referred to science as "nothing but trained and organized common sense."

This book is concerned with a small section of science, that of applied microbiology. Are there special aspects of experimental design required for assuring discoveries in the applied phases of microbiology? To answer, one must examine the subject.

Chapters, treatises, even complete books have been written on the distinctions between various forms of research. Is a program fundamental or applied, basic or developmental, pure or directed? The National Science Foundation of the United States (1962), in attempting to bring order out of chaos, has applied a series of definitions: Basic research is directed toward increase in knowledge in science; applied research is directed toward practical application of science; developmental research is the systematic use of scientific knowledge directed toward the production of useful materials, devices, systems, or methods, including design and development of prototypes and processes. Compare these definitions with the objectives expressed in the various chapters of this book. As an example, analytical microbiology clearly

---

[1] "The World Book Encyclopedia" (1955), p. 7621, Field Enterprises, Inc., Chicago, Ill.

concerns devices, systems, and designs and is truly developmental by definition. Its practice, however, may lead to significant basic discoveries. For instance, the attempt to provide an assay for measurement of the bactericidal action of penicillin in 1942 resulted in the discovery that the antibiotic failed to kill nongrowing cells. This observation, in turn, aided the elucidation of the separate biochemical paths of cell wall synthesis and cell protein synthesis. Recognition of the biochemical mechanism of killing of cells by penicillin permitted design of new methods for the enrichment of mutants in a microbial population. Penicillin killed the rapidly growing prototrophs, whereas nongrowing mutants proved resistant. The mutants quickly became the dominant survivors and could be isolated with ease.

Truly, applied microbial science expands as a chain reaction. Developmental research leads to basic discoveries, which generate new methodology and new applied approaches; these lead to discovery of new products, to further development, and to further basic discoveries. A similar course could be traced through any one of the fourteen special-subject-matter chapters of this book. Basic research requires an observation which must be explained, an object for investigation. Through the practice of applied microbiology, the objectives are provided.

Because of the intermingling of the basic, the applied, and the developmental aspects of microbiology, it is not possible to provide an experimental design suitable for one facet alone. The design that is fashioned must be applicable to all microbial science. No better outline can be chosen than that presented by Claude Bernard (1865) over 100 years ago:

> A true scientist is one who combines in himself both theory and experimental practice.
>
> 1 He discovers a fact.
> 2 An idea, connected with this fact, is born in his mind.
> 3 On the basis of this idea he reasons, makes an experiment, and conceives and implements the necessary material conditions.
> 4 From this experiment, new phenomena appear, which he must then observe, and so forth.

THE DISCOVERY OF A FACT

The discovery of a new fact of interest to the scientist can result from fortuitous laboratory coincidences arising during regular experimentation. A keen and appreciative observer is required. The story of Alexander Fleming's appreciation of the significance of the lytic activity of *Penicillium notatum*, a chance contaminant on a staphylococcus smear plate, is a classic of our time. For the purposes of applied microbiology, however, the discovery of a fact often is synonymous with the recognition of a need. In this recognition, the applied microbiologist has the advantage of the contributions of nonscientists: the politician who recognizes the concern of his constituents about the pollution of a river, the commercial development specialist who recognizes the desirability of a fabric which resists mildewing, the agricultural specialist who sees signs of a vitamin deficiency, the physician who is concerned with antibiotic resistance. Our daily lives are replete with examples of needs which can be fulfilled through research in applied microbiology. We thereby arrive at the first principle of experimental design. *The objective of research must be clearly and precisely expressed.*

Certainly, the purpose of experimental design is to increase the possibilities of success in the performance of research. Success is by no means guaranteed. Each laboratory worker knows of associates who move from topic to topic, occupying their time fully, yet never attaining a full explanation for a phenomenon or a practical achievement. Ask these workers the goal of their effort and the answer is apt to be as vacuous as their efforts.

The true scientist's need for clarity has been stated succinctly by Albert Einstein (1952), the master scientist:

> The formulation of a problem is often more essential than its solution, which may be merely a matter of mathematical or experimental skill. To raise new questions, new possibilities, to regard old problems from a new angle, requires creative imagination and marks real advance in science.

Applied microbiology exists within the context of an industrial society. Not only do problems exist; financing must be provided for their solution. A secondary benefit of a clearly stated objective for a research program is the advantage gained in communication with nonscientists, especially those in management or in government granting agencies responsible for the funding of research. Thus, the clear statement of an objective not only prepares the mind of the scientist for the birth of an idea, it also provides the basis for flow of information from scientist to financier needed to succor the idea.

THE BIRTH OF AN IDEA

We now attain the realm of the scientist, the field in which he alone can operate. An idea for solution of a problem must be generated. This is creative work. One thus reaches the second principle of experimental design. *To achieve success on a research project, one must assemble competent scientists who are dedicated to the objective.* At our present stage of educational development, a person is not called a scientist until he has accumulated much specialized knowledge. Ability to organize is essential for success in scientific endeavor. Fortunately, an educated scientist has this ability; otherwise he would not have been capable of assimilating the essential facts and conclusions required to achieve his academic degree. Years of formal training, along with the skills developed during laboratory practice, are an essential ingredient in matching people with projects. Proper matching provides the raw material necessary for the birth of ideas necessary for solution of problems.

A high rate of successful discoveries in the area of applied science is favored especially when the assemblage of scientists includes those with a prepared mind and those who practice disciplined experimentation. Fleming's reaction to his observation of the lysis of staphylococci often is quoted as an example of the prepared mind (Fig. 1-1). Previous experimentation with therapeutic agents had prepared him to recognize the chemotherapeutic potentiality of the substance he called penicillin. Within 8 months of his original observation, Fleming (1929) expressed his ideas on the significance of his observation in a scientific publication for the scientific world to judge. A previous observation of a similar laboratory phenomenon by the biologist John Tyndall led to naught. The objective, application in chemotherapy, was missing. Tyndall was concerned with spontaneous generation and, although he described his observations, their significance was missed and they remained buried from the sight of others who may have had the interest.

An even purer example of recognition of need, followed by the birth of an idea by a trained scientist with a prepared mind, and of the clear expression of his conclusion, is the recognition by Charles Nicolle (1932) of the transmission of typhus by fleas, as told in his own words:

Figure 1–1 Photograph of Fleming's original petri dish culture which led to the discovery of penicillin.

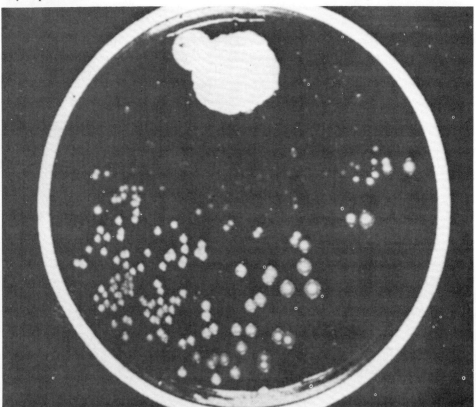

This shock, this sudden illumination, this instantaneous self-certainty of a new fact—I know of it, I have experienced it in my own life. It is in this way that the mode of transmission of exanthematic typhus was revealed to me. Like all those who for many years frequented the Moslem hospital of Tunis, I could daily observe typhus patients bedded next to patients suffering from the most diverse complaints. Like those before me, I was the daily and unhappy witness of the strange fact, that this lack of segregation, although inexcusable in the case of so contagious a disease, was nevertheless not followed by infection. Those next to the bed of a typhus patient did not contract the disease, while, almost daily, during epidemic outbreaks, I would diagnose contagion in the douars (the Arab quarters of the town), and amongst hospital staff dealing with the *reception* of patients. Doctors and nurses became contaminated in the country, in Tunis, but never in the hospital wards.

One day, just like any other, immersed no doubt in the puzzle of the process of contagion in typhus, in any case not thinking of it consciously (of this I am quite sure), I entered the doors of the hospital, when a body at the bottom of the passage arrested my attention.

It was a customary spectacle to see poor natives, suffering from typhus, delirious and febrile as they were, gain the landing and collapse on the last steps. As always I strode over the prostrate body. It was at this very moment that the light struck me. When, a moment later, I entered the hospital, I had solved the problem. I knew beyond all possible doubt that this was it. This prostrate body, and the door in front of which he had fallen, had suddenly shown me the barrier by which typhus had been arrested. For it to have been arrested, and contagious as it was in entire regions of the country and in Tunis, for it to have remained harmless once the patient had passed the Reception Office, the agent of infection must have been arrested at this point. Now, what passed through this point? The patient had already been stripped of his clothing and of his underwear; he had been shaved and washed. It was therefore something outside himself, something that he carried on himself, in his underwear, or on his skin, which caused the infection. This could be nothing but a flea. Indeed, it was a flea. The fact that I had ignored this point, that all those who had been observing typhus from the beginnings of history (for it belongs to the most ancient ages of humanity) had failed to notice the incontrovertible and immediately fruitful solution of the method of transmission, had suddenly been revealed to me.

As a complement to a prepared mind, disciplined experimentation is a desirable adjunct fostering the birth of an idea. Scientists are individualists. A group working on an unfulfilled need may be thought to receive flashes of genius. Closer examination of the great leaders of our field, however, will demonstrate their close involvement with laboratory experimentation. Pasteur drew many of his notable successes from the laboratory experience gained on previous projects. Nicolle's flash of discovery in recognizing the flea as a carrier of typhus would not have been possible without years of experimentation on disease transmission by more conventional means.

The two principles of experimental design enumerated above, a clear statement of an objective and the gathering of creative experimentalists, are too often neglected in the applied phases of microbiology. The practical needs are so clear that there is a desire to get on with the job, to engage immediately in laboratory activities, yet nothing is less productive. Possibly too much time is spent in discussing the history of the successes in science. The all too obvious lack of planning and design which is evident in the history of failures is seldom examined.

EXPERIMENTATION AND IMPLEMENTATION

No phase of experimental design for applied microbiology receives more attention than experimentation and implementation. Read the remaining chapters of the book! In developing general approaches to fulfilling these essentials of design, emphasis must be given to the need for organization. As a deterrent to the waste of research effort which results from disorganization, the next principle will be stated succinctly. *Develop a hypothesis.*

Throughout their educational lives, scientists have been taught a pattern for scientific endeavor: to assemble observations, to classify them, to develop a hypothesis for explaining the observations, to design experiments to test the validity of the hypothesis, and, if validated, to present the hypothesis as a fact. The progression is not as straightforward when applied to the topics in this book. The target of the applied microbiologist is often a new creation which never before existed: a new bactericide, a new sterilization procedure, a new sewage treatment process. The existing observations are unlikely to be adequate for development of a full-fledged hypothesis subject to simple verification. The specialist group, assembled according to the second principle stated above, is required to draw upon the limited existing observations having relation to the stated objective (first principle) and to define those additional experiments which are necessary before a hypothesis can be developed. The development of a hypothesis, therefore, becomes a three-step process. These are (1) the recognition of those observations which are missing in the literature for the development of a reasonable hypothesis, (2) the experimentation to collect the essential data, and (3) the formulation of the hypothesis. Additional experimentation to validate or reject the hypothesis then becomes possible.

There is a companion principle which has special significance for the experimentation directed at collection of essential data needed to formulate a hypothesis. The principle is well known to the engineer who has had broad experience with unit processes, but it is less familiar to the microbiologist. The principle may be stated as follows: *In the designing and implementation of laboratory experiments, consideration must be given to all parameters. These include the biological, the chemical, and the physical.*

The breadth of topics within applied microbiology makes impossible the discussion of this principle in general terms. Each scientist must apply his or her own skills to the examination of the specific project to be certain that no critical aspect of research has been overlooked. To provide an illustration of the principle, an experience of the author as a participant in a cooperative research program on cyanocobalamin, and later as a director of the microbiological phases of the project, will be presented.‡ With deviations as necessary, the approaches will be similar in most fields of microbiology.

In the early phases of the research which led eventually to the discovery of vitamin B_{12}, an approach for control of the disease pernicious anemia was formulated. Isolation of a factor postulated to be present in liver, extracts of which were known to prevent expression of symptoms of the disease, was to be undertaken. Only new patients, or those in relapse due to inadequate therapy,

---

‡Throughout the text, this symbol will be used to call attention to an author's practical example of a microbiological principle. For further information see the Preface.

were available for assay of the isolated fractions. This greatly inhibited progress in achieving the objective of the research project.

Past experience with the application of microorganisms as indicators of nutritional deficiencies caused the research personnel to be receptive to a possibility that a microorganism could show a nutritional deficiency correlative with pernicious anemia in human beings. This idea was held in spite of the best reviews and monographs of the time which had discarded the approach.

After several experiments resulted in failure, a microbiological assay became available from a cooperating university which showed the possibility of correlation. Results were so variable, however, as to be nearly noninterpretable. An objective was formulated: to establish whether human beings and the microorganism respond to the same factor in liver. Microbiologists skilled in developing microbial assays were gathered for the project.

The microorganism involved was *Lactobacillus lactis.* First consideration was given to biological factors. Microorganisms often are variable because of wide genetic diversity in the mass culture population. Single-cell isolates were made (Shorb, 1947). Stable strains differing in responsiveness from the parent population were recovered and preserved by lyophilization.

A chemical approach was then initiated. The initial surveys had shown that the culture grew best if both liver extract and tomato juice were present in a casein hydrolysate medium. The requirements of all known amino acids, vitamins, and nucleosides were established, and eventually a medium was devised in which every ingredient was a known substance except for very small amounts of liver extract (Table 1-1). Paper chromatography of the liver extract indicated that a single supplemental entity was adequate for full growth.

Even at this stage of development, the microbiological assay was not suitably quantitative for practical application. The organism still varied greatly in response. In approximately two trials out of three, maximum bacterial growth occurred even in the absence of liver factor. The procedure was inadequate to guide purification of the growth factor of liver or to establish whether microorganism and human being were correlative in response. This status existed in three pharmaceutical research laboratories and one university all working toward similar goals.

Because such requirements had not been significant in their past experiences, microbiologists working on the project had neglected one aspect of research, the physical factors. A chance observation provided the breakthrough at a time much later than would have been the case had the physical parameters been researched. It was recognized that a technician who routinely shook racks of assay tubes after removing them from the autoclave, prior to inoculation, produced the most consistent assays. The receptive mind of the laboratory supervisor led to investigation of the oxygen tension in the assay tubes, and the data showed the cyanocobalamin requirement of *L. lactis* to be a direct function of oxygen tension. In fact *L. lactis*, considered by all assay experts to be a microaerophile, grew well under highly aerobic conditions, and even in the presence of hydrogen peroxide when adequate amounts of cyanocobalamin were added (Fig. 1-2). Through application of unexpected technology, an aerated microbiological assay method, a pharmaceutical research laboratory succeeded in isolating pure vitamin B_{12}, in demonstrating that the microbe and pernicious

Table 1-1 Minimal nutrient requirements of *Lactobacillus lactis* (ATCC 10697)

Nutrient	Requirement, µg/ml	Nutrient	Requirement, µg/ml
DL-Glutamic acid*	460	Nicotinic acid	0.5
DL-Threonine	390	Pantothenic acid	0.1
DL-Aspartic acid	260	Riboflavin	0.06
DL-Leucine	230	Pyridoxamine PO$_4$	0.0002
L-Cystine	200	Biotin	0.0002
DL-Isoleucine	140	Vitamin B$_{12}$†	0.00002
DL-Phenylalanine	130	Uracil	10
L-Lysine	110	Guanine	10
DL-Methionine	70		
L-Arginine	70	**Supplemental factors**	**mg/ml**
L-Tyrosine	50	KH$_2$PO$_4$	1.0
DL-Tryptophan	50	K$_2$HPO$_4$	1.0
DL-Valine	50	MgSO$_4$ · 7H$_2$O	0.2
L-Histidine	40	FeSO$_4$ · 7H$_2$O	0.01
DL-Serine	30	NaAc	6
		Glucose	10

*DL-Amino acids were used because of commercial availability. The L-enantiomorph at half concentration will provide maximum growth of *L. lactis*.

†Vitamin B$_{12}$ is contained in liver extracts which are used for treatment of pernicious anemia. When vitamin B$_{12}$ is not added to the medium, the growth of *L. lactis* correlates directly with the vitamin B$_{12}$ content of the liver extract added to the medium.

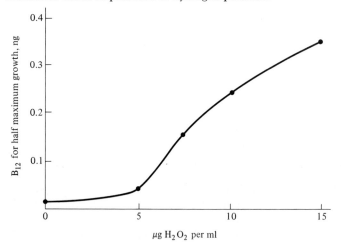

Figure 1-2 Vitamin B$_{12}$ (cyanocobalamin) utilization by *Lactobacillus lactis* in presence of hydrogen peroxide.

anemia patient were, in fact, responsive to the same entity, and in developing a valuable commercial product.

EXTENSION TO NEW PHENOMENA

A key factor for success in science is formulation of a series of goals, the solution of which leads to new formulations and new successes. The applied microbiologist must be engaged constantly in formulating such new goals, modifying them as necessary as data develop. We come to the final principle of experimental design: *Each experimental result must be examined not only for the extent to which it solves the objective for which it was designed, but also for the new problems which are raised.* Successful operation of this principle leads to chain discoveries. The assay system previously described permitted not only the purification of cyanocobalamin from liver but, in addition, the discovery that many bacteria produce cyanocobalamin and the isolation of crystalline material from this source. The commercial need for B_{12} was fulfilled by using *Streptomyces griseus* fermentations, in which it was recovered as a by-product of streptomycin manufacture. The assay system, thereafter, permitted discovery of higher-yielding cultures which were cultivated for cyanocobalamin production alone. The availability of ample B_{12} from microbial sources permitted screening for other utilities, and a discovery that vitamin B_{12} is an animal protein factor required for optimum growth rate of farm animals. This greatly increased the market volume for the compound. The ability of *Lactobacillus lactis* to grow with thymidine as a substitute for B_{12}, and of other assay organisms to grow with methionine, led to significant advances in defining essential metabolic pathways of aerobic and anaerobic bacteria and of animals. With cyanocobalamin available as a growth requirement, the possibility of discovery of antagonists existed and has been realized in the isolation of descobaltocobalamin, using the inverse of the normal assay system. This sequence could be extended greatly, quoting additional basic discoveries and new opportunities for applications derived from the original fulfilled goal.

On a larger scale, the revisions of textbooks of physics and chemistry which resulted following the discovery of x-rays and their many applications in biology and medicine are common knowledge. So are the developments which resulted from successfully controlled nuclear fission. That minor discoveries also have a chain-discovery effect is not so clearly recognized, nor are the opportunities provided by these discoveries always fulfilled.

ADDENDUM: EFFICIENCY

The factor of efficiency has been placed purposely as an addendum. It must be considered within the scope of experimental design. It must not substitute for experimental design. The statistical approach to experimentation is, without doubt, time-saving. The statistical principles enumerated in the early period of research on penicillin, when the wartime effort placed time at a premium, with too few laboratory workers available, remain applicable today (Brownlee, 1950). Textbooks are available and should be consulted, or, even better, statisticians should be included in the task force gathered to attack problems of major

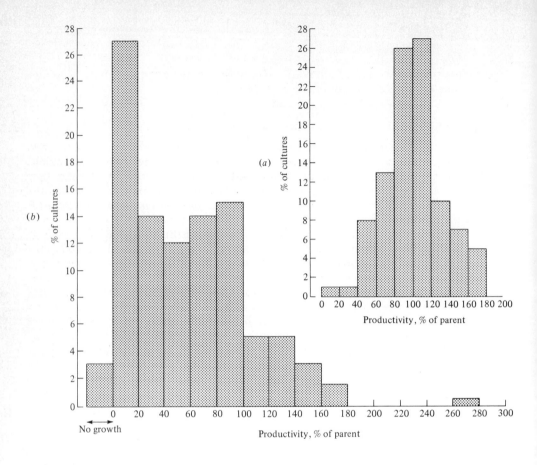
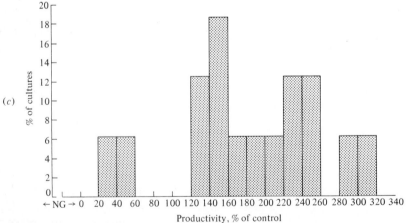

Figure 1-3 (*a*) Spread in tetracycline productivity of 100 isolates recovered from a culture of *Streptomyces viridifaciens*. (*b*) Spread in tetracycline productivity of 100 isolates recovered from a culture of *Streptomyces viridifaciens* after ultraviolet light treatment to greater than 99 percent kill. (*c*) Spread in tetracycline productivity of prototrophic mutants recovered from a homocysteine-requiring auxotroph of *Streptomyces viridifaciens*.

significance. Great care must be taken, however, to prevent use of statistical approaches as a substitute for creative research thought.

One example may suffice. The mutational techniques, whereby strains of microorganisms of continually increasing potency have been selected for antibiotic, amino acid, or vitamin production, are a natural target for the statistical approach. The obvious experimental limitations are those in which efficiency of design and determination of significance of collected data are determinant. Yield improvements through mutation have been reported to occur in no more than 1 in 3000 cultures examined. Statisticians have been successful in improving design so that less than one-third as many technicians now accomplish the same rate of discovery of improved cultures as in the initial mutation programs. No number of statistical approaches applied alone, however, can be expected to match the order of magnitude of improvements which have resulted from creative research applied to the problem. New techniques, in which only those cultures already established to be mutants are evaluated for tetracycline productivity, have permitted discovery of improved cultures within days (Dulaney and Dulaney, 1967), in contrast to the weeks of painstaking searching previously required (Fig. 1-3). The statistical approach can be applied as a supplement to the new methodology. In experimental design, however, the statistics must always be kept as the servant, never allowed to become the master.

CONCLUSIONS

The elements of experimental design are present throughout each of the chapters on applied microbiology contained in this book. The experienced microbiologist may scan through the chapters for new ideas. The student will need to search out these elements and classify them as part of the essential background for becoming a scientist.

There are certain essential principles which sometimes are overlooked in designing the approach to an applied problem—or to any problem in science. These principles are vital for success, and restating is justified.

1 The objective of a research project must be clearly and precisely expressed.
2 Competent scientists who are dedicated to the objective must be assembled.
  a A competent scientist has a prepared mind.
  b A dedicated scientist is a disciplined experimentalist.
3 A hypothesis subject to laboratory verification should be formulated.
4 In the designing of the laboratory experiments, consideration must be given to all parameters including the biological, the chemical, and the physical.
5 Each experimental result must be examined with a view toward new, relevant research opportunities.

LITERATURE CITED

Bernard, Claude (1865): "Introduction à l'étude de la médicine expérimentale," Flammarion, Paris.
Brownlee, K. A. (1950): A Plant-Scale Planned Experiment in Penicillin Production, *Ann. N.Y. Acad. Sci.*, **52**:820–826.

Dulaney, E. L., and D. D. Dulaney (1967): Mutant Populations of *Streptomyces viridifaciens*, *Trans. N.Y. Acad. Sci.*, **29**:782–799.
Einstein, A., and I. Infeld (1952): "The Evolution of Physics," p. 95, Simon and Schuster, New York.
Fleming, A. (1929): On the Antibacterial Action of Cultures of a Penicillium, with Special Reference to Their Use in the Isolation of *B. influenzae*, *Br. J. Exp. Pathol.*, **10**:226–236.
National Science Foundation (1962): Reviews of Data on Research and Development, *Rep. NSF* 62-9, p. 8, Government Printing Office, Washington, D.C.
Nicolle, Charles (1932): "Biologie de l'invention," Alcan, Paris.
Shorb, M. S. (1947): Unidentified Growth Factors for *Lactobacillus lactis* in Refined Liver Extracts, *J. Biol. Chem.*, **169**:455–456.

GENERAL REFERENCES

Hill, K. (ed.) (1964): "The Management of Scientists," Beacon Press, Boston.
Seyle, H. (1964): "From Dream to Discovery—On Being a Scientist," McGraw-Hill, New York.
Taton, R. (1957): "Reason and Chance in Scientific Discovery," Philosophical Library, New York.

2
ANALYTICAL MICROBIOLOGY

Frederick Kavanagh

Analytical microbiology is the use of microorganisms in performing quantitative analytical tasks. It is a part of the general subject of analysis and not simply a minor branch of bacteriology. Microorganisms as reagents in quantitative analysis could be considered the theme of this chapter. They are used when they provide more specific, more sensitive, or more efficient assays than chemical methods. They also have the advantage, contrary to general opinion, of great consistency. In the well-developed methods that have been used for the large-scale assays (antibiotics, amino acids, and vitamins) during the past 30 years, demonstrating error caused by microbiological variation is very difficult.

The test organism selected for an assay depends upon the kind of compound to be assayed. Lactic acid bacteria such as *Lactobacillus plantarum* or *L. delbruekii* are used in most vitamin assays. Usually a gram-positive organism such as *Staphylococcus aureus*, *Bacillus subtilis*, or *Sarcina lutea* is selected to assay for the antibiotics active primarily against gram-positive bacteria. Either a gram-positive bacterium listed above or a gram-negative organism may be used to assay for the broad-spectrum antibiotics. The most popular gram-negative bacteria are *Escherichia coli*, *Klebsiella pneumoniae*, *Bordetella bronchiseptica*, and *Streptococcus fecalis*. *Staphylococcus aureus* has been used more than any other of 21 bacteria in an FDA laboratory (Arret et al., 1971), performing 60 different microbiological assays for antibiotics approved for human use in the United States. The penicillins, cephalosporins, erythromycin, tylosin, and Vancomycin may be assayed by *S. aureus* and *B. subtilis* plate methods and an *S. aureus* turbidimetric method. The official method is an *S. aureus* plate assay. Europeans prefer *B. subtilis* and *B. pumilus* as their plate assay organisms.

No more than a hint of the complexity of the subject can be given in this chapter. Anyone who expects to use the methods seriously should prepare by studying standard works on the subject. A good place to start is with two articles by E. E. Snell (1948, 1950). He is one of the authors of the first practical quantitative microbiological assay for a vitamin. The method, for riboflavin, is still used after more than 30 years.

Those who need more details about theory and practice of microbiological assaying as well as about specific methods for antibiotics, amino acids, and vitamins should consult the appropriate chapters in "Analytical Microbiology" (Kavanagh, 1963a, 1972a). Pearson (1967) discusses briefly the principles of microbiological assaying as applied to vitamins and amino acids. The little book by E. Barton-Wright (1962) gives the methods for vitamins and amino acids found to be practical by one experienced analyst. Other books are "Vitamin Assay" by Strohecker and Henning (1965), describing methods used at E. Merck A.G. (Darmstadt); "Methods of Vitamin Assay" (Freed, 1966), prepared by the Association of Vitamin Chemists and giving methods applicable to foods; and "The Vitamins," vol. VII (György and Pearson, 1967). Methods having official standing must be used to assay products bought under federal or state standards. These methods are found in "Official Methods of Analysis" (AOAC, 1975), "The Pharmacopoeia of the United States of America" (USP, XIX), and methods published in the *Federal Register*. These official methods are the work of committees and may differ in minor or significant detail from the most recent methods. They rarely are designed for large-scale assays and may not be the most accurate. However, they are official and must be used on occasion.

RESPONSES
INTRODUCTION

Several responses of microorganisms can be used in assaying. Requirements of the responses are capability of accurate measurement, small variation of sensitivity within the population, rapid development of the response, and show of gradation with graded concentrations of the active substance.

Some responses used in assaying are cell multiplication, production of metabolic products such as carbon dioxide and acids, responses of subsystems within the cell such as leakage of salts from the cell through a damaged cell membrane, and production of light. Cell multiplication and production of acid are used more frequently than any of the other responses. Vitamins and amino acids may be measured by their capacity to increase cell population and the concomitant production of acids. Antibiotics reduce the rate of cell division. Certain antibiotics quench light emission by photobacteria. A few antibiotics cause the yeast cell to lose salts.

CELL MULTIPLICATION

Amino Acids

Certain species of lactic acid bacteria have absolute requirements for several vitamins, purines, pyrimidines, and amino acids. When supplied with a medium lacking only one required substance, the limiting cell population is proportional to the amount of substance supplied over a limited range. If this is not true, then the medium is deficient in some other required substance, or limiting growth was not obtained for some other reason. Quite often the reason is too short a growth period which, of course, is related to incubation temperature. An indication of the upper range of concentration for 10 amino acids assayed by *Streptococcus faecalis* is given in Table 2-1. The total volume of liquid in the tubes was 6 to 10 ml. Usually four tubes are used and the sample assayed at two or more concentrations. Therefore the smallest sample taken for assay of histidine, for example, would contain about 100 µg of histidine. This would be a sample of about 3 mg for casein. By scaling the volumes down and using only two tubes at two concentrations, an assay could be obtained on a 60-µg sample of casein. The technique of manipulating such small samples is more difficult than for macro size samples and would be done only when necessary.

At least 19 amino acids can be assayed for by one or more species of lactic

Table 2-1 Amino acid assays with *Streptococcus fecalis* (upper limit of concentration in the assay medium is given)

Amino acid	Assay range, µg/ml	
	Turbidimetric	Titrametric
L-Arginine	4.5	10
L-Histidine	2	5
L-Isoleucine	7	10
L-Leucine	8	10
L-Lysine	15	20
L-Methionine	3	5
L-Threonine	5	10
L-Tryptophan	—	2
L-Valine	7	10
L-Glutamic Acid	12	

SOURCE: G. D. Shockman (1963), in F. Kavanagh (ed.), "Analytical Microbiology," p. 649, Table XIII, Academic Press, Inc., New York.

acid bacteria. The specificity of the method is so great that each amino acid can be quantitatively measured in the presence of varying amounts of the other 18. This is possible only because the organism will not grow if any one of the required amino acids is missing from the assay medium. Also, the medium is designed so that the total amount of amino acids is so great that amino acids added in the sample are without influence. For example, the basal medium contains 3 mg/ml of total amino acids before the sample is added. In the example of the assay of casein for histidine, the maximum amount of sample would be 0.18 mg/ml. Therefore only 6 percent of the total amino acids would come from the sample. The medium also contains a large excess of vitamins, purines, pyrimidines, and minerals.

Vitamins

The vitamins are essential constituents of intermediary metabolic systems and are discussed at length in texts on biochemistry; their function need not be considered here. See Guirard and Snell (1962) for a brief discussion of nutritional requirements of microorganisms and Koser (1968) for an extensive account of vitamin requirements of bacteria and yeasts.

Many fungi, bacteria, algae, and protozoa require one or more vitamins for growth in a medium otherwise complete. A complete medium contains sugar, essential minerals and nitrogen sources, and the essential vitamins. Species of all the organisms listed have been used. The amount of any one vitamin required usually is very small, and for this reason vitamins were known at one time as accessory growth factors. They are essential to all organisms. The vitamins not synthesized by the organism must be supplied by its food or environment. The vitamins required by these lower forms of organisms are biotin, folic acid complex, riboflavin, pyridoxine complex, thiamine, inositol, vitamin B_{12} complex, pantothenic acid, nicotinic acid, and p-aminobenzoic acid. These vitamins are also known as the B vitamins for historical reasons. An organism requiring one or more of these compounds can be found. The most useful for assay purposes have been the lactic acid bacteria. Since they possess multiple requirements, one basal medium can be used for assaying for several vitamins. For example, *Streptococcus fecalis* ATCC 8043 requires vitamin B_{12}, riboflavin, folic acid, pyridoxine, biotin, nicotinic acid, and pantothenic acid. Each vitamin is assayed in a medium complete except for that vitamin. Thus the vitamin assay system is analogous to that of the amino acid system. Although there are obvious advantages to using one organism for assays of many different essential substances, practical assays are done with a number of organisms.

Growth, measured by increase in cell mass, is proportional to the amount of limiting vitamin present, as Snell showed for *Lactobacillus casei* growing in the presence of riboflavin. Thus the dry weight of bacterial cells harvested from the tubes of an assay is a measure of the riboflavin in the tubes. The cells can be counted by an automatic electronic or optical counter. These methods of measuring growth are time-consuming and usually are replaced by a measurement of turbidity.

Although the example of the assay of vitamins is by means of a lactic acid organism, the dry weight of the mycelial mat produced by the appropriate fungus could be the measure of the vitamin. Methods employing fungi generally are used only for special investigations because bacterial methods are quicker and more convenient. Protozoa and algae are not popular for routine assays either.

Sensitivities of microbiological assays for the B vitamins are given in Table 2-2. The sensitivities are high except for the inositol (a growth substance for certain fungi) method which has a sensitivity in the amino acid range. Sensitivities generally are considerably higher than for typical chemical methods. An indication of the distribution of the water-soluble vitamins in nature is given in Table 2-3. Most of the B vitamins, nicotinic acid, biotin, thiamine, riboflavin, folic acid, are of universal occurrence whereas vitamin B_{12} has limited distribution. Vitamin B_{12} seems to be a bacterial product that is required by bacteria and animals including protists.

Generally microbiological assay methods are more specific than chemical assays. For example, the Ninhydrin method of assay for amino acids measures the amino group without distinguishing between D and L forms. The microbiological method usually measures only the L form of amino acids. The specificity for vitamins is a little different from the specificity for amino acids because a slightly different chemical structure may have equivalent or nearly equivalent activity in satisfying growth requirements of the assay organisms. Microbiological methods measure function, not structure. A few of the vitamins, riboflavin for one, seem to have unique structures, and only one naturally occurring form is known. For the other vitamins, activity of a particular sample can be reported only as its equivalent of the standard substance. Other evidence, paper or thin-layer chromatography, must be obtained to identify the active substance. Where a number of molecular structures have equivalent function, folic acid and

Table 2-2 Approximate upper limit of concentration of vitamin per milliliter of assay medium

Vitamin	Organism	Upper limit
p-Aminobenzoic acid	*Neurospora crassa*	50 ng*
Biotin	*Lactobacillus plantarum* (ATCC 8014)	200 pg†
	Saccharomyces cerevisiae (ATCC 7754)	40 pg
Folic acid	*Lactobacillus casei* (ATCC 7469)	200 pg
	Streptococcus fecalis (ATCC 8043)	800 pg
Inositol	*Saccharomyces carlsbergensis*	1 µg
Nicotinic acid	*Lactobacillus plantarum* (ATCC 8014)	50 ng
Pantothenic acid	*Lactobacillus plantarum* (ATCC 8014)	10 ng
Pyridoxine	*Saccharomyces carlsbergensis* (ATCC 9080)	3 ng
Riboflavin	*Lactobacillus casei* (ATCC 7469)	20 ng
	Streptococcus sp. (ATCC 10100)	1 ng
Thiamine	*Lactobacillus viridescens* (ATCC 12707)	3 ng
	Saccharomyces cerevisiae	200 ng
Vitamin B_{12}	*Lactobacillus leishmannia* (ATCC 7830)	10 pg
	Ochromonas malhamensis (ATCC 11532)	20 pg
	Euglena gracilis (ATCC 12716)	10 pg

*ng = 10^{-9} g
†pg = 10^{-12} g

Table 2-3 Occurrence of vitamins in natural products as $\mu g/100$ g

Substance	Th.*	Rib.	Biotin	F.A.	N.A.	P.A.	B_6	B_{12}
Serum, human	1–9	2–4	1–2	1–5	30–150	6–40	1–20	14–100
Yeast, dried	2800	6200	200	700	28000	9500	3400	0
Beef, lean	50	200	4	15	4000	1100	77	2
Heart	580	890	...	3	7800	2100	120	11
Kidney	370	2500	...	58	6400	3400	400	30
Liver	260	3000	100	290	13000	7000	800	80
Eggs, whole	100	290	25	5	100	2700	22	8
Peanuts	300	130	39	57	16000	2500	300	0
Vegetables, fresh								
Beans, lima	210	110	...	10–60	1400	470	500	
Beets	80	180	...	13	400	150	50	
Cabbage	60	50	...	6–14	300	200	290	
Celery	50	40	...	7	400	400	60	
Spinach	110	200	2	50–100	600	300	200	
Potatoes, peeled	100	40	...	4–12	1200	500	160	
Tomatoes	60	40	2	2–16	500	300	100	
Apples	40	30	...	0.5	200	100	26	
Bananas	...	...	4	10	...	260	500	
Strawberries	20	50	...	5	200	300	55	
Wheat	...	...	5	27–51	...	1300	210	
Milk, whole	40	1700	5	10	100	290	6	0.4

*The abbreviations in the column heads refer to thiamine, riboflavin, biotin, folic acid activity, niotinic acid, pantothenic acid, vitamin B_6 activity, and vitamin B_{12}. Vegetables and fruits do not contain Vitamin B_{12}.

vitamin B_{12}, for example, the samples usually are mixtures. The proper assumption in assaying an unknown sample is that any possible mixture that can occur will occur. Sometimes organisms with high specificity for a particular molecular structure can be found. A good example is the response of lactic acid bacteria to the folic acid complex.

The responses of three bacteria to some of the many forms of folic acid (Table 2-4) are illustrative of the responses to the multiform vitamins. *Lactobacillus casei* responds to all except highly conjugated forms. It is used to measure "total folic acid." The response of *Streptococcus fecalis* is seen to be slightly more selective than *L. casei*. The differences between responses of the two organisms are slight but enough to have caused much confusion in the early days of research on folic acid, before it was known to be one of the multiple-factor vitamins. *Pediococcus cerevisiae* responds only to some of the reduced forms of folic acid. It is used to assay the reduced forms and compounds convertible to them by treatment with reducing agents such as ascorbic acid. By combining chemical treatment of the sample with specificity of response of these three bacteria, much can be learned about the folic acid composition of a sample. Similar procedures are employed in assaying for the several forms of the other vitamins (György and Pearson, 1967).

Table 2-4 Relative responses of assay organisms to several forms of folic acid.

Compound	Lactobacillus casei (ATCC 7469)	Streptococcus fecalis (ATCC 8043)	Pediococcus cerevisiae (ATCC 8081)
Folic acid (PteGlu)	+	+	−
Pteroyldiglutamic acid	+	+	−
Pteroyltriglutamic acid	+	−	−
Pteroylheptaglutamic acid	−	−	−
H_4PteGlu	+	+	+
5-CHO-H_4PteGlu	+	+	+
10-CHO-H_4PteGlu	+	+	+
5,10-CH≡H_4PteGlu	+	+	−
5-CH_3H_4PteGlu	+	−	−
Pteroic acid	−	+	−

SOURCE: E. L. R. Stokstad and S. W. Thenen (1972), in F. Kavanagh (ed.), "Analytical Microbiology," vol. 2. pp. 387–408, Academic Press, Inc., New York.

Miscellaneous Assays

Such compounds as sugars, purines and pyrimidines, available nitrogen, and essential inorganic elements can be and have been assayed by microbiological methods. For most of these, chromatography and specific chemical reactions make unnecessary the use of microbiological methods.

Bioassay for Fe, Cu, Mn, Mo, Zn, and Co in biological materials may be more sensitive than chemical methods. It is a direct method involving a minimum of sample preparation. *Aspergillus niger* has been used for many years to measure traces of minerals in soils and biological minerals. Table 2-5 gives the range of amounts of elements measured with *A. niger*. These sensitivities are in the range of vitamins assayed by lactic acid bacteria.

In addition to substances discussed above, presumably anything essential to cell multiplication of a microorganism could be assayed. Some of these substances are oxygen, carbon dioxide, organic acids, hematin, and a fairly large number of organic compounds. Light and temperature could be measured by systems which, in effect, would be biological integrators of their biological effectiveness.

Table 2-5 Trace metal assays by means of *Aspergillus niger*

	Range, μg/50 ml medium
Fe	0.1 − 10
Cu	0.05 − 4
Zn	0.5 − 10
Mo	0.0005 − 0.010
Mn	0.01 − 5

SOURCE: D. J. D. Nicholas (1966), *Ann. N.Y. Acad. Sci.*, **137**:217.

Antibiotics

All the examples given earlier are of the classes of substances that stimulate cell multiplication. The antibiotic substances, on the contrary, decrease cell multiplication and do so by interfering with some essential metabolic system. Penicillin and Vancomycin interfere with synthesis of cell wall material. Erythromycin, chloramphenicol, and tetracycline inhibit synthesis at the ribosomal level. Streptomycin and neomycin cause misreading of the genetic code, thereby inhibiting synthesis of functional proteins. Whatever the mechanism, the end result for analytical purposes is reduction in the rate of cell division.

Microbiological methods of assaying for antibiotics have been the methods of choice ever since antibiotics were discovered; in fact, that is how antibiotics were discovered. Microbiological methods are the official methods. The methods are capable of an accuracy as high as chemical methods, when one exists, and are considerably more sensitive. As examples of sensitivity, penicillin G can be measured in blood when it is present at less than 0.1 µg/ml, and tylosin can be assayed with acceptable accuracy when present in animal feed at 20 g/ton even though the feed also contains penicillin and organic arsenical compounds. The method selected to illustrate the principles of microbiological assaying will be the turbidimetric and plate methods for penicillin.‡

METABOLIC PRODUCTS

The principal metabolic products of value in assaying are acid and carbon dioxide. The former is produced by the lactic acid bacteria used in vitamin assays. Reduction in carbon dioxide output is a measure of the concentration of antibiotic in certain assay systems.

Acid may be measured directly by titration or indirectly by pH. The latter is much more convenient than the former and probably is as accurate and precise. The pH range of a vitamin assay standard curve is from about pH 5.5 to about 4. Examples of the response of standard samples are given in Fig. 2-1. The wide range of tenfold or more of the pH method is of advantage when assaying samples of unknown and widely ranging potency such as blood of people receiving vitamin therapy.

The assay methods illustrated in Fig. 2-1 have been in use for many years. These curves illustrate an all too common fault in the design of microbiological assays as practiced. The defects of these curves are too little slope of the vitamin B_{12} and nicotinic acid curves for high precision; range shorter than it need be for all three; and region of little responses (less than 1.5 ng) of the folic acid curve used in obtaining potency of samples.

Most vitamins assayed by means of lactic acid bacteria may be titrated to measure total acid produced after 48 to 72 h of growth or pH measured after 24 h of growth.

A respirometric method was developed for measuring the antifungal agents nystatin and amphotericin B in turbid samples (Gerke et al., 1963). These agents reduced carbon dioxide output of *Candida tropicalis* during a 2 to 4 h incubation period.

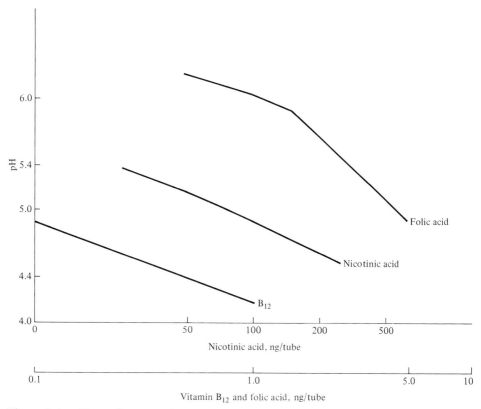

Figure 2-1 pH as a function of concentrations of vitamin B_{12}, nicotinic acid, and folic acid assayed by means of lactic acid bacteria.

RESPONSES OF SUBSYSTEMS OF THE CELL
Leakage
Certain antibiotic substances cause the cell membrane of yeast to leak accumulated salts (Isaacson and Platt, 1968). The response is rapid and may be carried out at relatively high temperature. A suspension of bakers' yeast in distilled water is treated with nystatin, and the concentration of salts in the external medium is measured by a conductimetric method. Although this method probably could be applied to other organisms and antibiotics, it is practical when the test organism is easily obtained in large quantity, as is bakers' yeast. A specific ion electrode could also be used.

Light
Certain marine bacteria luminesce in the presence of oxygen. One of these, the gram-negative organism *Photobacterium fisheri*, was used briefly in early antibiotic research when a rapid assay for penicillin was needed. The tests were

made in artificial seawater and read in a dark room. The test was fast. Some substances quenched the luminescence immediately (10 min after addition); several of these were gliotoxin, hydrogen peroxide, 1,4-naphthoquinone, spinulosin, and p-toluquinone. The immediately effective substances were quinones or extremely toxic substances. Penicillins, streptomycin, sulfonilamide, and tyrothricin were inactive. The tests were also read at 1, 2, 3, and 24 h. Some compounds were active after an hour or two and others only after 24 h. Patulin was an example of a slowly active antiluminescent substance. Streptomycin showed activity at 24 h, presumably because it killed the bacteria. Thus both antiluminescent and antibacterial activity could be obtained in one test. The method was not used because the substance of interest then, penicillin, was inactive. Details of the method and applications were given by Kavanagh (1947, 1963c).

OBTAINING AND MEASURING RESPONSES

pH

Acid produced by a lactic acid bacterium responding to graded amounts of a vitamin or amino acid may be measured titrametrically or by the pH of the medium. The latter is favored because of the ease and rapidity of measurement. The only piece of equipment is a good expanded-scale pH meter. The expanded-scale meter is needed because the range of pH is about 1.5 units over the entire standard curve. If the range is not restricted, the low pH obtained in the presence of the larger concentrations of active compound will limit production of acid and cause excessive curvature of the standard curve. The object of selecting medium, time, and concentration of substance being measured is to obtain a nearly straight standard curve. The absolute value of pH is of no significance; however, reporting pH to two or three decimal places is not absurd. The same solution measured with a good expanded-scale pH meter with a low-resistance glass electrode and a reference electrode with a glass sleeve junction will replicate readings within ±0.002 pH unit. Such equipment properly used will cause insignificant error in assay values of samples. All the tubes in a test must be at the same temperature during the measurement of pH. Examples of standard curves are given in Fig. 2-1.

LEAKAGE

Salts leaking from the yeast cells as in the nystatin assay are measured with a conductivity bridge. Again the temperature must be constant if the answers are to be accurate. Standards and samples are measured and potency of samples interpolated from the standard curve.

TURBIDITY

Introduction

The photometric method of assay is very simple in principle: The test substance is added to a suspension of the test organism in a nutrient medium, the mixture incubated, and the response of the test organism measured. The test organism may be any that will give a uniform suspension. Bacteria, protozoa, fungi, yeasts, and algae have been used. Just the barest outline of the many factors that affect the assay can be given here. For details see Kavanagh (1963b, 1972a).

Cell multiplication may be measured directly by cell count and by turbidity, or indirectly (chemically) by measuring total nitrogen, total protein, RNA, and specific cell wall constituents such as muramic acid. Chemical methods, although possible, are not practical in large-scale assays nor is cell counting even by modern semiautomatic equipment. Measurement of turbidity is easier and simpler than the chemical procedures and is the one generally employed. It can be measured automatically and rapidly with good precision.

Turbidity is generally measured photometrically, not nephelometrically. The latter is the method used in special procedures employing very small concentrations of cells where high sensitivity is required. Nephelometric methods are more sensitive to extraneous suspended material in the medium than photometric methods. Also, there are no accurate semiautomatic instruments, which is a necessary requirement for large-scale use in assaying.

Concentration of Cells

The bacteria commonly employed in assaying have cell dimensions ranging from 0.5 to 3 μm. The angular distribution of light scattered from particles of this size is anisotropic, with much greater scattering in the forward direction (direction of propagation of the irradiating light) than in either the backward or the 90° directions. The usual photometer used to measure the optical density of a suspension responds to scattered light as well as to collimated light, making the measured absorbance less than the true value.

The relationship between absorbance A and the concentration of bacteria, N, for most photometers and spectrophotometers is expressed approximately by the equation

$$A = HN - KN^2 \tag{2-1}$$

Size of Cells

The assumption has been made in discussing the measurement of turbidity that size and contents of the cells in the population of a given species do not appreciably affect the measurement of concentration.

Young cultures have larger and more varied sizes of cells than old cultures. Size becomes more uniform and smaller as the cultures age so that the cultures present typical textbook pictures after 24 h. Cells of *Staphylococcus aureus* at 4 to 6 h were twice the diameter of the old cells. Cells of *Escherichia coli* from a 2-h culture were more than three times as long as those from a 24-h culture. The length of many of the rod forms changed much more than the diameter.

The results of many investigators indicate that the photometric method, when suitably corrected for instrument response, measures cell mass, not cell number. The method gives an accurate measure of cell number only when all the cells are of the same size, as in the older cultures (and, perhaps, when grown in the presence of certain antibiotics). Large cells scatter more light than small cells. The influence of cell size on apparent concentration measured photometrically must always be kept in mind in interpreting curves constructed on the basis of such measurements.

The inherent bias caused by different cell sizes distorts some of the dosage-response curves but does not cause practical difficulties in assaying antibiotics. The growth curves obtained in assays of vitamins and amino acids will not be subject to distortion caused by cell sizes if the population density is limited by exhaustion of the factor assayed and not by a short time of growth.

Air Bubbles and Shaking Errors

The air bubbles formed and suspended in a broth suspension of bacteria by the usual vigorous shaking needed to completely suspend the bacteria interfere seriously with accurate measurement of turbidity. The measuring instrument responds to air bubbles as well as to bacteria.

Shaking the tubes causes three types of errors. One error results if the tubes are not shaken vigorously to suspend bacteria collected on the walls of the tube; the turbidity is low by about 10 percent when *Staphylococcus aureus* is the test organism. A second and larger error comes from air bubbles in a vigorously shaken tube when it is measured immediately after shaking. The third and smallest error is caused by sedimentation of the bacteria in a tube vigorously shaken and then allowed to stand until air bubbles rise. All errors are small if the shaken tubes stand about 15 min before being measured. Yeast settles very rapidly and must be resuspended before measuring turbidity.

Flow Birefringence

Turbidity of suspensions of rod forms of bacteria is more difficult to measure accurately than turbidity of suspensions of cocci because of flow birefringence resulting from movement of the rods. The movement is caused by motion imparted to the cells during filling of the cuvette and from thermal agitation caused by absorption of heat from the light beam. The combination of these two sources of motion causes the photometer to reach its balance point slowly and erratically if at all. The largest error in a vitamin assay may be that caused by flow birefringence.

The impracticability of accurate, quick measurement of turbidity of long rod forms of bacteria in the usual photometer is one reason for preferring cocci as the test organism, or, at worst, short rod forms.

Turbidity of suspensions of rod-shaped organisms can be measured without error caused by flow birefringence (Kuzel and Kavanagh, 1971a, 1971b).‡ This is done by measuring absorbance of suspensions flowing through an Arthur H. Thomas Co. 9120-NO5 flow cell at a rate of about 1.5 to 2 ml/s. No other flow cell has been found to be as satisfactory. The cell is fitted with polyethylene tubing and an 8-in length of 19-gauge stainless steel tubing as a sample probe. The vacuum is adjusted to obtain the proper flow rate. Frequently a small air bubble becomes trapped in the flow cell and causes an erroneous reading. The bubble can be removed by means of a water hammer produced by momentary shutting of the vacuum line between the flow cell and vacuum source. The line should be closed about 1 s after the start of the flow. A pinch clamp on a piece of rubber tubing is a suitable device for controlling flow rate.

Calibration of Photometers

For some problems in bacterial physiology absolute magnitudes of bacterial populations are not required, and the necessary information is conveyed by relative figures which are directly proportional to the quantity of bacteria. The general reference for this section is Kavanagh (1972b).

The important calibration is the establishment of the relation between the instrument response and the relative concentration of cells. The general shape of the calibration curve seems to be independent of the organism. The shape of the curve was the same for living and heat-killed *Staphylococcus aureus.*

The calibration curve is used to convert instrument scale readings either into relative concentrations of bacteria or into absolute numbers. The relative numbers are sufficient for constructing growth curves, for computing log-probability response curves, and for vitamin and amino acid assays. An absolute calibration for live bacteria is prepared by making a plate count on a suspension of the organism in the growth phase of interest. Bacteria in the log phase give very nearly the same total number and viable number. Since the bacteria are growing in a nutrient broth, the instrument is set to 100 percent transmittance (T) with broth to compensate for broth absorption.

The absolute calibration is not the same for all organisms. For one instrument, relative concentration of 100 represented 500 million of living *S. aureus* in the log phase, 800 million of living *Klebsiella pneumoniae*, or 1300 million of living *Salmonella gallinarum* cells per milliliter.

The photometer used to measure turbidity of bacterial suspensions should be calibrated if growth curves or the theoretical dose-response lines are to be constructed. The procedure is easy and direct.‡ Prepare a suspension measuring less than 10 percent transmittance, and kill by heating to 80 to 90°C. Centrifuge the suspension. Remove the broth, and wash the precipitate twice with 0.85% saline. Suspend the precipitate in 0.85% saline, and dilute to about 0.1 of the original volume of broth; let the suspension stand in a cylinder for an hour, and decant the upper 80 to 90 ml and use for preparing the calibrating suspensions. Dilute the suspension until it measures 10 percent transmittance. Assign the value of 100 to this stock suspension. Dilute it further with the aid of good volumetric pipets and volumetric flasks. Use saline solution as the diluent. Prepare a series of dilutions such as 10, 20, 40, 60, 80, and 100 of the stock suspension. Measure the turbidity of each as percent transmittance (T) under the condition of an assay after first setting the instrument to 100 percent transmittance with saline. Use a wavelength between 510 and 580 nm if a filter instrument is used or at about 550 nm in a spectrophotometer. Shorter wavelengths increase response to color of the medium faster than to the suspension. Measurement in the red (more than 600 nm) reduces both the influence of color of the medium and response to the suspension.

If the measurements of the standardizing suspension are made in terms of absorbency (optical density), plot the reading against concentration of suspension in the arbitrary units assigned before. If the measurements are in terms of percent transmittance, plot on semilog paper. Connect the points by a smooth curve. Use this calibration curve to convert instrument reading into numbers proportional to the concentration of bacteria (if all are of the same size). Use these relative concentrations of bacteria when preparing growth curves and log-probability plots of antibiotic dose-response lines (Kavanagh, 1963b). The instrument must be set to 100 percent transmittance with the same medium containing the suspension. In preparing dose-response lines of vitamin, amino acid, and antibiotic assays, this solution would be a sample of uninoculated medium.

A factor for converting turbidity of a live culture of each species of organism used in assaying into concentration of cells should be obtained. Thereafter, the concentration of cells in a living culture can be obtained by the rapid and convenient operation of measuring turbidity.

Concentration of Antibiotics and Turbidity

An equation (Kavanagh, 1968, 1972b) may be written to relate the concentration of a certain antibiotic substance and subsequent turbidity developed in the tubes of the turbidimetric assay. The equation has the form

$$\log N = F - BC \qquad (2\text{-}2)$$

where N = concentration of bacteria at end of incubation period
C = concentration of antibiotic
F, B = constants

The value of N will depend upon the initial concentration of bacteria in the broth, the growth rate in the absence of antibiotic, the concentration of antibiotic and its specific activity, and the incubation time. In an assay, all the factors except C and N are the same in all tubes. A dose-response line is constructed from the values of N corresponding to values of C. In an assay for penicillin G by *Staphylococcus aureus*, the dose-response line is very nearly straight. The best dose-response line is the one nearest to a straight line because errors of construction will be smaller than for a strongly curved line, especially if the latter is approximated by straight line segments.

Note well that response of bacteria, not of an instrument, is used in constructing the dose-response line, as was described earlier (Kavanagh, 1963b). The values of N or numbers proportional to N areobtained from the turbidity by means of the calibration curve constructed as described in the above subsection on calibration of photometers.

Equation (2-2) applies to those systems (and concentration) of antibiotic and organism in which the antibiotic reduces the growth rate of the organisms but does not kill them. Such systems are *S. aureus* and cephalosporins, erythromycin, penicillins, and tylosin; and *Escherichia coli* and tetracycline and chloramphenicol (Kavanagh, 1968). This equation can be used to guide selection of the range of concentrations of antibiotic to be assayed with minimized errors caused by errors in measuring turbidity. These concentrations are those near the upper concentration limit of a practical assay. The concentration of bacteria in

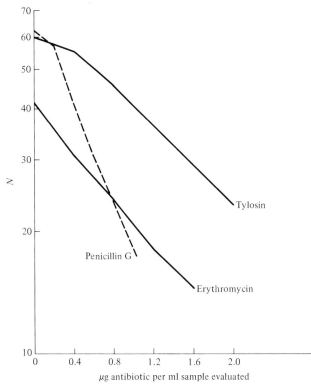

Figure 2-2 Examples of the dose-response line of three antibiotics using *Staphylococcus aureus* as the test organism. N = concentration of bacteria in the broth.

tubes containing these concentrations of antibiotic will be between 25 and 45 percent of the bacteria in the tube without antibiotic. Dose-response lines for three antibiotics are given in Fig. 2-2.

Antibiotic Standard Line

Dose-response lines relate concentration of organism to concentration of antibiotic whereas a standard curve or a calibration line relates instrument reading to concentration of antibiotics. A form of standard curve more convenient than Eq. (2-2) for antibiotic assays will be given (Kavanagh, 1972b). It is called the inverted logarithmic plot and is illustrated by Fig. 2-3. The data are in Table 2-6. The equation of the line has the form

$$\log (D - T) = E + BC \tag{2-3}$$

where B, E, D = constants
T = percent transmittance
C = concentration of antibiotic

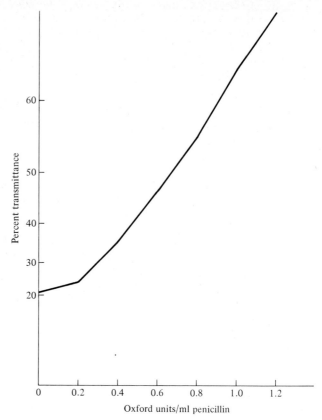

Figure 2–3 Penicillin G standard curve. (See Table 2–6 for details.)

Table 2-6 Penicillin G standard curve
Concentration C as units per milliliter of penicillin G in standard solutions; 0.5 ml of solution per assay tube. T is percent transmittance and A is the corresponding absorbance of the suspension of *Staphylococcus aureus* in the assay tubes after incubation.

C, units/ml	% T		Mean % T	A
0	20.8	20.8	20.8	0.682
0.2	24.0	24.2	24.1	0.618
0.4	35.9	35.7	35.8	0.446
0.6	46.7	46.9	46.8	0.330
0.8	55.6	55.4	55.5	0.256
1.0	63.4	63.5	63.4	0.197
1.2	68.0	67.8	67.9	0.168

The value of D will be between 70 and 100, depending upon details of the assay. The line ranges in shape from slightly sigmoid to straight when the value of D is selected to minimize curvature. Such values of D can be found if a computer is available to make the very large number of computations. When graphing by hand, values of D are most conveniently changed by one numbered division of the log scale on the graph paper.‡

Another convenient calibration line may be derived from Eq. (2-2) by substituting A for N to give

$$\log A = G + HC \qquad (2\text{-}4)$$

The line will be slightly more curved than that representing Eq. (2-2). Equation (2-4) is needed when the photometer scale is graduated in absorbance and not in transmittance or when the data are processed by a computer. The analyst must select the value of D in Eq. (2-3) before the computer can make the calculation. The computer does not need such instruction to calculate from Eq. (2-4).

pH in Assays for Antibiotics

The pH of assay media can have a profound influence upon antibiotic assays. A small change in the pH of the medium can cause a large change in the activity of the antibiotic. The direction of the change in activity depends upon the chemical characteristics of the antibiotic. The general rule is that the activity of an acidic antibiotic, such as the penicillins and cephalosporin derivatives, is less at the higher pH. Basic antibiotics (erythromycin, tylosin, streptomycin, and capreomycin) are always more active at the higher than the lower pH. The activity of a neutral antibiotic (chloramphenicol) is independent of the pH of the medium. The property of pH sensitivity can be used to regulate the range of an antibiotic assay because the free acids and free bases are the active compounds and their concentrations are a function of the pH of the solution.

Most antibiotics are diluted to assay level in pH 6, 7, or 8 phosphate buffers. The pH 6 buffer is used for the acidic antibiotics such as the penicillins and cephalosporins and the pH 8 buffers for the basic antibiotics. The assay medium also has a pH approximating that of the pH 7 buffer. The pH of buffers and media were selected to increase the sensitivity of the assays. For example, the penicillin G turbidimetric assay is about 1.5 times as sensitive at pH 6 as at pH 7 and the erythromycin assay about 5 times as sensitive at pH 7 as at pH 6. The tetracyclines are diluted in pH 4.5 buffer for stability reasons and assayed in broth at pH 7. To obtain maximum sensitivity when assaying a stable acidic antibiotic, as low a pH as will permit growth of a sensitive organism should be used.

Other diluents must be used when potassium ion interferes, as in the monensin assay, or phosphate as in the hygromycin B assay. Possible interference with the assay by buffer and medium components should always be kept in mind. Usually the interference shows as a reduction in sensitivity.

Ordinarily, the dilution of the sample in buffer is so great that the initial pH of the sample does not influence the pH of the diluted sample. However, there are situations where the dilutions are not very large; here initial pH can be very important and should be adjusted to that of the buffer before dilution is made.

Another class of samples in which pH is important is fairly low concentrations of antibiotics in highly concentrated salt solutions of high buffer capacity such as eluates from chromatographic columns. Unless the sample can be diluted sufficiently to reduce the buffer capacity of its salts to that of the buffer used as diluent, an accurate answer will not be obtained. When this situation occurs, the best that can be done is to realize the possibility of an inaccurate answer and not place too much dependence on it. The point of all this discussion is that sample and standard must be assayed in solutions of the same pH and buffer capacity if the answers are to be reliable. Fortunately, this problem is usually taken care of by the extent of the dilution of the sample.

Turbidimetric Assay for Vitamins

If the growth of bacteria is limited only by the amount of vitamin, the number of bacteria in the volume of medium is proportional to the amount of vitamin initially present. Twice as much vitamin gives twice as many bacteria, etc. No increase in number of bacteria over the inoculum occurs if the inoculum and medium are free from vitamin. This relation between amount of vitamin and the number of bacteria is linear, not logarithmic. An apparent logarithmic relation may be obtained when the medium is inadequate or the time too short for the growth to be limited solely by the amount of vitamin.

The usual published dose-response line seems to be a many-point curve of percent transmittance or optical density plotted against the logarithm of vitamin concentration. These curves, if the graph is smooth, can be used in practical assaying. For disadvantages of such lines, see Kavanagh (1972b).

Defects in medium and operations (incubation period, for example) are detected much more easily from the theoretically correct linear plot than from the empirical logarithmic graph. Responses of standard and samples when plotted as concentration of bacteria versus concentration of vitamin form a series of straight lines with a common zero point and different slopes. The same measurements plotted as percent transmittance versus logarithm of concentration form a series of approximately straight lines with nearly the same slope.

The equipment available to most who do vitamin assays measures turbidity of suspensions of rod-shaped bacteria, such as those used in vitamin assays, with such large error that deviation of the lines from linearity and parallelism goes undetected. Even such data properly graphed could reveal gross inadequacies of medium and technique.

The logarithmic presentation of the data loses part of the information and requires an unnecessarily large number of points. This will be illustrated by referring to vitamin B_{12} assays (Kavanagh, 1963b; Skeggs, 1963). If the basal medium is adequate and not contaminated by the vitamin, a graph of concentration of bacteria against concentration (or amount) of vitamin in the tube should be a straight line through the origin. This has been obtained in vitamin B_{12} assays. A line above the origin (growth in the zero tube) indicates contamination of the medium. A line concave upward near the origin and showing curvature indicates destruction of vitamin B_{12} by reason of inadequate reducing agent. Since the line is straight, two points and the zero tube establish the line and its straightness. Therefore, the theoretical plot is more efficient than the logarithmic plot because many fewer tubes give the same information. An indication of quality of medium can be obtained as a bonus from using the theoretical relationship and an additional set of tubes with only a small amount, say one-tenth, of the maximum level.

DIFFUSION METHODS

Introduction

The three diffusion methods in general use for quantitation of antibiotic potency are the disc, the cylinder plate, and the cup plate. In these methods, a small amount of a solution of active substance confined to a small area of an agar plate seeded with the test organism is incubated for 16 to 18 h. The diameter of the zone of inhibition is related to the logarithm of the concentration of the antibiotic. The theory of the formation of the zone of inhibition developed by Cooper and his associates in publications beginning in 1946 will be followed here. Diffusion assays can be modified to provide a test for any antibiotic. Special tests can be made very sensitive, and, if a chromatographic step is included, can be specific. Thin-layer chromatograms may be converted into the very sensitive assay needed for tissue residue studies (Kline and Golab, 1965).

The extensive review by Cooper (1963) should be consulted for the many details that are not given here. All that can be done is to give the theoretical equation derived by Cooper for the cup-plate method and to explain the meaning of the terms of the equation.

$$X^2 = 4DT_0 \ln \frac{m_0}{m'} = 9.2DT_0 \log \frac{m_0}{m'} \tag{2-5}$$

where D = diffusion coefficient of antibiotic in water at incubation temperature
T_0 = time required to form edge of zone of inhibition
m_0 = initial concentration in cup (hole in layer of agar)
m' = critical concentration of antibiotic
X = distance diffused

Zone diameter = $2X$ + diameter of hole, inside diameter of cylinder, or diameter of disc. The concepts of critical time T_0 and critical concentration of antibiotic, m', are crucial to the theory of the formation of diffusion zones. The value of m' is a function of the composition of the medium and may or may not be independent of incubation temperature. Concentration of antibiotic at the zone edge is m'. The critical time, when the zone is determined, is also related to N', the critical concentration of bacteria,

$$T_0 = L + 3.32G \log \frac{N'}{N_0} \tag{2-6}$$

where L = lag time
G = generation time
N_0 = initial concentration of bacteria

The critical time is the time required for the population to increase from N_0 to a particular value N', which is called the critical population. Critical time will thus vary with factors affecting lag time (L), generation time (G), size of inoculum (N_0), or the critical population (N'). For a particular strain of bacteria, incubation temperature and composition of the medium influence G.

The values of N_0 and G can be controlled easily. Therefore, T_0 should have very little daily variation in a well-run laboratory.

Equations (2-5) and (2-6) may be combined into the general equation (2-7) which can be used to explain the influence of temperature, medium composition, and inoculum size upon zone size (Kavanagh, 1974).

$$X^2 = \left(9.2D \log \frac{m_0}{m'}\right)\left(L + 3.32G \log \frac{N'}{N_0}\right) \tag{2-7}$$

Zone size is increased by a small inoculum (N_0), anything increasing generation time, increase in concentration of antibiotic in the cup or cylinder, and reduction in thickness of the agar in the cylinder-plate method. All these factors can be controlled in practical assays so that the same concentration of antibiotic will give nearly the same zone size each time (usually daily). For application of the Cooper equation to diffusion assays, see Kavanagh (1974, 1975b).

Selecting the Dose-Response Line

For the purposes of this discussion the theoretical dose-response line (Cooper, 1963) can be put into the form

$$\log C = F + BX^2 \tag{2-8}$$

where C is concentration of active substances and X is the distance diffused. Over a short range of C, the equation approximates as

$$\log C = H + Bd \tag{2-9}$$

where d is the zone diameter. When the range of concentrations is more than fourfold, the quadratic equation is more likely to fit the points than the linear equation. Nonetheless, there are numerous examples in the literature and in practice of the "best straight line" being drawn through a set of points following the quadratic equation. This practice can put a serious bias in the potencies of samples of almost any concentration. If a long standard curve must be used because the samples have truly unknown potencies, the curve should be drawn point to point. The best practice is to use the shortest dose-response line that will encompass the samples. For control of purified antibiotics, pharmaceutical preparations, and the like, a twofold curve is adequate because potencies of the samples are known to within ±25 percent of the true values before the assay.

The spacings of the concentrations of the standard curve form a geometric series in diffusion assays in contrast to the arithmetic series of the turbidimetric assay. The concentrations of the standard curve are related by a constant multiplier which, in practice, may be any value from 1.25 to 4. In the penicillin assay by the diffusion method given in the later section Methods of Assay, the multiplier is 2.

Applying statistical analysis to potencies calculated from the linear equation when the quadratic equation is the true representation of the data is a waste of time.

Measuring the Response

Response in diffusion assays is the appearance of circular translucent areas or zones surrounding the cylinder or disc containing the antibiotic. The zones are bounded at their periphery by an intensely opaque area of bacterial growth. Any device which will accurately measure the diameter of these circular zones will suffice in this phase of the assay. An assortment of calipers, millimeter rules, scale projectors, and automatic and semiautomatic measuring devices have been utilized for this purpose. The selection of this equipment will undoubtedly be governed by the number of samples being assayed, the degree of precision needed, and the allowable cost.

The use of "zone readers" specifically designed for this purpose normally requires the careful and symmetrical placement of cylinders or discs on the plates in the earlier stages of the assay. This is to ensure that, upon rotation, the zone will be in the field of view of the measuring instrument. A simple template placed under the plates serves to do this.

With anything short of fully automatic equipment, the principal error in measuring zone diameter is that of the human eye in distinguishing the boundary between the contrasting areas. In fact, even a fully automated system may not be capable of perfect consistency in this regard. The error involved is usually rather small except in those assays where there is very poor definition of zones or where a "halo" effect is seen. Many times these undesirable effects are the result of trying to extend the lower limit of sensitivity of the assay. A valid assay is possible only when standard and sample give the same type of zones.

Computation of Answers

Construct the standard dose-response curve by plotting the average corrected response value (millimeters) for each of the standard levels against the logarithm of the respective doses and by joining these plotted points. Normally this is simplified somewhat by plotting this curve on semilogarithmic graph paper, with the doses represented by the logarithmic scale.

Correct the mean sample responses for plate variations and interpolate from the standard curve to obtain the potency of the diluted sample. Details of this procedure are given in the discussion of the assay for penicillin in the later section Methods of Assay. Multiply the reading obtained by the dilution factor for the particular sample to obtain the potency of that sample.

Validity of Assay

The degree to which this concept is pursued will probably hinge upon the purpose of the assay, the time and equipment available for computation, and the statistical talents of the analyst. With only slight modification, the one-dose assay method described can be converted to a two-dose technique which allows for ascertaining the parallelism of the standard and sample response lines. The sample should always be assayed at two concentrations when the materials are very complex or when a new sample type is presented to the analyst. If assay conditions are altered, or if a new diluent is proposed, it would certainly be well to demonstrate that the response line is truly linear over the range of doses being considered and that standard and sample dosage-response lines have the same slope. Many other applications of statistical methods present themselves in these methods but are beyond the scope of this discussion. Those interested in pursuing this matter further will find the book by Hewitt (1976) to be an excellent guide to applications of statistical procedures to microbiological assaying.

The assumption is made that the standard and sample dose-response lines have the same slope and curvature in a valid assay. How much they may differ and how much the assay may be considered to be valid is difficult to answer and can be ascertained only by experience (see Hewitt, 1976).

COMPARISON OF PLATE AND TURBIDIMETRIC METHODS

More than 30 years of experience in assaying for antibiotics has shown that one of the growth responses, turbidimetric or diffusion, is applicable to most samples. Other responses are used comparatively infrequently.

The turbidimetric and plate (diffusion) methods are equally satisfactory when applied to assay of pure substances and can be about equally accurate. The two methods may give somewhat different answers when applied to impure materials. The impurities may consist of substances enhancing or decreasing response of the organism to the compound being assayed. Such substances are particularly troublesome in the turbidimetric assay (Kavanagh, 1975a). Unless compensation for the effect of the impurities can be achieved or their influence swamped, assays in their presence will be biased. Often, but not always, a diffusion method is less sensitive to the presence of interfering substances. This is true if the compound being assayed diffuses faster than the interfering materials. An assay is assumed to be free from interference if the potency of the sample is independent of the concentration assayed and if small concentrations of standard added to the several concentrations of sample are recovered within the limits of the error of the assay.

A particularly troublesome situation is the one in which standard and samples contain several active components of a complex in different ratios. Examples are bacitracin and neomycin. To add to the confusion, regulatory agencies may employ standards of different composition, which also may differ from those used by the manufacturer to assay the same sample; further, the composition of the sample may differ substantially from the standards. The result is three different values for the same lot of bacitracin, the purchaser being the third party. The three parties also may use different plate methods. Who has the correct answer? Each answer is correct because a microbiological assay measures *activity* in terms of the standard. Use of a common standard and method would make possible agreement as to potency. The same argument applies to neomycin and other antibiotics occurring as mixtures.

Method is not important when the standard and samples are essentially pure as, for example, penicillin G and erythromycin. Disagreements about their potency between supplier and customer or either and a regulatory agency are rare.

Although either the plate or turbidimetric method usually may be used, which one is selected may depend upon the preference of the analyst and the equipment available. A laboratory equipped to employ the large-plate method and experienced with it will tend to employ it to the exclusion of other methods. Modern turbidimetric methods are accurate and require less work than plate methods to achieve the same precision. A laboratory with semiautomatic or automatic turbidimetric equipment will do as many as possible of its samples turbidimetrically. It will use a plate method only when turbidimetric methods are not applicable, because of the much greater efficiency of turbidimetric methods. Most American pharmaceutical companies use a semiautomatic assay system (Kuzel and Kavanagh, 1971b) to assay for antibiotics and vitamins. It is also used by food manufacturers to assay for vitamins and amino acids.

METHODS OF ASSAY
SERIAL DILUTION
Introduction

Cell multiplication as an assay response is quantitated usually by some form of turbidimetric measurement because it is more convenient for large numbers of samples than a direct count of numbers or measurement of dry weight of cells.

The simplest form of the turbidimetric method is the end-point test where the response is recorded as all or none. This is the serial dilution test used so extensively in serology. It was one of the tests used in the early antibiotic work and was adequate where the isolation steps were essentially all or none, as in many solvent extraction procedures. It was economical of time and material.

Several dilutions of the antibacterial substance in small tubes are inoculated with a test organism and incubated, and the lowest concentration of the substance which causes apparently complete inhibition of growth of the organism is taken to be the minimum inhibitory concentration. The activity of the compound is computed from the minimum inhibitory concentration and may be reported as micrograms of substance per milliliter. For many purposes, the answers so obtained are as useful as those given by the more laborious plate or turbidimetric assays. One good operator can do all the work required to put on 60 samples in a working time of about 4 h, including reading and recording the answers. No elaborate equipment is needed. The details for doing three variants of the serial dilution method with different accuracies are given by Kavanagh (1947, 1963a).

The simple twofold serial dilution method gives a geometric series of dilutions in which the answer lies between a concentration equal to and one-half as much as the one reported. Supposedly it could be any value in between. Thus the serial dilution method has such a large inherent uncertainty that it is not suited to precise determination of quantity. It should never be used for control work where an error of 10 percent would be considered large. Nonetheless, it is a useful method for guiding chemical operations of the all-or-none sort and in hospitals for measuring sensitivity of pathogens to antibiotics.

PHOTOMETRIC ASSAYS
Introduction

Photometric methods are used for assay of growth-promoting and growth-inhibiting substances. An example of the former assay is given in Chap. 3, and only the assay of inhibitory substances will be considered in this chapter.

Although the example of assay to be given here uses penicillin G assay as a model, the assay is easily modified to be applicable to other antibiotics. Most substances active against *Staphylococcus aureus* can be assayed in the penicillin system by changing the standard and its range of concentrations. The data in Table 2-7 were obtained by just one change in the penicillin system of changing the active substance. The range of concentrations of most of the antibiotics would be different if the pH of the buffer and medium were changed, as discussed on page 29.

Antibiotics active against gram-negative bacteria usually should be assayed with the aid of one such species. Streptomycin, for example, has very little activity in the penicillin system and is best assayed with a strain of *Klebsiella pneumoniae* (ATCC 10031) sensitive to it. This strain was selected because its population is uniformly sensitive to streptothricin in contrast to the Heatley strain of *S. aureus*. Selection of assay strains is discussed by Kavanagh (1947, 1963b). Such antibiotics as streptomycin, chloramphenicol, tetracyclines, cephalosporin C, neomycin, and viomycin may be assayed by species of gram-negative bacteria.

Assays for penicillin G will serve to illustrate general principles and methods.‡ The photometric method has a range from 0.4 to 2 units/ml of penicillin G solution. The method is applicable to samples not grossly contaminated with bacteria, and free from penicillinase. The solutions should be clear and uncolored when diluted to the assay concentration. Bacteria and suspended solids, but not penicillinase, can be removed by Seitz filtration. The assay method is capable of good accuracy, and the answer can be obtained within 5 h after the sample is submitted for assay. It is not suitable for assay of blood because of interference by proteins.

Table 2-7 Response of *Staphylococcus aureus* to several antibiotics in the system used for measuring penicillin. The concentrations are those giving 20 to 80 percent growth and 50 percent growth.

Antibiotic	pH 7	
	80–20%	50%
Erythromycin, µg/ml	0.5–2	0.9
Monensin, µg/ml	0.6–80	14
Penicillin G, µg/ml	0.3–1.4	0.5
Tylosin, µg/ml	0.7–3	1.5

Test Organism

STOCK CULTURES

The assay organism for penicillin is the Heatley strain of *Staphylococcus aureus* (ATCC 9144). It forms a good suspension without lumps and does not give a precipitate of extraneous material when the suspensions are heated at the end of the incubation period.

Fresh slant cultures should be prepared each week. The slants should be kept in a refrigerator, and the set used for a week. Freshly slanted tubes of Penassay seed agar are satisfactory. Never transfer from liquid to liquid.

STANDARD INOCULUM

Inoculum can be prepared in several ways. The important point is to have an adequate concentration of bacteria. Transfer a loop full of bacteria from a slant to 100 ml of nutrient broth in a 250-ml flask. Incubate the inoculated flask for 12 h at 36°C without shaking.

Add 20 ml of this inoculum to each liter of assay broth just before the broth is added to the assay tubes. Do not shake the inoculum flask before decanting the suspension into a graduated cylinder for measurement of volume.

Daily preparation of inoculum may be neither necessary nor advantageous. A suspension stable for a week can be made by washing the bacteria from a large agar surface (flat bottle) with pH 7.0 phosphate buffer. To grow the bacteria, inoculate an agar surface with a heavy suspension of bacteria, grow for from 4 to 6 h at 37°C, and wash off with buffer. Chill the suspension and keep it in the refrigerator where the viability may decrease by only 10 percent in a week. *Staphylococcus aureus* dies more rapidly in nutrient broth than in buffer.

Standard Solutions

Make the primary stock solution in sterile pH 6.0 phosphate buffer. Dissolve a sample of about 25 mg of standard in the buffer and dilute to 25 ml in a volumetric flask. Compute the concentration in terms of units per milliliter and make further dilutions as indicated later. Some prefer to make the stock solution of exactly 1000 units/mg by weighing 62.7 mg of the potassium salt and diluting to 100 ml with sterile buffer.

The stock solution of penicillin G is fairly stable if prepared without gross contamination and kept in the refrigerator at 5°C. The loss in activity in a week may be no more than 5 percent. It is good practice to prepare the stock solution at least every other day.

Make working standards by diluting the stock solution with pH 6 buffer to the concentrations needed for preparing the standard curve. Prepare immediately before use and discard after the test is set up.

Preparation of Samples for Assay

TREATMENT OF SAMPLES

Samples submitted for assay may be liquids, solids, or semisolids such as tissues. The liquids may be nearly pure solutions, whole beers from a fermentation area, or process samples with widely varying purities. The solids may be pure salts, mixes, tablets from a manufacturing area, or agricultural grade mixes. Examples of several classes of samples will be considered.

Penicillin G is sensitive to acid, alkali, high temperature, and penicillinase and needs to be handled rapidly at low temperatures if marked loss is to be avoided. These sensitivities of penicillins must be kept in mind when devising extraction procedures. Freezing of solutions can cause considerable loss of penicillin.

Fermentation beers Filter the samples through filter paper (Whatman No. 1) to remove mycelium, calcium carbonate, and oils. Dilute the slightly turbid or clear filtrate with sterile pH 6 phosphate buffer to approximately 1 to 1.5 units/ml. These filtered samples are not sterile and will become grossly contaminated if allowed to incubate at room temperature for several hours. The lack of strict sterility is of no disadvantage so long as the dilutions are kept cool and are assayed promptly. Should the samples be contaminated with a penicillinase-producing bacterium, appreciable loss of penicillin occurs rapidly.

Powders Dissolve about 50 mg of penicillin powder in pH 6 buffer and dilute to 50 ml in a volumetric flask. Make further dilutions in buffer as needed for the particular assay. If the sample contains the procaine salt of penicillin G, dissolve it in 5 ml of absolute methanol and then dilute immediately with buffer. Remove starch and other methanol and water-insoluble materials by filtration, or just let the solids settle to the bottom of the flask and remove the sample (1 to 2 ml) from the clear portion.

Mechanics of the Assay

The range of the assay is from 0.3 to about 2 units of penicillin G per milliliter of test solution; the best part of the calibration curve lies between 0.4 and 2 units/ml. (The concentration of penicillin G in the assay broth is about $\frac{1}{21}$ of these concentrations.) Prepare the standard response curve by pipetting 0.50 ml of standards of concentrations 0 (buffer), 0.5, 1, 1.5, 2 units/ml into clean, sterile, 18 × 150 mm culture tubes. Use a 1-ml measuring pipet. Make at least four of the 0 tubes and two of each of the standard solutions. Dilute the sample to an estimated concentration of 1 to 1.5 units/ml and pipet 0.5 ml into each of two tubes. Usually the potency of the sample will be known to lie between rather narrow limits. If not, then estimate the minimum concentration, dilute to the estimated 1 to 1.5 units/ml, and prepare further dilutions in steps of 4 (4, 16, 64, 256). At least one of the dilutions will fall on a usable portion of the calibration curve.

If pipetting is well done with accurate pipets, two tubes of each level of standard and sample are enough. Although there is no statistical advantage to replicating assay tubes more than twice, there is an advantage to including a set of standards in each rack. A rack of 72 (6 × 12) tubes contains 5 pairs of tubes of standards and 31 pairs of tubes of samples. Each rack constitutes an assay.

When all the samples have been pipetted, add 10 ml of inoculated penicillin assay broth rapidly with a pipetting machine. As soon as all the tubes have received broth, place the racks of tubes in a well-stirred 36 to 38°C water bath. The stirring must not be vigorous enough to cause the tubes to move even the slightest. The exact volume of broth added is not important as long as the same volume is added to each tube. Incubate the test for at least 3 h. At this time, measure the turbidity of the 0 tubes, and continue to do so every 15 min until the population density reaches about 350 million cells per milliliter. At this point, transfer the racks quickly to an 80°C water bath for 1 min to kill the bacteria. Put the racks of hot tubes into a water bath at about 25°C (tap water) to cool the solutions to room temperature.

The incubation period, at least 3 h, should be long enough so that additional incubation changes the sensitivity of the assay but not the slope of the log-probability dose-response line. In routine assays, the conditions are held constant, and incubation is the same day after day.

Influence of extraneous materials upon the penicillin assay generally is not of concern because the dilution is usually large enough to eliminate such influences. If the concentration should be so low that only small dilutions (50×) are possible, then the sample should be assayed at several dilutions (0.4, 0.6, 0.9, 1.2 units/ml). The concentration in the undiluted sample as estimated from the two extreme concentrations should be within 10 percent of that computed from the 0.9 unit/ml dilution, and there should be no trend to the figures.

Response and Answers

The turbidity of the solutions should be measured with a photometer or turbidimeter. All the commonly used instruments show a nonlinear relation between scale reading and concentration of bacteria above a rather small concentration. The instrument should be fitted with a quick-emptying cuvette.

Shake the tubes several times by inversion to resuspend bacteria that settled on the bottom of the tube during the incubation and killing periods. Let the tubes stand at least 5 min before reading the test. Set the instrument to 100 percent transmittance with uninoculate broth. Read all tubes of the standard curve in order of decreasing penicillin concentration. Then measure the sets of sample tubes. Readings made sooner than 10 to 20 s after filling the cuvette of the photometer may be in error by unknown and variable amounts because of the presence of air bubbles. The readings on the pairs of tubes should agree within 1 percent of the full scale (as represented by the reading of the 0 tube).

Plot the mean instrument readings according to Fig. 2-3. Draw the standard curve through the points obtained by averaging the two readings for each concentration. Average the readings for the two tubes of a sample and obtain the concentration from the calibration curve. Multiply the concentration so obtained by the dilution of the sample to obtain the concentration of penicillin in the undiluted sample.

Meaning of the Assay

The answers are obtained in terms of units of penicillin G activity. The meaning of the answer depends upon the composition of the sample because a bioassay measures activity, not quantity. If only penicillin G is present, the answer is in units of penicillin G. If other penicillins are present (F, dihydro F, and K, as in fermentation samples), the answer is in terms of penicillin G activity which cannot be precisely computed into total units of penicillin because the several penicillins are present in unknown amounts and each has a specific activity against *Staphylococcus aureus* different from that of penicillin G. If the sample contains much K, then the assay in terms of G will be higher than it would have been had the K been replaced by a molecular equivalent of G.

DIFFUSION METHOD OF ASSAY FOR PENICILLIN
Introduction

The plate method is more difficult to use accurately than the photometric method. It is applicable to nonsterile samples (free from penicillinase), to bloods, and to samples containing slowly diffusing substances which would interfere with the photometric method.

The range of the plate assay is from 0.025 to 0.1 unit/ml when *Bacillus subtilis* is the test organism. The time interval between submission of sample and computation of answer is about 24 h.

Although a number of different organisms have on occasion been utilized by various workers for the bioassay of penicillin by the cylinder-plate technique, only two have achieved widespread recognition and acceptance. The first from the standpoint of use in the United States is *Staphylococcus aureus* (ATCC 6538P). The second organism commonly used (in Europe) is *B. subtilis*. This organism is somewhat more sensitive than *S. aureus* to penicillin and has found particular application in the assay of biological fluids, animal tissues, feedstuffs, and the like. In these materials, the antibiotic is frequently present in microgram or even submicrogram amounts and is often accompanied by active nonpenicillin factors. These extraneous materials, whether inhibitory or stimulatory to the test organism, must be removed or adequately diluted to obtain an accurate estimation of the penicillin content of the material being assayed. The greater susceptibility of *B. subtilis* to penicillin will, in most cases, permit dilution of these factors to such an extent as to eliminate the need for their removal. Also *B. subtilis* forms a much better-defined zone edge than does *S. aureus*.

The methodology of typical assays employing *B. subtilis* will be presented in detail as a classic example of antibiotic plate assays.‡

Many factors such as pH, temperature, time, concentration of agar, and diffusion constants play vital roles in the successful conduct of these assays. The device (paper disc, cylinder, etc.) for applying the test solutions may affect the diffusion of the antibiotic and thus alter the range of the dose-response curve.

Test Organism

The test organism is *Bacillus subtilis* (ATCC 6633) used in the form of a spore suspension. Two methods for producing the spores will be given.

Prepare a slant culture by incubating the inoculated slant of Trypticase soy agar for 7 days at 36 to 37°C to permit spore formation. Refrigerate the covered slant and transfer every 6 months to fresh medium.

Spore suspension may be prepared in liquid if a shaker is available. Put 800 ml of Trypticase soy broth containing 30 mg/l of manganese sulfate monohydrate into a 2-l Erlenmeyer flask, plug, and sterilize. Inoculate the broth with the surface growth of one slant and incubate at 37°C overnight. Then place the flask on a reciprocal shaker at 37°C for 5 to 7 days or until a stained smear shows spores in 80 to 90 percent of the cells. Harvest the spores by centrifuging in sterile bottles. Decant the supernatant liquid, add a few sterile beads and 100 ml of 0.85% NaCl solution, shake to suspend the spores, and centrifuge. Repeat. Decant the supernatant and add about 100 ml of saline to each bottle, suspend the spores, and pour the contents into a 500-ml plugged flask. Pasteurize by heating in the water bath at 65°C for 30 min. Make a viable spore count, and measure the turbidity with a calibrated photometer. Such a spore suspension, if kept in the refrigerator, may be used for 5 years or more.

The spores may be produced on an agar surface. Fill a toxin bottle about one-third full of a special agar (add 5 mg of $MnSO_4 \cdot H_2O$ to 1 l of Trypticase soy agar), sterilize, and lay on a flat side. Wash the material from a slant culture of *B. subtilis* with 10 ml of saline solution and transfer to the agar surface in the toxin bottle. Incubate for 7 days at 37°C. Make a smear, and harvest if more than 80 percent of the cells contain spores. Wash the growth from the agar with 75 ml of saline solution. Pasteurize at 65°C for 30 min, centrifuge, decant, resuspend in 75 ml of saline, and repeat twice or until the supernatant liquid is clear. Transfer the suspension to a stoppered flask and store in the refrigerator. Measure turbidity and make a viable spore count. As a first approximation to the proper amount of inoculum, use 0.2 ml of a 1:10 dilution in saline to inoculate 100 ml of assay agar.

There is no fixed rule governing the concentration of spores that is best to use in a diffusion assay. This can be determined only by trial. The edge of the inhibition zones in the range of 14 to 20 mm should be well defined and smooth. Test several dilutions of the spore suspension to find the range that gives an easily read zone. The concentration of spores can be calculated from the viable spore count and the volume of inoculum used. Use this level of inoculum in subsequent assays.

Each lot of spore suspension must be standardized. Once the best concentration of spores to use in a test has been determined, subsequent lots can be standardized by means of measured turbidity and the calibration obtained on the first lot of spores.

Standard Solutions

Prepare a stock solution of the appropriate penicillin working standard as described above in the subsection on photometric assays. Use the stock solutions for not more than 2 days. Dilute samples of the stock standard solution in sterile pH 6.0 phosphate buffer immediately prior to use on the day of the assay. Prepare solutions with concentrations of 0.025, 0.05, and 0.1 unit/ml.

Not all penicillins or penicillin derivatives exhibit identical dose-response curves even under identical assay conditions and, therefore, the standard must be identical to the compound being assayed.

In preparing the standard response curve it should be remembered that the buffer diluent mentioned above applies only when samples can be similarly diluted. If the activity is present in biological fluids, organic solvents, etc., the standards must be diluted to contain identical concentrations of these materials or the assay may be invalid. All too frequently, the erroneous assumption that 5 to 10 percent extraneous materials would not affect the assay has caused considerable embarrassment to the analyst.

Sample Preparation

In general, penicillin solutions and bulk powders may be diluted directly in pH 6.0 phosphate buffer prior to assay. If the particular compound is not sufficiently water-soluble, a small quantity of solvent such as formamide or methanol may be required to solubilize the sample prior to dilution in buffer.

Although space does not permit a detailed description of the many techniques of preparing samples, a few precautions are in order. Since a variety of organic solvents may be used in preparing the samples, the possible effect of these solvents on the assay system must be considered. If the solvents are known not to interfere, use the same concentration of solvents in preparing the standard curve (a compensated standard). If the samples vary widely in pH, adjust the pH to that of the standard curve even though subsequent dilutions are to be made in buffer. This is particularly important if dilutions are small in magnitude (1:5) as the capacity of the buffer may not be sufficient to achieve the required pH.

Mechanics of the Assay

DESIGN

Like other facets of the assay, the design itself may be varied somewhat to fulfill the needs of the assayist. Rather elaborate statistical models may be devised to answer specific questions regarding the reproducibility of the method, the effect of environmental changes, or the homogeneity of the materials to be assayed. In routine practice, however, the common approach is to accept the slightly greater variability of the simpler designs and to attempt to achieve the desired precision by means of replication. Perhaps the main drawback in the design is that it represents a one-dose assay; that is, only one concentration of the unknown is measured against the response curve. Obviously, reading at a single point on a line or curve does not enable any predictions to be made regarding the parallelism of the sample dose-response line with that of the standard. Samples likely to contain contaminants with microbiological activity, either stimulatory or inhibitory, should be assayed at two or more dilutions. If this is done and parallelism is demonstrated, the assayist may have considerably more faith in the validity of the results than one can have in a one-dose assay.

MEDIUM

Pour a layer of 10 ml of inoculated Penassay seed agar into the plate and allow it to harden. One of the most important steps in the entire assay is the pouring of agar layers of uniform depth, and great care should be exercised in ensuring that the plates sit on a level surface for this operation.

Two-layer plates may be used. Pour a bottom layer of 20 ml of uninoculated Penassay base agar. When it is solidified, carefully spread a 5-ml layer of inoculated Penassay seed agar on it. One advantage of using *Bacillus subtilis* spore inoculum is the possibility of inoculating and pouring a hot seed layer (~60°C). If a heat-sensitive organism were used, *Staphylococcus aureus* for example, the seed agar would have to be cooled to about 46°C before inoculating. The agar is viscous at this temperature and solidifies quickly when put on to the lower layer. Use of a thicker layer of agar (total 15 to 25 ml) reduces sensitivity of the test. Compensation for this should be made when selecting the lowest concentration of standard. For best zones and lines, the lowest concentration should give a clear zone at least 4 mm wide around the cylinder or disc.

INCUBATION

Incubate the *Bacillus subtilis* plates at 30 to 35°C overnight. Stacks of petri plates, whether glass or plastic, are relatively poor conductors of heat, and considerable plate-to-plate variation will occur if the plates are incubated in stacks. Incubate the plates in a layer one plate deep.

Obviously, in a cylinder-plate assay (as opposed to the paper disc), considerable care must be taken in transporting plates to the incubator and in placing them on the shelves. Spillage from the cylinders causes irregularly shaped zones which cannot be interpreted.

CALCULATION OF POTENCY

Measure the zone diameter and proceed as described in the discussion of diffusion methods in the section Obtaining and Measuring Responses.

Operational Details in Petri Dish Methods

TWOFOLD STANDARD CURVE

Operational details of petri dish assays will be illustrated by a slight modification of the single-dose assay used for 30 years by the Food and Drug Administration.

The two standards span a twofold interval, such as 1 and 2, and the sample is diluted to be between the standards. There are three plates of standards and three for each sample. Each plate has six cylinders (or discs) on it.

Fill three cylinders on the standard plates with the low standard and three with the high standard. The odd-numbered positions receive the lower concentration.

On the sample plates, fill three cylinders (even-numbered positions) with the high standard solution (reference) and the other three with the sample solution.

As soon as the cylinders on a set of three plates have been filled, put them into the incubator in a layer one plate thick.

Add 0.2 ml of solutions to the cylinders with an automatic pipet. Discs may be loaded by dipping into the solutions. However, more precise results will be obtained if the discs are loaded by means of an automatic pipet. A 12.5-mm-diameter disc can be loaded with about 0.1 ml of solution.

After overnight incubation, measure the diameter of the clear zones around the cylinders (or discs) as accurately as possible with the equipment available.

Find the mean of the nine zones of the low standard and the mean of the nine zones of the high standard. Plot the two points on semilogarthmic single-decade paper with concentrations on the log scale. Draw a straight line connecting the two points.

Measure the zones on the sample plates and find the mean of the nine reference zones and the mean of the nine sample zones. Differences in prediffusion time, incubation temperature, etc., cause plate-to-plate variation in zone size. As has been shown (Kavanagh, 1974), most of such variations can be removed by applying what is known as the FDA correction. The high standard and the reference are the same solution; therefore the zone sizes should be the same in the absence of operational variations. The correction to be applied to the sample zones is the number that must be added (algebraically) to the size of the reference zone to make it identical with the high standard. This correction usually is less than 2 mm. The corrected mean sample zone diameter is entered on the calibration line to find the concentration of standard that would give a zone of the same size. The number is taken to be the concentration of the sample.

The simple, two-standard short-range assay is adequate for samples with fairly precisely known potency but not for true unknowns. For these samples, a wider-range standard curve is required.

If a fourfold standard would be adequate, place three standards in ratios of 1, 2, 4 on each of four plates and treat as for the two-standard test. Use three plates for the sample, with the middle concentration of standard as the reference. Draw the standard line from point to point.

If a wider range of standard is needed, put five concentrations with ratios of 1, 2, 4, 8, and 16 on six plates. Use two plates for each sample with the 4 standard as reference. Draw the standard curve from point to point. The same number of zones is obtained for each standard and each sample. This assay should be used where highest precision is not needed or possible. If used as a range-finding assay, the sample should be assayed again after dilution to fall within the range of the twofold assay.

MEDIA AND BUFFERS

Concentrations are given in grams per liter of water.
A. Media

Penicillin assay broth, G & R No. 3	
Peptone	5
Yeast extract	1.5
Beef extract	1.5
Dextrose	1
NaCl	3.5
KH_2PO_4	1.32
K_2HPO_4	3.68

G & R No. 3 + 0.5% dextrose
Add 5 g of dextrose to a liter of G & R No. 3 broth.

Penassay seed agar, G & R No. 1

Peptone	6
Pancreatic digest of casein	4
Yeast extract	3
Beef extract	1
Dextrose	1
Agar	1.5

Penassay base agar, G & R No. 2

Peptone	6
Yeast extract	3
Beef extract	1
Agar	1.5

The media in prepared form are available from several manufacturers.

LITERATURE CITED

AOAC (1975): "Official Methods of Analysis of the Association of Official Analytical Chemists," 12th ed., Association of Official Analytical Chemists, Washington, D.C.

Arret, B., D. P. Johnson, and A. Kirshbaum (1971): *J. Pharm. Sci.,* **60**:1689.

Barton-Wright, E. C. (1962): "Practical Methods for the Microbiological Assay of the Vitamin B-Complex and Amino Acids," 52 pp. United Trade Press, London.

Cooper, K. E. (1963): in F. Kavanagh (ed.), "Analytical Microbiology," pp. 1–86, Academic, New York.

Freed, M. (1966): "Methods of Vitamin Assay," 3d ed., prepared and edited by The Association of Vitamin Chemists, Inc., 424 pp., Interscience, New York.

Gerke, J. R., J. D. Levin, and J. F. Pagano (1963): In F. Kavanagh (ed.), "Analytical Microbiology," pp. 406–409, Academic, New York.

Guirard, B. M., and E. E. Snell (1962): In I. C. Gunsalus and R. Y. Stanier (ed.), "The Bacteria," vol. IV, pp. 33–93, Academic, New York.

György, P., and W. N. Pearson (1967): "The Vitamins," vol. VII, 354 pp., Academic, New York.

Hewitt, W. (1976): "Microbiological Assay," Academic, New York.

Isaacson, D. M., and T. Platt (1968): Abstracts of Annual Meeting, Detroit, American Society of Microbiology, 1968.

Kavanagh, F. (1947): *Bull. Torrey Bot. Club,* **74**:303.

——— (ed.) (1963a): "Analytical Microbiology," 707 pp., Academic, New York.

——— (1963b): In F. Kavanagh (ed.), "Analytical Microbiology," pp. 142–217, Academic, New York.

——— (1963c): In F. Kavanagh (ed.), "Analytical Microbiology," pp. 125–140, Academic, New York.

——— (1968): *Appl. Microbiol.,* **16**:777.

——— (1972a): In F. Kavanuagh (ed.), "Analytical Microbiology," vol. 2., p. 631, Academic, New York.

——— (1972b): In F. Kavanagh (ed.), "Analytical Microbiology," vol. 2, pp. 43–121, Academic, New York.

——— (1974): *J. Pharm. Sci.,* **63**:1459.

——— (1975b): *J. Pharm. Sci.,* **64**:844.

——— (1975b): *J. Pharm. Sci.,* **64**:1224.

Kline, R. M. (1968): *Antimicrob. Agents Chemother.,* **1967**:763.
—— and T. Golab (1965): *J. Chromatogr.,* **18**:409.
Koser, S. A. (1968): "Vitamin Requirements of Bacteria and Yeasts," 663 pp., Charles C Thomas, Springfield, Ill.
Kuzel, N., and F. Kavanagh (1971a): *J. Pharm. Sci.,* **60**:764.
—— and —— (1971b): *J. Pharm. Sci.,* **68**:767.
Pearson, W. N. (1967): In P. György and W. N. Pearson (eds.), "The Vitamins," vol. VII, pp. 1–26, Academic, New York.
Skeggs, H. R. (1963): In F. Kavanagh (ed.), "Analytical Microbiology," pp. 552–565, Academic, New York.
Snell, E. E. (1948): *Wallerstein Lab. Commun.,* **11**:81.
—— (1950): In P. György (ed.), "Vitamin Methods," vol. I, pp. 327–505, Academic, New York.
Strohecker, R., and H. M. Henning (1965): "Vitamin Assay," 360 pp., Verlag Chemie, Weinheim, Germany. (Trans. by D. D. Libman.)
USP (1975): "The Pharmacopoeia of the United States of America," 19th ed., United States Pharmacopoeia, New York.

3
VITAMINS
Helen R. Skeggs

A vitamin is defined as an organic compound present in minute amounts in natural foodstuffs that must be available to the body from dietary or other sources in order that a specific function of physiologic maintenance or growth may proceed normally.

The role of vitamins in human nutrition has been well established. The discovery of the dramatic curative effect of natural foods or extracts of natural foods in diseases such as beriberi, pellagra, and scruvy led to their elucidation early in the twentieth century. The name "vitamine" was proposed by Casimir Funk in 1912[1] as descriptive of the substance, thought to be an amine, present in rice polishings that was curative in beriberi and thus vital for life. Although it subsequently developed that extracts of natural foods contained not one but several substances capable of exerting profound physiologic effects in minute quantities, and that few were amines, the name "vitamin" has been retained to define the class of substances rather than the single entity. The substances in rice polishings became known as the vitamin B complex.

The relationship between the growth of microorganisms and these substances vital in human nutrition was observed by Roger J. Williams[2] who wrote as follows:

---

[1] Casimir Funk (1912), *J. Stak Med.*, **20**:341.
[2] Roger J. Williams (1919), *J. Biol. Chem.*, **38**:465.

> From the cumulative evidence offered, we believe we are justified in concluding that as far as present knowledge is concerned, the substance or substances which stimulate the growth of yeast is or are identical with the substance or substances which in animal nutrition prevent beriberi or polyneuritis.
>
> If this conclusion is true, the water soluble vitamine must be a most fundamental nutritional requirement playing an indispensable role for a great variety of organisms. We have accidentally observed that some species of mold are able to produce it.
>
> As it is apparently possible to cause a single yeast cell to produce from 20 to several thousand cells in 24 hours by varying the vitamine content of the culture medium, we hope the method may be valuable both as a qualitative, and ultimately as a quantitative test for vitamine.

This observation that the *growth* of an organism might be so dependent on a single nutrient that growth could be used to define and quantitate the presence of that substance represented the birth of the *microbiological assay*. Coupled with it was recognition of the concept of fundamental biochemical unity that was to result in the contribution of knowledge from studies of microbial nutrition to the fields of nutrition and biochemistry. Further, there was recognition of the synthetic capabilities of microbial species in supplying nutritional factors.

Historically it is of interest that Dr. Williams traced his investigations back to an original observation by Louis Pasteur who had noted that yeast growth in simple medium was a function of the size of the inoculum, or a carryover of broth.

Although it was 20 years after Williams' observations before the first workable microbiological assay for a vitamin was described by Snell and Strong[1], these two decades saw the emergence of thiamine, riboflavin, pyridoxine, and nicotinic acid as components of the vitamin B complex, essential in the nutrition of various animal species, and the existence of other factors as yet unidentified had been recognized as falling into the vitamin category.

In the same era, studies in biochemical laboratories were revealing that these essential nutrients were in some manner related to enzyme activities. In 1932 Warburg and Christian[2] isolated a yellow respiratory enzyme from yeast that was found to consist of a combination of a protein and riboflavin phosphate. Neither component of the isolated enzyme was of itself active, but the two components could be recombined to re-form the active enzyme.

Subsequently, a yeast enzyme system that catalyzed the decomposition of pyruvic acid to acetaldehyde and carbon dioxide was found to consist of a protein, magnesium, and pyrophosphate ester of thiamine, cocarboxylase. Thus the fields of nutrition, biochemistry, and microbiology were converging to give a concept of the fundamental biochemical unity of living matter. By this concept, life in all species

---

[1] E. E. Snell and F. M. Strong (1939), *Ind. Eng. Chem., Anal. Ed.*, **11**:346.
[2] O. Warburg and W. Christian (1932), *Biochem. Z.*, **254**:438.

is thought to proceed along essentially the same metabolic pathways. Thus strains of organisms that propagate in completely inorganic media have the capability of synthesizing de novo the vast array of carbohydrate, lipid, and protein molecules, as well as vitamins and nucleic acids, that make up the structures of their cells. Other organisms and higher animals are more dependent on exogenous supplies of organic complexes in their diet. Once a given chemical structure enters the metabolic cycle, whether from dietary sources or by de novo synthesis, it follows the same pattern or patterns. The enzymatic capabilities with which the organism is genetically endowed determine whether a given nutrient can be synthesized by the organism or must be supplied in its food.

In considering the relationship between vitamins and their coenzymes, it is important to bear in mind that the vitamin serves as the precursor or center of the coenzyme. The coenzyme serves as the catalyst (frequently in conjunction with specific ion requirements) activating a variety of specific enzyme systems. Thus, in a vitamin deficiency, not one but several metabolic pathways may be blocked, and therefore in deficiency diseases a variety of symptoms may be encountered, depending on the individual susceptibility of the host to specific enzyme derangement.

The vitamins, their related coenzymes, and some of the metabolic functions in which they currently are known to participate are listed in Table 3-1. No coenzyme forms have yet been defined for the "fat-soluble" vitamins A, D, E, and K. These are so distinguished because of their occurrence in conjunction with body fat and their solubility characteristics. Vitamins of the B complex and vitamin C, on the other hand, are described as water-soluble since they are found in aqueous extracts of natural foods. Chemically these water-soluble vitamins range in complexity from the relatively simple structure of nicotinic acid, pyridine-β-carboxylic acid, to the complicated corrinoid structure, with a molecular weight of over 1300, of vitamin B_{12}. Activation to the coenzyme form often is achieved by phosphorylation, frequently in conjunction with a nucleoside.

The microbial population provides a wide spectrum with respect to organisms with requirements for specific vitamins. At one end are the "manufacturing" classes, capable of synthesizing from inorganic chemicals the whole array of complex organic intermediates necessary to complete the metabolic cycle. At the other end are the "consumer" classes which require preformed organic substance of a varying degree of complexity to survive. Somewhere in the course of evolution these organisms have dispensed with the genes directing the synthesis of specific nutrients, probably for ecological reasons. Just as today's housewife does not bake bread because it is readily available at the corner store, so the microorganism normally living in a plethora of

Table 3-1

Vitamin	Coenzyme	Enzyme reaction
Thiamine (B_1)	Thiamine pyrophosphate (TPP, cocarboxylase)	Oxidative decarboxylation of α-keto acids; transketolase
Riboflavin (B_2)	Flavin mononucleotide (FMN) Flavinadeninedinucleotide (FAD)	Cytochrome c reductase; amino acid oxidases; xanthine and glycolic oxidase; succinic dehydrogenase
Pantothenic acid	Coenzyme A	Transfer of acyl groups. 2-carbon metabolism; fatty acid synthesis; biosynthesis of aromatic rings, terpenes, and steroids
Nicotinic acid (B_4)	Diphosphopyridine nucleotide (DPN, NAD) Triphosphopyridine nucleotide (TPN)	Dehydrogenase reactions
Pyridoxine (B_6)	Pyridoxal phosphate Pyridoxamine phosphate	Transaminases; decarboxylase; phosphorylase
Folic acid (B_c)	Tetrahydrofolic acid	Synthesis of purine; single-carbon transfer
Cyanocobalamin (B_{12})	Coenzyme B_{12}	Isomerization of methylmalonic acid to succinic acid; 1-glutamic acid to 1-β-methyl aspartate; possibly in synthesis of deoxyribosides
Biotin	1'-N-carboxybiotin (?)	Carboxylation; decarboxylation systems; synthesis of fatty acids
Lipoic acid	Lipoic acid	Pyruvic and α-ketoglutaric dehydrogenase
Ascorbic acid (vitamin C)	....	Tyrosine oxidase; hydroxylation reactions
Vitamin D	25-Hydroxycholecalciferol (?)	Metabolism of calcium and phosphate oxidation of citrate
Vitamin A	Retinene	Visual process
Vitamin K (menadione)	....	Possibly in respiratory chain and oxidative phosphorylation
Vitamin E (α-tocopherol)	....	Possibly in respiratory chain and oxidative phosphorylation

vitamins and amino acids has given up their manufacture. Thus it is not surprising that many of our most useful assay organisms are commonly associated with food and dairy products, and that the earliest assay organisms were from the *Saccharomyces, Lactobacillus,* and *Leuconostoc* genera. Protozoa and algae strains, as representative of the aquatic ecology, also have become prominent for their specific nutrient requirements and usefulness in assay systems.

These "consumer" organisms have proved useful, as Williams predicted, as analytical tools in the microbiological assay of vitamins in that the amount of growth they can achieve is limited, over a defined range, by the amount of the required nutrient present.

GENERAL CRITERIA FOR VITAMIN ASSAY

Microbiological assay systems are useful and informative only when the limits of their capabilities are thoroughly understood. An assay system that may be completely adequate where evaluation of pure chemical potency is involved may be completely inadequate for the evaluation of the potency of the same vitamin in natural foods. Various factors enter into the choice of an assay system for a given set of conditions.

SPECIFICITY OF RESPONSE

It is important to know the limitations in the specificity of the response of the assay organism. Will it synthesize the required vitamin from direct precursors? The use of yeast such as *Saccharomyces cerevisiae* for the determination of thiamine, for example, never gained much popularity because the organisms could use either the "thiazole" or the "pyrimidine" moiety of the vitamin to achieve growth. Consequently, assay of natural products for thiamine activity with *S. cerevisiae* could give erroneously high results. Specificity of the assay could be greatly improved by carrying out a preliminary absorption of the thiamine in extracts of natural foods on Decalso, followed by elution and assay of the eluate, but were still slightly higher in most cases.

Will end products of the metabolic reactions of the vitamin satisfy its growth? Deoxyribosides will support growth of *Lactobacillus leichmannii* (ATCC 4797 or ATCC 7830; preferably the latter because growth is heavier) in the absence of vitamin B_{12}, for example, and results obtained with this organism on extracts of tissue, such as kidney, which is high in such nucleic acid derivatives would be erroneously high. Differential assay with *L. acidophilus* R 26 (ATCC 832; *Thermobacterium acidophilus* is one synonym) which responds only to deoxyribosides provides a means of differentiating between deoxyriboside and true B_{12} activity in this instance. The problem of deoxyriboside interference can be circumvented by using *Ochromonas malhamensis* as the assay organism.

Can the coenzyme or bound forms of the vitamin be utilized? In some instances, coenzyme forms of the vitamins have less activity per mole than they should in terms of the true vitamin content. In other cases, the coenzyme forms are fully active on a molar basis. To the best of this author's knowledge, the coenzyme form has never been shown to be more active than an equivalent amount of free vitamin. The relative activity of the coenzyme and other bound forms of the vitamin should be known, since frequently the extraction procedures used to release the total vitamin content present in the sample are dictated by the nature of the coenzyme complex. The answers to all these questions determine the usefulness of an assay organism, but to some extent these factors are within the control of the analyst.

ADEQUATE MEDIUM

The medium should contain all the nutrients necessary to support optimal growth of the organism, except for the one factor being measured.

In general, assay media are composed of a source of amino acids, usually a casein hydrolysate, glucose, mineral salts and buffers, purines and pyrimidines, other vitamins, and perhaps a source of long-chain fatty acids. The concentrations of the individual ingredients have been arrived at by trial and error to give the optimal growth response when the missing factor is supplied in a variety of substances. Most media, especially those for the lactobacilli, contain either casein hydrolysates or amino acid mixtures based on the concentrations present in casein, so that the balance which is natural to the organism is maintained.

Many lactic acid organisms require oleic acid, but the free acid is inhibitory just above the growth range. Therefore the nontoxic sorbitan monoleate (polysorbate 80, Tween 80) is used to provide a source of fatty acid.

The balance of the salt mixture used in the medium for the assay usually allows an optimal rate of growth. Excess quantities of inorganic salts in the sample preparation can, in some instances, interfere with the rate of growth or the buffering capacity of the medium. Usually this would result in a low value for the sample. Where salt-balance interference is suspected, assay of a known sample in the presence of a simulated salt mixture (or other base which could cause problems) is called for to determine the extent of the problem.

Ideally, the medium allows no growth in the absence of the vitamin being assayed so that the "blanks" or zero tubes remain clear. However, a medium adequate for an evaluation of pure vitamin products may fail to provide all the extraneous substances that might affect the growth rate of the organism were food samples to be evaluated. Conversely, a medium adequate for evaluation of the vitamin content of foods or other natural products should certainly serve the purpose in analyzing pure samples.

One must bear in mind that the organism being used is being constantly maintained in the artificial environment of the laboratory, and that its needs may change over the course of time as it adapts (or mutates ecologically) to its environmental conditions. The availability of lyophilized cultures from original stocks may serve to correct abnormalities that arise in this instance. Supplementation of the medium with substances that could be present in the maintenance medium or low or absent in the assay media represents another approach to the solution, if the time and interest are available.

Commercially prepared, dehydrated media are available for most of the commonly used vitamin assays. These media are checked for adequacy when they leave the manufacturers' hands, but it must be appreciated that they contain potentially unstable components and storage under improper conditions can cause deterioration that results in poor performance. Medium for vitamin B_{12} assay with lactobacilli, for example, contains ascorbic acid which is notoriously unstable. Once opened, a jar of media will readily pick up moisture, which accelerates the decomposition of ascorbic acid, unless stored under the dry, cool conditions recommended by the manufacturer.

Distilled water can play an important role in the control of assay media. Most of the media formulas were developed in an era when distilled water came from a scarce supply or laboratory still, instead of a laboratory tap. The piping of distilled water throughout large laboratory facilities, as is now common, can prove troublesome because, regardless of the use of stainless steel or glass pipelines, joints are required and the ubiquitous class of manufacturing organisms can set up production centers. It may be necessary to collect freshly distilled water, sterilize it, and keep it on hand for successful assay determination. Water problems can be suspected when the blank tubes of an assay contain variable growth from day to day.

Maintenance of assay organisms usually consists of carrying stock cultures which are transferred monthly and starting fresh daily cultures from the stocks at various intervals. Recommendations for maintenance media vary, but generally they should be rich in nutrient. Skim milk media supplemented with 1% tryptose has been used successfully in these laboratories for years to maintain stock cultures of lactic acid organisms. Most published assay methods include details for maintenance of specific organisms.

CHEMISTRY AND OCCURRENCE OF THE VITAMIN

Any successful analysis, chemical or microbiological, depends on the analytical system being compatible with the chemical behavior of the compound. The analysis of vitamins is considerably hampered by the fact that, as part of coenzyme and biological systems, they frequently occur in conjugated or bound form. Criteria that can be established for the crystalline material in a test tube do not always apply to vitamins in natural products. When one is concerned only with the analysis of pure substances, as in a pharmaceutical formulation, knowledge of the chemical stability of the pure compound and of the interrelationships with the other components of the formulation is the major concern, and the effects of assay conditions such as autoclaving time and extracting buffers can be evaluated by the use of aseptic addition of samples and recovery of known amounts of added vitamins.

In natural products, however, elaborate extraction procedures sometimes are necessary, and, unhappily, procedures that work for one type of sample may be too strenuous or yield incomplete extraction in another. For example, 6 N sulfuric acid at 120°C for 1 h is successfully used to extract biotin from liver tissue. This treatment destroys biotin in serum samples. Crystalline cyanocobalamin is quite stable to autoclaving for an hour; vitamin B_{12} activity in liver can be destroyed under the same conditions. On the other hand, crystalline cyanocobalamin is completely destroyed by autoclaving at alkaline pH. Destruction of cyanocobalamin activity in liver samples by alkaline hydrolysis is erratic and frequently incomplete.

It is important therefore to know as much as possible about the chemical behavior of the vitamin being analyzed, as a crystalline compound and as it occurs naturally. The safe way to approach extraction problems is to carry out recovery experiments, but even here an element of doubt persists, since one cannot be certain that crystalline compounds will equilibrate with coenzyme or tissue-bound material. No analytical procedure, chemical or microbiological, is free from these problems, and the same rule of thumb applies here as elsewhere: A good extraction procedure aims, without loss of activity, to extract all the active principle present and convert it, if necessary, into the form of the standard used for comparison.

PROPER LABORATORY FACILITIES

Most microbiological assay systems for vitamins function in the nanogram and picogram range and should be considered in the realm of microchemical techniques. Glassware employed must therefore be scrupulously clean; reagents and chemicals including water should be of the highest purity. Accurate analytical balances should be available for weighing standards, and careful analytical technique should be employed. An autoclave should be available, and a constant-temperature incubator or water bath is a necessity. The fastidious microorganisms used in these procedures as master chemists are hard taskmasters and will not tolerate sloppy techniques.

EXAMPLE OF AN ASSAY: PANTOTHENIC ACID‡

The microbiological assay of pantothenic acid using *Lactobacillus plantarum* per se is straightforward and uncomplicated. It is an interesting vitamin in that its existence as an essential nutrient for microorganisms was discovered prior to the demonstration of its role in animal nutrition. Its need in human nutrition has yet to be defined, and yet its coenzyme, coenzyme, A, is one of the busiest of the biological catalysts in the metabolic cycle, transporting acyl groups in carbohydrate, fat, and sterol metabolism. It exists in nature, in addition to coenzyme A, as pantetheine (β-mercaptoethanolamine-pantothenic acid), which is, for some microorganisms, the essential form, as pantothenic phosphate, or as pantetheine-phosphate. None of these complexes is as active for the assay organism *L. plantarum* as it should be in terms of pantothenate content. A schematic representation of coenzyme A is given in Fig. 3-1, which shows the relationship of pantothenic acid tucked in the middle of the larger molecule, from which it must be disengaged in order to be available for *L. plantarum*. Alkaline phosphatase breaks the pyrophosphate linkage, and the liver enzyme breaks the peptide bond to release pantothenic acid from the β-mercaptoethanolamine complex. Unfortunately the enzyme preparations necessary to effect the release of pantothenate from bound forms frequently contain appreciable amounts of free pantothenate although resin treatment is used to remove such activity. It should be a fundamental rule, however, especially when fresh enzyme preparations are put into use, to run an enzyme blank as a sample in the assay procedure.

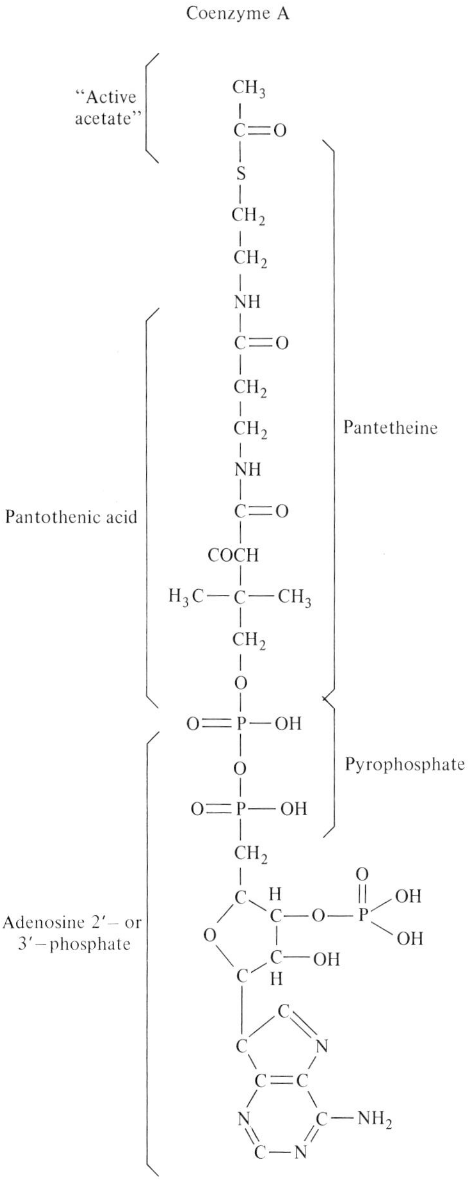

Figure 3-1 Structural form of coenzyme A.

Commercially available alkaline phosphatase is satisfactory for use. Bird[1] describes the preparation of the liver enzyme as follows:

---

[1] O. D. Bird (1963), Pantothenic Acid and Related Compounds, in F. Kavanagh (ed.), "Analytical Microbiology," pp. 497–517, Academic, New York.

Weigh and mince chilled chicken liver and transfer to chilled Waring Blendor. Add 20 ml chilled acetone per gram of liver. Homogenize 2 minutes. Filter through a Buchner funnel with coarse paper. Wash residue in funnel with chilled acetone, then peroxide free ethyl ether. Dry the powder in a vacuum dessicator over fresh phosphorous pentoxide. Remove connective tissue by passing the pink powder through a 40-mesh sieve. Rub the sieved powder into 10 volumes of ice-cold 0.02 M potassium bicarbonate solution to give a smooth suspension. Centrifuge at 1500× g for 30 minutes in the cold and store supernatant at $-20°$.

Wash Dowex-1 (X10 cross linkage, 200–400 mesh) twice with 10 volumes of N HCl and decant after centrifugation. Wash the acid-treated resin 8 to 10 times with 10 volumes of water until the supernatant is pH 5.0. Leave the resin in a slurry that can be pipetted with a 10-ml serological pipette.

To 1 volume of acid-washed Dowex-1 add sufficient M Tris buffer at pH 8.3 to bring to pH 8.0. Add 1 volume of ice-cold liver enzyme and stir for 5 minutes in an ice bath. Centrifuge the suspension at 4200× g in the cold. Decant the supernatant and repeat the treatment with acid-washed Dowex-1. Centrifuge, decant the supernatant, and store at $-20°$.

An amount of sample containing 5 to 15 micrograms of conjugated pantothenic acid is treated in a flask or test tube as follows:

1 ml of intestinal phosphatase preparation containing 20 mg alkaline phosphatase, purified (Pentex, Inc., Kankakee, Ill.) in water suspension
2 ml Dowex treated liver enzyme
1 ml 0.2 M Tris buffer adjusted to pH 8.3 with NaOH

Dilute to 10 ml with water, add 2 drops of toluene and incubate 3 hours at 37° in a water bath.

Immerse in boiling water to stop enzyme reactions, cool and dilute to approximately 0.03 mcg per ml with distilled water.

Once the extraction procedure is completed, the assay procedure is relatively simple.

A standard preparation of calcium pantothenate (108.8 mg equivalent to 100 mg pantothenic acid) is prepared by diluting an accurately weighed sample to a concentration of 0.04 μg of pantothenic acid per milliliter. This solution is dispensed into chemically clean 18 × 25 ml test tubes as follows: 0, 0.5, 1.0, 1.5, 2.0, 3.0, and 5.0 ml. Unknown solutions are similarly dispensed in 1, 2, 3, and 5 ml quantities. All tubes, standards and unknowns, are diluted to 5 ml with distilled water, and 5 ml of the double-strength medium shown in Table 3-2 is added to give a final volume of 10 ml. Tubes are then stoppered, sterilized by autoclaving at 121°C for 15 min, and cooled to room temperature.

The assay organism, *Lactobacillus plantarum* (ATCC 8041), is maintained in stock culture in agar stabs.[1] The inoculum is prepared from a 24-h culture grown at 30 to 33°C in 1% yeast extract, 1% glucose broth, or in microinoculum broth (Difco or BBL). The culture is centrifuged, supernatant discarded, and the cells resuspended in sterile physiologic saline to wash off residual pantothenic acid. After recentrifugation, the cells are resuspended in 10 ml of fresh saline, and further diluted 1 to 100 for the inoculum. Each assay tube is aseptically inoculated with 1 drop (0.05 ml) of the final diluted suspension.

The test is incubated at 30°C. Results may be measured turbidimetrically

---

[1] 1% yeast extract, 1% Bacto-tryptose, 0.1% glucose, 0.2% K_2HPO_4, 0.3% $CaCO_3$ (anhydrous), 1.5% agar.

Table 3-2 Double-strength basal medium

	Per 100 ml
Casein*	1.0 gm
Cystine	20 mg
dl-Tryptophan	40 mg
Sodium acetate (anhydrous)	1.2 gm
Adenine	1 mg
Guanine	1 mg
Uracil	1 mg
Xanthine	1 mg
Inorganic salts A†	1 ml
Inorganic salts B‡	1 ml
Glucose	4 gm
Thiamine chloride	0.2 mg
Riboflavin	0.2 mg
Nicotinic acid	0.2 mg
Pyridoxine HCl	0.4 mg
p-Aminobenzoic acid	0.02 mg
Biotin	0.5 mcg
pH adjusted to 6.6–6.8	
Oleic acid§	10 mg

*Hydrochloric acid hydrolyzed, Norit-treated, vitamin-free casein. Method of preparation is described in W. Horwitz (ed.) (1975), "Official Methods of Analysis," 12th ed., Association of Official Analytical Chemists, Washington, D.C. Commercial preparations are available.

†Salts A: Dissolve 10 g of K_2HPO_4 and 10 g of KH_2PO_4 in 100 ml of H_2O. Store under toluene in the refrigerator.

‡Salts B: Dissolve 4 g of $MgSO_4 \cdot 7H_2O$, 200 mg of NaCl, 200 mg of $FeSO_4 \cdot 7H_2O$, and 200 mg of $MnSO_4 \cdot 4H_2O$ in 100 ml of H_2O. Solution becomes cloudy.

§Optional; see text.

with any suitable instrument at 540 nm after 24-h incubation. Tubes should be chilled prior to reading to stop growth and prevent "drift" during reading time.

Alternatively, acid production after 72 h may be determined by titration to neutrality with 0.1 N NaOH, using bromthymol blue as an indicator, or by a pH meter.

A standard concentration-response curve is prepared by plotting either the turbidity values or the titrametric values against the level of pantothenic acid in the standard solution. A typical standard curve is shown in Fig. 3-2. The quantity of vitamin present in each level of unknown solution is read from the curve and the average concentration per milliliter determined. Values for each level of unknown tested should agree within 10 percent.

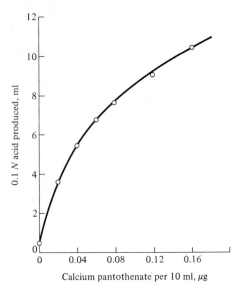

Figure 3-2 Titrametric response of *Lactobacillus plantarum* (ATCC 8014) to calcium pantothenate.

In the event that the sample contains materials not present in the medium that are capable of stimulating the response of the organism in the presence of pantothenic acid (but not in its absence), the sample may exhibit "drift." This is evidenced by increasingly high values in comparison with the standard as the level of sample increases. In the case of pantothenic acid, this phenomenon can occur in a sample containing fatty acids. Oleic acid, inactive in itself, will enhance the effect of a small amount of pantothenic acid and give erroneously high results. Addition of oleic acid to the medium will equalize the effect in the standard sample.

Uneven responses in tubes scattered throughout the assay usually means dirty glassware. The use of disposable tubes, when economically feasible, usually is quite satisfactory. Tubes that are reused are best soaked in dichromate cleaning solution, thoroughly rinsed in tap water, and finally with distilled water.

High blanks, accompanied by diminution of the response, usually indicate that the medium is contaminated with pantothenic acid.

In general, microbiological assay techniques tend to be replaced with chemical procedures when such methods become available. The primary reason is the time consumed, and the many steps involved in producing a final result. The microbial systems have an advantage, however, in that they define biologically active material, and, in spite of criticism to the contrary, frequently they are more specific than a chemical assay. For the reader who is interested in pursuing the subject further, the General References list several publications that give specific methods and detailed information about individual vitamins.

GENERAL REFERENCES

Association of Vitamin Chemists (1966): "Methods of Vitamin Assay," 3d ed., Interscience, New York.
Bird, O. D. (1963): Pantothenic Acid and Related Compounds, in F. Kavanagh (ed.), "Analytical Microbiology." pp. 497–517, Academic, New York.
Devlin, T. M. (1964): The Relation of Diet to Oxidative Enzymes, in M. G. Wohl and R. S. Goodhard (eds.), "Modern Nutrition in Health and Disease," 3d ed., pp. 522–533, Lea & Febiger, Philadelphia.
Oser, B. L. (1964): Vitamins and Deficiency Diseases, in B. L. Oser (ed.), "Hawk's Physiological Chemistry," 14th ed., chap. 21, McGraw-Hill, pp. 573–758, New York.
Snell, E. E. (1950): Microbiological Methods in Vitamin Research, in P. György (ed.), "Vitamin Methods," vol. 1, pp. 327–505, Academic, New York.
Wagner, A. F., and K. Folkers (1964): "Vitamins and Coenzymes," Interscience, New York.

4
ANTIBIOTICS
John N. Porter

The history of the application of scientific principles to chemotherapy in the treatment of infectious diseases has been relatively brief. Paul Ehrlich, often called the "father of chemotherapy," took the first of three significant steps in this field when he announced in 1910 that compound 606, Salvarsan, was effective in the treatment of syphilis. The second came in 1935 when Domagk reported that Prontosil, a red dye whose active ingredient was a short time later found to be sulfanilamide, was capable of curing infections caused by streptococci. The third and most significant development was the discovery of penicillin by Alexander Fleming in 1929. This discovery was to go virtually unnoticed until 1939, but it ushered in the antibiotic era which was spurred on by the needs of World War II and had a meteoric development over a span of 25 years. During those years some 1500 antibiotics were announced.

The term "antibiotic" was proposed by Waksman and in its presently accepted form may be defined as a product of microbial metabolism which is capable of antagonizing the growth and/or survival of microorganisms and which is effective in low concentrations. It should be pointed out that some liberties have been taken with this definition by many investigators. Semisynthetic antimicrobial metabolites are usually referred to as antibiotics. Furthermore, it is generally accepted that microbial products which act as antineoplastic agents come under a broad definition of the subject. Nevertheless, it is not correct to refer to unrelated synthetics, such as the sulfonamides, or to inhibitory substances derived from higher plants or animals, as antibiotics.

HISTORICAL DEVELOPMENTS

In the years prior to World War II Florey and Chain in England were interested in a variety of microbial antagonists. This led them in 1939 to initiate a study of penicillin in relation to the possible treatment of staphylococcal infections. Their work, together with that of Heatley and Abraham, in producing and purifying penicillin culminated in the first human trials in 1941. By 1944 there was sufficient penicillin to satisfy the demands of the Allied armed forces and within 10 years of the first trials over 20 tons of crystalline antibiotic were being produced each month.

Tyrothricin, a polypeptide antibiotic complex produced by *Bacillus brevis*, was described by Dubos in 1939. This complex is composed of two major components, gramicidin and tyrocidine, of which the former accounts for 20 percent and the latter 80 percent of the whole.

Penicillin and tyrothricin are primarily effective against gram-positive bacteria. Streptomycin, announced in 1944, has a somewhat wider spectrum of activity, being capable of inhibiting some gram-negative bacteria and mycobacteria. The discovery of chloramphenicol in 1947 and chlortetracycline, the first of a family of tetracyclines, in 1948 initiated the era of broad-spectrum antibiotics. This term was applied because of effectiveness against an unusually wide range of pathogenic microorganisms, namely, gram-positive and gram-negative bacteria, rickettsiae, large viruses, and certain protozoa.

The most recent major advance in antibiotics has been the discovery of synthetic and semisynthetic penicillins, tetracyclines, and cephalosporins. In several cases these "molecular modifications" have given products with some properties superior to those of the parent compounds.

SCREENING FOR NEW ANTIBIOTICS

Evidence that highly efficacious antimicrobial agents could be produced by fungi, true bacteria, and actinomycetes prompted the introduction of numerous screening programs to find new antibiotics. At first, with the success of penicillin, fungus culture collections were combed for leads but this approach did not turn out to be particularly promising. Investigators turned to isolating their own cultures which were then subjected to screening procedures designed to elicit and detect antibiotic activity. Major emphasis shifted to new streptomycete isolates, and these became the source of most of the commercially important antibiotics.

In the United States most of these programs were carried on by scientists working for pharmaceutical companies but significant discoveries were also made by scientists in government and academic laboratories. Overseas, similar programs were instituted, notably by the Japanese, British, and Russians.

As contrasted with chemical or biological modifications of known antibiotics, screening procedures to find new agents are empiric in nature and involve in vitro detection of activity, in vivo testing to establish whether there is detectable activity in the animal against induced infections, and antibiotic identification to determine whether the agent is novel.

IN VITRO TEST SYSTEMS

Two types of methods are available. On the one hand, the potential inhibitor may be grown on agar plates which are subsequently inoculated with one or more organisms against which suppressive activity is sought. On the other, fermentation fluids may be assayed by any of several available procedures.

In the agar method the organisms to be screened are grown for several days as colonies or streaks on plates which are then flooded and swabbed or cross streaked with selected test organisms. Because of their compact, nonspreading, confluent manner of growth, this method is particularly adaptable to screening streptomycete isolates. In routine procedures these actinomycetes are grown from 5 to 7 days at 28°C on a suitable medium, such as asparagine-dextrose agar, and then plates containing them are cross streaked with bacteria or fungi. The degree of inhibition and the type of organisms against which this is expressed serve to give an initial assessment of the inhibitor.

A useful modification is to make pinpoint inoculations of several actinomycetes on an agar plate and then, after a period of incubation, to make replications from the original or master plate onto a series of new plates, each of which can be flooded with a different test organism.

The various modifications on agar plates have certain drawbacks, however, and one may prefer to test cultures for antibiotic activity by using liquid media at the very first stage of the investigation. For one thing, agar results do not correlate perfectly with those in fermentation flasks. For another, tests for antiviral, antiprotozoal, or antitumor activity are mainly limited to fermentation fluids, and if one wishes to test for more than bacterial or fungal antagonism the agar test method is an unnecessary intervening step.

Agar diffusion, agar dilution, and broth dilution detection and assay procedures are available for testing fermentation fluids for antibacterial, antifungal, or antiprotozoal activity. These, as well as chemical assay methods, are described in detail in Chap. 2. Initial screening procedures for antiviral or antitumor activity must be carried out in systems using living cells, although not necessarily in the intact animal. Various tissue culture or egg embryo tests may be employed but it must be emphasized that many leads that are accepted as active by these methods cannot be substantiated in animals.

IN VIVO TEST SYSTEMS

Screens to detect substances with potential utility in human or veterinary medicine must at some point involve the treatment of infections in living animals. Crude culture filtrates can be employed for such tests, but it is more common practice to use preparations with some degree of purification and fractionation. Such preparations, in which the amount of inactive substances are decreased and the potency of the active fraction increased, are less likely to be toxic unless the toxicity resides in the active fraction. Furthermore, fermentation fluids frequently comprise mixtures of antibiotics, the components of which may or may not be related. Insofar as feasible it is desirable to test individually the components of such mixtures.

Antibiotic screening programs usually involve testing in mice, but other animals are used with some infections. Most frequently the route of administration of the drugs is oral, intraperitoneal, or subcutaneous although other types of treatment may be employed. In most animal tests the results are judged by the degree of the reduction of mortality, but in some, such as those involving superficial mycoses, the object is a suppressive effect in treated animals as compared with the severity of infection in inoculated untreated controls.

ANTIBIOTIC IDENTIFICATION

The initial identification of antibiotics in fermentation fluids and crude preparations may be carried out by utilizing paper chromatography and electrophoresis and, to some extent, light absorption, solubility, stability, and antimicrobial spectra, including strains resistant to known antibiotics. Newer methods include the use of preparative thin-layer chromatography and mass spectrometry. A complicating factor at this stage is the ubiquitous presence of antibiotic mixtures, which affects the reliability of some of the data obtained.

Because of mixtures and because of the fact that antibiotics frequently occur in nature as closely related chemical structures, the surest conclusions as to identity or novelty can be derived only from chemical and physical data on highly purified, preferably crystalline, single components. Additional data are obtained from studies of light absorption in the ultraviolet and infrared regions, solubility, stability, crystal structures, melting point, optical rotation, and various color reactions. Definitive proof of structure will, of course, provide the ultimate identification.

ANTIBIOTICS OF MAJOR IMPORTANCE

Of the approximately 2000 antibiotics hitherto described the vast majority are of no economic value. Toxicity or ineffectiveness in vivo has eliminated most from consideration as agents in chemotherapy. Some of these, however, such as puromycin and the actinomycins have proved to be valuable tools for investigating molecular and genetic processes within the cell. With the exception of a few marginal cases those antibiotics of clinical importance are listed in Table 4-1. Most of these were discovered as the result of screening programs for new entities or as modifications of known antibiotics.

EFFECTIVENESS AND LIMITATIONS OF AVAILABLE ANTIBIOTICS

In treating infectious diseases modern physicians have available to them numerous antimicrobial agents from which to make a choice, including the antibiotics tabulated in Table 4-1. On the basis of their breadth of activity, these can be classified as having narrow, intermediate, or broad spectra of activity.

The narrow-spectrum antibiotics are primarily limited to those active against gram-positive organisms, although some are also active against neisseriae and spirochaetes. Penicillin G, for example, is highly effective in most

Table 4-1 Antibiotics of clinical importance

Generic name	Type	Source	Antimicrobial spectrum and some important properties
Amphotericin B	Amphoteric heptaene	*Streptomyces nodosus*	Fungi
Bacitracin	Weakly basic polypeptide	*Bacillus subtilis*	Gram-positive bacteria
Cephalosporins	Derivatives of 7-aminocephalosporanic acid	*Cephalosporium* spp. and *Emericellopsis* spp.	Generally penicillinase-resistant but inactivated by cephalosporinase
Cephalexin		Semisynthetic	Gram-positive and some gram-negative bacteria; orally absorbed
Cephaloglycin		Semisynthetic	Like cephalexin; orally absorbed
Cephaloridine		Semisynthetic	Like cephalexin; not orally absorbed
Cephalothin		Semisynthetic	Like cephalexin; not orally absorbed
Chloramphenicol	Aromatic structure with a nitro group; neutral	*Streptomyces venezuelae*; synthesis	Gram-positive and gram-negative bacteria; rickettsiae; large viruses; *Entamoeba*
Colistin	Basic polypeptide	*Aerobacillus colistinus*	Primarily gram-negative bacteria
Erythromycin	Basic macrolide	*Streptomyces erythreus*	Gram-positive and a few gram-negative bacteria; some protozoa
Gentamicin	Strong base; aminoglycoside	*Micromonospora* spp.	Gram-positive and gram-negative bacteria
Griseofulvin	A spirocyclohexenone benzofuranone; neutral	*Penicillium* spp.; synthesis	Fungi
Kanamycin	Tetraacidic base; aminoglycoside	*Streptomyces kanamyceticus*	Gram-positive and gram-negative bacteria; some protozoa
Lincomycin	Basic glycoside	*Streptomyces lincolnensis*	Gram-positive bacteria
Neomycin	Complex of B and C, which are isomeric; basic; aminoglycosides	*Streptomyces fradiae* and other *Streptomyces* spp.	Gram-positive and gram-negative bacteria; mycobacteria
Novobiocin	Dibasic acid	*Streptomyces* spp.	Primarily gram-positive bacteria

Nystatin	Amphoteric tetraene	Streptomyces spp.	Fungi
Penicillins	Strong monobasic carboxylic acids; derivatives of 6-aminopenicillanic acid		
Ampicillin		Semisynthetic	Gram-positive and gram-negative bacteria; penicillinase-sensitive; acid stable
Carbenicillin		Semisynthetic	Gram-positive and gram-negative bacteria including *Pseudomonas* strains; penicillinase-sensitive; not orally absorbed
Cloxacillin		Semisynthetic	Gram-positive bacteria; penicillinase-resistant; acid stable
Dicloxacillin		Semisynthetic	Like cloxacillin
Methicillin		Semisynthetic	Gram-positive bacteria; penicillinase-resistant; acid labile
Nafcillin		Semisynthetic	Like cloxacillin
Oxacillin		Semisynthetic	Like cloxacillin
Penicillin G		*Penicillium* spp.; *Aspergillus* spp.	Gram-positive bacteria; penicillinase-sensitive; acid labile
Penicillin V		Biosynthetic	Gram-positive bacteria; penicillinase-sensitive; acid stable
Polymyxin	Basic polypeptide	*Bacillus polyomyxa*	Primarily gram-negative bacteria
Rifamycin	A complex of which derivatives of rifamycin B, containing an ansa-naphthohydroquinone group, are the most important	*Nocardia mediterranea*	Primarily gram-positive bacteria and mycobacteria
Rifampin		Semisynthetic	Gram-positive and gram-negative bacteria; mycobacteria; some viruses; orally absorbed
Spectinomycin	Basic aminocyclitol	*Streptomyces* spp.	Gram-positive and gram-negative bacteria, specifically *Neisseria gonorrhoeae*
Streptomycin	Strong base; aminoglycoside	*Streptomyces* spp.	Gram-positive and gram-negative bacteria; *Mycobacterium tuberculosis*

Table 4-1 Antibiotics of clinical importance (*Continued*)

Generic name	Type	Source	Antimicrobial spectrum and some important properties
Tetracyclines	Amphoteric		Gram-positive and gram-negative bacteria; rickettsiae and large viruses; coccidia; amoebae and balanthidia; mycoplasms
Chlortetracycline		*Streptomyces aureofaciens*	
Demeclocycline		*S. aureofaciens* mutant	
Doxycycline		Semisynthetic	
Methacycline		Semisynthetic	
Minocycline		Semisynthetic	Also inhibits staphylococci resistant to other tetracyclines
Oxytetracycline		*Streptomyces rimosus*	
Tetracycline		Catalytic hydrogenation of chlortetracycline; *S. aureofaciens* mutant	
Tyrothricin	Mixture of polypeptides	*Bacillus brevis*	Gram-positive bacteria
Vancomycin	Amphoteric glycopeptide	*Streptomyces orientalis*	Gram-positive bacteria

infections caused by streptococci, pneumococci, and sensitive staphylococci, and several of the newer penicillins, because they are unaffected by penicillinase, are useful against resistant staphylococci. However, since cross allergenicity has been reported from among all the penicillins in clinical use, it is necessary, when treating patients allergic to penicillin, to turn to other antibiotics such as erythromycin, lincomycin, or the related clindamycin. Vancomycin is used in certain instances but is more toxic than the other antibiotics. The polypeptide antibiotics, bacitracin or tyrothricin, may be employed topically.

In addition to gram-positive cocci, intermediate-spectrum antibiotics are active against some gram-negative bacteria and some mycobacteria. They are perhaps best exemplified by the aminoglycoside antibiotics, including gentamicin, kanamycin, neomycin, and streptomycin. These antibiotics are all clinically limited by renal and ototoxicity following parenteral use, which is directly related to the levels employed and the length of therapy. The cephalosporins may also be considered intermediate-spectrum antibiotics.

The broad-spectrum antibiotics—the tetracyclines and chloramphenicol—are used in a variety of infections caused by both gram-positive and gram-negative bacteria, although chloramphenicol must, in every case, be employed with caution because of the occasional occurrence of aplastic anemia. They are also excellent for use against the various rickettsial infections. The tetracyclines are useful in trachoma.

Colistin, polymyxin, and the aminoglycoside antibiotics, e.g., kanamycin, play a specific role in the therapy of gram-negative infections. However, colistin and polymyxin are more effective in infections caused by *Pseudomonas* spp; kanamycin is more useful against *Proteus* spp. Gentamicin is a newer antibiotic which has supplanted some of the others in the treatment of serious gram-negative bacterial infections. For infections caused by *Pseudomonas* spp it is frequently administered in combination with carbenicillin. Streptomycin is used mostly in combination with other antimicrobials such as penicillin, the tetracyclines, and chloramphenicol in gram-negative infections. In the treatment of tuberculosis it is commonly combined with isoniazid or *p*-aminosalicylic acid.

There is currently an antibiotic available for each of the three general types of fungus infections of human beings. Griseofulvin is effective in many superficial infections caused by some species of *Trichophyton*, *Microsporum*, and *Epidermophyton*. Amphotericin B, which has limited utility because of its toxicity, is employed in deep mycoses, particularly North American blastomycosis, histoplasmosis, coccidioidomycosis, cryptococcosis, and candidiasis. Candida infections of the alimentary tract or localized vaginal or skin infections usually respond to nystatin.

In spite of the impressive advances in antimicrobial therapy there is a need for additional agents to fill the gaps in the currently available armamentarium or to supplement the use of drugs already in existence. For example, new agents are needed to combat refractory gram-negative bacteria, especially strains of *Pseudomonas*, *Proteus*, *Aerobacter*, and *Salmonella*. Many mycobacteria, especially the atypical ones, resist treatment with available drugs. Safe, effective agents for the treatment of neoplasms and virus diseases have not yet made their appearance. As already indicated, certain antibiotics have limited usefulness because of their toxicity. And finally, there is always the problem of the development of resistant strains, some of which succumb to the newer antibiotics only to have other strains appear in their stead; it is probable that this process will be a continuing one. Thus, it is important that the search for new antimicrobials that can be used in human medicine be continued, for the need for more effective agents is still great.

No general discussion of antibiotics would be complete without mentioning the agricultural and nonmedical uses of antibiotics. There are at least six:

1. Disease therapy in livestock, poultry, and plants
2. Prevention of disease in livestock, poultry, and plants
3. Improvement of weight gains and feed conversion of livestock and poultry
4. Food preservation
5. Biochemical tools
6. Selective agents in culture media

At the present time only the first three are commercially important. The use of antibiotics in food preservation is no longer permitted in the United States. The therapy of bacterial diseases of livestock and poultry developed rapidly when the price decreased to levels which made it economically feasible. It was also learned that intermediate levels (50 to 250 g) of chlortetracycline, oxytetracycline, or tylosin per ton of feed given to animals at certain times in their life when they are particularly susceptible to disease (sometimes referred to as times of stress) would aid in the prevention of disease. This prophylactic use of antibiotics in animal feed, particularly at the time of weaning, vaccination, or after shipment, at present is of the greatest economic importance, especially where there is intensive livestock production. Monensin, the major component of a group of closely related monocarboxylic acids, is a more recently described antibiotic useful in the prevention of coccidiosis in chickens.

It has been recognized for more than 20 years that small amounts (1 to 50 g) of antibiotic per ton of feed for young animals and poultry would increase weight gains and improve feed conversion, i.e., number of pounds of feed per pound of gain. This effect appears to be produced as a result of the modification of the intestinal flora of treated animals. The antibiotics approved for this purpose in the United States include chlortetracycline, oxytetracycline, bacitracin, procaine penicillin G, streptomycin, lincomycin, and, most recently, moenomycin.

Although streptomycin has been used successfully against pear blight and tetracycline has been used against a yellows disease of mulberries, these do not represent important economic applications.

MODE OF ACTION

In the early days of the antibiotic era there was no clear conception of how any of the then known antibiotics actually exerted their antimicrobial effects. Today the modes of action of many newly described antibiotics are studied soon after discovery, and there is now a large body of knowledge available about all the antibiotics of commercial importance.

In general, antibiotics may either prevent the formation of DNA or RNA, inhibit protein synthesis, interfere with the formation of the cell wall, or impair the integrity of the cell membrane. For some antibiotics the site of action has been pinpointed with considerable accuracy. For others the precise locus is unknown. The antibiotics listed in Table 4-1 may be arranged according to the modes of action as follows:

1. Interference with nucleic acid synthesis
    griseofulvin
    novobiocin
    rifampin
2. Impairment of the translation of genetic information into protein synthesis
    chloramphenicol   neomycin
    erythromycin   spectinomycin
    gentamicin   streptomycin
    kanamycin   tetracyclines
    lincomycin
3. Impairment of the synthesis and function of the cell wall
    bacitracin   penicillins
    cephalosporins   Vancomycin
4. Impairment of the function of the cell membrane
    amphotericin B   polymyxin
    colistin   streptomycin
    nystatin   tyrothricin

Novobiocin interferes with DNA polymerization and also inhibits DNA-dependent RNA polymerase, as does rifampin. The locus of action of griseofulvin has not been precisely worked out but the antibiotic may prevent the assembly of purine nucleotides.

The antibiotics in group 2 manifest their inhibitory effects at the ribosome. The site of action of spectinomycin is at the 30S ribosomal subunit. Chloramphenicol interferes with the binding of mRNA to the ribosome; the tetracyclines inhibit the formation of the necessary complex between ribosome, mRNA, and aminoacyl-sRNA, probably by inhibiting the amino acid–tRNA binding reaction. Both types of antibiotics are bacteriostatic. Streptomycin, on the other hand, distorts the ribosome site where tRNA and mRNA pairing occurs and thus may induce defective protein molecules. The related aminoglycosides kanamycin, gentamicin, and neomycin behave similarly. All four are bactericidal.

Vancomycin and bacitracin inhibit the enzyme glycopeptide synthetase, which is responsible for condensation of the glycopeptide backbone of the cell wall. The penicillins and cephalosporins prevent the necessary cross linking of the glycopeptide.

In group 4 the component parts of tyrothricin, gramicidin and tyrocidine cause a leakage of amino acids from the cell by uncoupling oxidative phosphorylation and decreasing respiration. The other antibiotics alter the osmotic barrier function of the cell membrane in two other ways. Polymyxin and colistin (polymyxin E) act as cationic detergents with affinity for phosphate radicals. Amphotericin B and nystatin, both polyene antifungals, bind to sterol groups which apparently reorient the lamellar structure of the membrane.

THE TETRACYCLINES: ANTIBIOTICS OF INDUSTRIAL IMPORTANCE‡

The tetracycline family of broad-spectrum antibiotics will serve to illustrate the principles of producing antibiotics on a commercial scale.

Duggar and coworkers discovered chlortetracycline in 1948, which was closely followed by the discovery of oxytetracycline in 1950 by Finlay et al. Both antibiotics were found by means of screening programs involving actinomycetes isolated from soil samples. Tetracycline was mentioned by Stephens in 1952, and its properties were described in separate papers in 1953 by Boothe and by Conover. A fourth member, demeclocycline (demethylchlortetracycline), was reported from a mutant strain of *Streptomyces aureofaciens* in 1957 by McCormick et al.

Structure and Physicochemical Properties of Four Tetracyclines

The structure of these four early tetracyclines are given below, and their physicochemical properties in Table 4-2.

	R_1	R_2	R_3
Tetracycline	H	CH_3	H
Chlortetracycline	Cl	CH_3	H
Oxytetracycline	H	CH_3	OH
Demeclocycline	Cl	H	H

Chlortetracycline and oxytetracycline are unstable in strongly acidic solutions and are particularly unstable at pH values above 7.0. Tetracycline is more stable at alkaline pH than either, and demeclocycline is the most stable of all, being relatively resistant to degradation by acid or alkali. In addition to stability, tetracyclines can be distinguished by paper chromatography, agar diffusion, countercurrent distribution, ultraviolet and infrared absorption spectra, and by certain color reactions.

Following the introduction of the four tetracyclines described above, three additional tetracycline antibiotics, which were produced by semisynthetic methodology, found their way to clinical acceptance and have assumed a share of the antibiotic market. The first two of these were methacycline (6-demethyl-6-deoxy-5-hydroxy-6-methylenetetracycline) and its reduction product, doxycycline (α-6-deoxy-5-hydroxytetracycline), both of which were announced in 1963 by scientists of Charles Pfizer & Co. The third semisynthetic tetracycline, minocycline (7-dimethylamino-6-demethyl-6-deoxytetracycline), was reported by Lederle Laboratories, Division of American Cyanamid Company, in 1966.

Table 4-2 Physicochemical properties of the tetracyclines

	Tetracycline	Chlortetracycline	Oxytetracycline	Demeclocycline
Empirical formula	$C_{22}H_{24}N_2O_8$	$C_{22}H_{23}N_2O_8Cl$	$C_{22}H_{24}N_2O_9$	$C_{21}H_{21}N_2O_8Cl$
Molecular weight	444.2	478.5	460.2	464.9
Melting point (°C)	170–173	168–169	185	174–178 (sesquihydrate)
UV_{max} (nm) in 0.1 N HCl	268–355	230–262.5–367.5	267–357	
$[\alpha]_D^{25}$ in 0.1 N HCl	−257.9	−240	−196	−258* (sesquihydrate)
Rf in n-butanol, ammonium hydroxide, water (4:1:5)	0.39	0.47	0.27	

*0.5 percent in 0.1 N H_2SO_4

BIOLOGICAL PROPERTIES

Befitting the term "broad-spectrum antibiotic," the tetracyclines inhibit a wide variety of microorganisms, including many species of gram-positive and gram-negative bacteria, mycoplasma, rickettsia, and certain large viruses and protozoa. Fungi and true viruses are unaffected.

Chlortetracycline is somewhat more active against most sensitive microorganisms than are the other tetracyclines, but the degree of activity depends upon the conditions under which the test is carried out. For example, in one study employing *Staphylococcus aureus* in which measurements were made after a relatively short incubation period, oxytetracycline was 24 percent, tetracycline 25 percent, and demeclocycline 75 percent as active, respectively, as chlortetracycline. It should be noted, however, that gram-negative bacteria show more variability in response than do gram-positive. In therapy, sensitivity differences may be minimized or even reversed because of the ability of some analogs to produce and maintain high blood levels. The incidence of side effects also has a bearing on dosage levels and routes of administration.

Among members of a structurally related family of antibiotics a high degree of microbial cross resistance is ordinarily manifested to the various analogs. This is also true among the tetracyclines with the notable exception of minocycline. Important exceptions also occur among the penicillins.

INDUSTRIAL PRODUCTION

The successful commercial development of an antibiotic production process depends fundamentally upon two things: high fermentation yields of the antibiotic in large fermentors and an economical purification process. Pharmaceutical manufacturers of antibiotics maintain pilot-plant facilities whose purpose is to upgrade yields and to bring about improvements in processing procedures. Studies are continually being made on strain improvement, inoculum conditions, fermentation conditions, and various combinations of these factors. For example, improved mutant strains almost always require adjustments in fermentation conditions in order to achieve the high yields in fermentors that are obtainable in shaken flasks.

STRAIN DEVELOPMENT

The usual way to carry out a strain selection program is to plate out the organism on solid media and then select individual colonies and test them for potency. The process may involve a simple, empirical selection of isolates but more commonly the organism is subjected to prior treatment with a mutating agent such as x-ray or ultraviolet radiation or nitrogen mustard. Tests for potency can be made directly on the isolation plates by flooding with a sensitive organism or by growing colony transplants in liquid nutrient media which are then assayed.

At times phenotypic expressions such as color or colony appearance are found to correlate with antibiotic activity. In such cases selection is greatly facilitated. It has been reported, for example, that the highest yields of chlortetracycline appear to be produced by sterile (nonsporulating) more intensely orange-yellow pigmented colonies. McCormick et al. report that a simple procedure for selecting strains capable of producing the 6-demethyltetracyclines is to select those of *Streptomyces aureofaciens* which have dark-maroon colonies when grown on corn-steep agar medium. On the other hand, no correlation has been reported to have been established between productivity and other properties of *S. rimosus* strains which produce oxytetracycline.

Strain selection programs may expose qualitative as well as quantitative antibiotic differences. Mutant strains of *S. aureofaciens* have been discovered which are capable of producing either tetracycline or demeclocycline. Thus, in carrying out studies on strain selection, one should be alert for the production of additional components which may prove to be useful antibiotics in their own right.

CULTURE MAINTENANCE

The maintenance and propagation of a particular culture depend upon the conditions under which the culture can grow, remain viable, and retain its antibiotic-producing properties. Some strains show very little change with repeated transfer whereas others lose their productive capacity so readily that retention without repeated transfer is a necessity. In fact, the culture may have to be plated out and new colonies selected at frequent intervals.

The length of time which fungi and actinomycetes remain viable is a function of their spore-producing capacity. This has an important bearing on the selection of media for culture maintenance. However, some organisms are more satisfactorily preserved in the vegetative phase because of segregation into productive and nonproductive offspring upon the production of spores. Loss of antibiotic productivity upon repeated transfer is most often attributed to this cause.

Cultures may be maintained as agar slants frozen in a commercial freezer or preferably in liquid nitrogen at a lower temperature, as slants overlaid with sterile mineral oil, in tubes of sterile soil, or lyophilized after suspension in serum, gelatin, or other colloidal material. The freezing method is simple and effective but lyophilization is probably the most popular because of the small storage space taken up by lyophilized ampules and the fact that freezer space is not normally required.

INOCULATION PREPARATION

In addition to their role in promoting culture viability, agar media which promote good sporulation are generally preferred for growing cultures to be used in the preparation of inocula. The medium constituents differ, of course, with the sporulation requirements of the organism to be used. Published media for penicillin production with *Penicillium chrysogenum* variously include such ingredients as glycerol, molasses (both cane and sugar beet), corn-steep liquor, peptone, and several inorganic salts. With streptomycetes good sporulation is usually obtainable on starch-containing media although a variety of other media have proved satisfactory for this purpose. The pH of the medium may also be an important factor in spore production. In the case of cultures which sporulate poorly or not at all the vegetative mycelium may be blended in a Waring Blendor to produce a uniform inoculum.

Because of the large volumes of liquid media employed in commercial antibiotic fermentations an inoculum preparation of at least two stages is required. Shake flasks are inoculated from slants and are in turn used as inocula for larger glass vessels or for small tanks. Depending upon the size of the production fermentors the latter may be employed directly or an intervening inoculum tank may be needed. In practice, there are a number of variations in the types and sizes of the various vessels used for inoculum buildup purposes.

The composition of the liquid media used for inoculum preparation can have a profound effect on the amount and type of mycelial growth. In general, it is desirable to use media which produce fine centers of growth rather than large pellets. The latter are commonly observed in fungus fermentations, and again it may be necessary to homogenize the inoculum in a blender to obtain satisfactory results.

In the case of the tetracyclines, much of the specific information dealing with inoculum and fermentation media and conditions is to be found in various patents. *Streptomyces aureofaciens* is relatively unique among the streptomycetes in that virtually all strains readily utilize sucrose. One representative inoculum medium from a Lederle Laboratories patent on the 6-demethyltetracyclines is as follows:

Sucrose	3%
Corn-steep liquor	1.65 vol%
$CaCO_3$	0.7%
$(NH_4)_2SO_4$	0.2%

The inoculum medium described in the Charles Pfizer & Co. patent for the production of oxytetracycline by *Streptomyces rimosus* differs considerably.

Enzymatic digest of casein (N-Z-amine)	1%

Cerelose	1%
Yeast extract	0.5%
NaCl	0.5%
$CaCO_3$	0.1%

FERMENTATION CONDITIONS

The media used in industrial antibiotic production originate from laboratory and pilot-plant studies whose purpose is to elicit maximum yields from a particular strain of organism. In fact, such studies may have the additional purpose of obtaining significant yields of one particular component from among two or more components which a culture is capable of making. It is necessary to carry out studies with every strain which is receiving serious consideration as a production strain. In general, the more productive cultures seem to require more concentrated media than the strains from which they were derived.

A wide variety of substrates from which to choose is available. Carbon sources may include such carbohydrates as the various sugars, starch or glycerol, and a number of plant oils. Commonly used nitrogen sources are amino acids, casein and milk products, animal and plant meals, corn-steep liquor, meat and liver extracts, and various peptones. Growth substances such as distillers' solubles or yeast extract may be added. Sodium chloride, di- and monopotassium phosphate, magnesium sulfate, sodium nitrate, and other salts may be included for supplementary or buffering purposes. However, many of the natural products contain enough minerals to satisfy any requirements for trace elements.

Calcium carbonate is frequently added to the medium to counteract excess acidity, but an abundance may interfere with extraction procedures. In chlortetracycline fermentations the amount varies between 0.25 and 1.0 percent, and the value is determined experimentally according to the amount of corn-steep liquor and carbohydrates in the medium. Foaming may be controlled by the addition of any one of several antifoam agents, often being added as needed to fermentors while the fermentation is in progress. One frequently used agent is octadecanol at a level of 0.3 to 3 percent in lard oil.

In designing acceptable media for fermentations in large volumes it must be kept in mind that excessive costs must be avoided where it is possible and feasible to do so.

Two representative formulas for use with *S. aureofaciens* are listed below. The first is from the Petty patent for producing chlortetracycline. The materials are added in the order listed.

Corn-steep liquor, 50% solids	3.0% w/v
$CaCO_3$	0.9% w/v
Sucrose	3.0% w/v
$(NH_4)_2SO_4$	0.33% w/v
NH_4Cl	0.1% w/v
$MgCl_2 \cdot 6H_2O$	0.2% w/v
$FeSO_4 \cdot 2H_2O$	0.0041% w/v
$MnSO_4 \cdot 4H_2O$	0.005% w/v
$ZnSO_4 \cdot 7H_2O$	0.01% w/v
$CoCl_2 \cdot 6H_2O$	0.0005% w/v

The second formula is from the McCormick patent for producing 6-demethyltetracyclines and essentially substitutes corn starch for sucrose.

Corn starch	5.5%
$CaCO_3$	0.7%
$(NH_4)_2SO_4$	0.5%
NH_4Cl	0.15%
$FeSO_4 \cdot 7H_2O$	0.004%
$MnSO_4 \cdot 4H_2O$	0.005%
$ZnSO_4 \cdot 7H_2O$	0.01%
$CoCl_2 \cdot 6H_2O$	0.0005%
Corn-steep liquor	3.0% w/v
Cottonseed meal	0.2%
Lard oil	0.2% v/v

A very different medium was recommended for the production of oxytetracycline by *S. rimosus* as given in the Sobin patent on this antibiotic.

Soybean meal	3%
Corn starch	0.5%
N-Z-amine B	0.1%
$NaNO_3$	0.3%
$CaCO_3$	0.5%
Vegetable oil	0.4%

Media may be manipulated by additions or deletions to give proportionally increased yields of a desired component or to encourage the formation of a particular component. A series of biosynthetic penicillins was produced by incorporating various acyl side chains as suitable precursors. Some have been of clinical interest, particularly phenoxymethyl penicillin, penicillin V.

Numerous examples have been reported among the tetracyclines. Yields of chlortetracycline are increased when chlorides are added to the medium, but when tetracycline is desired the conditions must be reversed. Most mutant strains capable of producing tetracycline produce a certain amount of chlortetracycline unless conditions unfavorable to the production of the latter antibiotic are employed. These include fermentation on chloride-free synthetic media or, in natural media, the use of chloride inhibitors or the elimination of chlorides from such media by chemical means.

Bromide, iodide, thiocyanate, and 2-(2-furyl)-5-mercapto-1,3,4,-oxadiazole have been shown to be effective chlorination inhibitors. Dechlorination of medium ingredients may be carried out either by passage through ionic exchange resins or by precipitation by treatment with silver or mercury salts.

A patent by Bristol Laboratories describes a fermentation process in which both tetracycline yield and the proportion of tetracycline to chlortetracycline is increased by using cottonseed endosperm flour as one nitrogen source in the medium, another being dechlorinated corn-steep liquor.

In addition to media, other fermentation conditions must be studied for each new production strain in order to obtain optimum results. Considerable latitude has been reported from fermentation to fermentation among the tetracyclines. Generally the optima fall within the following limits:

Amount of inoculum	0.1 to 5.0 vol%
Age of inoculum	24 to 36 h
Aeration	0.3 to 1.5 vol/(vol)(min)
Agitation	Wide variation from nothing to 750 r/min or more
Temperature	26 to 28°C (one report recommends 28° for the first 24 h and then 25° for the remainder of the fermentation)
Harvest times	From 1 to 4 days
Changes in pH	These rise or fall between pH 4 and 8, depending upon the fermentation, and are normally controlled by appropriate buffers, notably calcium carbonate

Temperature optima for fungus fermentations are usually lower.

ANTIBIOTIC RECOVERY

At the time they are harvested, culture fluids contain a very small amount of antibiotic and a large proportion of cells and nonantibiotic materials. The purpose of extraction procedures is to obtain the antibiotic in a very pure state free from contaminating substances. The methods used are determined by the physicochemical properties of the antibiotic, particularly its solubility and its ionic character.

Some antibiotics are to be found predominantly within the cells of the producing organism but most, including those of industrial importance, are located in the extracellular fluids. Consequently, the first step in the purification process is usually to remove the cells by filtration or centrifugation (some procedures bypass filtration and go directly to extraction). The clear fluids are then treated by solvent extraction, adsorption and elution, or precipitation. All these methods may be necessary at some stage before a purified product is obtained.

To illustrate differing methodologies, brief descriptions are given below for extraction processes used with erythromycin, streptomycin, and the tetracyclines. It should be kept in mind that various alternative techniques are available.

Erythromycin is readily soluble in most organic solvents but less so in water. Filtrates are extracted with amyl acetate at pH 9.4 and back-extracted into water at pH 5.1. After adjustment to pH 8.0 the aqueous phase is concentrated; then upon further adjustment to pH 11.0 the antibiotic precipitates from solution. Recrystallization is carried out from either acetone-water or petroleum ether.

Streptomycin is a water-soluble basic substance scarcely soluble or insoluble in most organic solvents. One method of purification is to adsorb the activity onto activated charcoal at neutrality and elute with acid alcohol. The antibiotic can then be precipitated with acetone and further purified by chromatography on an alumina column. In a concentrate of the eluate from the column streptomycin is converted to a salt-picrate, reineckate, helianthate, etc., and finally to the trihydrochloride.

The solubility and ionic characteristics of the tetracyclines make possible a choice of several procedures.‡ One that is commonly used is to precipitate the antibiotics from broth by means of high-molecular-weight quaternary ammonium compounds at pH 8.5 to 10.0. The precipitates are dissolved in a solvent such as methylisobutylketone, butanol, or a water-chloroform mixture. The activity is then extracted from the organic phase with aqueous acid or precipitated from the solution.

Another method, described for chlortetracycline, is to adsorb the antibiotic on a diatomaceous earth or activated carbon column and elute under ultraviolet light with acid acetone or alcohol. A first blue band that is eluted is discarded; a second, yellow band possesses the active material and is retained. This is concentrated, taken up in butanol, which in turn is concentrated, and the antibiotic precipitated with absolute ether.

The extraction of tetracycline may be carried out by a third representative procedure. The filtrate is extracted at pH 8.5 with butanol or methylisobutylketone. The solution is concentrated and the antibiotic precipitated as the free base upon the addition of Skellysolve C. The precipitate is extracted with ammonium hydroxide and the solution passed through a Florisil column from which the activity is eluted with dilute acid.

PRECLINICAL AND CLINICAL DEVELOPMENT OF A NEW ANTIBIOTIC

Once it has been established that an antimicrobial agent is both new and capable of suppressing infections in animals it must still traverse a lengthy road of study and evaluation before approval will be given by the Food and Drug Administration (FDA) for use in human beings. At a minimum, 2 to 3 years are required for this to happen. The Food, Drug and Cosmetic Act of 1938 forbade the introduction of a new drug into interstate commerce until the producer had submitted to the FDA acceptable evidence for the safety of the drug under the recommended conditions of use. The Kefauver-Harris amendments of 1962 added the requirement that substantial evidence should also be presented for the effectiveness of the drug in respect to all therapeutic claims made for it.

Initially, pharmacology and toxicology testing is carried out on large numbers of animals of several species. Upon completion of these studies, which include 30-day observations in the animals, an investigational new drug application (INDA) is filed with the FDA. The INDA includes results of the various tests plus dosage forms to be used in forthcoming human studies.

Only in human beings can all the hazards of a new drug be known. The first human studies are for ascertaining toxicity and tolerance plus absorption and excretion and are carried out on volunteers. This leads into carefully controlled clinical trials for efficacy, which involve a small number of doctors and a small number of patients. Finally, the antibiotic is tested on a much wider group of patients, usually several hundred.

During these several stages long-range pharmacology and toxicology studies are continued in animals. Upon the satisfactory conclusion of all these investigations, a new drug application (NDA) is filed, with the hope of approval by the FDA.

LITERATURE REFERENCES ON ANTIBIOTICS

The student of antibiotics will soon find that there is an overwhelming amount of literature pertaining to virtually every phase of the subject. There is a substantial annual increase in the number of published papers dealing not only with various phases of antibiotic research and with the introduction of new antibiotics, but also with specific antibiotics of commercial importance. For example, the number of papers published in 1971 on only the tetracyclines was in excess of 2000. Fortunately, excellent reviews appear periodically; these are very helpful as a starting point in bringing current knowledge completely up to date.

The General References are intended to be illustrative of the source material for this chapter but by no means cover all aspects of the subject.

GENERAL REFERENCES

Animal Health Institute (1970): "Proceedings of the Antibiotic Presentations to the U.S. Food and Drug Administration Task Force on the Use of Antibiotics in Animal Feeds," Washington, D.C.

DiMarco, A., and P. Pennella (1959): The Fermentation of the Tetracyclines, *Prog. Ind. Microbiol.,* **1**:45.

Duggar, B. M. (1949): Aureomycin and Preparation of Same, U.S. Patent 2,482,055.

Evans, R. C. (1968): "The Technology of the Tetracyclines," Quadrangle, New York.

Hash, J. H. (1972): Antibiotic Mechanisms, *Ann. Rev. Pharmacol.,* **12**:35.

Hatch, A. B., G. M. Harty, G. H. Buelow, D. H. Phillips, R. M. Hofstead, H. J. Palocz, L. F. Sawmiller, and P. A. Hahn (1965): Notes on the Technology of Antibiotic Fermentation Development, *Simp. Ferment. Assoc. Brasil. Quim., 1st,* Sao Paulo, Brazil, **1964**:15.

Kiser, J. S., G. O. Gale, and G. A. Kemp (1971). Antibiotics as Feedstuff Additives: The Risk-Benefit Equation for Man, *CRC Crit. Rev. Toxicol.,* **1**:55.

McCormick, J. R. D., U. Hirsch, E. R. Jensen, and N. O. Sjolander (1959): 6-Demethyltetracyclines and Methods for Preparing the Same, U.S. Patent 2,878,289.

McGhee, W. J., and J. C. Megna (1957): Process for the Production of Tetracycline, U.S. Patent 2,776,243.

Noyes Development Corporation (1969): "Tetracycline Manufacturing Processes," Park Ridge, N.J.

Petty, M. A., Jr. (1955): Production of Chlortetracycline, U.S. Patent 2,709,672.

Sobin, B. A., A. C. Finlay, and J. H. Kane (1950): Terramycin and Its Production, U.S. Patent 2,516,080.

Weinstein, L., and K. Kaplan (1970): The Cephalosporins. Microbiological, Chemical, and Pharmacological Properties and Use in Chemotherapy of Infection, *Ann. Intern. Med.,* **72**:729.

White, A. I. (1966): Antibiotics, in C. O. Wilson, O. Grisvold, and R. F. Doerge, "Textbook of Organic Medicinal and Pharmaceutical Chemistry," 5th ed., p. 318,

Wolinsky, E., and F. M. Calia (1972): Outline of Antimicrobial Agents and Therapy, *Mod. Treatment,* **9**:277.

5
MICROBIOLOGY OF STEROIDS
Herbert C. Murray

The sterols and steroids have been interesting to biochemists for over 100 years. Estrogenic and androgenic activities were found to be the property of some natural steroids in the 1920s. The finding that extracts from animal adrenal cortex could maintain life in an adrenalectomized animal and victims of Addison's disease led to the great activity in the isolation and identification of these frequently active, chemically related hormones. An illustrious group of names has been associated with this work, not the least being T. Reichstein and E. C. Kendall, in the field of adrenal-cortical steroids. Ringold (1963) lists 41 steroids isolated and identified by 1959; only 7 of these could be considered active corticosteroids. The most highly active natural glucocorticoids are cortisone (Kendall's compound E: $17\alpha,21$-dihydroxypregn-4-ene-3,11,20-trione) and cortisol (Kendall's compound F: $11\beta,17\alpha,21$-trihydroxypregn-4-ene-3,20-dione).

Chemical synthesis had been developed by scientists of Merck & Co. on a large scale by 1949. This work relied upon desoxycholic acid from ox bile and was based upon the work of several groups.

When the dramatic results of the effects of cortisone upon rheumatoid arthritis and acute rheumatic fever became known, the possibility of more widespread use of these hormones gave impetus to a search for other sources for 11-oxygenated steroids.

The announcement by the Upjohn group in 1952 that *Rhizopus nigricans* (formally *R. stolonifera*) could convert progesterone to 11α-hydroxyprogesterone in good yield introduced a commercial process involving microorganisms in the synthesis of glucocorticoids and eventually furnishing material for the synthesis of analogs. This also stimulated a lagging interest in the microbiology of steroids.

Considerable activity took place in the following years, until today it may be readily said that one microorganism or another will oxygenate various steroids in one position or another, until all the feasible positions of the basic nuclei have been reported as having been oxygenated, even position 10 in a 19-nor steroid and position 3 in a 3-deoxy steroid (Fig. 5-1).

In 1970, the reactions which remained of most importance to industry were 11 oxygenation by fungi, 16α hydroxylation by streptomyces, the dehydrogenation of the A ring of glucocorticoids and analogs by *Arthrobacter simplex*, mycobacteria, nocardias, some fungi (De Flines, 1969), and the 3-hydroxy dehydrogenases which occur in many microorganisms (Fig. 5-2).

STRUCTURE OF STEROIDS IMPORTANT IN MICROBIOLOGICAL DISSIMILATION

The steroids used in commercial steroid hormone processes are, for the microbiologist, only part of the widely occurring, mostly saturated, colorless natural products possessing a tetracyclic carbon structure. The natural steroids and those at present involved in the commercially important microbiological processes have only one basic structure out of a possible 64. This very fortunately simplifies many problems arising from microbial processes. This basic form is (trans, anti, trans, anti, trans) according to the coupling of the four carbon rings of the carbon chains into chair shapes.

The most common steroids have methyl groups at 13 and 10 (C-18 and C-19). The special arrangement of substituents on the carbon skeleton is distinguished in the case of cis position toward angular methyl (above the plane of the carbon skeleton) as β, that is, a full line (Fig. 5-1); in the case of trans, it is indicated as α by a dotted line. Table 5-1 gives examples of several steroids by common and chemical name.

Figure 5–1 Basic steroid structure. (Solid lines to 18 and 19 represent angular methyl groups 17-β configuration.)

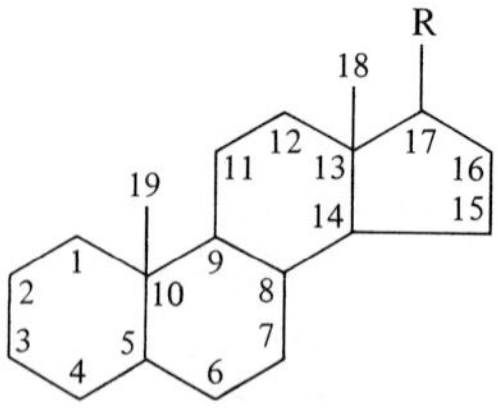

Figure 5–2 Sites of important microbial activity: dehydrogenation at 1 and 2, oxidation of alcohol to ketone at 3, and hydroxylations at 11 and 16.

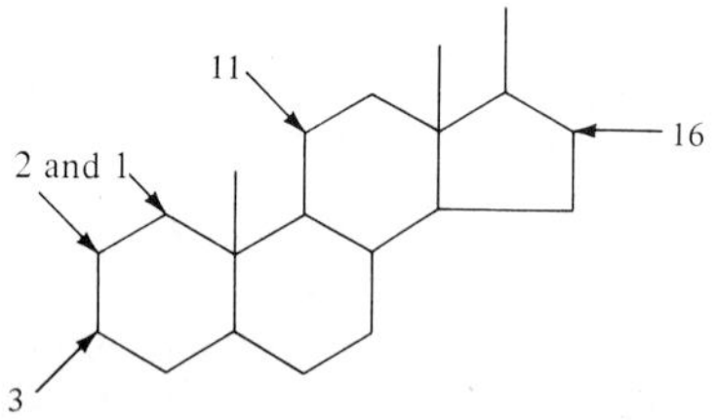

Table 5-1 Some examples of steroid nomenclature

Generic name	Chemical name
Androstenedione	Androst-4-ene-3,17-dione
Testosterone	17β-Hydroxyandrost-4-en-3-one
Progesterone	Pregn-4-ene-3,20-dione
11-Deoxy-17α-hydroxycorticosterone (Reichstein's) compound S*	17α,21-Dihydroxypregn-4-ene-3,20-dione
Cortisone (Kendall's) compound E*	17α,21-Dihydroxypregn-4-ene-3,11,20-trione
Cortisol, hydrocortisone (Kendall's) compound F*	11β,17α,21-Trihydroxypregn-4-ene-3,20-dione
Prednisone Δ-1-E*	17α,21-Dihydroxypregna-1,4-diene-3,11,20-trione
Prednisolone Δ-1-F*	11β,17α,21-Trihydroxypregna-1,4-diene-3,20-dione

*Trivial names.

A fuller description of the structure and configuration of steroids and sterols may be found in the monumental monograph of Fieser and Fieser (1959), as well as in other references (Florkin and Stotz, 1963; Shoppe, 1964). The nucleus of the steroids may also be referred to as the cyclopentanoperhydrophenanthrene skeleton by comparing it with phenanthrene, a three-ring aromatic hydrocarbon which is then reduced (perhydro) and to which a five-membered cyclopentane ring has been added.

The rings have the thermodynamically stable chair shape. There are exceptions to this description, especially with the estrogens (estrone and estradiol) which have no angular methyl group at 19 and have an aromatic A ring. There is also a class of synthetic 19-nor (methyl) steroids which have no methyl group on C-10.

Steroids involved in steroid hormone synthesis may be classed as:

1. Sterols: cholesterol, stigmasterol, sitosterol, etc., containing a 3-hydroxy and C_{28} or C_{29} atoms; C_8 or C_9 chain attached at C-17 (Fig. 5-3).
2. Bile acids having C_{24} or C_{27} and C_{28} carbons frequently hydroxylated in the nucleus and having a terminal carboxy on the side chain (Fig. 5-4).
3. Androgens having 19 carbons, most frequently a ketone or hydroxy group at 3 and 17 and unsaturation of C-4, 5 (Δ4) (Fig. 5-5). Some synthetic androgens may have other substituents such as a methyl group at 17.
4. Estrogens: generally C_{18} compounds having an aromatic or phenolic A ring and oxygen at 17 and sometimes 16 (Fig. 5-6). Synthetic estrogens may have other substituents.
5. Progestins: generally of 21 carbon atoms and having oxygen at the 3 position with unsaturation at 4,5 or 5,6 and a 20-ketone and 21-methyl group. Synthetic progestins, of course, may have a variety of substituents (Fig. 5-7).
6. Adrenal cortical hormones: the adrenal cortical hormones are 21-carbon steroids having a 3-ketone, 4-5 unsaturation, 20-ketone and 21-hydroxy group, and an 11-ketone or 11-hydroxyl group. These and substituted analogs

Cholesterol

Stigmasterol

Ergosterol

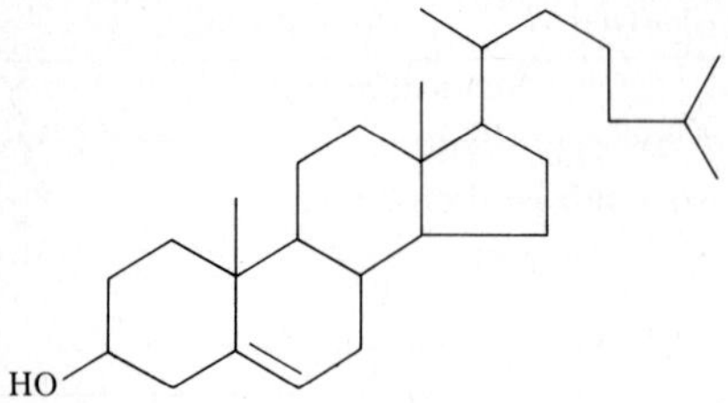

Figure 5–3 Structural form of sterols: cholesterol, stigmasterol, and ergosterol.

Desoxycholic acid

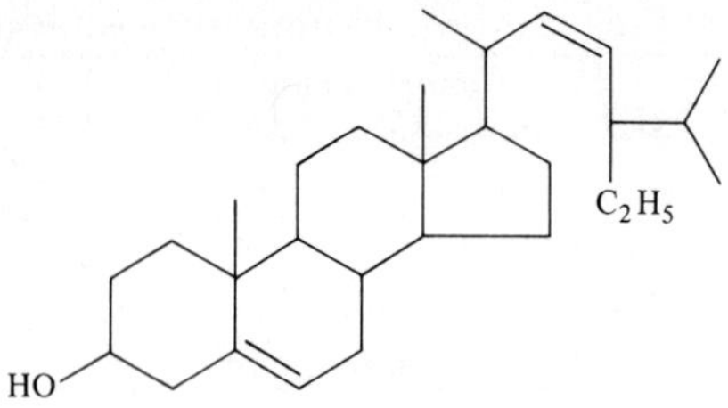

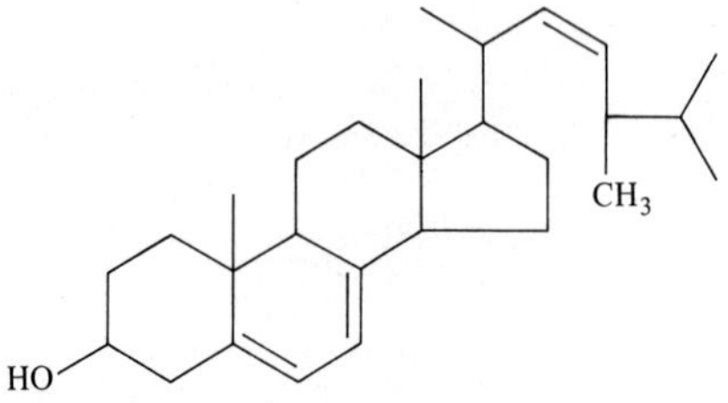

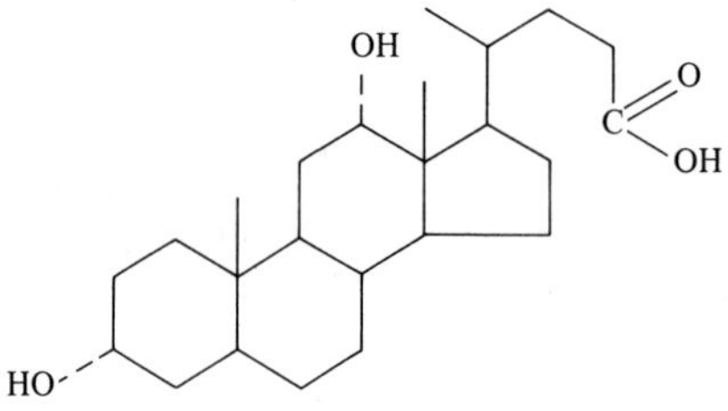

Figure 5–4 Structural form of a bile acid: desoxycholic acid.

Androstenedione

Testosterone

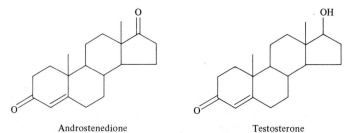

Figure 5–5 Structural form of androgens: androstenedione and testosterone.

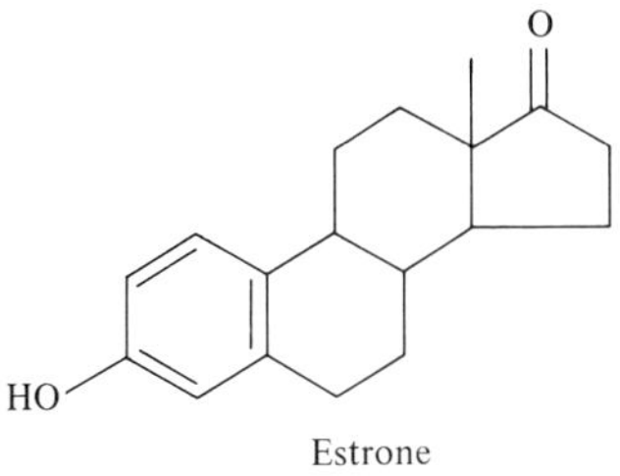

Figure 5-6 Structural form of an estrogen: estrone.

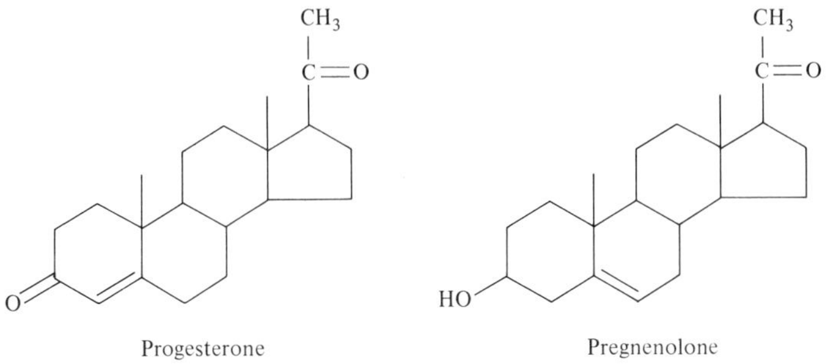

Figure 5-7 Structural form of natural progestins: progesterone and pregnenolone.

are the steroid hormones to which microbiological processes are most important.

ANALYTICAL METHODS FOR THE STEROIDS

It would be hard to overemphasize the influence of modern chemical techniques upon the successful studies of microbial dissimilation of steroids and other compounds.

Paper chromatography (PC) has been a very satisfactory tool for the analysis of fermentations. Thin-layer chromatography (TLC), generally on silica gel, has such versatility that it can be used in almost any of these studies. In many specific studies, vapor-phase chromatography (VPC) can give rapid results of extreme sensitivity. For identifications of other nature, nuclear magnetic resonance and mass spectrometry are extremely useful.

Each method has its advantages. Thin-layer chromatography has been developed to such an extent that microscope slide chromatograms may be run in a very few minutes. Samples may be applied to TLC in some cases directly from beers, or with rapid extraction of a few milliliters of beer with a solvent. The laboratory worker can follow the course of a conversion without elaborate equipment. One can change solvent systems according to one's preference and use any of an unlimited number of tests to detect the results. Of course, other methods taking advantage of the chemistry of the steroid molecule can be used.

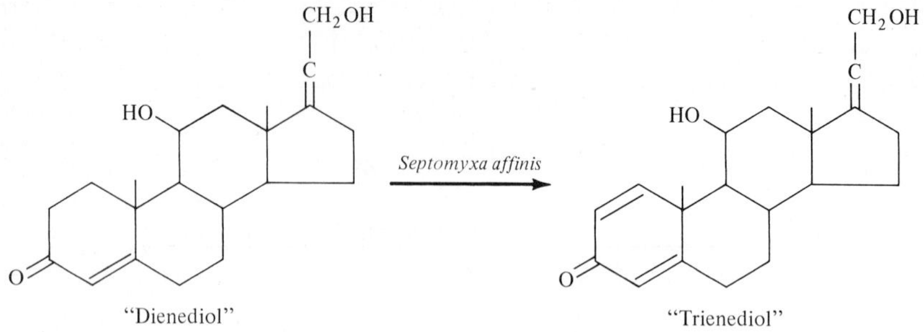

Figure 5-8 Conversion of "dienediol" to "trienediol."

A dehydrogenation reaction can be followed in extracts by the uv shift which occurs when a C-4,5 unsaturated 3-keto steroid is converted to a C-1,2 and C-4,5 unsaturated 3-keto steroid. Koepsell (Chen et al., 1962) used such a technique in following the conversion of a cortisol intermediate to a prednisolone intermediate. In this case, the absorbance decreased slightly at 243 nm and increased greatly at 268 nm (Fig. 5-8).

ECONOMICS OF STEROID PROCESSES

In 1950, only a year after Merck first introduced cortisone, cortical hormone sales were $13 million; by 1954 they were $50 million, and they reached $120 million in 1959. In 1965, domestic sales in the United States reached about $160 million.

The sex steroids and progestational hormone sales were about $12 million in 1950, and their properties as antifertility agents, in addition to other uses, increased the sales to somewhere around $150 million in 1965.

The use of microorganisms in the partial synthesis of these hormones has had significant effects. In 1949, progesterone cost about $15 per gram. Its commercial use as a basis for cortical steroids brought costs to under $0.15 per gram in 1967. Cortisone was priced at about $30 in 1949. In 1967, cortisone or hydrocortisone (cortisol) bulk prices were listed as being slightly less than $1.00 per gram.

Commercially, the glucocortical hormones may have one or more microbial steps, and the exact amounts produced by these routes are not published. The production of 11α-hydroxyprogesterone by fermentation is quite large, but the bile salts route is reportedly still being used for some analogs. The dehydrogenation of cortisone to prednisone by *Corynebacterium (Arthrobacter) simplex* is important (Fig. 5-9), and the 16 hydroxylation by streptomycetes for preparation of triamcinolone is also important. The last two reactions could be replaced by direct chemical routes, but for sensitive steroids, the microbial steps are economical (Fig. 5-10).

Synthetic progestins and estrogens are increasingly used as antifertility agents. Synthetic analogs are in demand for fattening beef and for estrus control in farm animals. Estrogens, progestins, and androgens are also therapeutic agents for human and animal disorders.

Microbial processes are available for production or modification of C_{19}

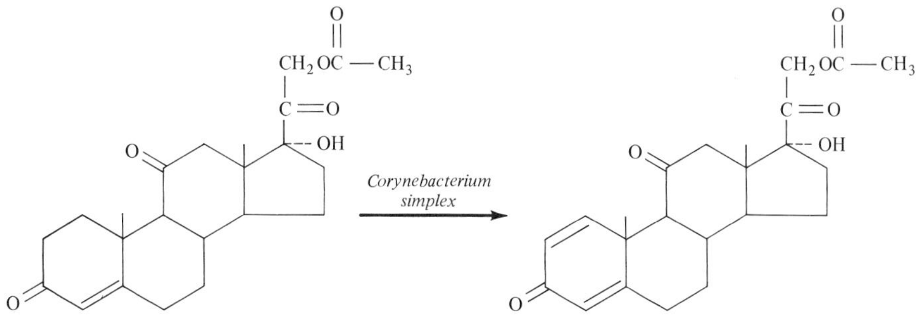

Figure 5–9 Dehydrogenation of cortisone acetate to prednisone.

steroids and are used to a limited extent in the production of substituted androgens or estrogens.

The degradation of progesterone (C_{21}) to a C_{19} steroid with or without dehydrogenation at the 1,2 positions furnishes a potential supply of testosterone or, by further chemical modification, estrogenic hormone. This use is controlled by the price of progesterone. Strictly chemical methods are available for producing C_{19} steroids.

The modification of C_{19} steroids directly by microbial methods is used for producing 11-oxygenated testosterones. The androgens can be oxygenated and dehydrogenated, as are progesterones.

Estrone and steroids with aromatic A rings are not oxygenated in the same pattern as is progesterone. Steroids with 3-hydroxy-5-ene or androst-5-ene configuration are not hydroxylated in positions one would expect from analogy to progesterone.

Methyl- and fluoro-substituted estrogens can be produced from substituted 19-nor methyl steroids. When the 19-methyl group is absent on a steroid with a 3-ketone and 4,5 dehydrogenation, dehydrogenation at 1,2 leads to aromatization of the A ring (Fig. 5-11).

MICROBIAL 11α HYDROXYLATION OF PROGESTERONE

If the structural form of progesterone (pregn-4-ene-3,20-dione) is compared with the natural corticoids cortisone (17α,21-dihydroxypregn-4-ene-3,11,20-trione) or

Figure 5–10 16 Hydroxylation of cortical steroid.

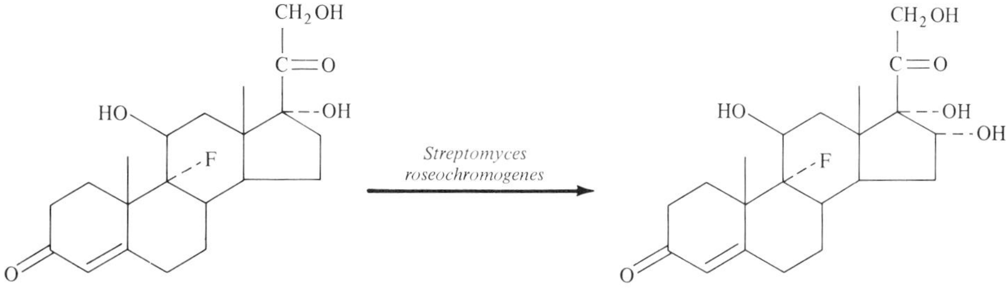

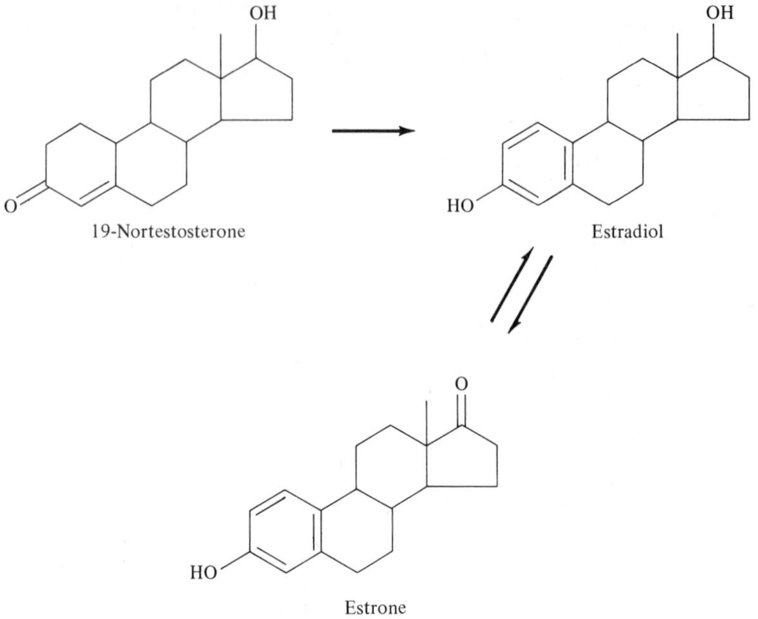

Figure 5-11 1 Dehydrogenation of 19-norsteroid.

cortisol (11α,17α,21-trihydroxypregn-4-ene-3,20-dione) (Fig. 5-12), it is seen that the glucocorticoids have additional oxygen at 11, 17, and 21. We have mentioned before that although there are reliable synthetic methods for introducing the 17- and 21-hydroxy groups, the partial synthesis of the 11-oxygenated steroids by other than microbiological processes is difficult.

11α Hydroxylation of progesterone by *Rhizopus nigricans* was developed as a timely solution to the problem of producing glucocortical hormones.‡ The action of *R. nigricans* on progesterone is a good example of the commercial feasibility of hydroxy steroid production. The mechanics of the same hydroxylation by other molds (aspergilli, etc.) resembles it, as do a number of other hydroxylations at other positions by several organisms: 15α hydroxylations by fusaria or penicillia, 6β,11α dihydroxylation, etc., by *R. arrhizus*, *Aspergillus niger*, etc. (Fig. 5-13).

Progesterone is readily available from several sterol sources by chemical procedures: stigmasterol from soya sterols, diosgenin from inedible yams (*Dioscorea*), and ergosterol from fungi. The commercial bulk product normally needs no upgrading for microbiological processing.

Rhizopus nigricans (ATCC 6227b) is the microorganism currently used. Several common mycological media are satisfactory for storage. Agar containing 5% malt extract and 0.5% peptone in tap water can be used, and the culture will be viable for several months in the refrigerator. Slants less than 2 weeks old are preferred for seeding liquid cultures, since the spores seem to be more viable. Rhizopus does not thrive on Czapek-Dox and other synthetic media. The culture spores may be stored in soil cultures and be viable for prolonged periods.

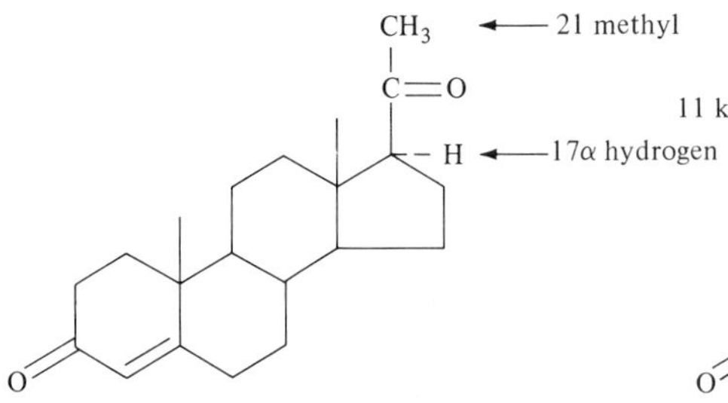

Progesterone (pregn-4-ene-3,20-dione)

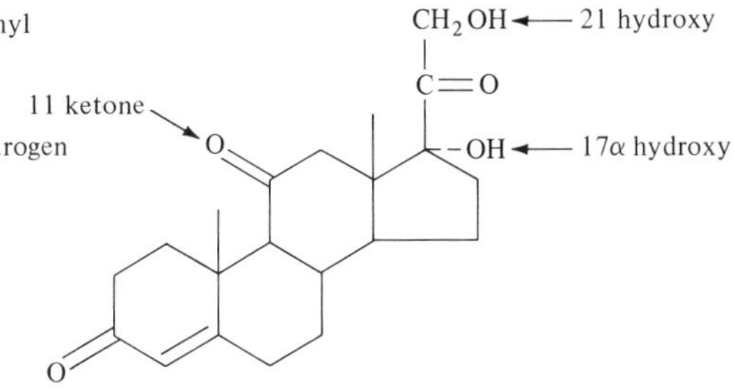

Cortisone (17α-hydroxy-pregn-4-ene-3,11,20-trione)

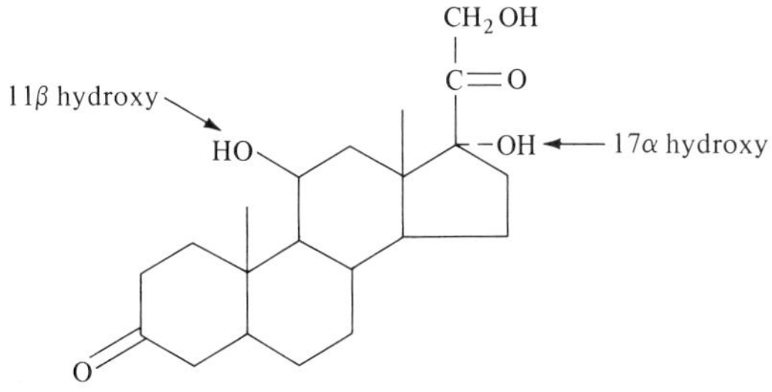

Cortisol (11β,17α-dihydroxy-pregn-4-ene-3,20-dione)

Figure 5–12 Structural form of progesterone and natural corticoids cortisone and cortisol.

Rhizopus nigricans has an optimum growth temperature of about 28°C and does not grow above 32°C.

Fermentation flasks and even large tanks may be directly inoculated with aqueous suspensions of spores harvested from tube or bottle slants. Sensibly, one uses vegetative growth by inoculating heavily from agar; a flask of liquid medium thinly seeded with rhizopus slowly forms a large whitish solid ball or pellet which does not disperse. Several media can be used for seed or fermentation; the exact constituents might be critical when the highest substrates are obtained. An initial pH of 4.8 or above is satisfactory, and slightly acid pH will contribute to a thorough sterilization without undue darkening of the medium. A medium containing about 10 g of corn-steep solids and 10 g commercial dextrose in 1 l of tap water adjusted to pH 5 with sodium hydroxide has been used in the process. Commercially, the most economical media change with costs of crude nitrogen and carbohydrate sources.

A heavy growth of rhizopus appears in 24 h in a flask shaken at room temperature. This seed can be used for several days if stored in the refrigerator. If the incubation is continued for longer than 48 h at 28°C, the growth becomes so thick that transfer and pipetting of the culture are difficult, rather resembling the manipulation of wet cotton. Contamination and variation of the mold have not been significant, and flask experiments, after the culture has grown, can be made without aseptic precaution. No critical ratio of inoculation to fermentation medium has been reported. A heavy inoculum is advised since this should overcome minor strain variation and contamination. It is good

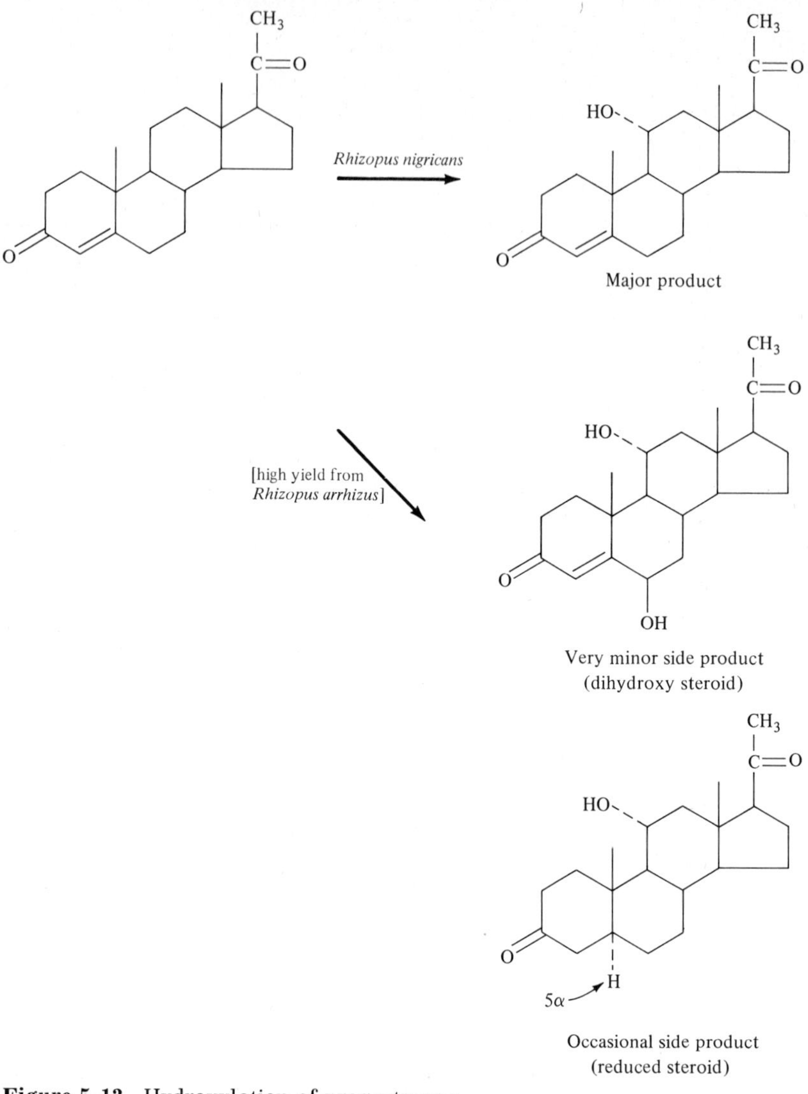

Figure 5–13 Hydroxylation of progesterone.

practice to incubate the culture until a heavy growth is apparent (obvious initial abnormalities may be detected by smell or appearance); this will be 11 to 18 h with even moderate aeration. An aeration rate of 5 l of air per 100 l of medium is about minimal; higher may be used, but excessive aeration can require an antifoam, such as lard oil, which interferes with the purification of the crystalline product. For flask experiments, almost any commercial shaker is satisfactory. Routinely, rotary shakers with 1½ - to 2½ -in displacement have been used to make conversions in 250-ml Erlenmeyer flasks containing 100 ml of medium. A speed of about 250 to 300 r/min is satisfactory. Surface cultures do not convert the progesterone to any extent.

Agitation disperses the oxygen, progesterone, and culture. This accelerates the various heterogeneous phase reactions involved with substrates and products of limited water solubility, the cells of microorganisms, and oxygen in a water dispersion.

The hydroxylation has been described as a first-order reaction; that is, after a short period of induction, there is a linear decrease of substrate and increase of product. The residual progesterone is probably in a mixed crystal formation with the product. A similar observation of reaction rate has been observed with dehydrogenations, and also mixed crystals formed from substrate and product can cause a high amount of unconverted starting material (Chen et al., 1962). Substrates dispersed as large insoluble crystalline masses are not rapidly transformed by microorganisms.

Tanks may be inoculated with rhizopus by pouring the culture from the flask into the fermentor or by pipetting with special large tubes or siphons.

Progesterone dissolved in acetone is added to the fermentation at any time; 19 h is satisfactory. No sterile precautions are needed with the solution; sterilizing an acetone solution by filtration is not necessary. Open flames are hazardous. After growth is apparent, experiments have been made with flasks and even tanks open to the air.

As the process is described here, up to 2.0 g of progesterone per liter of culture may be added and will be hydroxylated satisfactorily within 48 h. The product is extracted from the beer with methylene chloride. If less than 400 mg/l of progesterone is added, most of the product will be in the beer; at higher levels, some 11α-hydroxyprogesterone will be filtered out with the mycelium as well as most of the remaining progesterone. This is a practical description of early processes. Developments have since led to higher levels of steroids, i.e., use of surfactants, suspension of steroids, etc.

EXTRACTION

Methylene chloride and a variety of nonpolar solvents (ethyl acetate, amyl acetate, ethylene chloride, chloroform) are used for extracting the products of this conversion. Petroleum products, ether, and benzene are not satisfactory. On the laboratory scale, methylene chloride has desirable properties since it is readily evaporated or distilled at a low temperature and the residue does not have to be heated unnecessarily. Several extractions of the beer with solvent are made (for example, three or four times with a quarter volume) and the amount of extraction followed by TLC with uv detection. Continuous extraction processes are used commercially. Extracting the mycelial cake with acetone recovers much or all of the steroid, and a moderate volume of methylene chloride will recover the rest. After removal of the solvent from the extracts, a light-brown-colored mass is obtained. This can be decolorized with carbon and recrystallized from acetone-methanol or methylene chloride. A large amount of the desired product may be obtained by dissolving the solid mass in methylene chloride, filtering through filter paper, and evaporating to a small volume upon a steam bath. The product will readily cool, crystallize upon cooling, and can then be recrystallized from acetone, etc. Yield is expected to be 90 percent or more of weight of added progesterone.

SIDE REACTIONS OF PROGESTERONE HYDROXYLATION

With escessively low aeration, and high carbohydrate media, *Rhizopus nigricans* produces 11α-hydroxy-5-pregnane-3,20-dione in up to 10 percent of the product. Another product which is almost always present is 6β, 11α-dihydroxyprogesterone. The reason that *R. nigricans* does not produce more than a very small percentage of dihydroxyprogesterones in this fermentation is not clear. *Rhizopus nigricans* may produce the enzymes for 6 hydroxylating steroids in fair amount—even 11α-hydroxyprogesterone may be hydroxylated slightly; but, fortunately, the product from the progesterone conversion is mostly monohydroxylated and untouched progesterone. *Rhizopus arrhizus* can be used to make 11α-hydroxyprogesterone, but it will almost quantitatively convert this product to dihydroxyprogesterone unless some care is exercised.

Rhizopus nigricans does not grow above 32°C, and essentially no hydroxylation will take place. *Rhizopus arrhizus* grows well up to 37°C, and the higher temperature accelerates the production of 6β,11α-dihydroxyprogesterone. A variety of analytical methods can be used for this conversion. The simplest is TLC or PC. For example, a TLC silica gel system with 10% methanol-benzene can be used. The steroids are detected by uv absorption at about 340 nm, and the reduced steroids having no conjugated ketone are detected by sulfuric acid, iodine vapor, dinitrophenylhydrazine reagent, etc.

DISCUSSION OF THE 11α HYDROXYLATION PROCESS

Termination of the conversion is not critical. After the progesterone is converted to 11α-hydroxyprogesterone, slight further reaction occurs, and so the extraction of product can be made when convenient. A large number of fungal hydroxylations may be halted by stopping aeration or cooling. Generally no necessity is seen for poisoning or pasteurizing the culture. This is not true of the reactions with streptomycetes, corynebacteria, and nocardias, since further degradation or change of the steroid nucleus may follow.

PRODUCTION OF 11-HYDROXY STEROIDS BY OTHER MICROORGANISMS

Many mucoraceons fungi introduce oxygen into the 11 position of progesterone and other steroids. The *a* strain of *Rhizopus nigricans* is equivalent to the *b* strain. Almost all rhizopus species resemble *R. nigricans*, i.e., producing mainly monohydroxyprogesterone, or *R. arrhizus* readily yielding 6β11α-dihydroxyprogesterone. *Absidia* and *Cunninghamella* strains may hydroxylate progesterone to an 11α- and an 11β-hydroxy steroid, generally with one or more of other positions—6β, 7α, 14α, 9α—also hydroxylated.

Curvularia lunata has exceptional ability to make 11β-hydroxy steroids upon some 17α,21-dihydroxy steroids. *Curvularia lunata* may also hydroxylate at 7 and 14 and even dehydrogenates at 1,2.

Curvularia lunata is used to produce cortisol from Reichstein's compound S (Fig. 5-14). DeFlines (1969) indicated that 16 substitution or 17 acetylation will block the formation of 14-hydroxy compounds by *Cur. lunata* (Fig. 5-15). The process he described uses compound S 17,21-diacetate, and if the pH is below 6.5, hydrocortisone 17-acetate is obtained in very good yield at a high substrate level. The acetylation of the starting material makes the hydroxylation proceed more rapidly. The acetoxy group at the 17 position blocks hydroxylation at 14. Reichstein's compound S can be added as a solid and will disperse itself, but the free compound seems toxic to many fungi.

Curvularia Blakesleeana has been used to convert Reichstein's compound S to cortisol (Fig. 5-16), and under limited conditions, some may be prepared, apparently by poisoning the enzymes converting the 11β-hydroxy group to a ketone (Eroshin and Krasilnikov, 1961).

Aspergillus ochraceus, as well as several other aspergilli, yields 11α-hydroxyprogesterone and also 6β,11α-dihydroxyprogesterone. *Aspergillus ochraceus* will also cleave the side chain of progesterone. Weaver et al. (1960), using finely ground progesterone and Tween 80 as a surfactant, reached levels of 20 to 50 g of progesterone per liter of culture and still had a satisfactory yield of 11α-hydroxyprogesterone. Finely dispensed steroids and similar techniques are used with a variety of cultures, including *R. nigricans*. The solvents acetone, dimethylformamide, ethyl alcohol, propylene glycol, etc., show some toxicity to cultures at relatively low levels (2 to 5 percent) of the culture and interfere with the hydroxylations. All steroids or processes do not respond to the high-level techniques. Weaver et al. (1960) pointed out that the product from progesterone is not inhibitory to the 11α hydroxylation by fungi, but that the product from many other steroids can be inhibitory. The product from hydroxylating 16α,17α-epoxypregnene-3,20-dione at low levels by *R. nigricans* inhibited the hydroxylation of progesterone (Fig. 5-17) as well as the steroid from which it had been produced. On the other hand, during high-level conversions, undesired product formations were suppressed.

Vezina et al. (1963) studied the steroid conversions with spores, in the absence of complete media. They could not relate the 11α hydroxylation by *A. ochraceus* to an induced enzyme phenomenon. The addition of carbohydrate

Figure 5-14 Conversion of compound S to cortisol.

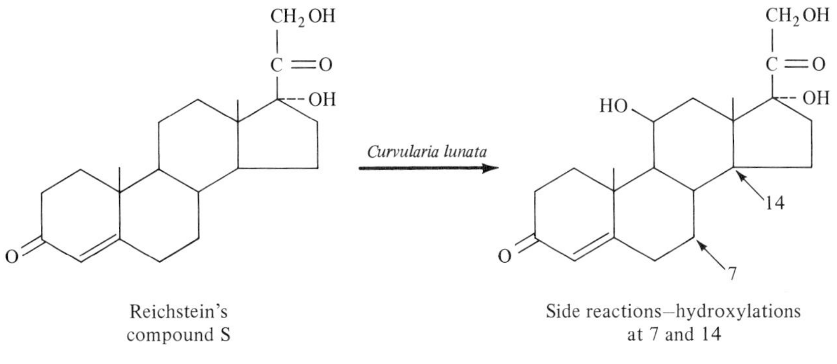

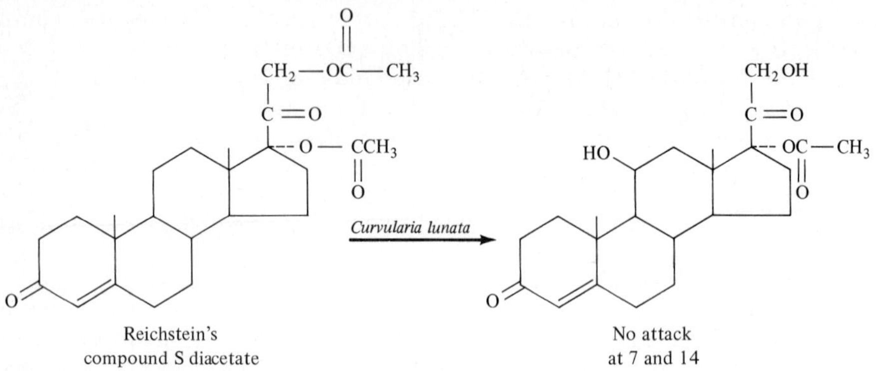

Figure 5-15 Conversion of compound S diacetate.

(Reichstein's compound S diacetate → Curvularia lunata → No attack at 7 and 14)

accelerated the hydroxylations by fungal spores, and there are other indications that the presence of sugar may enhance other hydroxylations. The dehydrogenations and side-chain cleavage discussed later are inhibited by the presence of available carbon. Spore conversions are useful for special conversions since

Figure 5-16 Conversion of compound S.

(Reichstein's compound S → Curvularia blakesleeana → Cortisol → Cortisone, also 17α-hydroxy, 11α-hydroxy, and 6β-hydroxysteroids)

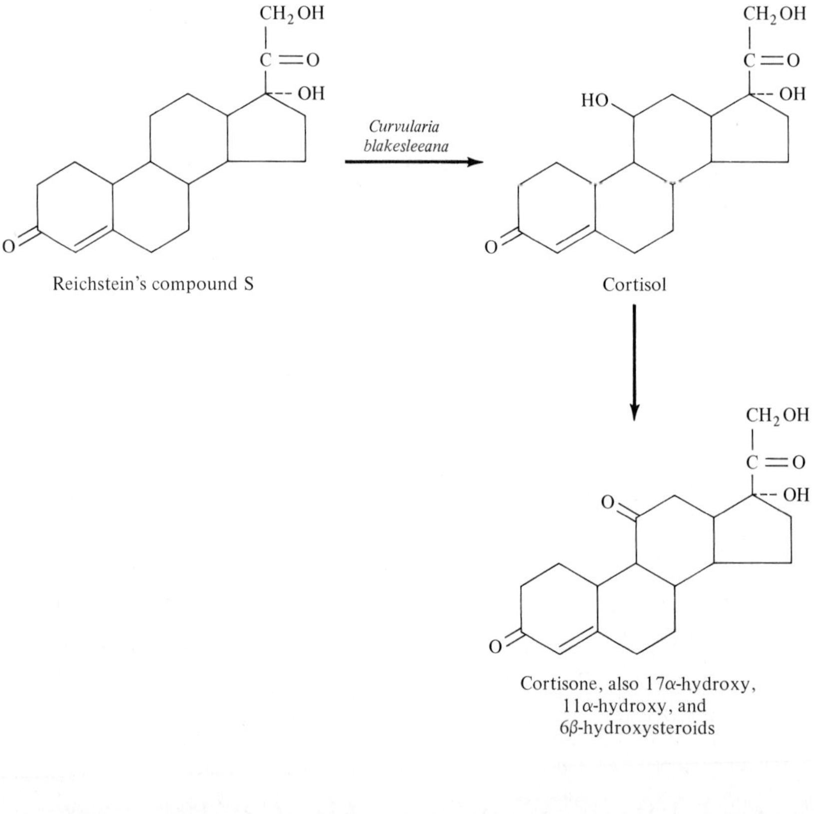

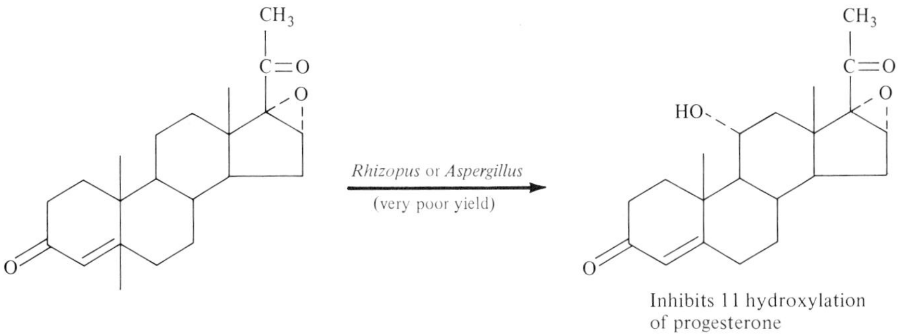

Figure 5-17 Product inhibition of conversions.

(diagram shows: steroid structure → Rhizopus or Aspergillus (very poor yield) → 11α-hydroxylated steroid; labeled "Inhibits 11 hydroxylation of progesterone")

adverse solvent effects do not readily affect the conversions. The isolation of products from a low-solids media (buffer solution) is very complete.

16 HYDROXYLATION OF 9α-FLUORO-11β,17α,21-TRIHYDROXYPREGNA-1,4-DIENE-3-ONE WITH *STREPTOMYCES ROSEOCHROMOGENES* (ATCC 11009)

Another hydroxylation important for production of a synthetic cortical steroid analog is the hydroxylation of the C-16 (Fig. 5-18). Steroids may be oxidized to 16-hydroxy steroids synthetically, but upon the highly substituted steroid 9α-fluorocortisol, etc., the hydroxylation with streptomycetes has advantages. In contrast to the 11α-hydroxylation with *Rhizopus* and *Aspergillus* species, the fermentation conditions are more critical. This apparently arises, to some extent, from the nature of the substrate as well as the organism.

Goodman and Smith (1960) studied the 16 hydroxylation of 9α-fluorohydrocortisone and 9α-fluoroprednisolone (Δ1 analog) extensively and also reported the nonenzymatic isomerization to D-homoannulated steroids during and after hydroxylation or during microbiological 1,2 dehydrogenation. They found that this undesired reaction was induced by excessive iron in the media.

A fermentation medium consisting of 4.0% starch, 25% corn-steep liquor, 0.5% calcium carbonate, and 0.2% lard oil by volume is inoculated with a 24-h vegetative growth of *S. roseochromogenes* (ATCC 3347) grown in a seed medium of 3% sucrose, 2% corn-steep, 0.5% calcium carbonate, and 0.2% ammonium sulfate. This medium is all dispensed in 50-ml amounts in 250-ml flasks. After 24-h aeration, a solution of 25 mg of 9α-fluorohydrocortisone (9α-fluorocortisol) in dimethylformamide is added to the flasks in less than 1 ml of dimethylformamide. After 48 h, aliquots or the whole flasks may be analyzed by chromatography.

Since Goodman and Smith (1960) showed the advantage of adding 0.5% monobasic hydrogen phosphate to prevent the iron effect, it is necessary to add this amount to the fermentation.

The aeration at 26.5°C upon a 2-in-throw rotary shaker at 185 r/min would correspond to an air rate of 0.2 to 0.3 l/(min)(l) of medium in a tank.

Iron has a demonstrably adverse effect on the product. It also has an

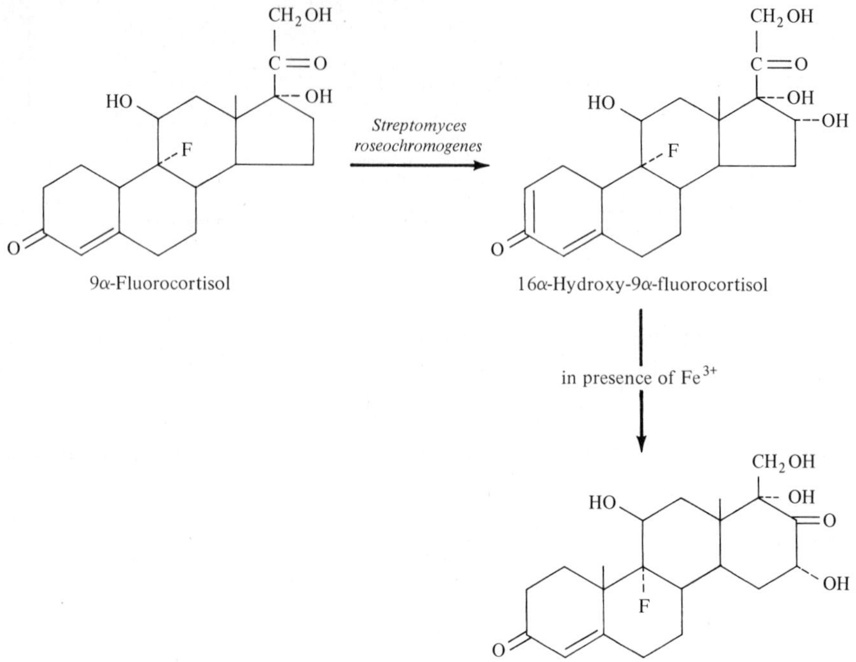

Figure 5-18 16 Hydroxylation of 9-fluorocortisol by *Streptomyces*.

adverse effect during the 1,2 dehydrogenation of the product to triamcinalone. If iron was present during the dehydrogenation with *Corynebacterium simplex*, the isomerization of the 16-hydroxylated compound occurred as well as dehydrogenation.

ENZYMOLOGY OF MICROBIAL HYDROXYLATION OF STEROIDS

The nature of the fungal enzymes which catalyze the hydroxylation of steroids has not been completely determined. Tamm (1962) noted that these must be very sensitive enzymes. Only a few reports have been made of successful steroid hydroxylations with cell-free extracts. Hayano and other workers (Block and Hayaishi, 1966) over the years made several fundamental observations with mammalian extracts and whole fungal cells.

On the basis of tracer studies, it is clear that the hydroxylations are direct replacement of the hydrogen on a carbon atom. The oxygen of the hydroxyl group comes from molecular oxygen and not water. The hydroxyl group is of the same configuration (α or β) as that of the hydrogen which is removed without inversion.

Not all the cofactors have been determined with these enzyme systems. All the mammalian steroid oxygenase systems require NADPH or cofactors which can be replaced by NADPH. These enzymes are called monooxygenases and mixed-function oxygenases. Kimura and Estabrook (Block and Hayaishi, 1966) simultaneously reported a ferrodoxin-like nonheme iron protein; adrenodoxin was found with the 11β-steroid hydroxylase of adrenal glands. Kimura summarized the electron transport system from NADPH to molecular oxygen as consisting of adrenodoxin reductase (a flavoprotein), adrenodoxin (a nonheme iron protein), and cytochrome P-450.

Zuidweg (1968) published results of a cell-free hydroxylation with extracts of *Curvularia lunata*. The steroid hydroxylation studied was the conversion of 17α,21-dihydroxy-pregn-4-ene-3,20-dione (or Reichstein's compound S) to cortisol and 14α-hydroxy compound S. These studies were a 5-year intensive effort.

The enzyme is induced in the whole cells by addition of compound S. The mycelium is washed free of steroid and homogenized in a Potter-Elvehjem homogenizer at 4°C. The extraction is successful only if EDTA, glutathione, and Tween 80 are present. When incubations were made at 25°C, no activity was measured. The hydroxylations were most successful at lower temperatures. The enzyme was stable as a precipitate at −30°C. Solutions were very labile. The inducers could be C_{18}, C_{19}, or C_{21} steroids not substituted at 11 or 14. Two exceptions as inducers were 11-ketoprogesterone and 14-dehydroprogesterone. Both 14α and 11β hydroxylating activities were found to be in the extracts. This technique has not been performed successfully in many laboratories. Frequently the hydroxylases seem to be attached to cell membranes and are soluble with difficulty.[1]

The hydroxylation, like that of mammalian enzymes, was NADPH- and O_2-dependent. Some other characteristics of the Kimura and Estabrook enzymes were not found (Zuidweg, 1968).

The specificity with which both fungal and mammalian enzymes hydroxylate in one or another position is not understood, and so the existence of another factor has been postulated.

Sih and Whitlock (1968) studied the 9 hydroxylation and 9 epoxidation of steroids with cell-free extracts from nocardia. This enzyme was more stable than that from *C. lunata*, since incubations for 5 min at 25°C produced 9α-hydroxy steroids. The enzyme was in a particulate fraction from sonicated cells, and the supernatant contained an inhibitor. The nocardia are well known to utilize steroids, and the 9 hydroxylation is part of the mechanism by which the molecule is disrupted.

OTHER STEROID HYDROXYLATIONS

Hydroxylations of many positions of specific steroids can be obtained in high yield.

Mucor griseo-cyamas produces 14α-hydroxy progesterone readily. *Aschochyta linicola* makes 9-hydroxylated steroids as almost the only compound from several C_{19} and C_{21} steroids. Fusaria, calonectria, and some penicillia are a readily available source for 15α-hydroxy steroids. Some fusaria and *Calonectria decora* also produce 12β, 15α-dihydroxy steroids.

An interaction of organism and steroid is readily observed. *Rhizopus nigricans* produces mainly 11α-hydroxy progesterone as given in the example.

---

[1] D. P. Wallach, personal communication.

Testosterone and methyl testosterone are converted by *R. nigricans* to 11α-hydroxy, 6β-hydroxy, and dihydroxy analogs. *Sporotrichum sulfurescens* (and other *Sporotrichum* strains) can convert progesterone to 11-hydroxy testosterone in high yield and primarily hydroxylates testosterone at the 11 position. However, *S. sulfurescens* does not easily cleave the side chain of 17α,21-dihydroxypregn-4-ene-3,20-dione (Reichstein's compound S) but produces 11α, 17α,21-trihydroxypregn-4-ene-3,20-dione (11-epicortisol). A large number of these interactions could be found. In almost all hydroxylations, more than one position may be affected.

Hydroxylations at 14 may be accompanied by hydroxylations at 6,7,9, and hydroxylations at 17 by hydroxylations at 6 and 11.

The proof must be accomplished by unequivocal physicochemical methods. It is rash to conclude that some specific hydroxylation is the exclusive property of one group of microorganisms. Some bacilli have made 15-hydroxy steroids. At least one fungus, *Sepedonium chrysospermum*, may produce 16α-hydroxy steroids from saturated pregnanes.[1] The 3 hydroxylation and the 1 hydroxylations of steroids lacking oxygen in the A ring have been accomplished with *Cal. decora*. This could also be considered another example of interaction. Sir Ewart R. H. Jones (1973) and his many collaborators studied the action of microorganisms on many "unusual steroids." Their goal was to determine points of attachments or attack by enzymes. This can be risky business because certain substituents on a molecule may make extraordinary changes in the rate and position of attack. Jones found that *Cal. decora* produced the 11α hydroxy product in one particular steroid. J. A. Campbell and D. M. Squires found that a related compound when substituted at the 16 position with a cyano group was hydroxylated at 12β by *R. nigricans*; *R. nigricans* makes the 11α-hydroxy compound in the absence of the cyano group.[2]

It has been observed before, although these hydroxylations could be practical, that no extensive commercial hydroxylations are known except for those at 11 and 16.

DEHYDROGENATION OF STEROIDS

Most of the marketed anti-inflammatory steroids are unsaturated at the 1,2 position (Fig. 5-19). Several cultures are used for the dehydrogenation of a precursor to the final cortical steroid: prednisone, prednisolone, triamcinolone, 6-methylprednisolone, and others. Another process utilizes the dehydrogenation of a chemical intermediate which then is converted chemically into prednisolone.

In contrast to the hydroxylation, several bacteria can be more efficient than fungi in dehydrogenating steroids at the 1,2 carbons. The first dehydrogenated anti-inflammatory steroid marketed, prednisone, was from cortisone dehydrogenated by *Corynebacterium (Arthrobacter) simplex* (Fig. 5-9).

The dehydrogenation of steroids by microorganisms is directly in competition with chemical dehydrogenations. If, for instance, dehydrogenation at 4,5 is necessary as well as on a saturated 5α or 5β, a chemical method may be used.

---

[1]G. S. Eonken, personal communication.
[2]J. A. Campbell, personnal communication.

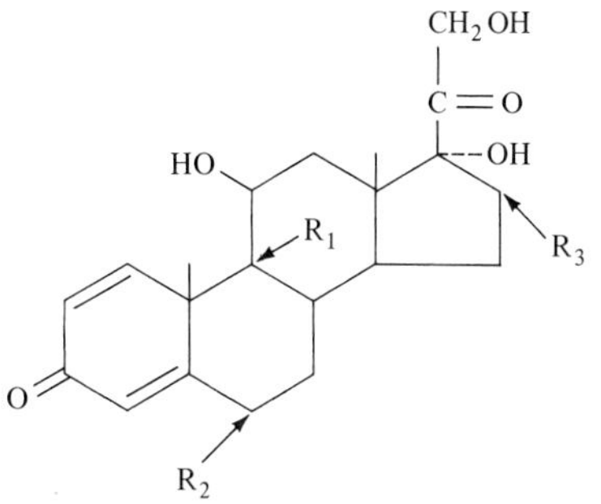

Figure 5-19 Corticoid structure with possible substitutions that enhance anti-inflammatory activity: $R_1 = 9\alpha$ F or H; $R_2 = $ H, 6α CH_3, 6α F; $R_3 = $ H, 16α CH_3, 16β CH_3, 16α OH.

The microbiological dehydrogenations are accompanied by side reactions which decrease the yield of desired products. Fortunately, manipulation of the several fermentations gives a high recovery of desired products (Fig. 5-20).

The extreme example of the application of such studies would be the "pseudo-crystallofermentation" of Kondu and Masuo (1961), which merits particular discussion because of novelty and extreme high levels of substrate. The culture was *C.* (or *A.*) *simplex* grown on a medium of 5 g of peptone, 5 g of corn-steep liquor, and 5 g of glucose per liter of water, buffered at pH 7.0. After 24 h of growth, up to 500 g/l of finely powdered cortisol was added and the fermentation continued.

The conversion was followed by observing the changes in crystal formation since both substrate and product were crystalline in the medium. The yields were extremely high. When 1 g of cortisol was added per 100 ml of culture, 0.985 g of crude prednisolone could be recovered in 24 h. When 50 g/ 100 ml of culture was added, 46.8 g of crude prednisolone could be recovered in 5 days.

This example of a conversion is on the level with an organic catalytic process. The change of cortisone to prednisone by *C. simplex* does not proceed as well at this extremely high level. *Corynebacterium simplex* does not remove the 21-acetate and, thus, prednisone acetate may be made from cortisone acetate.

Several other bacteria, *Bacillus sphaericus*, nocardias, and mycobacteria, dehydrogenate cortisol or cortisone readily. At a lower concentration than in the example, the degradation of the steroid may be observed. Most organisms which dehydrogenate steroids also have the ability to further degrade or change the molecule by removing the side chain, reducing the 20-ketone to an alcohol, and also by a 9α-hydroxylation mechanism bring about the breakage of the B ring of the steroid.

Many true fungi also dehydrogenate steroids at the 1,2 position, and, like some nocardia, many dehydrogenate at 4,5. For chemical purposes, the latter dehydrogenation has not been very important.

Studies with cell-free preparations indicate that these enzymes are induced. *Septomyxa affinis* clearly demonstrates this effect in the growing culture. It will not dehydrogenate cortisol to prednisolone unless another less

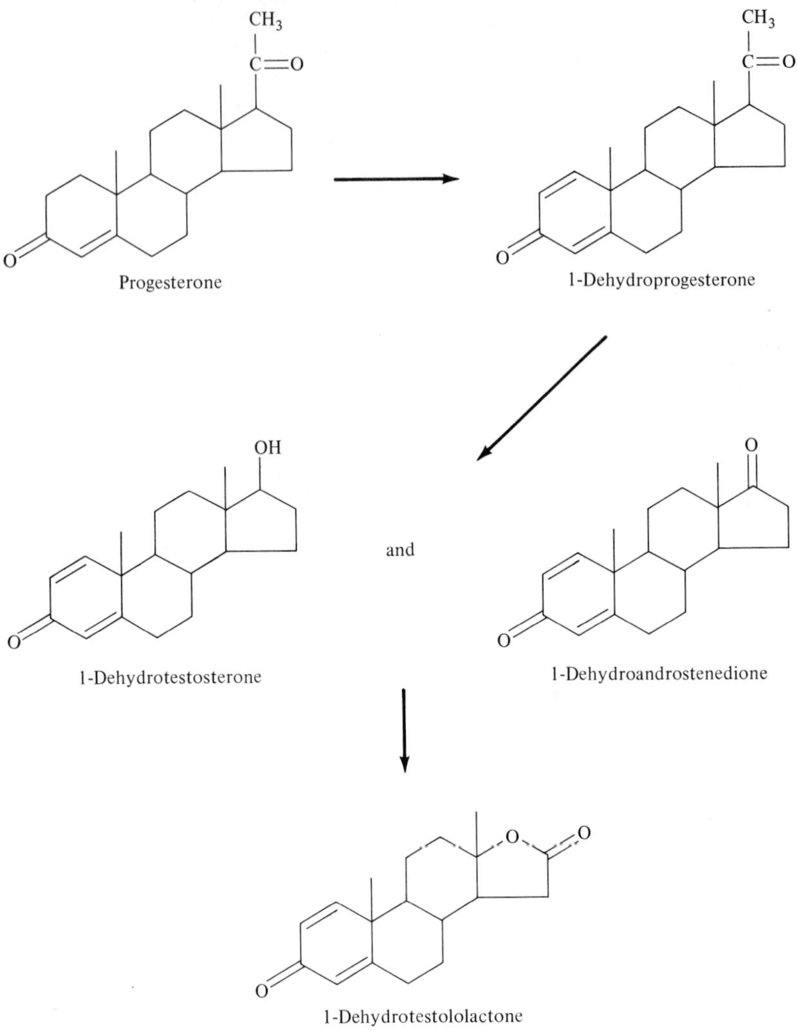

Figure 5-20 Dehydrogenation of progesterone by several microorganisms.

polar steroid is present, i.e., progesterone or a steroid of similar polarity. This effect has been used to convert 6α-methyl cortisol to 6α-methyl prednisolone. *Corynebacterium simplex* seems to degrade 6α-methyl cortisol as it dehydrogenates the A ring. The dehydrogenation of 6α-methyl cortisol by *S. affinis* is considerably slower than, for instance, the dehydrogenation of progesterone and, although pregna-1,4-diene-3,20-dione can be prepared with this fungus, the air rate must be carefully controlled and the conversion must be carefully monitored by sampling and TLC, or another method of controlling the dehydrogenation and side-chain removal (Fig. 5-21).

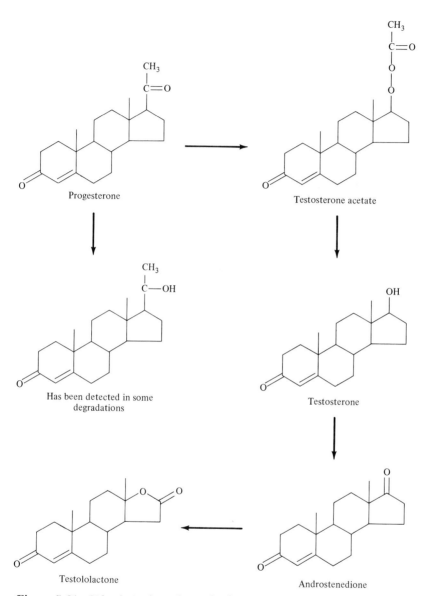

Figure 5-21 Side-chain degradation by fungi.

ENZYMOLOGY OF STEROID 1-DEHYDROGENASES

The steroid 1-dehydrogenases are readily obtained from microorganisms by the usual techniques (Sih and Whitlock, 1968).

The induced enzymes from nocardias, *Bacillus sphaericus* and *Pseudomo-*

nas testosteroni, have been studied in cell-free extracts and the enzyme which dehydrogenates at 1 can be separated from that dehydrogenating at 4 on saturated A-ring steroids. The dehydrogenation does not need molecular oxygen, and generally some hydrogen acceptor such as phenazine methosulfate, reazurin, or 2,6-dichlorophenal indophenol can be used.

Among the several 1-dehydrogenating cultures, there seems to be a difference in the enzymes. *Septomyxa affinis* dehydrogenates only at 1, 2 but dehydrogenates both 5α, 5β, and 4,5 unsaturated steroids and those with 11 or other hydroxyl groups. Nocardia and Sih's nocardial enzyme dehydrogenated at 1 and 4 of 5α and 5β saturated steroids, and dehydrogenated steroids with 4,5 unsaturation and 11-oxygenated steroids. *Pseudomonas testosteroni* enzymes did not seem to dehydrogenate 11-oxygenated steroids. No hydroxylation participates in the reaction. Evidence strongly points to the dehydrogenation being brought about by enolization and hydrogen removal.

MICROBIAL PRODUCTION OF C_{19} AND C_{18} STEROIDS

Many fungi cleave the side chain of progesterone, and the remaining steroid is testosterone or androstenedione. These two active androgenic hormones can be chemically treated to yield other active hormones (Fig. 5-21).

Fungi have been reported to be useful in this conversion. Several penicillia degrade progesterone: *Penicillium lilacinum*, *P. chrysogenum*, etc. Many aspergilli also accomplish this reaction. *Aspergillus flavus* and *A. tamarii* have been studied extensively. Most of these organisms will further degrade the steroid, yielding testolic acid or its lactone testololactone. It can be said that, if progesterone is inexpensive enough, this process would be economical for the production of testosterone. However, chemical methods might also be used.

The mechanism of this cleavage may not be the same for all side-chain cleaving organisms. Testosterone acetate has been obtained from aspergilli, cladospora, and streptomycetes. 20β-Hydroxypregn-4-en-3-one occurs in penicillia and other conversions, but it is not known why this might be a necessary intermediate. There is some evidence that 17α-hydroxypregn-4-ene-3-one, as well as testosterone acetate, may be an intermediate for *P. lilacinum*, analogous to the reaction in mammalian systems.

Chemically, this reaction would be a peroxidation similar to the Baeyer-Villiger reaction. If progesterone is substituted at other than the 17 position, a similar cleavage can be made so that substituted androgens can then be prepared.

PROGESTERONE TO ANDROSTA-1,4-DIENE STEROIDS

Many cultures which dehydrogenate steroids at the 1,2 positions also readily cleave the side chain (Fig. 5-20). The resulting steroids may be chemically converted to aromatic A-ring steroids such as estrone and estradiol.

The cultures which will accomplish the side-chain degradation and dehydrogenation include *Fusarium javanicum*, *Hypomyces haematococeus*, *Septomyxa affinis*, and *Cylindrocarpon radicicola*. The dehydrogenating bacteria are not as useful here since they seem to degrade the molecule by disrupting the

B ring, and the molds seem to further degrade the steroid by producing testolic acids or testololactones. Substituted progesterones may also be converted to $\Delta 1,4$-C_{19} steroids, and production of substituted estrogens then is a possibility.

STEROLS TO C_{19} COMPOUNDS

Microorganisms are known to ulitize the sterols cholesterol and sitosterol, which have saturated side chains. This potential source of steroids would be endless if a good way of controlling the further degradation of the steroid nucleus beyond the C_{19} carbon were found. Chemically, cholesterol by chemical modification and oxidation with chromic acid can be converted to 3-hydroxy-androst-5-ene-17-one. The yield is barely sufficient to be economical (15 percent or so). However, this process was a source of the C_{19} steroids for many years. Diosgenin stigmasterol and compounds having "unsaturated" side chains are economically converted to progesterone and androstene compounds by chemical means (Fig. 5-3).

Several laboratories have reported conversion of cholesterol to a C_{19} steroid by *Mycobacterium* or *Nocardia* strains. Sih suggested that 19-hydroxy cholesterol prepared chemically can be converted to estrone in good yield by a nocardia or mycobacteria.

Marschek, Kraychy, and Muir (1972) reported that mutants from *Mycobacterium* sp. would yield androstene dione and 1-dehydroandrostene dione in satisfactory amount from cholesterol (Fig. 5-3) and other sterols. Generally mycobacteria and sterol-utilizing microorganisms completely metabolize these compounds with no residual steroidal intermediate. This work should provide another source for steroids besides the ones from sterols readily degraded by purely chemical means.

OTHER MICROBIOLOGICAL EFFECTS ON STEROIDS

The earliest reported activity of microorganisms on steroids were the steroid ketone and double-bond reductions reported by Mamoli and Vercelloni in 1937. Since then, many similar reductions and hydroxy steroid dehydrogenations have been found. Specific bacterial enzymes have been prepared for oxidizing 17β or 3β and 3α-hydroxy steroids to ketones. These are of considerable interest for steroid metabolism research and can be used by organic chemists. Similar enzymes are available in fungi. Penicillia hydroxylates at 15 and converts pregnenolone to 15-hydroxy progesterone by oxidizing the 3-hydroxy group to a ketone. Anaerobic bacteria may reduce the 4,5 unsaturation of a steroid and also the 3- and 17-ketones of androstenedione to alcohols. Yeast may reduce 17-ketones. The hydroxy steroid dehydrogenases have very specific molecular requirements, and several studies have been made of the nature of these enzymes (Sih and Whitlock, 1968). Fungi, as indicated above, may also reduce ketones and the 4,5 unsaturation. Anaerobic bacteria seem to produce 5β-steroids by reduction; certain molds produce the 5α-steroids.

Calonectria decora or *Fursarium culmorum* hydroxylates some 11β-hydroxy steroids to 15α-hydroxy analogs. *Rhizopus nigricans*, *Cunninghamella baineri*, and other molds which carry out 11α-hydroxylations can oxidize the 11β-hydroxy group and also hydroxylate at 6, 14, etc.

At one time it was felt that *Curvularia blakesleeana* could introduce a ketone at 11 on compounds. Since then, several molds, including *Cur. blakesleeana*, have been shown to convert 11β-hydroxy steroids to 11-keto steroids (Eroshin and Krasilnikov, 1961). The direct introduction of oxygen as a ketone is improbable. *Flavobacterium dehydrogens* hydrolyzes various primary and secondary steroid acetates and some cyclic acetate esters. *Flavobacterium dehydrogens* specifically oxidizes only certain of the resulting alcohols and does not hydrolyze higher acetyl esters or benzoates nor methyl esters of cholenic acids nor long-chain fatty acids.

ESTERIFICATION

Trichoderma spp. and yeasts have acetylated certain specific steroids at 21. Other molds remove 21-acetates. The 17β-acetate of testosterone is readily removed.

SOME GENERAL CONSIDERATIONS OF EXTENSION OF THESE PROCESSES

It must be apparent that the microbial steroid reactions are a chemical economic problem as well as a microbial and biochemical research tool.

Microbial dissimilations of other organic molecules are the same type of problem and, in some cases, it should be concluded that a good chemical study of a desired reaction may be more rewarding than considerable screening and fermentation studies.

The specific oxidation of the 17β-hydroxy group of androstenediol preferentially over the 3-hydroxy group can be done chemically in better yield than the microbial oxidations. On the other hand, DeFlines (1969) reports that the conversion of 3β,17α,21-trihydroxy-pregn-5-ene-3-one to the 3-keto-Δ4-steroid (Reichstein's compound S) is accomplished with bacterial enzymes on a commercial level. The chemical methods do not differentiate between the several hydroxyl groups (Protiva et al., 1968).

FURTHER USE OF MICROORGANISMS IN TRANSFORMING COMPOUNDS FOR ORGANIC SYNTHESIS

The potential for using microorganisms for conversion of one steroid or organic chemical to another closely related analog seems to be unlimited.

Microorganisms have been found to hydroxylate many organic cyclic and aliphatic compounds. Others degrade the aromatic nucleus to interesting structures. Several authors have cautioned that, with the present state of the art of organic chemistry, the results from microorganisms are sometimes less useful than those from improved catalytic or stereospecific synthetic reactions. The microbial conversion is frequently more expensive than the chemical process, if the amounts of reactants which must be used are less and the costs of processing are higher. When the microbial transformation replaces several steps, it becomes practical. The microbial processes are also of utility to chemists when some unactivated carbon is to be hydroxylated or when a certain stereospecifici-

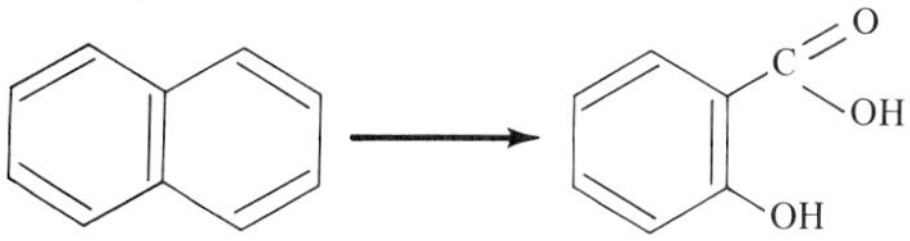

Figure 5-22 Conversion of naphthaline to salicylic acid by nocardias or mycobacteria.

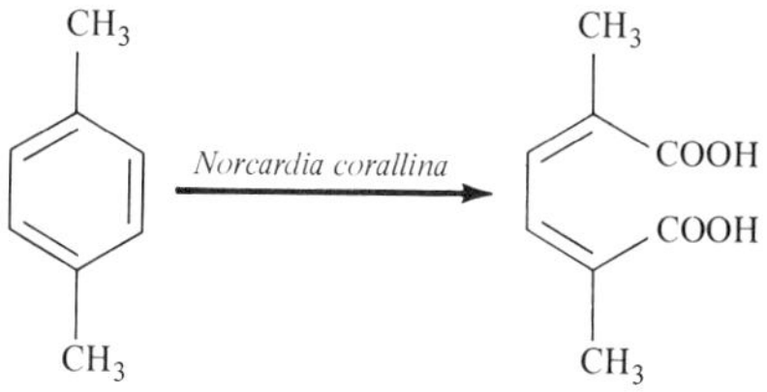

Figure 5-23 Conversion of p-xylene to α,α'-dimethyl-*cis*,*cis*-muconic acid.

ty is desired; for instance, the preferential oxidation of one of several hydroxyls to a ketone or the preferential removal of an acyl group from a sensitive ester.

Examples of hydroxylations of various terpenes, alkaloids, and glycosides are frequently reported. The degradation of aromatic compounds to organic acids are also noteworthy. Naphthalene can be converted to salicylic acid in quantities almost competitive with chemical synthesis (Fig. 5-22). A *Nocardia corallina* strain produces α,α'-dimethyl-*cis*-,*cis*-muconic acid from *p*-xylene which accumulates in high quantity (Raymond, 1969) (Fig. 5-23).

An example of a nonsteroidal conversion is the conversion of the xanthenone, lucanthone, to hycanthone described by Rosi et al. (1967) (Fig. 5-24). A study of the metabolism of lucanthone by screening against several microorganisms led to the selection of *Aspergillus sclerotiorum* for further studies. Upon scale-up, a hydroxy methyl analog was isolated and designated hycanthone. Since the lucanthone was more toxic than the product, it was added in portions, and substrate levels of 13.5 g/l were obtained if the conversion was extended to 9 days. Concentrations (8 percent) of sugar increased the hydroxylations and could be maintained by periodic addition of glucose as well as the lucanthone hydrochloride.

Hycanthone is apparently the cause of antischistosomal activity of lucanthone. The same authors studied the hydroxylation of other schistosomicidal compounds. Some of the hydroxy methyl compounds were readily produced by *A. sclerotiorum*. The hydroxylations compared favorably with chemical routes to the same products.

The lucanthone to hycanthone conversion illustrates the use of microorganisms for reinforcing metabolic studies. It also may lead to a preparative method for the product.

Figure 5-24 Conversion of lucanthone to hycanthon.

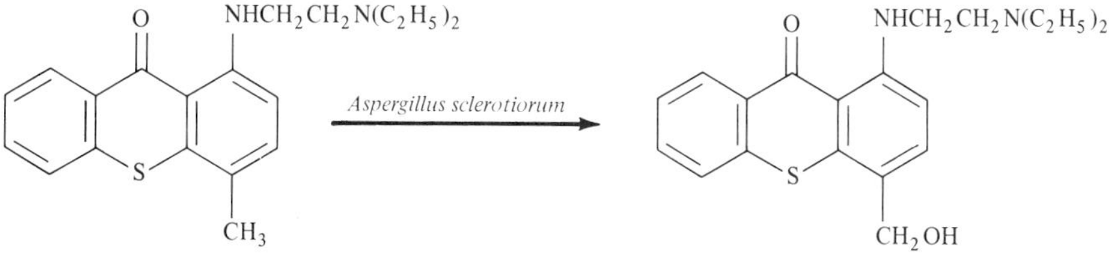

INTRODUCTION TO REFERENCES

The literature cited and the general references in the following pages are not exhaustive. The intention is to supply the reader with a few references which can be used for sources of information. Fieser and Fieser (1959) is the definitive monumental monograph on steroids.

Any one of these references has a uniqueness which merits its inclusion. Undoubtedly there are others to be found which might serve as well. Charney and Herzog (1967) supply comprehensive lists and references covering the field.

LITERATURE CITED

Block, C., and O. Hayaishi (1966): "Biological and Chemical Aspects of Oxygenases," pp. 179–194, Maruzen Co., Tokyo.

Carlstrom, K. (1967): Mechanism of Side Chain Degradation of Progesterone, *Acta Chem. Scand.,* **21**:1297–1303.

Charney, W., and H. Herzog (1967): "Microbial Transformations of Steroids—a Handbook," 728 pp., Academic, New York.

Chen, J. W., H. J. Koepsell, and W. D. Maxon (1962): Kinetics of 1-Dehydrogenation of Steroids by *Septomyxa affinis, Biotechnol. Bioeng.,* **4**:65–78

DeFlines, J. (1969): The Use of Biocatalysis in the Synthesis and Transformation of Steroids, in D. Perlman (ed.), "Fermentation Advances," pp. 385–390, Academic, New York.

Eroshin, V. R., and Krasilnikov (1961): Selective Microbiological Oxidation of the 11-Hydroxyl Group of Hydrocortisone, *Dokl. Akad. Nauk USSR,* **137**:968–969.

Fieser, L., and M. Fieser (1959): "Steroids," 945 pp., Reinhold, New York.

Florkin, M., and E. H. Stotz (eds.) (1963): Sterols, Bile Acids, and Steroids, "Comprehensive Biochemistry," vol. 10, Elsevier, Amsterdam.

Goodman, J. J., and L. L. Smith (1960): Effect of Medium Composition on Isomerization of 9α-Fluoro-16α-hydroxyhydrocortisone and 9α-Fluoro-16α-hydroxyprednisolone (Tramcinolone) during Microbiological Fermentations, *Appl. Microbiol.,* **8**:363–366.

Jones, Sir Ewart R. H. (1973): The Microbiological Hydroxylation of Sterols and Related Compounds, *Pure Appl. Chem.,* **33**(1):39–52.

Koneo, Eiji, and Eitaro Masuo (1961): Pseudo-Crystallofermentation" of Steroids: A New Process for Preparing Prednisolone by a Microorganism, *J. Gen. Appl. Microbiol. (Tokyo),* **7**:113–117.

Marschek, William J., Stephen Kraychy, and Robert D. Muir (1972): Microbial Degradation of Sterols, *Appl. Microbiol.,* **23**:72–77.

Protiva, J., V. Schwarz, and K. Syhora (1968): Steroid Derivatives. LIII. Substrate Specificity of *Flavobacterium buccalis* in Microbial Transformations of Derivatives of 3β-Hydroxypregn-5-en-20-one, *Collect. Czech. Chem. Commun.,* **33**:83–91.

Raymond, R. L. (1969): Biotransformations Using Hydrocarbons, *Process Biochem.,* **4**:71–74.

Ringold, H. J., and A. Bowers (1963): Steroids, in M. Florkin and E. H. Stotz (eds.), "Comprehensive Biochemistry," vol. 10, chap. 3, Elsevier, Amsterdam.

Rosi, D., G. Peruzzotti, E. W. Dennis, D. A. Berberian, H. Freele, B. F. Tullar, and S. Archer (1967): Hycanthone, a New Active Metabolite of Lucanthone, *J. Med. Chem.,* **10**:867–876.

Shoppee, C. W. (1964): "Chemistry of Steroids," 463 pp., Analytical, Butterworth, Washington, D.C.

Sih, C. J., and H. J. Whitlock, Jr. (1968): Biochemistry of Steroids, *Ann. Rev. Biochem.,* **37**:661–694.

Tamm, C. (1962): Conversion of Natural Substances by Microbial Enzymes, *Angew. Chem., Int. Ed., Engl.,* **1**:178–195.
Vezina, C., S. N. Seghal, and K. Singh (1963): Transformation of Steroids by Spores of Microorganisms, *Appl. Microbiol.,* **11**:50–57.
Weaver, E. A., H. E. Kenney, and M. E. Wall (1960): Effect of Concentration on Microbiological Hydroxylation of Progesterone, *Appl. Microbiol.,* **8**:345–348.
Zuidweg, M. H. (1968): Hydroxylation of Reichstein's Compound S with Cell-Free Preparations from *Curvularia lunata, Biochem. Biophys. Acta,* **152**:144–158.

GENERAL REFERENCES

Applezweig, N. (1959): The Big Steroid Treasure Hunt, *Chem. Week,* Jan. 31, 1959.
Goll. P. H. (1966): Microbiological Processing of Steroids, *Process Biochem.,* July, 1966, pp. 1–5.
Heftman, E. (1967): "Chromatography," 851 pp., Reinhold, New York.
Kieslich, K. (1969): "Präparativ anwend bare mikrobiologische Reaktionen: Synthesis." no. 3, pp. 120–134; no. 4, pp. 147–157.
Sanders, H. J. (1968): Arthritis and Drugs, pt. 3, *Chem. Eng. News,* **46**(34):53–94.
Wettstein, A., H. Hurlman, and E. Vischer (1964): Microbial Synthesis of Pharmacologically Active Substances, in "Global Impacts of Applied Microbiology," Wiley, New York.

6
MICROBIAL SYNTHESIS OF AMINO ACIDS

Donald R. Daoust

The microbial synthesis of amino acids on an industrial scale has developed rapidly in the past decade. The impetus for these advances originated chiefly from the interest in the nutritional applications of lysine and glutamic acid. Ever since the monosodium salt of L-glutamic acid was shown to be such an excellent flavor enhancer of food, its use in the commercial canning industry as well as in restaurant and home cooking has increased rapidly. L-Lysine long has been recognized as an essential amino acid in the nutrition of human beings and domestic animals, and its application as a diet supplement is sought particularly in the underdeveloped, overpopulated areas of the world where the chief staples are deficient in this amino acid. More recently the industrial production of L-lysine has received greater interest as a consequence of the encouraging results obtained in feeding trials with cattle, hogs, and poultry that are supplied a normal diet supplemented with this amino acid. The application of lysine for this purpose is prevalent especially in European countries. Experimental programs presently are under way in India and in several South American countries to feed undernourished children with bread prepared with lysine as a nutritional supplement. Kellogg Company, the large American cereal manufacturer, also employs lysine to fortify its Special K brand of cereal.

Currently, Japan leads the world in the microbial production of lysine and glutamic acid. Kyowa Fermentation Industry produces over 1000 tons of lysine annually, and its output of glutamic acid approaches 20,000 tons per year. The Ajinomoto Company of Japan, which produces close to 50,000 tons of glutamic acid annually, currently has plans for the construction of a large fermentation plant in South America. The facility is designed for the production of both lysine and glutamic acid.

Although lysine and glutamic acid are the only amino acids that have attained the status of microbial production on a commercial scale, there is considerable interest in the development of microbial processes for the nutritionally essential amino acids L-methionine, L-threonine, and L-tryptophan. Both the Ajinomoto Company and the Kyowa company reportedly are evaluating processes for the microbial production of the nutritionally essential amino acids threonine and tryptophan as offshoots of their lysine and glutamic acid processes. The commercial production of threonine by microbial synthesis is in process at the Kyowa Hakko Kogyo plant in Japan. This is a testimonial to the fact that an economical microbial process for any amino acid can be developed when the demand for it exists.

The remainder of this chapter is presented with the intention of pointing out the problems which are inherent in the development of a microbial process for the synthesis of an amino acid. It is the author's intention to draw freely for examples from the microbial processes for lysine and glutamic acid since these are the systems with which the greatest experience has been gained and about which the literature is replete.

SELECTION OF A MICROBE
GENERAL REMARKS

At the outset of a project designed to lead to the microbial synthesis of any amino acid the first and most important objective is the isolation or selection of a suitable microorganism with which to begin the studies. To this end one should give careful consideration to searching natural sources such as soils and waters. As an illustration of this point, it is interesting to note that the Japanese workers first uncovered the glutamic acid–producing cultures while screening the feces of various wild birds and mammals. A number of interesting alanine-producing cultures were isolated from samples of cheese by the same workers.

The scientific literature contains a considerable amount of information about the microbial production of all natural amino acids, albeit usually in very small quantities. Nonetheless, cultures which are known to produce some quantity of the amino acid are a suitable starting point with which to familiarize oneself with the various methods that will be employed to subsequently isolate more productive cultures.

Generally, one must determine if the cultures do, indeed, produce the amino acid which is sought, and the extent to which this production takes place. The critical steps in the study involve (1) the design of a culture medium which will permit the production of the amino acid and (2) the evaluation of methods of detection to select one which will prove most suitable under the given cultural conditions. Once a suitable culture is found, the effort is transferred to the evaluation of the various physical and chemical factors of the process in order to improve production even further. Each of these steps will be discussed in subsequent sections.

SCREENING MICROORGANISMS

During the initial phase of the screening, one is concerned chiefly with the rapid examination of a great many cultures to determine their capacity to synthesize the amino acid. Organisms from natural sources such as soil, water, or sewage can be isolated by plating them in the conventional manner on the surface of a solid medium. A portion of the sample is added to a volume of sterile distilled water or phosphate-buffered saline, and a number of serial tenfold dilutions are prepared in the same diluent. Each of these is plated on the surface of a general, all-purpose plating medium such as nutrient agar or Trypton glucose yeast extract agar. Usually a 0.1-ml aliquot of a given dilution in the series is added to the agar plate and distributed evenly over the surface with the aid of a sterile, glass spreading rod. Following incubation at 25 to 30°C for 2 to 3 days, those plates which contain a sufficient number (30 to 50) of discrete, well-isolated colonies are selected as the source of cultures to be evaluated for production.

It should be added that the plating of samples from nature gives rise to a wide variety of microorganisms. The large majority of superior amino acid–producing cultures are bacteria which belong to the genera *Arthrobacter*, *Microbacterium*, *Brevibacterium*, *Micrococcus*, or *Corynebacterium*. Consequently the search for amino acid producers might be limited to the bacteria and exclude completely the fungi. Pimaricin or cycloheximide at a concentration of 50 µg/ml in the initial plating medium suffices to restrict the growth of fungi.

The cultivation of all isolates on a maintenance or storage medium certainly is not justified at this stage since their potential to synthesize the amino acid is yet to be determined: However, a very good method of keeping the cultures intact for future use and evaluating them simultaneously for production is the replica-plate method devised by Lederberg and Lederberg (1952). The technique, which is discussed in greater detail by Elander and Espenshade in Chap. 9, enables one to transfer simultaneously all the colonies on a given agar plate to the same respective site on the surface of a second, third, or fourth agar medium.[1]

Although a culture may thrive on a given medium, one has no assurance that it will produce the amino acid under the same conditions. For example, in all the microorganisms found to produce substantial quantities of L-glutamic acid, the principal prerequisite for this synthesis has been a nutritional dependency of these cultures for biotin as well as the limitation of this vitamin in the production medium. Whereas 1 µg of biotin per liter limits growth and permits good

---

[1]The initial evaluation of culture isolates for amino acid production need not be carried out on a solid medium although this method is the most rapid. Broth cultures, which carry certain advantages in evaluating cultures, may be used; this aspect will be discussed in a subsequent section of this chapter.

glutamic acid production, the use of 15 μg of biotin per liter cuts off all but a small amount of glutamic acid synthesis at the expense of growth. Therefore a suitable production medium must be designed based on current knowledge of the biosynthetic pathways of the amino acids (see Umbarger and Davis, 1962).

The ideal medium should provide a readily available supply of nitrogen in the form of either inorganic salts such as ammonium sulfate, ammonium chloride, and ammonium nitrate, or organic compounds such as urea, peptones, meat extracts, and protein hydrolysates. A suitable carbon source is of the utmost importance; its choice is predicated primarily upon the amino acid being studied. The total synthesis of the amino acid from glucose, naturally, requires the inclusion of this sugar or of common disaccharides, such as sucrose, which contain it. Under certain circumstances the synthesis of an amino acid can be obtained by the transformation of precursor compounds. L-Threonine can be produced by microbial conversion of its precursor L-homoserine. Processes have been described for production of L-tryptophan by microbial conversion of either 3-indolepyruvic acid or anthranilic acid. L-Methionine is produced by a large number of microorganisms by conversion of β-methylmercapto-α-hydroxybutyric acid. The precursor substance comprises the carbon source for the final product in such cases. It is advisable, however, to include in the medium a low level of glucose in order to stimulate the physiologic activity of the culture.

The addition of yeast autolysate or yeast hydrolysate is recommended since each is a good source of vitamins and various cofactors. A well-balanced salt solution consisting especially of magnesium, manganese, and iron should be provided since these frequently are required as cofactors for enzymatic reactions. The use of a buffer system cannot be emphasized too strongly since, in its absence, dramatic changes in pH result following production and excretion of an amino acid. A medium designed for the evaluation of cultures which produce amino acids is given in Table 6-1. Wherever circumstances command, modifications must be made to accommodate the specific requirements of the microorganism or the biosynthetic pathway involved.

The search for novel producers of amino acids is a tedious and often frustrating task, and the evaluation of thousands of culture isolates frequently is performed in industrial laboratories before a suitable culture is obtained.

Although it may be assumed that the screening medium is satisfactory and the source of cultures abundant, the final task remains of determining which organisms produce the amino acid that is sought. To this end one must employ or devise assay methods which will detect the presence of the given amino acid.

ASSAY METHODS

The method of assaying the cultures for production of an amino acid should be merely qualitative in nature at the outset of such a program, since it is the easiest way to screen a large number of cultures in a relatively short period of time. The type of assay used depends primarily upon the way the culture was grown during production. Some assay methods lend themselves to the evaluation of cultures grown on solid medium whereas others are restricted to broth cultures. The remainder of this section is devoted to a brief discussion of qualitative and quantitative assay methods for amino acids.

Table 6-1 Medium designed for the evaluation of microorganisms synthesizing amino acids from glucose

Ingredient	Concentration (per liter)
Glucose	25 g
Yeast autolysate	0.5 g
Meat extract	0.5 g
$(NH_4)_2SO_4$	7.0 g
K_2HPO_4	7.0 g
KH_2PO_4	3.0 g
$MgSO_4 \cdot 7H_2O$	0.52 mg
$FeSO_4 \cdot 7H_2O$	0.014 mg
$MnSO_4 \cdot H_2O$	0.005 mg
NaCl	0.019 mg

pH 7.2

Sterilize by autoclave 15 min at 15 lb/in^2

NOTE 1: The salts can best be prepared in the form of a stock solution and added as a dilution to the main body of medium.
NOTE 2: The pH is adjusted with 0.2 N NaOH.
NOTE 3: For plating purposes agar is added to a final concentration of 2 percent.

DETECTION OF AMINO ACIDS PRODUCED ON SOLID MEDIUM
Bioautography

METHOD I

Plates of the production medium are stamped (inoculated) with a number of cultures by replica plating from the initial isolation medium. The replicated cultures are incubated at 28°C for 2 to 3 days to allow ample time for growth of the organism and production and diffusion of the amino acid into the medium. These colonies are exposed to a dose of ultraviolet light intensive enough to kill the cells and prevent overgrowth during the assay. Subsequently the agar plate is overlaid with approximately 10 ml of molten amino acid assay medium seeded with the specific assay organism. Following incubation at 37°C for 16 to 24 h the assay organism grows in the immediate area surrounding any colony that produced the amino acid. The density of this zone of growth is used as a relative measure of the quantity of amino acid released by the given culture.

It is at this stage that the value of the replica-plate method is most appreciated. Since the colonies on the production medium were sterilized with uv light to prevent spreading and interference with the assay, the cultures which produce the amino acid can be identified and retained by comparing the finished assay plate with the original master plate from which the production medium was inoculated. When the two plates are superimposed, the colony on the isolation plate which aligns with a halo of growth on the assay plate can be picked off and inoculated to a storage or maintenance medium.

The multiple amino acid auxotroph *Leuconostoc mesenteroides* (ATCC 8042) and the chemically defined lysine assay medium (Difco) are used for the detection of L-lysine. Since this assay organism requires a number of amino acids for its growth it can be employed to detect any one of them so long as all the others are supplied in the assay medium. Techniques for the microbiological assay of amino acids are described in the monograph Difco Supplementary Literature (1966).

The reader is cautioned in using this method of assay that the production medium alone should not contain a sufficient level of the amino acid to interfere with the assay. Should this problem arise, the source of the amino acid (meat extract, protein hydrolysate, etc.) in the production medium should be reduced in concentration or an alternative assay should be employed.

METHOD II

The following method consists of evaluating cultures individually rather than in groups, as in method I. The isolation from natural sources and the replica plating on to production medium should be done as described earlier. The colony and the production medium upon which it is growing are punched out of the agar plate with a sterile, metal cork borer (no. 3 or no. 4). The agar plug, as it is called, is transferred to the surface of a piece of chromatography paper (Whatman No. 1) which has been slotted to prevent intermixing of samples. The plugs are allowed to stay on the paper for a period of time to permit the amino acid to leach into the paper. The plug is removed, the paper is air-dried, and chromatographed in a solvent system such as 80% aqueous phenol or other recommended combinations of solvents (see Block et al., 1958) designed for chromatography of amino acids. Following adequate development time the chromatograms are air-dried to remove all traces of the solvent and then placed in a rectangular dish large enough to accommodate the paper strip. The dish should contain the solid amino acid assay medium seeded with the specific assay organism. The paper strip can be either left on or removed after about 1 h. The plate is incubated overnight at 37°C. The assay organism grows on the assay medium only in the area which corresponds to the spot to which the specific amino acid migrated on the paper strip during chromatography.

A suitable means of verifying the efficacy of this technique requires that plates of production medium be prepared to contain different levels (0.05 to 0.25 µg/ml) of the authentic amino acid. Plugs must be punched out of these plates and treated as described above. This permits an evaluation of the behavior of the amino acid under actual assay conditions and indicates the sensitivity of the assay as well. This method is applicable also for cultures grown in a broth medium. A sample of the broth is applied in microliter quantities, and the paper is treated exactly as described above.

METHOD III

Amino acids can be detected as a result of their capacity to reverse the inhibition of specific antimetabolites (inhibition reversal method). For example, the antimetabolite β-2-thienylalanine is an inhibitor of *Bacillus subtilis*. This inhibition is reversed if L-phenylalanine is added to the medium. Since the reversal is specific for phenylalanine, the test is ideal for the detection of cultures which produce this amino acid. In this particular example the production plate would be overlaid with phenylalanine assay medium (containing no phenylalanine) supplemented with the inhibitor and inoculated with the specific indicator organism (*B. subtilis*). A selected list of amino acids, their competitive analogs, and the assay organisms was compiled from a more complete source (Shive and Skinner, 1962) and appears in Table 6-2.

DETECTION OF AMINO ACIDS PRODUCED IN LIQUID MEDIUM

Microbiological Assays

The detection from broth culture of microorganisms that produce amino acids usually involves more time and labor than is required with agar plates. The cultures are isolated from natural sources as described previously. Discrete colonies are picked from these plates and cultivated on individual agar slants. The growth from these slant cultures is used to inoculate a starter or seed broth

Table 6-2 Reversal inhibition assays available for detection of amino acid-producing microorganisms

Amino acid	Inhibitor	Indicator culture
Alanine	1-Aminoethanesulfonic acid*	*Proteus vulgaris*
Arginine	Homoarginine	*Escherichia coli* B
Aspartic acid	β-Hydroxyaspartic acid	*Lactobacillus plantarum* (ATCC 8014)
Glutamic acid	Methionine sulfoxide	*Lactobacillus plantarum* (ATCC 8014)
Glycine	1-Aminoethanesulfonic acid	*Staphylococcus aureus*
Histidine	2-Thiazolealanine	*E. coli*
Isoleucine	2-Cyclohexeneglycine	*E. coli* (ATCC 9723)
Leucine	2-Amino-4-methylhexanoic acid	*E. coli* (ATCC 9723)
Lysine	4-Thialysine	*Pediococcus cerevisiae* (ATCC 8042)
Methionine	Methionine sulfoximine	*Pediococcus cerevisiae* (ATCC 8042)
Proline	3,4-Dehydroproline	*E. coli* (ATCC 9723)
Serine	α-Methylserine	*Pediococcus cerevisiae* (ATCC 8042)
Threonine	2-Amino-3-hydroxpentanoic acid	*Streptococcus fecalis* (ATCC 9790)
Tryptophan	5-Methyltryptophan	*Streptococcus fecalis* (ATCC 9790)
Tyrosine	5-Hydroxy-2-pyridinealanine	*Leuconostoc dextranicum* (ATCC 8086)
Valine	2-Cyclopenteneglycine†	*E. coli* (ATCC 9723)

*1-Aminoethanesulfonic acid merely retards the growth of *P. vulgaris*. The reversal must be sought early in the assay.
†Isoleucine must be added to the medium since it is required along with valine for the reversal of this inhibitor.

medium which consists frequently of a very rich mixture of ingredients. The purpose of the seed medium is to obtain a culture which is primed physiologically for the production phase. The seed culture usually is inoculated into the production medium at a level of 5 percent on a volume basis. Production cultures are incubated at 28 to 37°C for 3 to 5 days unless special circumstances demand differently. Assays are done on aliquots of 5- to 10-ml samples that are first exposed to flowing steam for approximately 10 min in order to release the free intracellular amino acids. The samples are cooled and centrifuged to remove the cells and other suspended solids, and the supernatant is used for the assay.

METHOD I: TURBIDIMETRIC ASSAY

The assay of culture broths by the turbidimetric as well as the subsequent assay (titrimetric) requires that the sample be freed of cells of the production culture. This is done by passing the sample through microbiological filters designed to sterilize liquids.

The assay of such samples for lysine, for example, calls for the addition of increasing volumes (0 to 3 ml) of the filtered broth to a triplicate set of clean, sterile test tubes, each containing 3 ml of double-strength lysine assay medium. The volume is raised to 6 ml with sterile distilled water. The same steps are followed with a control sample which consists of the uninoculated production medium which has been mixed with 15 µg of L-lysine per milliliter of medium. All tubes are inoculated with one drop of a standardized inoculum of the lysine assay organism *Leuconostoc mesenteroides* (ATCC 8042). The tubes are incubated at 37°C for 16 and 23 h at which time they are examined for growth.

The cell density or turbidity produced by the growth of the organism is dependent upon the amount of lysine present in the sample. An arithmetic plot of the turbidity (optical density or OD) versus the concentration of lysine in the known samples provides essentially a linear relationship which enables one to determine the lysine content of an unknown sample from its respective OD readings. Measurement of OD is made in instruments such as the Klett colorimeter, Bausch & Lomb Spectronic-20, and Beckman or Coleman spectrophotometers.

METHOD II: TITRIMETRIC ASSAY

The techniques involved in the titrimetric assay of amino acids are similar, essentially, to those used in turbidimetric assays. The major difference lies in the manner in which the response is measured. Whereas turbidimetric assays measure the response of the assay organism to the amino acid in terms of change in light absorption or turbidity, the titrimetric assay measures the response in terms of a change in physiologic response, specifically acid production. *Leuconostoc mesenteroides* (ATCC 8042) produces acid from glucose at a rate which is dependent primarily upon the concentration of lysine in the assay medium. These assay cultures must be incubated at 37°C for at least 72 h before being read. The results for unknown samples are calculated from a standard curve which is obtained by plotting the final pH attained by cultures grown in the presence of known and increasing levels of lysine. The sterility of unknown samples must be assured before they are assayed by either of these two methods.

It is not the purpose of this chapter to delve into the various facets of microbiological assays. However, the reader is referred to the excellent treatise by Shockman (1963) for a complete and comprehensive presentation of these techniques.

Manometric Assays

Amino acid decarboxylases are relatively specific enzymes produced by a variety of bacteria. The enzyme releases 1 mol of carbon dioxide (CO_2) for each mole of the specific amino acid with which it reacts. Lysine decarboxylase reacts only with lysine, and glutamic acid decarboxylase reacts only with glutamic acid. It is the combination of this specificity with the stoichiometry of the reaction which enables one to use these decarboxylases to assay for an amino acid. The technique requires the use of a Warburg respirometer, which consists of a temperature-controlled water bath and a series of reaction vessels (Warburg flasks) each coupled to an individual manometer. The principle of the instrument lies in its capacity to measure, in a closed system, the change in gas pressure in the reaction vessel which results from the utilization of the oxygen or, as in the case of decarboxylase reactions, the release of CO_2. The amount of CO_2 released is in direct proportion to the amount of the specific amino acid present. Hence a standard curve is constructed by plotting the volume (microliters) of CO_2 released enzymatically from samples containing increasing concentrations of the amino acid. The amount of amino acid present in unknown samples is determined by relating the volume of CO_2 released from these samples to the corresponding volume in the standard curve.

The method has been successfully employed for the assay of lysine and glutamic acid, and the availability of specific bacterial decarboxylases should extend its usefulness to assays for arginine, histidine, ornithine, tyrosine, and aspartic acid (see Fruton and Simmonds, 1958). The techniques of manometry are described and defined in the classic monograph written by Umbreit, Burris, and Stauffer (1959).

Automated Assays

The reaction of amino acids with their specific decarboxylase forms the basis of automated assays developed by the Technicon Corporation for evaluation of lysine and glutamic acid. Briefly, the Autoanalyzer, as it is called, is an intricate instrument composed of an automatic sampling device, a peristaltic pump, reaction unit, colorimeter, and chart recorder. The samples are prepared in a special buffer, placed in small sampling cups, and taken up individually by the automatic sampling device. The sample is pumped through a series of helical mixing coils to ensure thorough contact between the sample and the specific amino acid decarboxylase. The reaction releases CO_2 which proceeds to react with the phenolphthalein indicator also present in the mixture. The intensity of the pink color of the indicator is inversely proportional to the amount of CO_2 released.

The amount of amino acid present in a given sample is determined by comparing its light absorption (percent transmittance) with that from a standard curve. The latter is constructed by plotting the log percent transmittance of a given sample versus the concentration of the authentic amino acid in that same sample. The instrument is not difficult to operate and, when used properly, can deliver at least 60 quantitative assays in a single day.

DEVELOPMENT OF CULTURE
NUTRITIONAL STATUS

Once a large number of wild-type cultures has been screened and the choice of a production culture has been narrowed down to several of the better producers, one usually seeks ways of developing these to improve their performance even more. The avenues which are open to achieve this goal are (1) mutating the cultures; (2) improving the nutritional balance of the production medium; (3) regulating the physical and chemical factors of the process such as pH, temperature, and aeration until optimum levels are found; and (4) overcoming the effects of metabolic control mechanisms if these are inherent in the process.

MUTATION

The importance of mutation in the development of microbial processes cannot be emphasized enough. Mutation of cultures undoubtedly has played a major role in improving the productivity of almost every industrial microbial process ever employed to date. In order to emphasize this point, we shall draw upon the microbial production of the amino acids lysine and glutamic acid. The Japanese workers were the first to isolate microorganisms that release large amounts of free amino acid (glutamic acid) into the medium. They were quick to realize the potential of such cultures to produce other amino acids if the proper metabolic reactions could be controlled. During the course of a random mutation program a number of amino acid auxotrophs (mutants that require a given substance, in this case amino acids, in order to grow) were isolated from this original glutamate producer.

Several homoserine auxotrophs were found which produced a substantial level of L-lysine but little glutamate from a medium similar to that described in Table 6-1. A cursory examination of the metabolic relationship between homoserine and lysine (see Fig. 6-1) provides the explanation for this phenomenon. Lysine and homoserine are members of the aspartic acid family of amino acids. They are so placed because, like the other members of the group, diaminopimelic (DAP) acid, methionine, threonine, and isoleucine, they derive all or a major portion of their carbon skeleton from aspartic acid. It is readily apparent that mutating a culture to homoserine dependence, i.e., mutating to eliminate homoserine dehydrogenase (reaction 5), will divert into the lysine branch all the carbon which enters the aspartic acid pathway.

Essentially the same results are attained if the culture is mutated to methionine and threonine dependence by blocking reactions 8 and 6, respectively. One should note that these experiments were carried out in a medium with a sufficient level of biotin, a condition which, one will recall, is inhibitory to glutamate production. The biotin supply holds the key to amino acid production

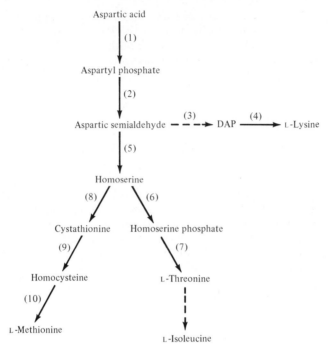

Figure 6-1 Metabolic pathway of the aspartic acid family of amino acids.

in these cultures. During biotin deficiency, glutamic acid is produced, whereas in biotin-sufficient cultures the carbon is diverted to aspartic acid from α-ketoglutaric acid (α-KGA), a glutamate precursor, by either the transamination of oxaloacetic acid or by amination of fumaric acid (Fig. 6-2).

The glutamate producers have been mutated to different amino acid requirements with different results. Threonine auxotrophs produce both homoserine and lysine; methionine-lysine auxotrophs accumulate threonine; tyrosine auxotrophs produce phenylalanine; isoleucine auxotrophs produce proline; lysine-dependent cultures accumulate diaminopimelic acid; leucine mutants produce valine; and arginine auxotrophs accumulate considerable ornithine. The list of mutants and the amino acids they produce is lengthy and should be left for the interested reader (see Dulaney, 1966; Daoust and Stoudt, 1966; Anderson et al., 1966).

The examples pointed out above demonstrate the versatility of mutation as a tool to develop new amino acid producers.

Glutamate production has been increased considerably in its own right by mutation. Some cultures produce considerable extracellular glutamate which they subsequently convert to proline or ornithine or combine with other amino acids to form glutamyl peptides. This waste of glutamic acid can be prevented usually by mutating the cultures to proline or ornithine dependence, or inhibiting the formation of peptides.

The activity of α-KGA dehydrogenase in most glutamate-producing cultures is either absent or very poor. In the case of the latter, a mutation to delete the enzyme entirely should suffice to improve the yields of glutamate to some extent.

The principles and techniques of mutation, particularly as they apply to industrial microbiology, are discussed fully by Elander and Espenshade in Chap. 9.

NUTRITIONAL REQUIREMENTS

Carbon Source

The carbon source employed in a microbial system for amino acid production is one of the most important factors to consider in determining the feasibility of that process. Experiments carried out in test tubes or in 250-ml or 2-l shake flasks can employ reagent grade glucose or sucrose as the carbon source. However, the cost of using these sugars becomes rather prohibitive when the process is scaled up to run in large fermentation vessels with thousands of gallons of medium. Consequently, such crude, inexpensive sugar substitutes as

Figure 6-2 Relationship of amino acid synthesis with metabolic schemes of glycolysis and biological oxidation.

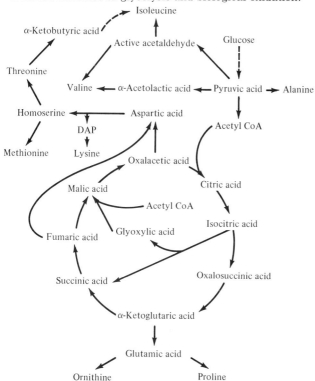

Table 6-3 Crude sugar substitutes used in industrial microbial processes

Source	Total sugar,* %
Brewers' maltose syrup	45
Dried whey	50†
Potato flour	80‡
Molasses	
Cane blackstrap	52
Beet	50
Corn sugar	50
Citrus	42
High test or invert	75

*Sucrose + invert sugar (glucose + fructose).
†As lactose.
‡Total carbohydrate.

molasses should be explored for their capacity to support the production of amino acid. An abbreviated list of the sugar substitutes is given in Table 6-3.

The various crude sources of sugar have to be evaluated in laboratory-scale shake flask studies to determine which is most suitable for the production of amino acid. During the early days of industrial-scale microbial synthesis of glutamic acid many of the batches of molasses used decreased the expected yield of the product. An inhibitory substance, present in some batches of the molasses, was shown to be the cause of the failure. Poor batches of molasses were identified by laboratory-scale studies and then made useful by diluting them with acceptable batches of molasses.

The concentration of sugar is a most significant factor in determining the success of an amino acid process. The most satisfactory way of determining the optimum level is to run a number of shake flask experiments with medium supplemented with different levels of the sugar.[1] An analysis of the amount of amino acid produced and the amount of sugar remaining after the process provides the data necessary to calculate the percent yield of the product. The sugar concentration which provides the maximum percent yield of product should be the optimum level to use for maximum efficiency. Shake flask studies frequently are done with sugar levels in the range of 5 to 15 percent whereas the charges of sugar employed in large fermentors total 20 to 25 percent but are added in increments over a period of time.

Nitrogen Source

The type of nitrogen source to use depends essentially upon the nutritional characteristics of the production culture. Auxotrophic microorganisms require little nitrogen other than nitrate or ammonium salts. In addition, specific requirements of mutant cultures must be considered.

---

[1] The concentration of the nitrogen source should be sufficient to satisfy the demand imposed by increasing the carbon level.

Hydrolysates of various protein meals such as soybean, cottonseed, peanut, etc., usually are an inexpensive, readily available source of amino acids. The cultures which produce glutamic acid and lysine (*Brevibacterium* sp., *Arthrobacter* sp., *Microbacterium* sp., and *Corynebacterium* sp.) all possess a potent urease activity and utilize urea very efficiently as a nitrogen source for the synthesis of these amino acids.

The optimum level of ammonium sulfate, urea, or protein hydrolysate must be determined experimentally. It is recommended that various levels be tried in shake flask experiments until the optimum conditions are found.

Several notes of caution should be interjected here with respect to protein hydrolysates and urea. The method of hydrolyzing the protein will determine, to an extent, the qualitative and quantitative nature of the hydrolysate. Acid hydrolysis destroys tryptophan and, to a lesser degree, serine and threonine. Arginine, serine, threonine, cystine, and cysteine are destroyed by alkaline hydrolysis. These factors should be borne in mind when selecting protein hydrolysates as a nitrogen source, particularly where a mutant culture requiring one of these sensitive amino acids is used. Urea is degraded when sterilized by autoclaving; therefore, it should be sterilized by filtration and added aseptically to the bulk of the medium whenever it is used as a nitrogen source.

OTHER FACTORS

The control of such factors as pH, aeration, and temperature is of the utmost importance in ensuring the success of any given industrial microbial process. Each of these parameters must be maintained at the optimum level if biosynthesis is expected to proceed at its most efficient rate.

pH

The conversion of sugar to amino acids results in a change in the reaction (pH) of the production medium. Failure to control the pH causes definite injury to the culture and reduces its productivity. The pH can be regulated manually or automatically; the choice of method depends upon the size of the vessel used for the production.

MANUAL pH CONTROL

Laboratory-scale shake flask processes (see Fig. 6-3) for lysine production have been controlled by the addition of approximately 15 mg of magnesium phosphate per milliliter of medium. The production of glutamic acid in shake flasks is controlled adequately by the periodic addition of sterile ammonium hydroxide. The range for most amino acid processes lies between pH 6 and 8, with the optimum level nearer to pH 7. The inclusion in the medium of an acid-base indicator such as phenol red provides an internal warning device which signals the need for addition of alkali. The method suffers from the obvious disadvantage that visual inspection must be made every 2 to 3 h around the clock to ensure proper control. Another disadvantage of controlling pH in this manner is the lack of physiologic synchrony of the cultures. Frequently, alkali additions are

Figure 6-3 Laboratory-scale shaker.

required in some flasks at times other than the routine inspection periods. If these are not controlled at the proper time, a serious discrepancy is introduced which could result in error in the interpretation of results. This method does suffice, however, for the routine laboratory studies of amino acid biosynthesis.

AUTOMATIC pH CONTROL

Instruments have been designed to control automatically the pH of a given process (Fig. 6-4). Briefly, such instruments consist of a pH meter, pH controller, timer, and strip chart recorder. The maximum and minimum limits of the desired pH range are preset before the start of the process. Whenever the pH drops to the minimum limit of the control range, an output relay activates a peristaltic pump which meters out alkali from a reservoir until the midpoint of the range is reached once again. A timer which is attached to the pump releases the alkali at a rate sufficient to ensure mixing and prevent overshoot and excessive pH cycling. The biggest advantages of automatic control are (1) the elimination of the time-consuming task of adjusting pH manually; (2) the accuracy with which pH is controlled; and (3) the permanent record of the pH profile for the process which is kept on the strip chart recorder. The instrument is used usually with 5-l fermentors or larger vessels.

Figure 6-4 Automatic pH control instrument.

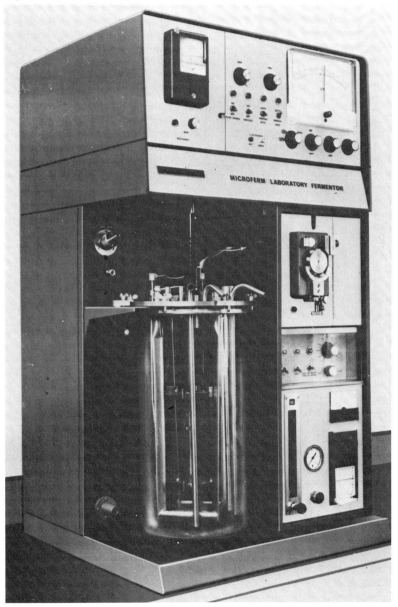

Temperature

The temperatures employed in amino acid processes have ranged between 25 to 37°C. Again, the optimum temperature for a given process must be determined experimentally. The lysine and glutamic acid processes proceed very nicely at 28°C but deteriorate markedly at 25 or 37°C.

Laboratory-scale shake flask experiments are carried out in thermostatically controlled incubators or water baths. Five-liter and larger fermentor vessels have built-in temperature control units which regulate the temperature of the medium, either indirectly (5-l fermentors) by heating the water in which the vessel is immersed or directly (larger fermentors) by passing steam-heated water through a built-in recirculating system in the jacket of the vessel.

Aeration

The microbial synthesis of amino acids proceeds best when the medium is well aerated. The use of indented (baffled) shake flasks (Fig. 6-5) usually provides all the aeration that is required for a laboratory-scale process. These vessels are indented at three equilateral sites along the lower outside edge of the flask. During incubation the flasks are placed on a rotary shaking apparatus with an adjustable shaking speed. The most common shaking conditions used for 250-ml flasks which contain 50 ml of medium are 220 rotations per minute with a 2-in stroke. Large 2-l baffled flasks are incubated on similar shaking tables adjusted to 150 r/min with a 1- to 2-in stroke. The rotary motion of the shaker plus the indentations in the flask causes the broth to splash vigorously to the center where it is aerated. Although cotton closures have proved adequate, some workers prefer to use special gauze pads (Topper sponges)[1], since these are less restrictive to the entrance of air but still assure sterility of the contents. Therefore, the degree of aeration in shake flasks is a function of the shaking speed, the stroke or displacement of the shaker, the degree of baffling of the flask, the type of closure used, and the volume of the medium in the flask.

Fermentors of 5-l capacity (Fig. 6-6) and larger receive air by means of lines that enter the vessel through a sparger located at the bottom of the vessel. The amount of air is controlled by means of a pressure regulator, flow meter and valve, and an air filter. In addition, these vessels are equipped with lateral stationary baffles and a number of impellers or blades located on an agitator shaft. The speed of the agitator can be adjusted to deliver the desired turbulence and aeration required.

The unit of measurement of aeration efficiency in microbial processes is the oxygen transfer coefficient or K_{dw}. It is expressed as the gram moles of oxygen per atmosphere per hour per liter of medium. K_{dw} is measured by agitating a given volume of a solution of sodium sulfite under the same conditions that one would the production medium. Two samples of this solution are taken from the vessel at different times, mixed with separate volumes of a solution of potassium iodide and iodine, and back-titrated separately with a solution of 0.1 N sodium thiosulfate. The principle of the test lies in the fact that air oxidizes the sodium

---

[1]Registered trademark of Johnson and Johnson.

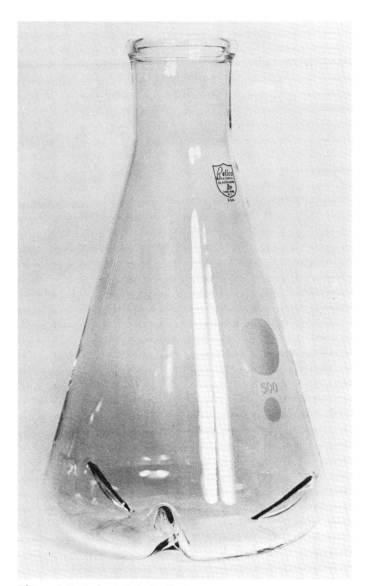

Figure 6-5 Indented shake flask.

sulfite at a rate which depends upon the conditions of aeration. The sodium sulfite, when mixed with the solution of $KI-I_2$, reduces the latter in proportion to the amount of unoxidized sulfite that remains. Consequently, the iodine which remains is determined by titration in the presence of starch with a standard solution of sodium thiosulfate.

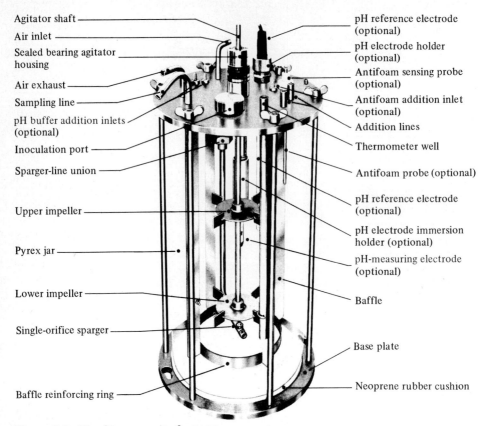

Figure 6-6 Five-liter-capacity fermentor.

The K_{dw} is calculated from the following formula:

$$K_{dw} = \frac{N \, \Delta R}{(4000)(V)(0.21)(\Delta t)}$$

where N = normality of sodium thiosulfate used to titrate iodine
ΔR = difference in volume of sodium thiosulfate used to titrate first and second samples
Δt = time between first and second samples
V = volume of sample of sodium sulfite (usually 5 ml)
0.21 = atmospheric pressure due to oxygen
4000 = factor which expresses relationship between millimoles O_2 and milliequivalents SO_3^{2-} expressed on a molar basis

A K_{dw} in the neighborhood of 5.5×10^{-4} has been reported by Japanese workers as satisfactory for the microbial production of glutamic acid.

Foaming, which frequently occurs during vigorous agitation, interferes with oxygen transfer into the medium. The problem is overcome, however, by adding to the medium any of a number of antifoam agents such as silicone (GE-66) or polyglycol 2000 (Dow Chemical Co.).

METABOLIC CONTROL MECHANISMS

Bacterial cells are equipped with natural controls which prevent both the overproduction of metabolites and the waste of carbon and nitrogen present in the medium. Two of these control mechanisms are end-product inhibition (retroinhibition) and repression. In end-product inhibition the *activity* of the first enzyme in a given metabolic pathway is inhibited as a result of the accumulation of the final product of that pathway. The sensitive enzyme binds the end-product inhibitor at a site different from that used to bind the normal substrate. This union of inhibitor and enzyme alters the structure of the enzyme and causes the latter to lose the stereospecificity required to bind the normal substrate.

Repression is the phenomenon whereby the *synthesis* of an early enzyme of a pathway is inhibited as a result of the accumulation of the final product of that pathway. This mechanism is more complex and less understood than the first. The end product, or effector, as it is called, combines with an inactive aporepressor (normal product of a gene) to form an active repressor. The latter prevents formation of the enzyme protein, possibly by interfering with transcription of the genetic message from DNA to messenger RNA.

There are a number of variations of these control mechanisms that fulfill the same purpose but in a different fashion. For example, the branched-chain amino acids isoleucine, leucine, and valine and the vitamin pantothenic acid share three common enzymes in their total synthesis from pyruvate (leucine, valine, and pantothenate) or α-ketobutyrate (isoleucine). The enzymes in these two pathways are repressed by a phenomenon called concerted feedback control or multivalent repression only if all four of the end products are in excess simultaneously.

The aspartic acid family of amino acids (Fig. 6-1) is a classic example of metabolic control in amino acid biosynthesis. The first step in this pathway, the phosphorylation of aspartic acid by aspartokinase, is mediated, in some organisms at least, by three isoenzymes of aspartokinase. The activity of these isoenzymes is governed by a phenomenon called differential feedback control. This means that one isoenzyme is controlled specifically by lysine, a second is controlled by threonine, and a third is regulated by methionine. The aspartic acid pathway contains a large number of other enzymatic reactions controlled by these same amino acids individually (see Fig. 6-1). Threonine regulates the phosphorylation of homoserine (reaction 6) which is the first step in the threonine branch of the aspartic pathway. Lysine exerts metabolic control at several points (reactions 3 and 4) along its branch from aspartic semialdehyde. Methionine regulates three reactions (steps 8, 9, and 10) along its synthetic branch from homoserine.

Control mechanisms such as these operate in most bacteria. However, seldom are all of them present in any single strain. For example, the control sites found in *Escherichia coli* may or may not be found in *Micrococcus glutamicus* or even in other strains of *Escherichia*. Obviously the microbial production of 40 mg of lysine per milliliter of medium would be impossible unless the culture lacked lysine-directed controls.

Although these control mechanisms are advantageous to the microorganism, they present a serious obstacle to the microbiologist who seeks the production of large quantities of an amino acid. There are a number of alternatives available to overcome the effects of these controls. Since this is not the purpose of this chapter, the reader is referred to the contribution by Elander and Espenshade in Chap. 9 or to the reviews by Stadtman (1963) and Demain (1966).

CULTURE STABILITY

Genetic stability is a very important characteristic to have in a culture that is mutated to a high level of productivity. Cultures which revert spontaneously to wild type during the production phase are undesirable and have no place in an industrial process. These revertants lose their capacity to produce the amino acid and, in addition, thrive on the ingredients of the medium at the expense of that population of cells that did not revert and is still productive. Frequently a mutant culture is made stable if a second metabolic mutation is imposed. An example is the biosynthesis of lysine by a homoserine auxotroph. According to the explanation offered earlier (Fig. 6-1), such a mutant produces lysine in rather high yield from sugar. If this homoserine mutation shows a tendency to revert spontaneously during production, a condition which would cause a shift in the amount of carbon entering the lysine branch, the mutant could be stabilized by adding a second mutation prior to threonine or methionine. Although the homoserine mutation would probably revert at the same rate, the revertant cultures could not overgrow the production culture because of their inability to produce threonine.

The only way to test the stability of a production culture is to examine the culture for its ability to grow in absence of the required metabolites, for example, homoserine and threonine in the example cited above. A culture which demonstrates a dependence for both is indicative of maximum stability and is most desirable. A culture which reverts at an extremely slow rate may not hamper significantly the yield of amino acid and may be tolerable under the usual circumstances.

LITERATURE CITED

Anderson, R. F., H. T. Huang, S. Singer, and M. H. Rogoff (1966): The Biochemistry of Glutamic Acid Fermentation, in Symposium: Microbial Production of Amino Acids, *Dev. Ind. Microbiol.,* **7**:7–15.

Block, R. J., E. L. Durrum, and G. Zweig (1958): "A Manual of Paper Chromatography and Paper Electrophoresis," Academic, New York.

Daoust, D. R., and T. H. Stoudt (1966): The Biosynthesis of L-Lysine in a Strain of *Micrococcus glutamicus*, in Symposium: Microbial Production of Amino Acids, *Dev. Ind. Microbiol.,* **7**:22–34.

Demain, A. L. (1966): Industrial Fermentation and Their Relation to Regulatory Mechanisms, in W. W. Umbreit (ed.), "Advances in Applied Microbiology," vol. 8, pp. 1–27, Academic, New York.

Difco Laboratories, (1966): "Difco Supplementary Literature," pp. 206–213, Detroit, Mich.

Dulaney, E. L. (1966): Microbial Production of Amino Acids: Methionine, Tryptophan, Homoserine, Ornithine, Citrulline, Isoleucine, Valine, Phenylalanine, Tyrosine, and Aspartic Acid, in Symposium: Microbial Production of Amino Acids, *Dev. Ind. Microbiol.,* **7**:47–58.

Fruton, J. S., and S. Simmonds (1958): "General Biochemistry," 2d ed., Wiley, New York.

Lederberg, J., and E. J. Lederberg (1952): Replica Plating and Indirect Selection of Bacterial Mutants, *J. Bacteriol.,* **63**:399.

Shive, W., and C. G. Skinner (1962): Amino Acid Analogues, in R. M. Hochster and J. H. Quastel (eds.), "Metabolic Inhibitors," vol. 1, Academic, New York.

Shockman, G. D. (1963): Amino Acids, in F. Kavanagh (ed.), "Analytical Microbiology," Academic, New York.

Stadtman, E. R. (1963): Symposium on Multiple Forms of Enzymes and Control Mechanisms. II. Enzyme Multiplicity and Function in the Regulation of Divergent Metabolic Pathways, *Bacteriol. Rev.,* **27**:170–181.

Umbarger, H. E., and B. D. Davis (1962): Pathways of Amino Acid Biosynthesis, in I. C. Gunsalus and R. Y. Stainer (eds.), "The Bacteria," vol. III, Academic, New York.

Umbreit, W. W., R. H. Burris, and J. F. Stauffer (1959): "Manometric Techniques," Burgess, Minneapolis.

7
MICROBIAL ENZYMES
Leland A. Underkofler

SOURCES OF ENZYMES AND THEIR IMPORTANCE

Almost without exception the chemical changes brought about by the cells of all living creatures—animals, plants, and microorganisms—are mediated by appropriate catalysts. Enzymes are the protein biocatalysts produced by living cells. They are, of course, produced by the cells to bring about and control the numerous biochemical reactions involved in the metabolic processes of the cells. Fortunately, most enzymes can be separated readily from the cells that produce them and can perform their catalytic activities entirely apart from the cells. A considerable number of enzyme preparations have found important applications both in research and in industry.

ANIMAL, PLANT, AND MICROBIAL CELLS AS ENZYME SOURCES

Since all living cells produce enzymes, it is possible to obtain useful enzyme products from animal tissues, plant tissues, and microorganisms. Very considerable quantities of enzymes are currently produced commercially from animal and plant sources, but for both technical and economic reasons microbial enzymes are increasingly important. Most higher animal and plant cells serve specialized functions and produce only enzymes required to catalyze reactions involved in their special functions.

On the other hand, a microbial cell must produce all the enzymes necessary for its complete metabolism. Microbial cells have the capacity of producing at least 2500 different enzymes, and probably more. There is a definite limitation in the supply to us of animal enzymes since they are a by-product of the meat-packing industry. Plant enzymes are limited by available acreage and labor and are subject to climatic, seasonal, and weather variations. Productive capacity for microbial enzymes may be expanded without limit to meet all demands. Developments in the production of microbial enzymes have assured potentially unlimited supplies and also have made available enzyme systems which cannot be readily obtained from plant and animal sources.

Advantages and Disadvantages of Microorganisms for Enzyme Products

The first major advantage of microorganisms as the source for useful industrial enzymes, as mentioned above, is the potentially unlimited supply. Fermentation production capacity can be expanded almost without limit to meet any level of demand.

The second principal advantage is the large number of enzymes which can be obtained economically from microorganisms. The only major limits are those imposed by the limits of possible enzymatic catalysis. A well-designed and intensive search among microbial strains can usually find an appropriate organism to produce almost any enzyme.

On the other hand, the multiplicity of enzymes produced by a single organism is sometimes a disadvantage. Often in an industrial process requiring only a specific enzymatic conversion, the presence of contaminating enzymes which will cause undesirable reactions is a distinct handicap. In such cases it is necessary to remove the undesirable enzyme contaminants, which may be difficult or costly. Fortunately, enzyme purification methods, such as differential inactivation, fractional precipitation, and column chromatography, have become applicable on a large scale, making it possible for the enzyme manufacturer to supply commercial enzyme products having the necessary performance characteristics.

INDUSTRIAL APPLICATIONS OF ENZYMES

Enzymatic processes have been used for centuries. For example, the fermentations for making beer, wine, bread, and cheese are older than recorded history. All fermentations are enzymatic conversions mediated through the metabolism of living organisms. When a system of several enzymes is necessary to produce desired changes, this is most conveniently done by intact cells in fermentation processes. A number of important fermentations are discussed in other chapters of this book. Where a single enzyme or a simple system of only two or three enzymes is involved in a desired reaction, cell-free enzyme preparations are usually preferable to intact cells.

The practical use of enzymes to accomplish reactions apart from the cell goes back into antiquity, long before the existence, the nature, or the functions of enzymes were understood. Malt for starch conversion to sugar for brewing, stomach mucosa for clotting milk in cheese making, papaya juice for tenderizing meat, and dung for bating hides in leather making are examples of ancient uses of enzymes. Later, crude enzyme preparations extracted from animal and plant tissues such as pancreas, stomach mucosa, malt, and papaya fruit found applications in the textile, leather, food, beverage, and other industries.

After the biocatalytic enzymes responsible for the action of the crude preparations became recognized and understood, a search began for better, less expensive, and more readily available sources of similar enzymes. The development and continual improvement of methods for large-scale production of enzymes, along with better understanding of and means for controlling enzymatic processes and applications, have resulted in a sizable number of commercial enzyme products and industrial uses of them.

A rather comprehensive tabulation of the important current uses for enzymes is given in Table 7-1. Selected examples for a few specific important applications will be discussed in some detail below. Other publications, such as the book by Reed (1975), should be consulted for the details of other uses.

Amylases

Amylases, the enzymes which act on starch, have the most numerous applications of all the commercial enzymes. Three types of amylases which hydrolyze starch in different manners are known:

$$\text{Starch} \xrightarrow{\alpha\text{-amylase}} \text{oligosaccharides} + \text{maltose}$$
$$\text{Starch} \xrightarrow{\beta\text{-amylase}} \text{maltose} + \text{dextrins}$$
$$\text{Starch} \xrightarrow{\text{glucoamylase}} \text{glucose}$$

α-Amylases are widely distributed in nature and are obtained from animal, plant, fungal, and bacterial sources. They all hydrolyze $\alpha(1\to4)$ linkages in large starch molecules in a random manner, first producing mainly dextrins and rapidly liquefying the starch. On prolonged action they may cause extensive saccharification, the hydrolysis products being mainly maltose and maltooligosaccharides. α-Amylases from different sources differ greatly in the extent of hydrolysis they bring about and in their thermal stability. For example, bacterial α-amylase can be used effectively at temperatures of 80°C or higher, whereas fungal α-amylase is very rapidly inactivated at 60°C.

β-Amylase is a saccharifying enzyme apparently produced only by higher plants. It splits maltose units progressively from the nonreducing ends of starch chains. Since it cannot hydrolyze $\alpha(1\to6)$ linkages, the enzyme ceases its action when it reaches the branch points of the amylopectin fraction of starch, leaving "beta limit dextrins."

Glucoamylase, frequently also called amyloglucosidase, produces glucose by progressive hydrolysis of glucose units from nonreducing ends of starch chains. Glucoamylase can hydrolyze $\alpha(1\to6)$ and $\alpha(1\to3)$ linkages, although more slowly than $\alpha(1\to4)$ linkages; hence it is capable of converting starch completely to glucose.

From Table 7-1 it can be noted that there are important applications for amylases in many industries and products. Among the most important are saccharification of starchy mashes for fermentation, in bread baking and numerous other food manufacturing processes, in modifying starch for paper and textile coating, in desizing (removing starch) from textiles, and as a pharmaceutical digestive aid.

All three of the types of amylase are used in syrup and dextrose manufacture from starch. Within recent years manufacturers have been taking ad-

Table 7-1 Important uses for enzymes

Industry	Application	Enzyme	Source
Analytical	Sugar determination	Glucose oxidase	Fungal
		Galactose oxidase	Fungal
	Glycogen determination	Glucoamylase	Fungal
	Urea determination	Urease	Plant
	Uric acid determination	Urate oxidase	Animal, fungal
Baking and milling	Bread baking	Amylase	Malt, fungal
		Protease	Fungal
		Lipoxygenase	Plant
Brewing	Mashing	Amylase	Malt, bacterial
		Glucoamylase	Fungal
	Chillproofing	Protease	Papain, bromelain, pepsin, fungal, bacterial
	Oxygen removal	Glucose oxidase	Fungal
Carbonated beverages	Oxygen removal	Glucose oxidase	Fungal
Cereals	Precooked baby foods	Amylase	Malt, pancreatin, fungal
	Breakfast foods	Amylase	Malt, fungal
	Condiments	Protease	Papain, bromelain, pepsin, fungal, bacterial
Chocolate, cocoa	Syrups	Amylase	Bacterial, fungal
Coffee	Coffee bean fermentation	Pectinase	Fungal
	Coffee concentrates	Pectinase, hemicellulase	Fungal
Confectionery, candy	Soft-center candies and fondants	Invertase	Yeast
	Sugar recovery from scrap candy	Amylase	Bacterial, fungal
Dairy	Cheese production	Rennin	Animal, fungal
		Lipase	Animal
	Milk, sterilization with peroxide	Catalase	Liver, bacterial, fungal
	Modifying milk fats for flavor	Lipase	Animal, fungal
	Milk, prevention of oxidized flavor	Protease	Pancreatin
	Milk, protein hydrolysates	Protease	Papain, bromelain, pancreatin, bacterial, fungal

Table 7-1 Important uses for enzymes (*Continued*)

Industry	Application	Enzyme	Source
	Evaporated milk, stabilization	Protease	Pancreatin, pepsin, bromelain, fungal
	Whole milk concentrate	Lactase	Yeast
	Ice cream and frozen desserts	Lactase	Yeast
	Dried milk, oxygen removal	Glucose oxidase	Fungal
Distilled beverages	Mashing	Amylase	Malt, fungal, bacterial
		Glucoamylase	Fungal
Dry cleaning	Spot removal	Protease, amylase	Bacterial, pancreatin, fungal
Eggs, dried	Glucose removal	Glucose oxidase, catalase	Fungal
	Mayonnaise and salad dressing, oxygen removal	Glucose oxidase	Fungal
	Improving whip of albumen	Lipase	Pancreatin
Feeds, animal	Pig starter rations	Protease, amylase	Pepsin, pancreatin, bacterial, fungal
	Poultry rations	Protease, amylase, hemicellulase	Bacterial, fungal
	Cattle rations	Protease, amylase	Fungal, bacterial
Flavors	Removal of starch, clarification	Amylase	Fungal
	Oxygen removal	Glucose oxidase	Fungal
Flavor enhancers	Preparation of nucleoside monophosphates	Ribonuclease	Fungal
	Flavor restoration	"Flavor enzymes"	Plant, fungal
Fruits and fruit juices	Clarification, filtration, concentration	Pectinases	Fungal
	Low methoxyl pectin	Pectinesterase	Fungal, plant
	Starch removal from pectin	Amylase	Fungal
	Oxygen removal	Glucose oxidase	Fungal
	Debittering citrus fruits	Naringinase	Fungal
Laundry	Spot removal	Protease, amylase	Bacterial
	Cold-soluble laundry starch	Amylase	Bacterial

Leather	Bating	Protease	Bacterial, pancreatin, fungal
	Unhairing	Protease	Bacterial, fungal, plant
Meat, fish	Meat tenderizing	Protease	Papain, bromelain, fungal, bacterial
	Tenderizing casings	Protease	Papain, bromelain, fungal, bacterial
	Condensed fish solubles	Protease	Bacterial, papain, bromelain
Paper	Starch modification for paper coating	Amylase	Bacterial, malt
Pharmaceutical and clinical	Digestive aids	Amylase	Fungal, pancreatin, bacterial
		Protease	Papain, bromelin, pancreatin, pepsin, fungal, bacterial
		Lipase	Pancreatin, fungal
		Cellulase	Fungal
	Wound debridement	Streptokinase-streptodornase, protease	Bacterial, animal, plant
	Treatment of bruises, inflammation, etc.	Streptokinase, protease	Bacterial, animal, plant
	Diabetic diagnosis and control	Glucose oxidase, peroxidase	Fungal, plant
	Varied clinical tests	Numerous	Animal, plant, microbial
Photographic	Recovery of silver from spent film	Protease	Bacterial
Starch and syrup	Corn syrups	Amylase, dextrinase	Fungal, malt
		Glucose isomerase	Fungal, bacterial
	Production of glucose	Glucoamylase, amylase	Fungal, bacterial
Textile	Desizing of fabrics	Amylase	Bacterial, malt, pancreatin
		Protease	Bacterial, fungal, pancreatin
	(See also Dry cleaning, Laundry)		

Table 7-1 Important uses for enzymes (*Continued*)

Industry	Application	Enzyme	Source
Vegetables	Liquefying purees and soups	Amylase	Fungal
	Dehydrated vegetables, restoring flavor	Flavor enzymes	Plants
	Maceration of dehydrated vegetables	Cellulase, macerating enzyme	
Wine	Pressing, clarification, filtration	Pectinases	Fungal
Miscellaneous	Drain cleaners and waste disposal	Lipase	Pancreatin, fungal
	High-test molasses	Invertase	Yeast
	Resolution of racemic mixtures of amino acids	Protease, acylase	Fungal
	Wallpaper removal	Amylase	Bacterial

vantage of different commercially available enzyme preparations, containing various proportions of α-amylase, β-amylase, and glucoamylase, either simultaneously or successively. Syrups of widely differing compositions, from high-dextrin syrups of low sugar content to high-sugar syrups of low dextrin content, are produced to meet the demands of different industries.

A quite recent, and perhaps the largest, application of amylolytic enzymes is in the production of dextrose from starch.‡ Replacement of the classic acid conversion process by the enzymatic method has made possible more than doubling the concentration of starch in the conversion slurry, has very materially increased the yields of recovered dextrose, and has considerably simplified handling and processing. The most efficient process appears to be as follows: A slurry of about 30% starch is made with water at pH 7.0. Best results are obtained in the essential preliminary liquefaction if calcium and sodium ions are present in concentrations of 0.01 M and 0.02 M, respectively. Then heat-stable bacterial amylase, equivalent in amount to 0.1 percent of dry solids weight of Miles Tenase, is added; the mixture is passed through a steam-jet heater at 90°C and is held with gentle agitation at 85 to 90°C for 45 min. The gelatinized and liquefied starch slurry is cooled to 60°C, adjusted to pH 4.0 with hydrochloric acid, and glucoamylase added, usually at a level of about 80 units per pound of starch. The saccharification is allowed to proceed to completion in tanks maintained at about 60°C, requiring about 72 h. The material is then filtered, purified by means of carbon and ion-exchange treatments, evaporated, and the dextrose crystallized.

Proteases

It can be noted from Table 7-1 that the various proteases have numerous applications. Because of the large number of different amino acids present and the resulting complex nature and high molecular weights of proteins, and the differing specificity of the various proteases toward particular peptide linkages, the action of proteases is quite complicated. Most of the practical applications of proteases have been developed by empirical methods for selecting the most suitable protease or combination of proteases and the most appropriate conditions of use. Description of the use of proteases in baking bread and crackers, in tenderizing meat, and in dry cleaning and laundry will illustrate their diversity and importance.

One of the largest uses of microbial proteases is in the baking industry. The gluten proteins of the flour are responsible for the unique and variable properties of bakery doughs. Necessary time for dough mixing as well as loaf characteristics including volume, symmetry, grain, texture, and compressibility of the baked bread depend upon the "strength" of the gluten which is quite variable in different flours. Fungal protease is added by the baker for probably more than half of the bread baked in the United States in order to control and reduce the mixing time and improve loaf characteristics. Either fungal or bacterial protease may be used in cracker, biscuit, and cookie doughs.

Another important use for proteolytic enzymes is in tenderizing meat, particularly beef. There is a great deal of variation in the toughness of beef, depending upon the condition of the animal and the part of the carcass from which it came. Surface tenderization of steaks by the individual consumer is usually accomplished by sprinkling with a powdered protease preparation or by immersion in a solution of the enzyme. Steaks for restaurants are dipped by meat packers into enzyme solutions and marketed as prepackaged, frozen steaks, which are broiled directly from the frozen state. Most of the tenderizing action occurs during the brief period of cooking before the temperature in the meat rises to the point where the enzyme is inactivated. A typical meat tenderizing composition contains 2% commercial papain or 5% fungal protease, 15% dextrose, 2% monosodium glutamate, and salt.

Antemortem tenderization, in which proteolytic enzyme solutions are injected into the vascular system of cattle before slaughter, is a recent development now practiced by meat packers on a fairly large scale. The vascular system effectively distributes the proteolytic enzyme throughout the tissues. This antemortem tenderization enables production and sale of a much higher percentage of tender cuts from all grades of beef. It also enables tenderization of large roasts which is not possible with surface application. The meat must be refrigerated until used to avoid overtenderization.

The single largest application for bacterial alkaline proteases is the recent incorporation of such enzymes in laundry detergents by all the major detergent manufacturers. Proteolytic enzymes have been employed for many years by dry cleaners as spotting agents. The enzyme solution is applied to spots of proteinaceous materials such as milk, egg, or blood, or the garment is immersed in a solution of the enzyme. The enzyme action degrades the proteinaceous materials to fragments which can be removed readily by the dry-cleaning solvents. The new laundry products now being widely marketed are of two kinds. Probably most efficient are those employed in a presoak treatment, which allows the enzyme to degrade the insoluble materials of the various stains on the soiled fabrics. These are then easily washed away in the subsequent detergent laundry step. The products containing enzyme with detergent for use in a conventional laundry cycle are also effective, as the enzymes employed have sufficient thermal stability to withstand the water temperature long enough to accomplish the necessary degrading action before they are inactivated.

Glucose oxidase

Glucose oxidase is of fungal origin and shows a high specificity for β-D-glucose,

$$C_6H_{12}O_6 + O_2 + H_2O \xrightarrow{\text{glucose oxidase}} C_6H_{12}O_7 + H_2O_2$$
$$\text{(Glucose)} \qquad\qquad\qquad\qquad \text{(Gluconic acid)}$$

Catalase, which is also present in commercial glucose oxidase, acts on the hydrogen peroxide,

$$2H_2O_2 \xrightarrow{\text{catalase}} 2H_2O + O_2$$

The net reaction for the glucose oxidase–catalase enzyme system is therefore

$$2C_6H_{12}O_6 + O_2 \xrightarrow{\text{glucose oxidase-catalase}} 2C_6H_{12}O_7$$

that is, 2 mol of glucose is oxidized for each mole of oxygen consumed. The glucose oxidase–catalase system is used commercially both for removing glucose and for removing oxygen. It is also used as an analytical reagent.

The most important application of glucose oxidase is for removing glucose from egg albumen and from whole eggs prior to drying. Powdered egg products are unstable and deteriorate during storage because of the nonenzymatic browning reaction between glucose and proteins. The best method for stabilizing dried egg products is removal of the glucose before drying. In commercial processing the enzyme is added, the batch held at about 90° F (32°C) with gentle agitation, and excess oxygen is supplied by continuously or periodically adding hydrogen peroxide. Completion of desugaring is determined by a negative test with Somogyi reagent, after which the egg product is dried.

Since oxygen is responsible for a wide range of types of deterioration of foods, glucose oxidase is effective in protecting certain foods and beverages by removal of oxygen from the foods and containers. Glucose oxidase is used to prevent color and flavor changes in bottled or canned soft drinks, particularly citrus drinks. Mayonnaise and salad dressings are outstanding examples of food products which can be protected against deterioration by addition of glucose oxidase. These products, being oil-in-water emulsions containing much air, packed in glass containers, and exposed to light and ambient temperatures, are very susceptible to rancidity. Removal of the oxygen by incorporating glucose oxidase can afford significant protection.

Because of its specificity, glucose oxidase is widely used as an analytical reagent for detection and quantitative determination of glucose in the presence of other carbohydrates. Both manual and Autoanalyzer procedures employ glucose oxidase for this purpose. It is of particular advantage in biological systems of interest to medical technicians and to food technologists. An important example is the use of test strips containing glucose oxidase and reagents which give a color reaction to determine the presence of glucose in the urine for detecting and monitoring diabetic conditions.

GENERAL CHARACTERISTICS OF ENZYMES

For an appreciation of microbial enzymes an understanding of simple enzyme theory is necessary: what enzymes are, how they are named and classified, how they work, and the factors which affect their action.

CHEMICAL NATURE OF ENZYMES

An enzyme is usually defined as a protein biocatalyst produced by a living cell. Enzymes are, therefore, rather special members of that broad class of substances known as catalysts. Catalysts influence the rates of chemical reactions without being used up; they take part in the reactions but reappear in their original form. Theoretically, a catalyst can convert an unlimited amount of reacting substance.

Like all catalysts, enzymes affect the velocities of chemical reactions, but not the extent of the chemical changes which occur. The enzymes only catalyze reactions which are thermodynamically possible, that is, are attended by losses of free energy. In order for most chemical reactions to occur, a certain amount of resistance must be overcome; the molecules must be activated by supplying energy. An enzyme lowers the amount of activation energy required by the reaction. Hence, enzymes are able to bring about, under mild conditions near room temperature, reactions which in their absence would require drastic conditions of high temperature or another high-energy source.

The enzymes produced by living cells are, of course, for the purpose of accomplishing specific metabolic needs. Since enzymes can be separated readily from the cells that produce them and can perform their catalytic activities entirely apart from the cells, they are available for useful practical applications. How many enzymes exist is unknown but certainly the number runs into many thousands. Perhaps about a thousand have been studied to some extent and have been somewhat purified. Some have been completely purified and isolated in crystalline form. All these highly purified enzymes have been found to be proteins, and without question all enzymes are proteins. In some enzymes nonprotein groups are also present, such as specific metals like calcium or zinc, or organic carbohydrate, heme, flavin, or other groups.

Enzymes differ from other catalysts in several respects, mainly because of their protein nature. Being proteins, the enzymes are denatured and inactivated when subjected to unphysiologic conditions, such as heat or strong chemicals. Hence, two distinctive properties of enzymes are their thermal lability and their sensitivity toward acids and bases. But the most important distinction of enzymes is their specificity of action. For example, chemical hydrolytic catalysts such as acids will catalyze the hydrolysis of many different kinds of substances including esters, acetals, glycosides (sugars), or peptides (proteins), whereas individual different enzymes are required for hyrolyzing each of these types of compounds and even of specific members of each class. Thus carbohydrases act specifically on glycosides and cannot hydrolyze ester or peptide bonds. Individual carbohydrases are necessary for each of the individual glycosides. Sucrose, lactose, and maltose are all disaccharides, but separate and distinct enzymes are necessary for the hydrolysis of each. The enzyme sucrase (invertase) which hydrolyzes sucrose has no effect whatever on lactose or maltose, and so on.

With enzymes two types of specificity are distinguished: substrate specificity and reaction specificity. Examples of substrate specificity are the hydrolysis of sucrose by sucrase and the oxidation of glucose by glucose oxidase. Examples of reaction specificity are the actions of proteinases which are capable of splitting particular peptide bonds of proteins. These peptide bonds are amide linkages between carboxyl and amino groups of the amino acids. Individual proteinases have narrow reaction specificity as to which peptide bonds they can split. For example, the action of trypsin is the rapid hydrolysis of only those bonds linking the carboxyl groups of the two basic amino acids, lysine and arginine, to other amino acids. Trypsin has little or no effect on peptide bonds formed by other amino acids.

Much research is under way to elucidate the exact composition of enzymes. For this purpose it is necessary that the enzymes be highly purified. Isolation and purification of enzymes are based upon stability, solubility, size, and charge of the enzyme molecules, involving sequences of procedures of selective inactivation, solvent or salt precipitation, chromatography, and electrophoresis. Using highly purified enzymes, it has been firmly established that they are proteins, with the molecular weights of individual enzymes ranging from about 13,000 to over a million. The molecular weights and amino acid compositions of many enzymes have been determined, as have been the amino acid sequences of some of the smaller enzymes. There is nothing in the amino acid analyses or sequences which clearly differentiates enzymes from other proteins. All proteins contain chemically reactive groups such as free amino, carboxyl, hydroxyl, sulfhydryl, and imidazole groups and frequently nonprotein prosthetic groups. However, the mere presence of reactive groups in protein molecules does not account for enzyme activity. Special structures in the molecules must be responsible for the specific catalytic functions of the enzyme proteins. It is known that the long peptide chains of native protein molecules are folded, arranged, and cross-linked into three-dimensional structures. In these conformational arrangements of enzyme molecules the reactive groups or combining sites are situated in such a manner as to make up active centers which fit the reactive groups of the substrates to enable substrate binding and catalytic activity. The molecular conformation and structure of the active centers for a few enzymes have been elucidated by laborious physical-chemical methods such as x-ray analysis. When enzymes are denatured and thus inactivated by heat or strong chemicals, it has been shown that the necessary tertiary arrangements for activity are destroyed by unfolding and rearrangement of the peptide chains. Often activity is regained by removal of the denaturing agents, thus permitting re-formation of the critical conformational structures.

An interesting development relating to enzyme composition and structure has been the demonstration of multiple molecular forms of an enzyme, designated *isoenzymes* or isozymes. These isoenzymes catalyze the same reactions but have very slightly different physical properties. Isoenzymes occurring together can be most readily differentiated by zone electrophoresis. The different electrophoretic mobilities indicate differences in electric charge associated with slightly different amino acid composition of the isoenzyme proteins. Isoenzymes appear to be quite common in animal tissues and have been demonstrated in plants and microorganisms as well. It has been clearly shown electrophoretically that commercial glucoamylase of fungal origin contains two isoenzymes, and these have been partially separated and purified. There appears to be no difference in their catalytic activities and so they have no importance in the practical application of glucoamylase in the commercial production of dextrose from starch.

Isoenzymes have important practical applications in clinical diagnosis. For example, electrophoresis of normal human serum shows five isoenzymes of lactate dehydrogenase. On the other hand, particular tissues have quite different patterns of occurrence of the lactate dehydrogenase isoenzymes. Heart has predominantly isoenzymes 1 and 2, whereas liver has almost entirely isoenzyme 5. In diseases leading to leakage of enzyme from damaged cells, the isoenzyme pattern of serum may change from the normal toward that of the particular tissue involved. Thus it is possible to distinguish between myocardial infarction and liver disease by electrophoresis of serum and determining the lactate dehydrogenase isoenzyme profile.

Although many microbial enzymes may readily be separated in soluble form from the cells or from the insoluble fragments of the cells when they are disintegrated, some enzymes are firmly associated with particulate matter in the cells. Such bound enzymes are localized in the cellular structure and undoubtedly play very important metabolic roles in such functions as nucleic acid replication, protein synthesis, and oxidative phosphorylation. These bound enzymes are of the greatest significance in cellular activities, but they have no industrial importance, at the present time at least.

NAMES AND CLASSIFICATION OF ENZYMES

Without some historical background and some knowledge of basic enzymology it may appear that utter chaos and confusion exist in naming and classifying enzymes. The names of early enzymes were completely unsystematic, such as diastase and ptyalin for starch-splitting enzymes from malt and saliva; pepsin, trypsin, and papain for certain proteinases; and catalase for the hydrogen peroxide decomposing enzyme. Long usage and familiarity make necessary the continued use of these trivial names. When it became clear that enzymes bring about specific reactions in specific compounds, more meaningful names became possible. The compound upon which an enzyme acts is known as its substrate. It, therefore, became customary, where possible, to name a hydrolytic enzyme after the substrate upon which it acts, with the ending "-ase"; thus lactase, amylase, urease, esterase, lipase, peptidase, proteinase. In other cases, enzymes were named from the reactions they catalyzed, such as dehydrogenases (oxidases), transferases, phosphorylases. Examples are lactate dehydrogenase, glucose oxidase, glucosyl transferase, glycogen phosphorylase. Schemes for enzyme classification were also based upon substrates acted upon and types of reactions catalyzed.

Since enzyme nomenclature and classification based partly on historical names, partly on names of substrates, and partly on names of reactions were unsatisfactory and confusing, the International Union of Biochemistry set up an Enzyme Commission in 1959 to devise a workable systematic plan. The proposed scheme was officially adopted in 1961 and slightly modified and republished in 1964 (International Union of Biochemistry, 1965). The enzymes are divided into six main classes which are further divided into subclasses and subsubclasses, according to the nature of the chemical reactions catalyzed. They are also coded on a four-number system intimately associated with this system of classification. The main classes are oxidoreductases, transferases, hydrolases, lyases, isomerases, and ligases. Oxidoreductases catalyze oxidations or reductions. Transferases catalyze the shift of a chemical group from one donor substrate to an acceptor substrate. Hydrolases catalyze hydrolytic splitting of substrates. Lyases remove groups or add groups to their substrates by means other than hydrolysis. Isomerases catalyze intramolecular rearrangements. Ligases (synthetases) catalyze the joining together of two substrate molecules.

The systematic name for an enzyme is derived in accordance with definite rules to identify the enzyme and indicate its action as precisely as possible. In general, the systematic name consists of two parts. The first part names the substrate, and the second, ending in *-ase*, indicates the nature of the process. The systematic rules for nomenclature are quite extensive but they cannot be applied if the substrate composition or the enzyme reaction is not fully understood, in which case only trivial names are possible. The Enzyme Commission also adopted more convenient trivial names for the enzymes for general use. In the majority of cases the trivial names are those already commonly employed. In biochemical publications it is now customary to give the classification number, systematic name, and trivial name for an enzyme when first mentioned, and subsequently to use only the trivial name. To illustrate, the familiar glucoamylase (trivial name) is EC 3.2.1.3 α-1,4-glucan glucohydrolase. The name indicates that the substrate is a glucan (starch), a glucose polymer having the glucose molecules joined by $\alpha(1\rightarrow4)$ linkages, and the reaction is the hydrolytic splitting off of glucose. Table 7-2 gives representative important enzymes, illustrating the classification and nomenclature. Where more than one trivial name is given, the first one is that adopted by the Enzyme Commission.

HOW ENZYMES ACT

Enzymes function as specific catalysts which increase the rates of chemical reactions. It is well established that the enzyme takes an actual part in the reaction by combining with the substrate; when the reaction products are formed, the enzyme reappears in the original form. Numerous physical and chemical factors profoundly affect enzyme action.

Enzyme-Substrate Combination

The equations for enzymatic action are

$$E + S \rightleftharpoons ES \tag{7-1}$$
$$ES \rightleftharpoons E + P \tag{7-2}$$

where E = enzyme
S = substrate
ES = labile intermediate enzyme-substrate complex
P = end product

The substrate is activated, that is, its molecules become more chemically reactive, when it combines with enzyme to form the complex, and the complex is in equilibrium with free enzyme and free substrate. The mass action equilibrium equation for this reaction is

$$\frac{(E - ES) \times (S - ES)}{(ES)} = K_M \tag{7-3}$$

where $(E - ES)$ = concentration of free enzyme
$(S - ES)$ = concentration of free substrate
(ES) = concentration of complex

The equilibrium constant has come to be known as the Michaelis constant since Michaelis was the first to express this concept of enzyme action involving

Table 7-2 Classification and nomenclature of some important enzymes

EC No.	Systematic name	Trivial name
	Oxidoreductases	
1.1.1.28	D-Lactate:NAD oxidoreductase	D-Lactate dehydrogenase
1.1.3.4	β-D-Glucose:oxygen oxidoreductase	Glucose oxidase
1.1.3.9	D-Galactose:oxygen oxidoreductase	Galactose oxidase
1.11.1.6	$H_2O_2:H_2O_2$ oxidoreductase	Catalase
1.11.1.7	Donor:hydrogen peroxide oxidoreductase	Peroxidase
1.13.1.13	Linoleate:oxygen oxidoreductase	Lipoxygenase, lipoxidase
	Transferases	
2.4.1.5	α-1, 6-Glucan:D-fructose 2-glucosyltransferase	Sucrose 6-glucosyltransferase, dextransucrase
2.4.1.19	α-1, 4-Glucan 4-glycosyltransferase (cyclizing)	Cyclodextrin glycosyltransferase, *B. macerans* amylase
	Hydrolases	
3.1.1.3	Glycerol-ester hydrolase	Lipase
3.1.1.11	Pectin pectyl-hydrolase	Pectinesterase, pectinmethylesterase
3.2.1.1	α-1,4-Glucan glucanohydrolase	α-Amylase
3.2.1.2	α-1,4-Glucan maltohydrolase	β-Amylase
3.2.1.3	α-1,4-Glucan glucohydrolase	Glucoamylase, amyloglucosidase
3.2.1.4	β-1,4-Glucan glucanohydrolase	Cellulase
3.2.1.15	Poly-α-1,4-galacturonide glycanohydrolase	Polygalacturonase
3.2.1.20	α-D-Glucoside glucohydrolase	α-Glucosidase, maltase
3.2.1.23	β-D-Galactoside galactohydrolase	β-Galactosidase, lactase
3.2.1.26	β-D-Fructofuranoside fructohydrolase	β-Fructofuranosidase, sucrase, invertase
3.4.4.1	.....	Pepsin
3.4.4.3	.....	Rennin
3.4.4.4	.....	Trypsin
3.4.4.10	.....	Papain
3.4.4.12	.....	Ficin
3.4.4.16	.....	Subtilopeptidase A, subtilisin, bacterial protease
3.4.4.17	.....	Aspergillopeptidase A, fungal protease
3.4.4.24	.....	Bromelain
3.5.1.1	L-Asparagine amidohydrolase	Asparaginase
3.5.1.5	Urea amidohydrolase	Urease
	Lyases	
4.1.3.6	Citrate oxaloacetate-lyase	Citrate lyase, citrase
4.2.1.3	Citrate (isocitrate) hydrolyase	Aconitate hydratase, aconitase
4.2.99.3	Poly-α-1,4-D-galacturonide lyase	Pectate lyase, pectate transeliminase
	Isomerases	
5.1.2.1	Lactate racemase	Lactate racemase
5.3.1.5	D-Xylose ketol-isomerase	Xylose isomerase, glucose isomerase

enzyme-substrate complex formation. By investigating reaction velocities at various substrate concentrations, the K_M for an enzyme can be determined. This is a fundamental constant in enzyme work since its value reflects the affinity of an enzyme for its substrate. The higher the affinity of an enzyme for its substrate, the lower is the value of K_M.

Factors Affecting Enzyme Action

Numerous factors affect enzyme activity. These must be taken into account in understanding the behavior of enzymes and their practical use. Among the most important factors are (1) time, (2) concentration of enzyme, (3) concentration of substrate, (4) temperature, (5) pH, and (6) presence or absence of activators or inhibitors.

TIME

Time is a most important factor in practical enzymatic applications. Initial velocities of enzyme reactions are usually very rapid. But as the reaction proceeds, the rate diminishes. A typical curve for the extent of reaction with time is shown in Fig. 7-1. The decreased velocity as the reaction proceeds may be due to a variety of reasons. The most important usually are exhaustion of the substrate and inhibition of the reaction by its end products. For many enzyme uses it may not be necessary or even desirable that the reaction go to completion. An example is the liquefaction and reduction of viscosity of heavy starch pastes by α-amylase. In such cases action for only a few minutes suffices. However, where completion of the reaction is desirable, several hours or even days may be required to allow the enzyme reactions to approach completion. An example is the saccharifying action of glucoamylase in the enzymatic process for industrial production of dextrose from starch. Depending upon the exact conditions employed, this requires a digestion period of 48 to 96 h.

CONCENTRATION OF ENZYME

For almost all enzymatic reactions, especially where substrate is present in excess, the reaction rate is directly proportional to the concentration of the enzyme, at least during the early stages of the reaction. Thus, for example, although velocity falls off during later stages of the reaction, in general, twice as much enzyme will complete the reaction in half the time.

CONCENTRATION OF SUBSTRATE

With very low substrate concentrations, enzymatic reaction velocity is proportional to the substrate concentration. Where substrate is present in large amounts, enzymatic reactions usually follow either zero-order or first-order reaction kinetics. Where substrate is present in considerable excess during early stages, the reaction conforms to zero-order kinetics, and the amount of product formed is proportional to time:

$$dP/dT = k_0 \tag{7-4}$$

For such reactions the amount of end product is doubled if the reaction time is

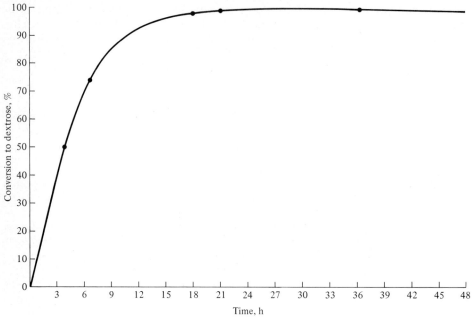

Figure 7–1 Enzymatic hydrolysis of maltose to dextrose by glucoamylase.

doubled. This is illustrated for the first 3 h in the curve of Fig. 7-1 where 20 percent conversion has taken place in 1.5 h and 40 percent after 3 h.

As previously mentioned and as shown in Fig. 7-1, the rate of reaction decreases with time. During the course of the enzymatic reaction, there is a continuing decrease in substrate concentration which is responsible for this decreasing rate. Most enzymatic reactions then follow first-order kinetics:

$$dP/dT = k_1 \times (S\text{-}P) \tag{7-5}$$

where k_1 is the first-order reaction constant and (S-P) is the concentration of substrate remaining at any given time. The rate of the reaction is proportional to the remaining substrate concentration, and equal fractions of the remaining substrate are transformed in equal time intervals. For example, if 50 percent of the substrate is converted in an hour, then in another hour 50 percent of the remaining substrate, or 25 percent of the original substrate, will be converted, and so on. This is illustrated in Fig. 7-1 for the period subsequent to the first 3 h. At that time 60 percent of the original substrate remained unconverted. During the next 3 h half of this, or 30 percent, was converted (total 70 percent), and in the succeeding 3 h half of the remaining 30 percent, or 15 percent, was converted (total 85 percent), and so on.

TEMPERATURE

Enzymes may be affected by heat in two ways. One effect may be inactivation. High temperatures cause denaturation and unfolding of the enzyme protein, resulting in loss of catalytic properties. There is considerable variation in the actual temperatures at which heat inactivation is substantial, depending upon the particular enzyme. For many enzymes thermal inactivation becomes appreciable and rapid at temperatures above about 50°C, but some enzymes are much more resistant. Glucoamylase is effectively employed at 60°C and bacterial amylases at 80°C or even higher. With most enzymes, thermal stability is better in the presence of substrate than in its absence.

The second effect of temperature on enzymatic reactions is in their rate. As with most chemical reactions, raising the temperature increases the velocities of most enzyme-catalyzed changes. A rough rule is that every 10°C increase in temperature approximately doubles the rate for many chemical reactions; that is, the temperature quotient Q_{10} is about 2. Actual Q_{10} values measured for individual enzyme reactions have been found to be in the range of 1.2 to 4.

In many practical applications of enzymes, the temperature is increased to speed up the reaction. However, a point may be reached where thermal inactivation of the enzyme will more than offset increased rate of reaction. A term frequently used is the "optimum temperature," which is the temperature for maximum velocity above which the rate decreases because of thermal inactivation. Caution must be exercised since numerous factors such as pH, enzyme concentration, substrate concentration, and particularly time have considerable effect on observed or useful optimum temperatures. The latter is apparent from the temperature-activity curves shown in Fig. 7-2. The apparent temperature optimum for hydrolysis of gelatin by bacterial proteinase differed by 10°C when 30- or 60-min incubations were used. This is, of course, explained by the greater inactivation at the higher temperatures with the longer incubation.

pH

The pH of the system in which enzymes act has a profound effect on enzyme activity. Each enzyme in the presence of its substrate has a characteristic pH at

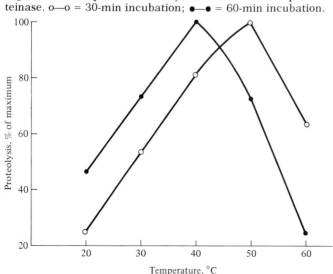

Figure 7-2 Temperature-activity curves for bacterial proteinase. o—o = 30-min incubation; ●—● = 60-min incubation.

which it is most active. This is known as its optimum pH. The optima for certain enzymes are quite sharp; other enzymes have rather broad optimum pH ranges. Typical pH-activity curves are given in Fig. 7-3. When the pH of an enzyme-substrate system is changed, the enzyme activity decreases rapidly on both sides of the optimum range until the enzyme is completely inactive. Usually change of the pH toward the optimum will reactivate the enzyme. However, holding at pH values above or below the levels of temporary inactivity will usually result in gradual denaturation of the enzyme and permanent inactivity. The pH optima for different enzymes vary widely. Pepsin and trypsin, two proteolytic enzymes, have pH optima of about 2 and 8, respectively. The pH range for best enzyme stability is not necessarily the same as for optimum activity.

ACTIVATORS AND INHIBITORS

Many enzymes need cofactors or activators for maximum effectiveness. So-called coenzymes are organic substances of this nature. Inorganic ions may be of particular importance in the practical use of enzymes. Although the activity of some enzymes is not noticeably affected by the presence or absence of salts, others are greatly influenced by the nature and concentration of the ions present. Certain ions are absolutely necessary for the activity of some enzymes whereas some others, such as Ag^+, Hg^{2+}, and Pb^{2+}, are highly toxic to nearly all enzymes.

Figure 7-3 pH-Activity curves for three commercial enzymes. (*From L. A. Underkofler (1968), Enzymes, in T. E. Furia (ed.), "Handbook of Food Additives," p. 59, Chemical Rubber Company, Cleveland. Reproduced by permission of the Chemical Rubber Company.*)

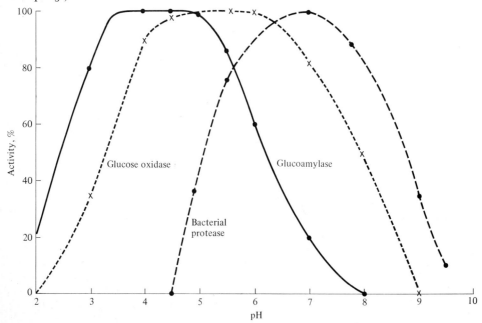

Some of the effects, particularly those of anions, are not very specific although the activity of some enzymes is greatly influenced by the presence of anions. For example, animal α-amylases are so greatly affected that chloride ion is usually regarded as the natural activator of these enzymes, and the presence of sodium chloride is indispensable for good activity of the amylase in pancreatin. Other monovalent, but not divalent, anions also have some effect on the activity of these amylases.

The activating effects of cations, on the other hand, are usually fairly specific. Fifteen different metal cations have been found to activate one or more enzymes, namely, Na^+, K^+, Rb^+, Cs^+, Mg^{2+}, Ca^{2+}, Zn^{2+}, Cd^{2+}, Cr^{3+}, Cu^{2+}, Mn^{2+}, Fe^{2+}, Co^{2+}, Ni^{2+}, Al^{3+}, as well as NH_4^+. Molybdenum compounds also activate certain enzymes, but the ionic form of the activator is obscure. The ions in the list are by no means interchangeable; in some cases only one particular metal ion is effective, whereas in other cases two or three can activate. For example, Mg^{2+} is the natural activator of the great majority (though not all) of the enzymes which act on phosphorylated substrates; in many of these cases Mg^{2+} can be replaced by Mn^{2+}, but not usually by any other metal. The metal ions may produce their activating effects on enzymes by a variety of mechanisms. For example, in some cases they may form an essential part of the active center of the enzyme; in other cases they may act as a binding link between enzyme and substrate. They may act by removing an inhibitor by forming a complex or precipitate with it; or they may affect equilibria between forms of an enzyme having different activities.

In some enzymes it has been shown that metal ions are essential constituents of the enzyme molecules. These are designated metalloenzymes. For example, pure α-amylase from *Bacillus subtilis* (*Bacillus amyloliquefaciens*) has been shown to contain 0.5 g atom of zinc and 2 to 3 g atoms of calcium per mole. The active enzyme is a dimer with the two units cross-linked by an atom of zinc. Nothing definite is known about how the calcium is bound. When the enzyme is treated in solution with chelating agents, such as ethylenediamine tetracetic acid (EDTA), the metals are removed, enzymatic activity is lost, and the metal-depleted amylase becomes susceptible to enzymatic proteolysis, in contrast with the very stable metalloenzyme. It is quite clear that α-amylase stability toward protease is totally dependent upon firmly bound calcium in the molecule.

In the true metalloenzymes the metal atom is firmly bound as a constituent of the enzyme itself. In other cases it is necessary to add some metal ion to activate the enzyme. The distinction may be only a matter of firmness of combination. Besides bacterial α-amylase, other well-known enzymes containing or requiring metal ions include certain peptidases, dehydrogenases, or oxidases.

Enzyme inhibition is also of great importance. An enzyme inhibitor is a substance which reduces or blocks an enzymatic catalysis. In practical applications the presence of inhibitors is often undesirable and care must be taken to exclude them. In other cases they are useful to stop, slow down, or control enzymatic reactions. Much use of inhibitors was made in elucidating the mechanism of glycolysis and fermentation and of the citric acid cycle. Specific inhibitors were used to block various successive enzymatic steps and allow intermediates to accumulate to be isolated and identified. Many poisons act by inhibiting a single enzyme in a major metabolic chain, thus having a profound and even fatal effect on the organism. An example of the complete inhibition and poisoning of a single enzyme is cyanide inhibition of cytochrome oxidase which results in cessation of aerobic oxidative processes and death in a very short time. The military "nerve gases" are essentially specific enzyme inhibitors. Related compounds have ever-increasing practical uses as insecticides.

There are many examples of nonspecific and specific, irreversible and reversible enzyme inhibitors. Enzyme proteins contain such groups as carboxyl, amino, and sulfhydryl, as well as other chemically active groups. The blocking of such groups which are involved in the sites necessary for binding or activation of the substrates gives rise to rather nonspecific inhibition. If the enzyme activity is restored by removing the inhibitor, as by dialysis, the inhibition is reversible. Poisons or inhibitors, such as certain heavy-metal ions, may result in irreversible inhibition by denaturation or destruction of the enzyme.

Specific inhibitors are those substances which block groups that confer specificity to an enzyme, that is, which react with combining sites of the active centers. Usually such reactions are reversible, and the inhibitor can be removed by dialysis or other means.

Substances which combine with the enzyme at the same site as the normal substrate produce competitive inhibition. The competitive inhibitors usually are substances similar in structure to the normal substrate, including in many cases products of the enzyme action. The enzyme is capable of combining with such substances but cannot activate them. The resulting hindrance of access of the normal substrate to the active centers inhibits the enzyme action.

Noncompetitive inhibitors are substances which combine with sites not essential for the substrate binding but which are essential for completing the enzymatic action. In noncompetitive inhibition the enzyme may combine with both inhibitor and substrate at the same time, thus causing interference with substrate activation and subsequent reaction. There are naturally occurring inhibitors of this kind, such as trypsin inhibitors found in serum, raw eggs, and legumes.

Methods of Enzyme Assay and Expression of Enzyme Activity

An important problem in enzymology is the quantitative measurement of the activity or concentration of an enzyme. Ordinary procedures of analytical chemistry are not applicable because the only way to detect an active enzyme is by what it does to its specific substrate. Hence, the potency or quantity of an enzyme must be determined by measuring how fast it changes its substrate under controlled conditions. As an example, the potency of an amylase can be determined by its action on starch. A solution containing a known weight of the enzyme preparation is added to a measured volume of starch solution. The pH, temperature, time, and other conditions of incubation are accurately controlled. The reaction is stopped by adding a suitable reagent, and the amount of starch hydrolyzed is determined. This can be done by measuring the change in viscosity of the starch paste, or the amount of starch not changed (starch-iodine color reaction), or the amount of sugar produced. In other words, enzymes are assayed by determining (1) the amount of substrate changed under standardized conditions of the procedure, (2) the change in physical nature of the substrate, or (3) the amount of reaction product. There are numerous ways of expressing the results of enzyme assays although the most common method is in terms of arbitrary units which are defined for the specific enzyme and substrate used. For example, amylase potency might be expressed in terms of saccharification units per gram, a unit being defined as the amount of enzyme which produces one gram of maltose when it acts on a 2% starch solution for 30 min at 30°C. Unfortunately, there is little uniformity between laboratories in defining units, or in the assay conditions of pH, temperature, and time. Hence, it is often impossible to compare assays of the same enzyme reported from different laboratories. Standardization of assay procedures for some of the important enzymes is being attempted by professional and trade associations.

Terms which are frequently encountered, especially in investigations of enzyme purification and use, are *specific activity* and *molecular activity*.

The specific activity of a pure enzyme is defined as the units of activity per milligram of enzyme. Where the enzyme preparation is not pure, specific activity is frequently expressed in units per milligram of protein, or in units per milligram of enzyme preparation.

The molecular activity is a useful term to describe the amount of substrate converted in unit time by a given quantity of enzyme if the molecular weights of the enzyme and of the substrate are known. The molecular activity represents the number of moles of substrate converted per mole of enzyme per minute. The molecular activities of different enzymes vary widely (100 to 5,000,000). The expression "turnover number" is sometimes employed instead of molecular activity.

PRODUCTION OF MICROBIAL ENZYMES

Although many useful enzymes are obtained from animal and plant cells, microorganisms are becoming the favored source for industrial enzymes because of their technical and economic advantages. There are limitations of supply of animal tissues as by-products from packing plants and also of plant materials, depending upon available acreage and labor. Considerable variability of quantity and quality of animal and plant enzymes occurs with changes in seasonal and climatic conditions. On the other hand, microorganisms provide sources for a wide variety of enzymes, and suitable strain selection makes possible high yields of the desired enzymes. Such enzymes can then be conveniently obtained by fermentation processes in unlimited quantities and in a very reproducible and controllable manner.

SOURCE ORGANISMS

Industrial enzymes are produced from selected strains from each of the microbial groups, molds, bacteria, and yeasts. Some of the source organisms currently employed and enzymes obtained from them are shown in Table 7-3. Of particular importance are strains of the *Aspergillus oryzae*, the *Aspergillus niger*, and the *Bacillus subtilis* groups (Beckhorn et al., 1965).

Table 7-3 Source organisms for some industrial enzymes*

Organism	Enzyme
Molds	
Aspergillus oryzae	Amylase, protease, ribonuclease
Aspergillus niger	Amylase, glucoamylase, cellulase, hemicellulase, pectinase, glucose oxidase, catalase
Rhizopus species	Amylase, glucoamylase, pectinase, lipase, protease
Mucor species	Protease, lipase
Trichoderma viride	Cellulase
Bacteria	
Bacillus subtilis (*B. amyloliquefaciens*)	Amylase, protease
Micrococcus lysodeikticus	Catalase, polynucleotide phosphorylase
Clostridium histolyticum	Collagenase
Streptococcus hemolyticus	Streptokinase-streptodornase
Yeasts	
Saccharomyces cerevisiae	Invertase
Saccharomyces fragilis	Lactase

*See Phaff (1959) and Davies (1963) for extensive compilations of enzymes produced by microorganisms.

Screening Methods

The number of individual enzymes produced by microorganisms is in the thousands since each microbial cell must possess all the enzymes necessary for its complete metabolism. The various enzymes function in hydrolyzing nutrient materials, in catalyzing metabolic reactions, and in synthesizing cellular materials. Hence, it is usually possible to find a microbial source for an enzyme to accomplish almost any desired reaction within the possible limits of enzymatic catalysis.

The kinds and amounts of various individual enzymes produced vary widely between strains of different organisms. Hence, the best source for a particular enzyme must be sought from appropriate microorganisms by suitable screening methods. Taxonomic descriptions in standard compendia of bacteria and fungi or original literature references often give leads to suitable organisms for investigation. Cultures for screening are obtained from culture collections and from new isolates from natural sources, particularly sources rich in the substrate upon which the desired enzyme must act. Initial screening may be conducted by plating the cultures on solid agar medium containing the potential substrate. Colonies selected for further study are those which show the largest zones of action on the substrate. They are then grown as pure cultures in liquid media, and further selections are based primarily on assay of the enzyme potencies produced. Alternatively, where the desired reaction cannot be easily detected on solid media in plate cultures, the initial screening may require growth in liquid media with enzyme assays on the culture liquids. The screening processes are employed not only for selecting original cultures for production of new enzyme products but also in finding new strains of higher yield potential.

Microbial strains which appear to have capacity for highest production of the particular desired enzyme are subjected to further study. Variations are made in medium composition and growth conditions to determine the maximum potential for producing the desired enzyme. Besides fermentation conditions for maximum enzyme yields, suitable procedures for recovering the enzyme from the culture medium also must be established. Finally, the fermentation and isolation methods must be scaled up through pilot plant to full manufacturing level.

Mutants

Mutant cultures are of great importance in the production of commercial enzymes. Frequently mutants can produce many times the yield of enzyme obtainable from the parent cultures. In other cases mutants may produce fewer undesirable enzyme contaminants or other metabolic products, thus facilitating enzyme recovery and purification. A detailed discussion of microbial genetics, including mutation, applied to industrial fermentations is given in Chap. 9. In brief, mutants for enzyme production are obtained by subjecting cultures to mutagenic agents to such degree that most of the cells are killed. The survivors are then plated on isolation media, colonies are picked, and the new isolates are screened for superior characteristics as compared with the parental strain.

GENERAL METHODS OF CULTIVATION

As with other fermentation processes, for enzyme production the microorganisms are cultivated by inoculating pure cultures into a sterile medium of suitable nutrient composition, followed by incubation at controlled temperatures with the necessary presence or absence of oxygen. Most commercial enzymes are derived from aerobic organisms.

Surface and Submerged Culture

Prior to World War II all commercial microbial enzymes were produced by surface culture methods. Submerged culture has now come into extensive use. For example, submerged culture has completely supplanted the process for producing bacterial enzymes by growing strains of *Bacillus subtilis* as a pellicle on the surface of thin layers of liquid medium (Wallerstein, 1939). However, several fungal enzyme products continue to be produced by the original Takamine moldy bran or koji process, in which the mold is cultivated on the surface of a solid substrate, moist wheat bran. Full details of this process are given below.

The submerged culture method for aerobic organisms is now well known and used. In this process the organisms are cultivated in liquid media in aerated tubes, flasks, or fermentors. Details of the process as applied to enzyme production are given below.

There are advantages and disadvantages to each of the semisolid surface and submerged culture methods, some of which are given in Table 7-4. The surface culture process has advantages in simplicity of operation, low power requirements, and freedom from contamination problems, but with the disadvantages of needing much space and costly hand labor. The submerged culture

Table 7-4 Comparison of surface and submerged processes

Surface	Submerged
Requires much space for trays	Uses compact closed fermentors
Requires much hand labor	Requires minimum of labor
Uses low-pressure air blower	Requires high-pressure air
Little power requirement	Needs considerable power for air compressors and agitators
Minimum control necessary	Requires careful control
Little contamination problem	Contamination frequently a serious problem
Recovery involves extraction with aqueous solution, filtration or centrifugation, and perhaps evaporation and/or precipitation	Recovery involves filtration or centrifugation, and perhaps evaporation and/or precipitation

method has advantages of low labor and space requirements, uniformity, and ease of control; its disadvantages are high power requirements and occasional serious contamination problems. Whether the surface or submerged culture method is employed for a particular enzyme is determined mainly by relative yields and convenience, although sometimes plant equipment availability and enzyme application are also factors.

Comparison of Laboratory and Large-Scale Operations

Although the general principles of microbial enzyme production in the laboratory on a small scale and in the industrial plant on a large scale are the same, the types of equipment and quantities of materials involved are very different. Frequently the problems encountered and the procedures employed in the laboratory and in the plant differ greatly.

MICROBIAL GROWTH CYCLES

All microbiologists are aware that microorganisms multiply by cell division and that there are five well-recognized phases—lag phase, logarithmic phase, stationary phase, declining phase, and survival phase—during incubation of an organism inoculated into a favorable growth medium.

During the initial lag phase the culture becomes established and begins to multiply. This lag phase may be long or short, depending upon the organism, the medium, and the conditions of incubation. Extended lag phases often are experienced when the growth medium differs considerably in composition from that used in growing the culture, when the culture is past the actively growing stage, or when spore inoculation rather than vegetative cell inoculation is used. The initial lag is usually of minor significance in the laboratory since maintenance of sterility in flasks presents no problems, and time of fermentation is not important. In the plant, however, minimizing or eliminating the lag is desirable since preventing contamination in large culture vessels is a constant problem, and time of fermentation is very important in the efficient utilization of plant equipment. By using inoculum medium of the same composition as used in the production fermentor and employing large inocula of actively growing seed cultures, lag in plant fermentors may usually be almost completely eliminated.

The logarithmic or exponential growth phase is one of accelerating multiplication. During this phase it can be assumed that the organisms are fully viable and all of equal vigor. However, restrictive mechanisms, such as accumulation of inhibitory products or exhaustion of essential nutrients, begin to come into play well before maximum population density is reached. In an ordinary batch fermentation the limit of growth is determined by the volume and the composition of the medium. However, if fresh medium is continuously added to the culture, it is possible to prolong the growth almost indefinitely in nearly continuous logarithmic phase. This is, of course, the basis of continuous fermentation procedures.

Following the logarithmic phase is the stationary phase during which the cell population remains almost constant. Some cells may continue to replicate, some remain static though viable, and others may be dying. There is considerable variation in the vigor of individual cells and in the type and rate of their metabolism.

In the declining phase, cell division still continues to occur but at an ever-decreasing rate, and the number of new cells being formed is far outnumbered by the number of cells dying off.

The last stage is the survival phase. Cell division ceases completely so that no new cells are being formed. Only those cells which were formed previously continue to survive and eventually die. Thus the curve of cell numbers versus time levels off and finally returns to the horizontal axis.

There is great variability in the growth phase during which enzyme accumulation occurs, depending upon the particular organism and enzyme. A desired enzyme may appear mainly during any phase except the lag phase. For example, certain bacterial proteinases are elaborated rapidly and almost entirely during the logarithmic phase, whereas production of bacterial amylases occurs mostly during the stationary phase. In some cases an extracellular enzyme may begin to appear during the early part of the logarithmic phase and continue to increase in amount during the later stages. Intracellular enzymes are probably produced mainly during the logarithmic phase but are released into the medium only as the cells undergo lysis during the declining phase. Therefore, recovery of intracellular enzymes is usually best accomplished by harvesting the cells at the end of the logarithmic phase, then releasing the enzyme into aqueous solution by lysis, mechanical grinding, or ultrasonication.

During fermentation the concentration of a desired enzyme being produced will increase to a maximum. Then, depending upon the particular enzyme, the concentration may remain constant for a considerable time or it may decrease either slowly or rapidly. When the latter occurs in a fermentation, there may be serious recovery problems involved in obtaining good enzyme yields.

The problems of optimum time of harvesting and effective enzyme recovery methods, both in the laboratory and in the plant, are affected by various factors of microbial growth and of enzyme growth and stability. In the laboratory usually somewhat longer fermentation times are required than in large-scale fermentations. Depending upon the organism and enzyme desired, laboratory fermentations of 2 to 14 days are common, whereas similar fermentations in the plant may require only 12 h to 6 days. In the laboratory small volumes are involved so that when maximum enzyme production is reached the cultures can be harvested and the enzyme recovered by simple methods, involving short times of minutes or a few hours at most. In the plant, processing a large batch may require many hours in each of the operations of filtering, concentrating, precipitating, recovering, and drying. In fact, the most difficult problems encountered in scaling up a process for producing a microbial enzyme are usually not in the fermentation but rather in the subsequent recovery steps.

STERILIZATION

The important topic of sterilization is covered in Chap. 14. The only satisfactory means for sterilizing plant fermentation media prior to inoculation with the desired cultures is by means of heat under pressure, and this is most commonly employed for laboratory media also. However, there are major differences in sterilizing conditions for laboratory and for plant fermentor media. In fact, sterilization is a major problem in plant scale-up from laboratory fermentations.

In the laboratory, media in test tubes, flasks, and fermentor jars are sterilized in steam autoclaves. Since small quantities of medium in tubes and flasks follow the temperature cycle closely, heating and cooling periods are quite brief so that commonly 20 min at 120°C ensures sterility. When the volumes of medium and sizes of containers are increased, there is a greater time lag of the temperature reached by the medium as compared with that of the steam-space temperature. A 30-l vessel of aqueous sugar solution requires about 2 h in an autoclave at an external steam temperature of 120°C for complete sterilization. Media in which heat penetration is poor, such as those containing undissolved solids or of viscous nature, require still longer times for sterilization. Heat-sensitive medium components further complicate sterilization. If all ingredients are soluble, in the laboratory small volumes of the complete media may be filter sterilized, or the heat-sensitive material in solution may be separately filter sterilized and added aseptically to the rest of the medium which has been heat sterilized.

In plant fermentors sterilization is often accomplished by heating the media in the fermentors by means of steam jackets or steam coils, then cooling by circulating cold water through the jackets or coils. Even with good agitation of the liquid in the fermentor, it takes many minutes or hours, depending upon volume, to reach the sterilizing temperature, say 120°C, and an equal or longer time to cool to inoculation temperature. The long heating cycles are effective in sterilization but at the same time can cause change or destruction of essential nutrients as well as reactions between medium constituents. Often such medium changes adversely affect growth of the fermentation organism, its production of enzyme, or recovery of the enzyme from the culture liquid. Sometimes it is possible to sterilize certain medium ingredients separately from the main batch and add these aseptically at fermentation temperature to minimize undesirable changes. Because of the volumes involved, filter sterilization of media or medium ingredients is not usually possible on the plant scale.

To avoid undesirable heat-caused changes in fermentation media, to approximate more closely laboratory sterilizing processes, and to make more efficient use of fermentor capacity, continuous high-temperature, short-time sterilization methods are commonly used for the media for large fermentors. Suitable continuous processing of free-flowing liquids permits very fast heating, close control of the holding time, and rapid cooling. The cool sterile medium flows continuously into the fermentor which has been previously sterilized by steam under pressure. Usually the culture is introduced into the partially filled fermentor, and growth begins while filling of the fermentor is being completed.

The problem of contamination of fermentations by undesired organisms is much greater in large-scale fermentors than in the laboratory. Contamination in laboratory flask fermentations is not a serious problem, assuming good laboratory procedures are followed. Preventing contamination in an industrial fermentation process is a constant battle. Fermentations can be contaminated through such routes as faulty agitator seals, leaks in cooling coils, inadequately protected inoculum, antifoam and sampling lines and valves, and failure of air-sterilizing filters.

AGITATION AND AERATION

Since most commercial enzymes are produced by aerobic organisms, aeration and agitation must be continuous during the course of the fermentation in submerged culture. In the laboratory the most common practice is to employ shaken cultures in flasks. Aeration in tubes or cylinders employing porous air-dispersing devices also may be used. Small jar fermentors provided with mechanical agitators and air spargers, similar in arrangement to industrial deep tank fermentors, have become common in well-equipped laboratories.

Plant fermentors for aerobic submerged culture fermentations were originally developed for production of antibiotics and now are widely used. For enzyme production they may vary between 1000- and 30,000-gal capacity. To provide agitation, a shaft, rotated by a powerful electric motor, runs from the top to the bottom of the fermentor, with an efficient seal at the top and a steady bearing at the bottom. Near its lower end the shaft carries a flat turbine impeller about two-fifths of the tank diameter with either straight or slightly curved blades. Sterile air is introduced by a sparger immediately below this impeller so that the air is intimately and continuously mixed with the fermentor contents. There may be one to four additional empellers higher up the shaft. Most fermentors have four vertical baffles inside the tank, about one-tenth of the tank diameter in width. For temperature control, cold water is circulated as necessary through jackets or coils. Foaming is controlled by introduction of sterile antifoam agents.

Agitation and aeration in laboratory shaken flasks are quite different from plant fermentors. Laboratory jar fermentors simulate plant equipment and are useful in developing successful fermentations. However, they provide little useful information about optimum plant fermentor operating conditions. To obtain maximum enzyme yields from any individual plant fermentation, optimum conditions of aeration, agitation, and power input must be determined rather empirically. When once established for a particular fermentation and vessel, these conditions are adhered to rigidly in plant operations.

GENERAL RECOVERY METHODS

After completion of the fermentations, further processing steps in obtaining the commercial enzymes are similar whether the aqueous enzyme solutions are derived from semisolid surface culture or submerged culture. These involve combinations of various standard operations, such as filtration, centrifugation, vacuum evaporation, precipitation, adsorption, drying, and blending, for concentrating, partially purifying, stabilizing, and formulating the enzyme materials into commercially useful products. Since these operations do not involve microbiology, they will not be considered further here. Details are available from other publications (Beckhorn et al., 1965; Underkofler, 1966a, 1966b).

INDUSTRIAL PRODUCTION

Although quite a number of publications have discussed the general procedures for producing microbial enzymes (Wallerstein, 1939; Hoogerheide, 1954; Underkofler, 1954, 1966a, 1966b; Beckhorn, 1960; Arima, 1964; Beckhorn et al., 1965), the specific details of current manufacturing processes have not been disclosed. The success of a microbial enzyme manufacturer depends on the selected cultures, the exact composition of media, and the cultural conditions. In order to maintain competitive positions, most manufacturers hold this information as confidential unless patented. Even in patents only laboratory or relatively small-scale examples are usually cited, and specific conditions for large-scale production may be quite different from laboratory operations. As much detail as possible, without using confidential information, for a few specific examples of industrial production methods will be given in this section.

Surface Culture

A general flow diagram of the semisolid culture or koji process is given in Fig. 7-4. In this process the fungus is cultivated on the surface of a solid substrate on trays in large chambers or in horizontal rotating drums. The chambers and trays, or drums, are washed with detergent; sanitized with sodium hypochlorite, formaldehyde, or quaternary ammonium disinfectants; and steamed between successive batches. The basic substrate material is wheat bran. This is mixed with water, and sometimes additional nutrient materials and acid or buffer salts may be added. The mass is sterilized in rotating vessels by means of steam under pressure. The mash is then cooled and inoculated with mold spores. These spores are prepared in the laboratory by incubating pure cultures on porous bran media in bottles until it is heavily sporulated. The inoculated bran mash is then spread in thin layers on trays and incubated in chambers. Alternatively, incubation may take place in slowly rotating drums. The temperature and humidity are controlled by circulating cool, moist air over the bran mass. The porous structure of the bran mash allows air to penetrate and the mold mycelium to grow throughout.

During the incubation period a considerable loss of weight occurs because of oxidation of organic matter to carbon dioxide and water. Periodically the

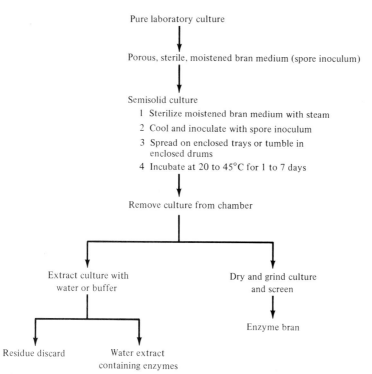

Figure 7-4 Flow diagram of a semisolid culture process. (*From L. A. Underkofler (1966a), Manufacture and Uses of Industrial Microbial Enzymes, Chem. Eng. Prog. Symp. Ser.*, **62** *(69): 11–20. Reproduced by permission of the Chemical Engineering Progress, American Institute of Chemical Engineers.*)

temperature, pH, moisture level, and enzyme content are determined. Temperature and moisture are controlled as necessary. The batch is harvested when the enzyme level has reached its maximum, which may vary from 1 to 7 days, depending upon the organism employed and other factors. The enzyme is extracted from the moldy bran (koji) either by batch extraction with water or buffer and filtration, or by percolation of water or buffer through the koji by a countercurrent extraction system which filters as well as extracts.

The semisolid surface culture method is now used extensively in producing such commercial enzyme products as fungal amylase (*Aspergillus oryzae*), fungal protease (*A. oryzae*), fungal glucoamylase (*Rhizopus* species), fungal cellulase (*A. niger, Trichoderma viride*), and fungal milk clotting enzyme (*Mucor pusillus*). The method was first developed by Takamine for large-scale production of fungal amylase (Takadiastase) both in the United States and in Japan. Processes employed today are only slightly modified from the original Takamine procedure and have been described by Shellenberger (1947), Underkofler et al. (1947), and Jeffreys (1948).

CULTURE

The organism employed is a selected strain of *A. oryzae* (ATCC 4814). The stock cultures are maintained on Czapek agar slants or wort agar slants.

SEED PREPARATION

Spore inoculum is required for the process. The medium for small seed bottles is prepared by mixing 15 g of wheat bran and 12 ml of water in each 250-ml wide-mouth bottle, covering the opening with a gauze and cotton filter disc, and sterilizing for 30 min at 120°C on 3 successive days. Each bottle is inoculated with spore suspension from a stock slant culture, the bran is distributed on one side of the bottle by gentle tapping, and the bottle is incubated at 30°C while lying on the side until the material is well grown, sporulated, and completely dry.

The medium for large seed bottles is a mixture in the proportions of 100 g of wheat bran, 10 g of ground corn, 100 g of glucose, 2.5 g of peptone, and 110 ml of water. Into each 2-qt wide-mouth bottle is placed 400 g of this mixture, the opening is covered with a gauze and cotton filter disc; it is sterilized for 30 min at 120°C on 3 successive days. Each bottle is then inoculated with a small quantity of dry material from a small seed bottle and incubated at 30°C while lying on the side until the material is well grown, sporulated, and dry.

MEDIUM PREPARATION, STERILIZATION, AND INOCULATION

Into the rotary sterilizing drum are charged 4000 lb of wheat bran and 375 gal of water. These ingredients are thoroughly mixed by rotating the drum, and the mixture is then sterilized by direct steaming at a pressure of 15 lb/in^2 for 3 h. Since there is condensation of steam into the mixture during sterilization, the quantity of water originally added must be varied slightly, as experience dictates, to obtain the desirable initial moisture content of about 51 percent in the mash when inoculated. Filtered cool air is then blown through the drum and cold water run over the outside until the mixture is cooled to 88°F (31°C). The mixture is inoculated by adding dry spore culture from 20 large seed bottles, and the drum is rotated for another hour to ensure uniform dispersal of the inoculum.

INCUBATION

The inoculated bran is spread on perforated trays in a layer about 1 in deep, and the trays placed on movable racks. The racks are wheeled into the incubation chamber which is equipped for circulating humidified air. The temperature of the air is raised to about 98°F (37°C) to initiate growth, and during the active growth period cooler air is circulated as needed to maintain the temperature between 100 and 105°F (38 to 41°C). During the incubation period the mold mycelium grows throughout the bran mass and forms a firmly matted cake, the koji. Maximum amylase potency is obtained after incubation for 24 to 36 h.

RECOVERY OF ENZYME

The koji is dumped from the trays, broken up into small pieces in a ribbon mixer, and loaded into a series of percolators with perforated false bottoms. The enzyme is extracted into aqueous solution by percolating water through the koji, pumping the solution from the bottom of the first percolator to the top of the next in the series, and so on. The first 500 gal of strong extract is processed further for enzyme recovery, and the second 500 gal of weak extract is employed to start extraction of the next batch of koji. An alternative method of extraction is to slurry the koji in water with agitation, then filter and wash the filter cake to obtain the aqueous enzyme extract.

The enzyme is precipitated by adding 3 volumes of cold 95% special denatured alcohol, agitating slowly for about an hour, and then allowing the precipitate to settle. The clear supernatant liquid is decanted off, an equal volume of cold alcohol is mixed with the precipitate slurry in the tank, and the enzyme precipitate recovered by filtration in a plate and frame press. The damp precipitate is transferred to cloth-bottomed trays and dried at 105°F (41°C) in an atmospheric drier. The dry enzyme concentrate is removed from the drier, ground to a fine powder, and transferred to moisture-proof containers.

ASSAY, STANDARDIZATION, AND FORMULATION

The enzyme concentrate is sampled and assayed in the laboratory for amylase activity. It is then formulated by intimately blending with a diluent to a standard potency, the diluent and potency depending upon the application. For example, fungal amylase for use in the bread-baking industry is standardized to 5000 SKB (American Association of Cereal Chemists, 1962) with wheat starch as diluent.

Submerged Culture

A general flow diagram for enzyme production by submerged culture is given in Fig. 7-5. The enzyme fermentations are carried out in closed, deep tank fermentors which may vary between 1000- and 30,000-gal capacity. These are equipped with agitators, spargers for introduction of sterile air, and jackets or coils for temperature control. The ingredients of the medium can be put into the fermentor and the medium batch sterilized in the fermentor by heating under pressure, or the liquid medium can be continuously sterilized as it is being pumped into the sterile fermentor. The laboratory pure culture is grown in shaken flasks of medium, which are used to inoculate the seed tank containing sterile production medium. Seed-tank operating and control conditions are analogous to those used in production fermentation. The cooled medium in the fermentor is inoculated with pure seed-tank culture. Appropriate aeration and agitation are supplied to the growing culture, and the temperature is maintained by circulating water through the jackets or coils of the fermentor.

The appropriate conditions for maximum enzyme production must be rather empirically determined for each individual fermentation process. Once optimum nutrient composition, pH, temperature, and aeration-agitation conditions are established for a given fermentation and vessel, these conditions are henceforth rigidly adhered to in plant operations in order to obtain consistently high enzyme yields. Continuous controls of temperature, pH, disappearance of mash ingredients, purity of culture, and enzyme level are maintained on samples taken from the fermentor periodically. The fermentation period may be from 1 to 5 days, depending upon the organism and enzyme being produced.

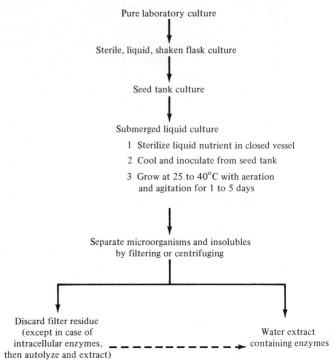

Figure 7-5 Flow diagram of a submerged culture process. (*From L. A. Underkofler (1966a), Manufacture and Uses of Industrial Microbial Enzymes, Chem. Eng. Prog. Symp. Ser.*, **62** *(69): 11–20. Reproduced by permission of the Chemical Engineering Progress, American Institute of Chemical Engineers.*)

When laboratory tests show that maximum enzyme production has been reached, it is the optimum harvest time.

The fermentor contents may then be cooled; preservatives, stabilizers, flocculants, or filter aids are added as necessary and the microbial cells and insoluble materials separated by centrifuging or, more commonly, by filtering. If the desired enzymes are extracellular, the microbial debris is discarded. In the case of intracellular enzymes, the cells are removed by filtration, the liquid is discarded, and the filter cake is slurried in water containing appropriate salts. After disintegration of the cells by grinding, ultrasonication, or lysis, the cell debris is removed by filtration or centrifugation, yielding a clear extract for further processing.

The submerged culture method is applied in the production of numerous commercial microbial enzymes among which are bacterial amylase, bacterial protease, fungal glucoamylase, pectinase, cellulase, hemicellulase, glucose oxidase, catalase, invertase, lipase, and lactase. Details of a process for the industrial production of bacterial amylase have been published (Underkofler, 1966a) and are again described here.‡

CULTURE

The organism employed is a selected strain of *Bacillus subtilis* (*B. amyloliquefaciens*). The stock cultures are maintained on nutrient broth agar slants. New transfers are made every 2 months, incubated at 90°F and stored under refrigeration until used. The stock agar cultures are used to inoculate seed flasks for plant fermentations.

MEDIUM PREPARATION AND STERILIZATION

The composition of the fermentation medium is given in Table 7-5. A volume of 1200 gal of the medium is employed in the fermentor.

The water is run into the fermentor and then the other ingredients added with the agitator running. After the fermentor is closed, steam is introduced into the jacket. With the agitator running, the medium is heated to 250°F (121°C) and held at that temperature for 35 min. Cold water is then circulated in the jacket and the fermentor contents cooled to 94°F (35°C), keeping tank pressure always above 5 lb/in^2 with minimum agitation and aeration to prevent excessive foaming.

SEED PREPARATION

Two 2-l Fernbach flasks each containing 1000 ml of nutrient broth are inoculated from stock agar cultures and incubated for 12 h at 90°F (32°C) on a reciprocating shaker. These flask cultures are used to inoculate the seed tank containing 40 gal of sterile medium similar in composition to that of the fermentor. The seed is grown for 10 h at 90°F (32°C), with agitation at 150 r/min and aeration at 5 ft^3/min at a tank pressure of 3 lb/in^2. At the end of this period the seed is examined microscopically for purity and is transferred aseptically to the fermentor.

FERMENTATION

The fermentation conditions maintained are temperature 94°F (35°C), tank pressure 5 lb/in^2, agitation 170 r/min, and aeration 35 ft^3/min. Samples for

Table 7-5 Percentage composition of medium for bacterial amylase production

Ground soybean meal	1.85
Amber BYF (autolyzed brewers' yeast fractions, Amber Laboratories)	1.50
Distillers' dried solubles	0.76
N-Z Amine (enzymatic casein hydrolysate, Sheffield Chemical Co.)	0.65
Lactose	4.75
MgSO$_4$•7H$_2$O	0.04
Hodag KG-1 antifoam (Hodag Chemical Corp.)	0.05
Water	90.40

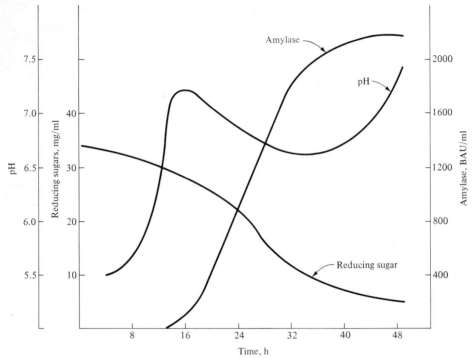

Figure 7-6 Control analysis curves for bacterial amylase fermentation. (*From L. A. Underkofler (1966a), Manufacture and Uses of Industrial Microbial Enzymes, Chem. Eng. Prog. Symp. Ser.*, **62** *(69): 11–20. Reproduced by permission of the Chemical Engineering Progress, American Institute of Chemical Engineers.*)

control analyses are taken aseptically immediately after inoculation, every 4 h for the first 24 h, and then every 2 h until the fermentation is completed. Assays are made on each sample for reducing sugar, pH, and amylase activity. The amylase assays are run by a method similar to that of the American Association of Textile Chemists and Colorists (1962) and expressed as bacterial amylase units (BAU). When two successive samples show no increase in amylase activity, the fermentation is terminated by stopping the aeration and cooling to about 70°F (21°C).

The curves in Fig. 7-6 were obtained by plotting the control analyses for a typical fermentation which was complete in 48 h. Subsequent processing of this particular fermentation was as follows:

RECOVERY OF ENZYME

To the 1200 gal of fermentor beer assaying 2159 BAU/ml was added 2 percent of diatomaceous earth filter aid, and the beer was filtered over an advancing-knife vacuum precoat filter. The volume of filtrate (including wash water) was 1100 gal assaying 2002 BAU/ml. The filtrate was transferred to a precipitation tank and 3000 gal of special denatured alcohol was added slowly with continuous agitation. The agitator was stopped, the precipitate allowed to settle, the clear supernatant liquid decanted, and 1000 gal of fresh alcohol added to the precipitate slurry with agitation. After stirring gently for a few minutes, the precipitate was again allowed to settle, the supernatant liquid was decanted, and the precipitate was recovered by filtration in a plate and frame press. The damp precipitate was transferred to a vacuum drier. After 8 h the dry enzyme concentrate was removed from the drier, ground to a fine powder, and transferred to moisture-proof containers. Weight of the product was 162 lb.

ASSAY, STANDARDIZATION, AND FORMULATION

The enzyme concentrate was sampled, assayed in the laboratory, and found to have a potency of 102,000 BAU/g. This material was formulated by intimately blending with powdered gypsum to produce a commercial bacterial amylase product standardized at 16,800 ± 5% BAU/g. For this purpose the 162 lb of concentrate was blended thoroughly in a mixer with 820 lb of powdered gypsum. A sample of the blend assayed in the laboratory showed 16,830 BAU/g.

Intracellular Enzymes

A few intracellular enzymes have industrial importance, including glucose oxidase, catalase, invertase, and lactase. Other intracellular enzymes are prepared, and some are marketed for use in research in molecular biology, or potentially for therapeutic applications. A few examples are polynucleotide phosphorylase, ribonucleases, deoxyribonucleases, DNA polymerases, and L-asparaginase.

The organisms producing the intracellular enzymes are cultivated in submerged culture, using media and conditions which give maximum yields of cells with the highest possible content of the desired enzyme. At the conclusion of the fermentation, the cells are removed by filtration, the liquid is discarded, and the cell filter cake is slurried in water containing appropriate salts and frequently appropriate preservatives and plasmolyzing agents. After disintegration of the cells by lysis, grinding, or ultrasonication, the cellular debris is removed by filtration or centrifugation, and the clear extract is further processed to recover the enzyme.

Full details are not available on any process for commercially producing an intracellular enzyme of industrial importance. For invertase, which has had the longest period of industrial use of any intracellular enzyme, patents and an early publication by Wallerstein (1939) give the general procedures employed commercially which are described below.

Invertase is prepared from bakers' or brewers' yeast or from special strains of *Saccharomyces cerevisiae* of exceptionally high invertase content. The yeast is cultivated at a controlled pH with continuous aeration and sucrose as the carbohydrate nutrient. After recovery by filtration, the yeast cake is subjected to a plasmolyzing process followed by autolysis. Chloroform, ethyl acetate, or toluene is usually used as a plasmolyzing agent. The following is an example: To 100 lb of yeast cake 10 lb of toluene is added and the mixture is allowed to liquefy for 3 h. Then 12 gal of water is added and the pH adjusted to 5.8. Autolysis is allowed to proceed for 24 h at 30°C. After completion of the autolysis, the pH is adjusted to 4.7, and the cell debris is separated from the autolysate by centrifugation or filtration. To obtain a product of high potency the autolysate is evaporated under vacuum at a temperature not exceeding 40°C to about one-fifth of its original volume. After enzyme assay, the product is standardized and stabilized by adding glycerol and water to give the desired enzyme potency and a final glycerol concetration of 60 percent. A somewhat purer product is obtained by precipitating the concentrated autolysate with 2 volumes of cold ethyl alcohol or isopropyl alcohol. The precipitate is separated by centrifugation or filtration. It is then mixed with sufficient 60% glycerol solution to dissolve the enzyme and give the desired potency, and is filtered to give the final product.

LITERATURE CITED

American Association of Cereal Chemists (1962): "Cereal Laboratory Methods," 7th ed., AACC Method 22-01, St. Paul, Minn.
American Association of Textile Chemists and Colorists (1962): "Technical Manual, vol. XXXVIII, p. B-132, Howes, New York.
Arima, K. (1964): Microbial Enzyme Production, in M. P. Starr (ed.), "Global Impacts of Applied Microbiology," pp. 227–294, Wiley, New York.
Beckhorn, E. J. (1960): Production of Industrial Enzymes, *Wallerstein Lab. Commun.*, **23**:201–212.
———,M. D. Labbee, and L. A. Underkofler (1965): Production and Use of Microbial Enzymes for Food Processing, *J. Agric. Food Chem.*, **13**:30–34.
Davies, R. (1963): Microbial Extracellular Enzymes, Their Uses and Some Factors Affecting Their Formation, in C. Rainbow and A. H. Rose (eds.), "Biochemistry of Industrial Microorganisms," pp. 68–150, Academic, New York.
Hoogerheide, J. C. (1954): Microbial Enzymes Other Than Fungal Amylases, in L. A. Underkofler and R. J. Hickey (eds.), "Industrial Fermentations," vol. 2, pp. 122–154, Chemical Publishing, New York.
International Union of Biochemistry (1965). "Enzyme Nomenclature. Recommendations. 1964 of International Union of Biochemistry," Elsevier, Amsterdam.
Jeffreys, G. A. (1948): Mold Enzymes Produced by Continuous Tray Method, *Food Ind.*, **20**:688–690, 825, 826.
Phaff, H. J. (1959): The Production of Certain Extracellular Enzymes by Microorganisms, in W. Ruhland (ed.), "Handbuch der Pflanzenphysiologie," vol. 11, pp. 76–116, Springer-Verlag, Berlin.
Reed, G. (1975): "Enzymes in Food Processing," 2d ed., Academic, New York.
Shellenberger, J. A. (1947): Commercial Production of Enzymes, *Chem. Eng.*, **54**(2):130–131.
Underkofler, L. A. (1954): Fungal Amylolytic Enzymes, in L. A. Underkofler and R. J. Hickey (eds.), "Industrial Fermentations," vol. 2, pp. 97–121, Chemical Publishing, New York.
———(1966a): Manufacture and Uses of Industrial Microbial Enzymes, *Chem. Eng. Prog. Symp. Ser.*, **62**(69):11–20.
———(1966b): Production of Commercial Enzymes, in G. Reed (ed.), "Enzymes in Food Processing," pp. 197–220, Academic, New York.
———(1968): "Enzymes," in T. E. Furia (ed.), "Handbook of Food Additives," pp. 51–105, Chemical Rubber Co., Cleveland.
———G. M. Severson, K. J. Goering, and L. M. Christensen (1947): Commerical Production and Use of Mold Bran, *Cereal Chem.*, **24**:1–22.
Wallerstein, L. (1939): Enzyme Preparations from Microorganisms. Commercial Production and Industrial Applications, *Ind. Eng. Chem.*, **31**:1218–1224.

8
ALCOHOLIC BEVERAGES AND FERMENTED FOODS

Gerhard J. Haas

Fermented foods were known to prehistoric peoples. Cheese (made from soured milk), mead, beer, and wine, all based on fermentations, and the baking of bread with yeast are mentioned in the earliest records, such as the writings of the Babylonians, Syrians, and early Egyptians.

For the purposes of this chapter, we shall define the term *fermented foods* as all foods (solids and beverages) which are obtained with the use of microorganisms. In the strict microbiological sense of the word, fermentation is anaerobic glycolysis, and the oxidative pathway is called respiration. Most microbial metabolic pathways used in food production are anaerobic and thus strictly fermentative; but there are exceptions such as the oxidation of ethanol to acetic acid in vinegar production.

Use of fermentation in food production was prompted by several reasons, the main objectives being (1) physiologic effect of alcohol, (2) improved stability (preservation), (3) acidulated flavor (such as in fermented milk), and (4) change in texture (rising of bread and curdling of cheese).

Only certain microorganisms yield food that is wholesome. Growth of many others makes the food either unpleasant or harmful [such as *Bacillus cereus* (food poisoning), *Staphylococcus aureus* (staph poisoning), or *Aspergillus flavus* (producer of aflatoxin, a carcinogen)]. The organisms used most commonly in foods are the yeasts

Table 8-1 Yeast Fermentations

Food	Organism
Beer	*Saccharomyces carlsbergensis* and *S. cerevisiae*
Wine, hard cider	*Saccharomyces cerevisiae* var. *ellipsoideus*
Hard liquor	*Saccharomyces cerevisiae*
Mead from honey	*Saccharomyces cerevisiae*
Bread	*Saccharomyces cerevisiae*
Saké from rice	*Saccharomyces sake* and *Aspergillus*
Shoyu (soy sauce)	*Aspergillus oryzae*, *Saccharomyces*
Kumiss from mares' milk (Russia)	Yeast, *Streptococcus lactis*, *Lactobacillus bulgaricus*
Kefyr from cows' milk (Turkey)	*Saccharomyces cerevisiae*, *Bacillus caucasi*
Taette from cows' milk (Sweden)	*Streptococcus lactis* (special rope-forming strain of saccharomyces)
Miso from soy and rice	*Saccharomyces rouxii*, *Aspergillus oryzae*
Kvass from rye (Russia)	*Saccharomyces*

of the genus *Saccharomyces*. The reason is the wide distribution of these organisms on fruits and grains, leading to the practice of fermenting steeped grain (prehistoric beer), mead (originally from honey of wild bees), and wine, using in each case the particular adhering flora.

The particular organisms involved and the respective fermented foods are shown in Tables 8-1 and 8-2. No attempt at completeness has been made in these tables; a rapid glance will show the reader that really very few genera are involved: *Saccharomyces*, *Aspergillus*, and various lactic organisms with few exceptions affect the fermentation.

However, the types of foods obtained by fermentation cover a wide range; alcoholic beverages, dairy products, vegetables, and high-carbohydrate foods are the most frequent. No attempt will be made in this chapter to describe in detail all the fermented foods. A large part of it will deal with brewing, as the brewing process incorporates many facets of the other processes. Although the oldest and one of the most important economically, brewing is one of the processes about which consumers know little. Other sections of this chapter will deal with wine, distilled beverages, and other fermented foods, but in a more abbreviated fashion.

BREWING

HISTORICAL DEVELOPMENT OF THE BREWING PROCESS

Among all the fermented foods the oldest are probably beer, bread, mead, and wine. Although there are differences of opinion about the country of origin, most

Table 8-2 Nonyeast Fermentations

Food	Organism
Vinegar	*Acetobacter*
Culture milk, butter, sour cream	*Streptococcus lactis, S. cremoris, Leuconostoc citrovorum*
Cheese	*Streptococcus lactis, L. citrovorum*, and other molds and bacteria, depending on variety of cheese
Yogurt	*Lactobacillus bulgaricus, Streptococcus thermophilus*
Acidophilus milk	*Lactobacillus acidophilus*
Sausages	*Leuconostoc citrovorum*
Pickles	*Lactobacillus plantarum*
Sauerkraut	*Leuconostoc mesenteroides, Lactobacillus plantarum*, and others
Tempeh	*Rhizopus oryzae, R. oligosporous*
Tofu	*Rhizopus, Aspergillus*

anthropologists believe that the Sumeric empire, the Babylonians, and the Egyptians all had a grain-based fermented beverage. This can be traced back to approximately 4000 B.C. Barley, millet, and other cereals were used; these were formed into loaves, baked, and then broken up, suspended in water, and fermented. Already in Babylonian and Egyptian times, several different kinds of beer existed, some sweet and some acidulated. Beer was known in ancient Israel and to the Greeks and Romans, although it was often considered inferior to wine; it was known also to the Arabs and to the Celtic and Germanic tribes.

All the ancient beers, however, were brewed without hops, and beer without hops would not taste like beer as we know it today. Many different spices and herbs were used in beer, particularly in Germany. Hops may have been known in ancient Egypt, but its first definite mention was documented in Bavaria in 736. Hops seem to have been used occasionally in those times; a document by the Holy Hildegard, an abbess in a South German convent, in 1079, on the use of hops in beer seems to have popularized and broadened the custom; from the fourteenth and fifteenth century on, hops were used for all beer brewed in Germany. In England, the widespread use of hops was adopted even later, about the sixteenth and seventeenth century.

Beer, as we know it today, is made of malt, as well as other sources of starch such as rice and corn; water, hops, and, of course, yeast as the fermentation agent constitute the rest of the basic brewing materials. The purpose of the brewing process is the conversion of the starch contributed by the malt and the adjuncts corn or rice to maltose and dextrins. This conversion is accomplished by the enzymes present in the malt. The solution of carbohydrates obtained (called *wort*) is then boiled with hops, cooled, and fermented to beer containing alcohol, carbon dioxide, and residual dextrins. The finished beer has as major constituents water, dextrins, alcohol, and carbon dioxide and, as minor though important constituents, unfermented sugars, proteins, and aromatics such as hop resins, fusel oil, etc.

MALT AND MALTING

The actual process of malting barley is carried out in the malthouse. Many breweries, particularly in Europe, have their own malthouses, whereas much of the American malt is produced by independent malting companies often located strategically near the barley growing areas. Malthouses not only make brewers' malt but also supply distillers' and bakers' requirements.

Malt contains the enzymes which degrade the starch of the malt itself and additional starchy adjuncts, such as corn starch and rice starch, to dextrins and maltose. The dextrins remain to give body to the beer, and the maltose is fermented to alcohol and carbon dioxide. The malt is also the source of the beer proteins which are important for foam and for much of the flavor typical of the beer. The malting procedure is outlined in Fig. 8-1.

Only certain types of barley are suitable for malting and brewing; these must contain the right proportion of constituents and also be viable, so that germination will take place during malting.

The terms two- and six-row barleys connote the arrangement of the kernels in the "ear" of barley. In six-row barley the kernels are arranged in six rows around the head. In two-row barley only two of the six are fertile, giving rise to two rows of developed kernels instead of six. These two different types of barley are often used together, depending on the total brewing formula; addition of adjuncts calls for a somewhat higher protein barley (with a greater enzyme content), the six-row barley. Six-row is the major barley raw material for most American beers. The two-row barley has less protein. All European beers are brewed from two-row barley malts alone, and this type is also used as a minor constituent in some American beers. In the particular wort and beer the protein level must be sufficient to supply enough yeast and nutrients for good foam but not so high as to give rise to protein hazes and bready taste upon pasteurization.

For processing into malt, the barley is cleaned and soaked in water so that the grain can take up the moisture needed for germination (sprouting). The water is then drained off and sprouting continued for 5 to 7 days, until the desired growth of the embryo has been obtained. During this time, the malt has to be turned and is thus aerated frequently to prevent local overheating and achieve uniformly controlled growth. At the end of the period necessary for optimum growth (measured by the length of the growth of the germ or acrospire and the hardness of the endosperm), the sprouts are killed by kilning the malt. Duration and temperature of this kilning step are very important for color, aroma, and level of enzymes. After kilning, the sprouts and rootlets are removed and the malt aged for several weeks. It is then ready for use in the brewery. The times and temperatures given in the flow diagram (Fig. 8-1) are average figures.

During malting, certain enzymes are increased and others are newly formed: β-Amylase is increased, and α-amylase is newly formed. α-Amylase is important for attack on the starch (liquefaction), and β-amylase for final sugar formation; α-amylase can degrade only straight chains (amylose) but not branch-chain amylodextrins; β-amylase attacks both straight and branch chains. A sort of tandem action occurs between the two enzymes, making possible the complete degradation of starch. This conversion starts in the malting process, increasing the proportion of lower carbohydrates, and it is even more important for production of the wort during mashing.

Other enzymes are also produced in malting, particularly proteases, which

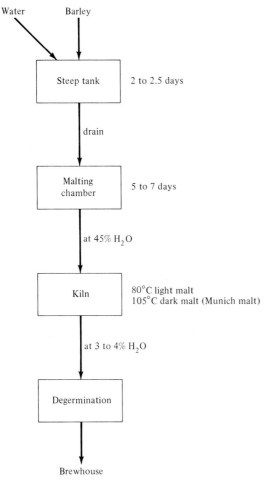

Figure 8-1 Malting procedure.

increase the soluble proteins; cytases, which degrade some of the pentosan gums; and phytases, which liberate phosphate and inositol (which is a growth requirement for many yeast strains). During kilning, browning reactions take place and melanoidins are formed which are important for flavor and color of the beer.

An average malt has the following composition, calculated on dry weight:

Starch	59%
Sugars	10%
Gums	10%
Cellulose	5%
Protein	10%
Fat	2.5%
Ash	2%

For dark beer, malts are kilned at higher temperatures; such malts are Munich malt, roasted malt, and caramel malt.

OUTLINE OF THE BREWING PROCESS

The malt is transferred to the brewery for grinding and use. Other raw materials for brewing include adjuncts (rice or corn) as additional sources of starch and hops (*Humulus lupulus*). The most prominent hop-growing regions in the United States are California, Idaho, Oregon, and Washington. In the brewery the flowers of the hops are boiled with the wort (many people mistake them for leaves because of their green color). The hop resins are contained in small yellow aggregates below the petals. They are a major flavor contributor, being responsible for the characteristic bitterness in beer and some of the aromatics. These resins have preservative properties, as will be more fully discussed. Hops also contain hop oil; the aromatics of hop oil probably also have some importance for beer flavor.

The other raw material of importance, often not thought of as such, is the water. Water should have the right hardness and calcium content for optimum enzyme activity and must not have off-flavors. Over and above this, the mineral content of the water influences the final beer flavor.

The brewing process for lager beer, as usually carried out in the United States, is outlined in Fig. 8-2.

The ground malt is mixed and gradually heated with water to let the amylases convert the starch in the malt to dextrins and maltose and the proteases convert the proteins to more soluble and simpler molecules. The brewer uses a mash diagram—holding the mash for certain periods of time at selected temperatures—to assure optimum conversion. Meanwhile, the adjuncts have been separately heated with a smaller portion of the malt and then finally boiled in the cooker to achieve gelatinization. The pregelatinized adjuncts are added to the main bulk of the brew in the mash tub where they go through the final stages of conversion until no free starch remains and the desired ratio of dextrin and maltose is obtained. The total mash, consisting of adjuncts and malt mash, including the barley husks, is then heated to about 72°C, when enzyme activity ceases, and is then passed through the lauter tub or mash filter where the so-called spent grain solids, the insolubles, mainly the barley husks, are removed by respectively siphoning off the wort or by filtration. The wort is then boiled with hops in the brew kettle for approximately 1½ h. This solubilizes the valuable hop constituents, while undesirable proteins and tannins are coagulated, forming the so-called break. At the same time the wort is sterilized.

From the brew kettle the wort passes via the hop strainer, where the insoluble part of the hops (about 70 percent of the original hop dry weight) is removed, to the hot wort settling tank. Here, some of the insolubles (the so-called trub) separate and are removed. The wort leaves the brewhouse via the wort coolers and is moved to the cellar, usually to the starting tank. In the starting tank there is further settling out of insolubles (cold break).

The brew is now moved to the fermentors. The yeast is added, and the wort is fermented to beer. Fermentation takes place at a low temperature (6 to 12°C) for a period of 8 to 10 days. At the end of the fermentation the yeast settles and

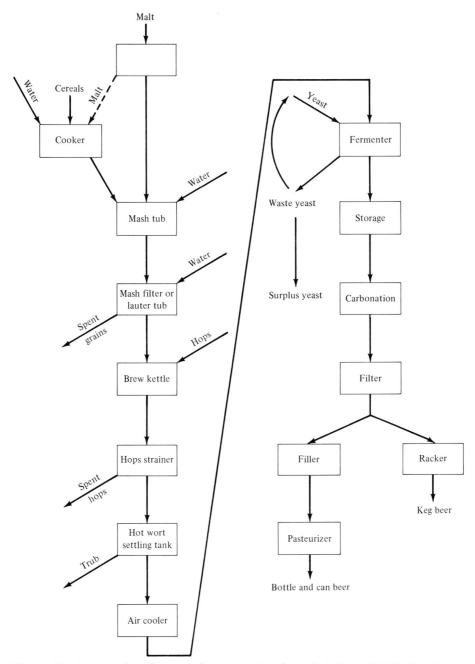

Figure 8-2 Brewery flow diagram. [By permission from G. J. Haas (1960), Microbial Control Methods in the Brewery, *Appl. Microbiol.*, **2**:115.]

the beer is removed and stored in tanks for further clarification, aging, and carbonation. The methods for carbonation vary. In many beers carbonation is carried out by returning, under pressure, the carbon dioxide produced and collected during the fermentation. Another system is the Kräusen process where the beer is fermented a second time with some fresh wort in closed tanks under pressure, thus saturating it with CO_2. In Europe fermentation is not completed in the fermenting cellar but in an "after fermentation" under pressure in closed tanks and the beer carbonated in this way. After adequate aging, the carbonated beer is filtered and then put into kegs, bottles, or cans.

Keg beer is sold unpasteurized, whereas bottle or canned beer is either pasteurized in the package or sterile filtered prior to filling. This is necessary to preclude further yeast growth, producing precipitates, or contamination by certain lactic organisms, causing precipitates or off-odors. Adequate pasteurization is assured by heating the beer for a least 5 min at 60°C. Many brewers use a longer heating time to have an adequate safety margin.

The final American lager beer (average, 1968) contains:

Alcohol	3.8%
Dextrins	4.3%
Protein	0.3%
Ash	0.3%
CO_2	0.4%

It also contains an appreciable amount of vitamins, particularly riboflavin, but is low in thiamine, because the yeast completely scavenges that vitamin. In addition, there are the aroma substances originating from hops, malt, and the yeast fermentation which are important for the taste. Much has been learned about beer flavor substances through the application of gas chromatography. Some of the minor constituents found in beer are enumerated in Table 8-3.

A term which is often applied to beer is *original gravity*. This represents the percentage of solids in the wort before the yeast was added and the wort fermented. It can be obtained by calculation from the analytical data for alcohol and total solids (called *extract* by the brewer) in the beer. The original gravity of the average American beer is 12 percent and about 16 percent for bock beers.

An important quality of beer is its head, or foam, which should cling to the side of the glass and not disappear too rapidly. The higher-molecular-weight constituents, particularly the protein, malt gums, and hop resins, together with the carbonation, are responsible for the degree of foam formation. The amount of foam a consumer likes varies widely with individual preference. For those who enjoy a good head, having clean and fat-free glasses is important.

Gushing is a fault of beer (and other carbonated beverages) when the carbon dioxide is insufficiently well bound and there are nuclei for carbon dioxide release. Tendency to gush is much reduced when the bottles and cans are stored in an upright position.

BIOCHEMISTRY OF BREWING

Alcohol and carbon dioxide are produced from maltose by the overall formula

$$C_{12}H_{22}O_{11} \cdot H_2O \longrightarrow 4C_2H_6O + 4CO_2$$

Table 8-3 Some minor compounds found in beer

Alcohols	Organic acids	Sulfur compounds	Esters
Methanol	Acetic acid	Hydrogen sulfide	Ethyl acetate
n-Propanol	Formic acid	3-Methyl-2-butene-1-triol and/or its mercaptans	sec-Butyl acetate
Isopropanol	Propionic acid		Isobutyl acetate
n-Butanol	Pyruvic acid		Isoamyl acetate
sec-Butanol	Succinic acid		Ethyl lactate
Isobutanol	Lactic acid		Isoamyl lactate
2-Methylbutanol	α-Ketoglutaric acid		Ethyl caproate
3-Methylbutanol	Malic acid		
Phenylethyl alcohol	Citric acid		
Glycerol	Tartaric acid		
	Oxalic acid		
	Caproic acid		
	Caprylic acid		

Carbonyl compounds	Hydrocarbons	Amino acids	Others	Sugars
Acetaldehyde	Myrcene	Proline	Indole	Maltose
Butyraldehyde	Farnesene	Alanine	Furfural	Glucose
Acetone	Caryophyllene	Tyrosine		Fructose
Methylnonylketone	Isocaryophyllene	Valine		Galactose
Diacetyl		Arginine		Arabinose
Formaldehyde		Leucine		Xylose
Propionaldehyde		Aspartic acid		Ribose
		Histidine		Maltotriose

The first step is the formation of two glucose units from maltose. Glucose is then fermented by the Embden-Meyerhof scheme, which is outlined in Fig. 8-3 very generally without naming all the enzymes and side reactions. Small amounts of glycerol are always formed by the side reaction shown. Fusel oil consisting of amyl, isoamyl, and isobutyl alcohols and their esters are formed from isoleucine, leucine, and valine; beer has to be processed for low fusel oil content, as excess would bring about possible lower palatability and even headaches. Certain other alcohols, aldehydes, and acids may also be formed, depending on the strain of yeast, the constituents of the medium to be fermented, and the temperature of fermentation. A very annoying constituent elaborated by certain yeasts is diacetyl ($CH_3COCOCH_3$). It has a butter aroma and flavor which is most undesirable in beers. Beer has to be brewed to prevent accumulation of undesirable levels of this agent.

MICROBIOLOGY OF BREWING

The yeast is one of the most important determinants of beer quality. Together with the raw materials, it determines pH, flavor, and final quality of the beverage. The vigor of the fermentation and yeast determines not only the flavor substances but also the amount of oligosaccharides that are not fermented. As

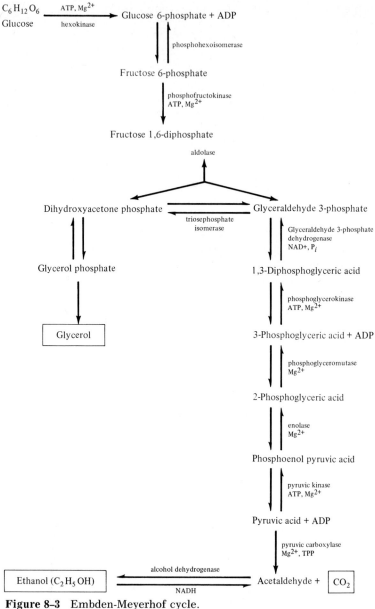

Figure 8-3 Embden-Meyerhof cycle.

already mentioned, the yeast may be responsible also for side reactions. If the fermentation is vigorous, spoilage microorganisms will be suppressed.

The yeast used to produce lager beer is a bottom fermenting, nonsporulating strain of *Saccharomyces carlsbergensis*, after the Carlsberg laboratories in Copenhagen, Denmark, where this strain was first grown in pure culture (Fig.

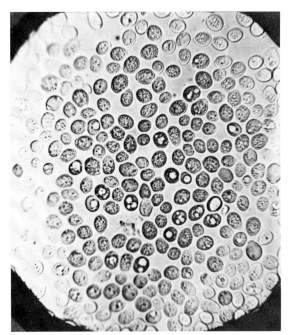

Figure 8-4 *Saccharomyces carlsbergensis.* (Courtesy of L. Saletan, Wallerstein Company, Division of Baxter Laboratories.)

8-4); it ferments melibiose to give acid and gas whereas *S. cerevisiae* does not.

Because the yeast is constantly reused in the brewery as the inoculum for subsequent batches, control of its quality is very important. Any mutation or degradation is bound to be magnified by repeated cycles of use. The same is true of contaminants which may outgrow the yeast.

The yeast is tested routinely for its purity and freedom from other organisms, by microscopy and by culturing on media which suppress yeast growth, so that one can determine small numbers of contaminants in the presence of large numbers of yeast. One of the media used most frequently is cyclohexamide (Actidione) containing agar. Cyclohexamide (Actidione) at low levels inhibits culture yeast without affecting bacteria and some "wild" (foreign) yeasts.

Before a discussion of the types of contaminants present, the quality of beer as a microbiological medium should be considered. Boiling in the kettle produces a virtually sterile medium into which the yeast is inoculated. Only very few organisms can thrive in beer for the following reasons: (1) The yeast rapidly uses up the oxygen and produces carbon dioxide. This restricts the potential contaminants to anaerobes. (2) Hops have a strong antimicrobial effect on most gram-positive organisms. The preservative value of hops is mainly due to the alpha resins or humulones. It is really a family of three compounds differing only by the side chain (Fig. 8-5). (3) Beer has an acid pH (3.7 to 4.5). (4) The alcohol has a slight antibacterial effect. (5) Beer is held at low temperatures. Because of these factors, only very few bacteria can grow in beer.

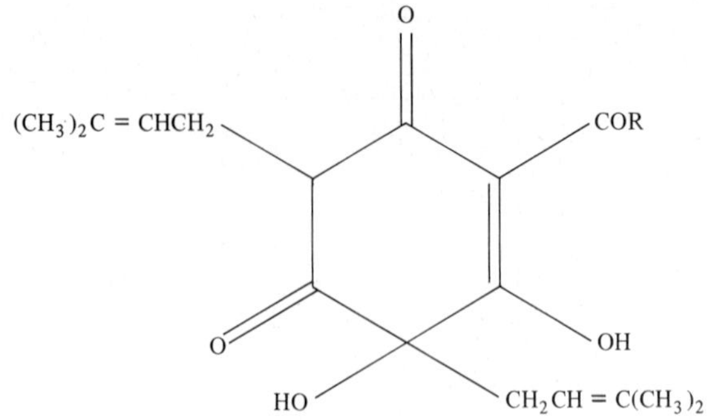

Cohumulone: R=CH(CH$_3$)$_2$
Humulone: R=CH$_2$CH(CH$_3$)$_2$
Adhumulone: R=CH(CH$_3$)CH$_2$CH$_3$

Figure 8-5 Structural form of the humulones. [By permission from G. J. Haas (1960), Microbial Control Methods in the Brewery, *Appl. Microbiol.*, 2:115.]

The most important potential contaminants are *Lactobacillus pastorianus* and *Pedicoccus cerevisiae*, which is considered a strain of *Leuconostoc citrovorum*. Also certain nonculture yeasts called "wild yeasts" can occur. Some beers at the early stage of fermentation, before the medium becomes anaerobic, may be contaminated by a gram-negative rod *Flavobacterium proteus*; this contaminant leaves an unpleasant taste in the beer, even though no live organisms remain. Another type which can contaminate the wort but not the beer is *Aerobacter aerogenes*, the so-called "termobacteria." No human pathogens have been found in beer; most die in a short time. *Escherichia coli* when added to beer at 200 cells per milliliter survive for 3 days.

Measures taken to fight contamination in the brewery include (1) good cleaning and disinfection throughout the brewery; (2) selection of the yeast from a batch which has shown the least degree of contamination, when reinoculating the next batch (repitching), and taking the yeast from the center of the strata which have settled out; and (3) introduction of fresh yeast, when this is necessary.

How often fresh yeast is introduced depends on the practice of the particular brewery. In some cases, large batches are grown in the so-called pure culture apparatus and introduced at approximately 3-week intervals. Other breweries grow batches large enough for introduction from liquid cultures maintained and continuously transferred in the laboratory. Some introduce this yeast once a year or even less frequently. Special techniques for pure culture development are employed by most breweries; such procedures include isolation of suitable single cells for propagation, using droplet culture and sterile transfer between Pasteur flasks (with sidearm).

Yeast is tested for its flocculant properties. Lager yeast should ferment vigorously in suspension during the early stages of fermentation and then settle down to form a heavy slurry. It must ferment at low temperatures, and the fermentation must be vigorous and go to completion without leaving an unfermented residue of mono- or disaccharides. Only a minimum of dead cells (usually determined by staining with methylene blue) may be present, and the yeast should exhibit the proper cytological appearance, slightly off round to ellipsoid but not too oblong. Other related factors which are sometimes evaluated are nutritional requirements and generation time. Foam properties, diacetyl content, taste, and pH of the beer may reflect conditions of the yeast.

RECENT TRENDS

During recent years, there have been several innovations. One is the use of liquid adjuncts. This is practiced quite widely, particularly where brewhouse space is critical. These liquid adjuncts are usually syrups of various degree of degradation (dextrose equivalent). They may be added in the mash tub or cooker. Another development is the use of hop extracts instead of hops. These are usually hexane extracts of hops containing the resins. Their use saves handling costs but is not very widespread yet.[1] Continuous processing practices have been tried: mashing as well as fermentation. Although this looked good in the small scale, problems occurred when scaled up to plant production, and there are only a few, if any, installations in commercial use employing such methods. The recent trend is away from in-container pasteurization to bulk sterilization and sterile filtration. Sterile filtered beer is often promoted as keg beer in the bottle because keg beer is not pasteurized.

BEER TYPES

Lager beer, for which the process is described here, is obtained by fermentation with a bottom fermenting, nonsporulating yeast (*S. carlsbergensis*) at a low temperature. It is usually lightly hopped and almost completely end fermented. Bock beer is a lager beer brewed from a wort of higher original gravity. To brew it, more malt and hops and fewer or no adjuncts are used, and usually bock beer is a dark beer. It is aged longer and has a higher alcohol content (5.5 percent or so by weight versus 3.8 percent). The origin of bock beer probably was in the north German town of Einbock. It became traditional during the Lenten season when other pleasures were not permitted.

Ale, porter, and stout all use top fermenting strains of *Saccharomyces cerevisiae*. They have a higher hopping rate and a higher gravity and are fermented at considerably higher temperatures. Some of the American ales, however, do not completely follow this description but are much closer to lager beer. Audit ales are ales brewed from very high-gravity wort, aged several years, and high in alcohol. They are popular in English colleges.

Pilsener type is lager beer made according to the beer prevailing in Pilsen, Czechoslovakia, light in color, dry (low in fermentable sugar), and with good hop aroma.

Malt liquor is usually somewhat higher in alcohol than regular beer.

"No carbohydrate beers" are beers brewed from wort, where all the dextrin has been hydrolyzed by enzymes (usually by the addition of fungal enzymes) to maltose and glucose, which are entirely fermentable by the yeast.

Beer consumption per capita was considerable in 1968. Total sales in the United States were approximately 110 million barrels (117 1 to the barrel) a year, amounting to a consumption of approximately 16 gal/year per person.

---

[1]Use of hop extract or preboiled (preisomerized) extract has increased recently.

WINE

PRODUCTION OF REGULAR WINE

Wine's history goes back just as far as that of beer. Wine making is a much simpler process, starting from fermentable sugars, i.e., fructose and glucose, instead of starch. The inoculum, the natural microflora from the skins, adheres to the grapes and, essentially, all that is necessary is to crush the grapes. The yeast is *Saccharomyces cerevisiae* var. *ellipsoideus* (Fig. 8-6). Its name, which is derived from its shape, distinguishes it from brewers' yeast; it cannot ferment melibiose, whereas brewers' yeast can. In spite of the convenience of this natural yeast inoculum, most American wines today are fermented by the addition of pure culture yeast. This assures rapid onset of vigorous yeast fermentation and uniformity of taste.

Only certain varieties of grapes make good wine, and only some grape-growing countries have an important wine industry. In the United States the most important wine-growing state is California, with New York State a distant second. In Europe, France, Germany, Spain, and Italy make famous wines; Greek, Portuguese, Hungarian, and Swiss wines are also widely appreciated. Chile and South Africa have important wine industries.

Of the many varieties of wine, if classified according to color, the two most basic types are white and red wine. There are also specialties such as rosé, sweet wines, champagne, and various wines made from fruits other than grapes.

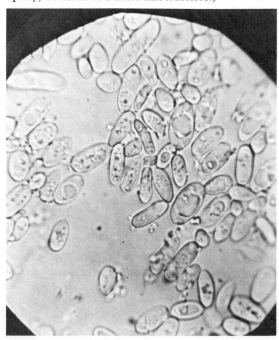

Figure 8-6 *Saccharomyces cerevisiae* var. *ellipsoideus*. (Courtesy of L. Saletan, Wallerstein Company, Division of Baxter Laboratories.)

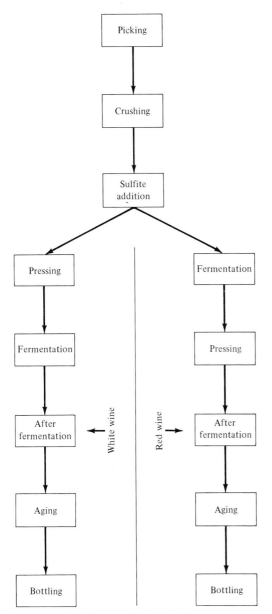

Figure 8-7 Wine-making flow diagram.

Within these groups there are a great number of varieties and types. The production of wines is outlined in Fig. 8-7.

The grapes are picked at the optimum degree of ripeness. This means that they must have not only optimum flavor but also the right sugar content and acidity. Grapes used for different varieties have a different composition.

The grapes are crushed carefully so as not to break the pits and stems which contain bitter components (tannins). Before filtration and fermentation, sulfite is added (approximately 100 ppm) for preservation. Most of the potential spoilage organisms are sensitive to sulfite whereas the wine yeast is not. In addition to its preservative power, sulfite is also an important antioxidant. The antimicrobial activity depends on the pH; the more free sulfurous acid is present, the greater is this activity. It also depends on other constituents, particularly aldehydes, with which the sulfurous acid combines.

For white wine the juice (also called "must") is removed immediately from the pomace by pressing. Composition of the must is approximately as follows:

Carbohydrates (glucose and fructose)	15–25%
Acids (mainly tartaric and malic)	0.9–1.5%
Tannins (for white wine)	0.02–0.04%
Tannins (for red wine)	0.1–0.25%
Pectins	0.12–0.15%
Ash	0.3–0.5%
Aroma substances	Trace
Nitrogenous substances	Trace
Higher alcohols	Trace
Aldehydes	Trace

The grape juice is now ready for fermentation.

For red wine, blue grapes are fermented with the skins, stems, and pits (this extracts the red color) and are only subsequently pressed. As an alternative, the juice may be heated in the presence of the skins to 110°F for 8 to 16 h, then cooled, pressed, and fermented. When neither method is employed, the resulting wine is not red but faint pink or rose-colored; this wine is marketed as "rosé." Red wine has a considerably higher tannin content than white or rosé wine, owing to the extraction of the tannin material from the skins.

The yeast starter is pregrown in sterile grape juice; approximately 1 to 3 percent of this starter is added as inoculum, and the fermentation sets in rapidly. The starting temperature is 12 to 15°C for white wine, and during fermentation the temperature is not allowed to exceed 23°C. Considerable heat is liberated during fermentation, and the amount of cooling necessary to control the temperature depends on the size of the batch and the type of equipment. A cool, slow fermentation favors a more aromatic, better wine. Red wine is fermented at a somewhat higher temperature than white wine. The yeast uses up the oxygen and to help maintain an anaerobic environment the containers are kept full by adding similar wine. Air is harmful as it will favor spoilage organisms, particularly acetic acid bacteria; also, it can give rise to discoloration and browning. Fermentation takes 5 to 8 days. At the end of this period, the sugar has been almost completely fermented, and yeast, some tannins, proteins, pectins, and tartrates have settled out.

The wine is moved from these precipitates and transferred to the storage cellars for completion of the "after" fermentation, clarification, and stabilization. Practices vary widely, but in general they encompass a constant check of the sulfite level and replenishment if necessary, the addition of certain clarification agents such as bentonite or activated carbon, and an aging and blending of the wine. Finally, the wines are filtered. Some types are pasteurized or sterile filtered. Wines should age at least 8 to 10 months before consumption, and flavor continues to improve over several years, especially for red wines. After 5 to 100 years most wines deteriorate.

Wine aging is really a science or may better be called an art of its own. Wine is unique in that aging continues in the bottle. The oxidation process continues slowly and takes the flavor and aroma through various stages. After a red wine has gone through its fermentation, it is harsh and almost undrinkable; then the color changes from purple to ruby red and in a mature wine to a tawny amber. The wine becomes mellow and well rounded and eventually hits its peak, where it plateaus for a while and then starts to deteriorate. Sometimes a red wine may still be good for close to 100 years. White wines do not last as long, rarely as long as 15 years, and some varieties and vintage years only 5 years or so.

For many wines which are high in acid, such as dry wines, certain micrococci (malolacticus, acidovorax) will reduce the acidity by fermenting the malic acid and producing lactic acid, which is a much weaker acid. This change (called malic acid fermentation) is desirable and is utilized by the winemaker (particularly for Bordeaux and Swiss wines).

Finished wines vary widely in their composition. Average table wines may contain:

Extract (Total solids)	2–3%
Carbohydrates	0.03–0.5%
Acids	0.5–1%
Ash	0.15–0.3%
Tannins	Trace
Amino acids	Trace
Aroma substances	Trace
Alcohol	6–9 wt %
	8–13 vol %

SWEET TABLE WINES (DESSERT WINES)

The high sugar and alcohol content of sweet table wines is achieved by various techniques:

1. Musts with higher sugar content are used, and the fermentation is checked by cooling before it has gone to completion. This procedure is used for sauterne.
2. Dry table wines may be blended with grape juice (must) to obtain sweet clarets.
3. The addition of brandy and sugar to wine stops fermentation and, at the same time, adds alcohol and sugar. This approach is employed to make port.
4. Grapes which have dried on the vines or even raisins are used as starting substrates.

Sherry is made by entirely different processes in Spain and in California. For Spanish sherry, grape juice from the palomino variety picked at full ripeness is treated with gypsum; this, by precipitating insoluble calcium tartrate, decreases the pH (by liberating sulfate ions) and increases the acidity and tartness.

The grape juice undergoes a double fermentation, first by the regular wine yeast and then by a special film-forming yeast, the so-called flor (flower) yeast, an alcohol-tolerant strain of *Saccharomyces cerevisiae* var. *ellipsoideus*, sometimes accompanied by strains of *Pichia, Hansenula,* and *Torulopsis*. Between the two fermentations sherry is aged about a year, and then is aged further during the second fermentation while it is under the flor yeast cover. This aging is carried out by having a number of barrels in series, drawing a portion off from the longest-aged barrel, and filling each barrel from the preceding one back to the one containing the newest wine. The sherry is then fortified with brandy to the desired alcohol content, clarified with clay, and blended with sherry from other barrels for uniformity of character.

In California, for most sherries the must is fermented, fortified, aged, clarified, and then baked at 120 to 140°F for up to 12 weeks, aged for a further short period (6 months to 1 year), then filtered and bottled.

Alcohol and extract data for some sweet wines are given as follows:

	Alcohol, g/100 ml	Extract, %
Sherry	14–17	3–7
Port	15–17	9–15
Sauterne	10–12	4–12
Samos	11–15	18–23
Tokay	9–13	7–15

Vermouth is a fortified wine to which an infusion of herbs has been added. The most important is the vermouth herb *Artemisia absinthium*. Further, thyme, coriander, cinnamon, nutmeg, cloves, and cinchona may be added. Italian vermouth is essentially white wine plus grape concentrate plus brandy plus herbs or herb extract plus caramel syrup; it has 12 to 19% sugar and 12 to 14 g/100 ml alcohol. French vermouth contains about 4% sugar and 15 g/100 ml alcohol.

CHAMPAGNE

To make sparkling wine or champagne, wine is fermented a second time in heavy glass bottles, to reach the desired CO_2 pressure (4 to 5.5 atm). After addition of sugar and yeast to the wine, the second fermentation is started at 15 to 18°C and continued at 8 to 10°C for several months. To remove the yeast the bottles are shaken at about 4°C for 2 weeks and then gradually inverted. The yeast collects in the neck of the bottle. The material forming the sediment and part of the liquid in the neck is frozen, the bottles put right side up and opened. The pressure propels (disgorges) the sediment from the bottle. The empty space created is refilled to taste, often with a solution of sugar in wine and/or brandy. The bottles are closed again and stored for several more months for improved quality, or they may be sold directly.

A tank procedure has been developed where this second fermentation is carried out in bulk; the cooled champagne is filtered under counterpressure and then filled at very low temperature into individual bottles.

WINE FROM NONGRAPE SOURCES

Preparation of wine is not confined to grapes; wines are prepared from a large number of fruits: hard cider (from apples), elderberry wine, blackberry wine, cherry wine, pear wine (poire), etc.

Mead made from honey goes back as far in antiquity as beer and wine from grapes. Mead was particularly popular among the ancient Germanic tribes. It is still produced occasionally in Scandinavia and Eastern Europe.

Saké (rice wine) is popular in Japan. It is made from polished rice, and the starch is saccharified by aspergilli through their amylases. The yeast used to ferment the mash after acidification by either lactic acid or lactic organisms is *Saccharomyces saké*, a variant of *S. cerevisiae*, but more acid-tolerant. Saké has 16.5% alcohol and 4.1% extract. It often is consumed hot.

Pulque from agaves is brewed in Mexico and parts of Latin America.

VINEGAR

As mentioned repeatedly in the discussion of both beer and wine making, the exclusion of oxygen (air) is important for these beverages. One of the major reasons is the potential conversion of alcohol to acetic acid by species of acetobacter. This spoils the alcoholic beverages, but these organisms are used for the production of vinegar.

The acetic acid bacteria (acetobacter) are gram-negative, motile rods occurring singly or in pairs and are obligate aerobes. They are film formers, forming a film on top of the tank. The biochemical reaction by which they form acetic acid from ethanol is as follows:

$$2CH_3CH_2OH + O_2 \longrightarrow 2CH_3CHO + 2H_2O$$
$$2CH_3CHO + O_2 \longrightarrow 2CH_3COOH$$

Some of the acetobacter species do not stop with acid production but continue the oxidation to carbon dioxide.

$$CH_3COOH + 2O_2 \longrightarrow 2CO_2 + 2H_2O$$

Thus, selection of proper organisms is important for vinegar fermentation. These must not be slime formers and should carry the reaction as far to completion as possible without destroying the acetic acid by oxidation. In addition they must be tolerant to ethanol. *Acetobacter xylinoides*, *A. orleanse*, and *A. acetigenum* are some of the organisms used; temperatures in commercial vinegar production range from 26 to 31°C.

Depending on the raw material, vinegars are differentiated as wine vinegar, cider vinegar, malt vinegar, and others. When the process is not started from an alcoholic beverage, the sugars are first fermented to alcohol by yeast, and the alcohol is then converted to acetic acid. The alcohol concentration should be between 10 and 12 percent. The basic methods of vinegar production are known as the slow process or Orleans method, the rapid, so-called generator process, and the more recent submerged fermentation in a so-called acetator. The latter development has also been worked out as a continuous process. With all methods, adequate aeration is, of course, of great importance for this oxidative process. The finished vinegar is aged in well-filled wooden barrels and subsequently filtered, mostly with the aid of diatomaceous earth. For preservation, the vinegar is usually pasteurized or sterile filtered.

In addition to the problems presented by acetic acid organisms which oxidize acetic acid and the slime formers, acetic acid manufacture may be plagued by vinegar eels which are nematodes. Pasteurization or sterile filtration will eliminate them.

Vinegar contains about 4 to 5% acetic acid, and the resulting product is diluted to this value if indicated. The other constituents depend on the source material from which the vinegar has been prepared.

Vinegar fermentation is not a competitive process for the production of acetic acid. Synthesis from ethylene or oxidation of synthetic ethanol is a more economical commercial method for the manufacture of this important acid.

DISTILLED ALCOHOLIC BEVERAGES

Yeast action is limited by the amount of alcohol present, and at about the level of 18 percent by volume its action ceases. For this reason, simple fermentation cannot yield alcohol contents exceeding 18 percent. For higher levels, to produce the so-called hard liquor, distillation is required. As alcohol has a boiling point of 78.5°C, the alcohol-water azeotrope (constant boiling mixture) a boiling point of 78.3°C, and water a boiling point of 100°C, the lower boiling fraction obtained by distillation is greatly enriched in alcohol. With an efficient distillation column, a 96 percent concentration, which is far higher than the 40 to 50 percent of most distilled beverages, can be obtained. In the distilling industry the alcoholic content is usually given in American proof figures, which are twice the alcoholic content, so that a beverage of 40% alcohol content is termed 80 proof.

Distilled alcoholic beverages may be divided into three major classes: (1) products starting from a starchy substrate and needing enzymes, usually in the form of barley malt, to convert the starch to fermentable sugars; (2) a type starting directly from a sugar substrate; depending on the kind of distillation equipment employed, a major or minor portion of the aromatics also are distilled and become part of the distillate; and (3) the type of liquor produced by adding flavor substances to quite pure ethanol which has been obtained by distillation and rectification. Starch-based distilled beverages include:

Name	Starch base
Scotch	All malt
Rye	Rye and malt
Bourbon	Corn, rye, and malt
Irish whiskey	Rye, wheat, and malt
Arrak	Rice

To make these beverages, about 15 percent of high-diastase malt is added to the other starch sources, water is added, and the combined mash is converted at 56°C under agitation. The mash is then heated briefly to 62°C after which it is cooled to 17 to 23°C and acidified with lactic or sulfuric acid to pH 4.7 to 5. As an alternative, the pH may be lowered by *Lactobacillus delbruckii* fermentation. The lower pH controls the contaminating organisms and favors the yeast. Sometimes ammonium fluoride is introduced to hold down contaminants, in which case an ammonium fluoride-resistant yeast is used. All distillery yeasts are top fermenting and should be alcohol-resistant and rapid fermenters. Fermentation of the acidified mash takes about 3 days. Temperature is controlled to remain below 32°C. A continuous method of fermentation is used by some distilleries.

At the end of fermentation the alcohol and aroma substances are distilled off. Different whiskeys are distilled to different final proofs and then diluted to the selected concentration. The type of distillation equipment and the final proof to which a whiskey is distilled determine very much the aromatic substances and the character of the final product. Bourbon, for instance, is usually distilled to give a 170 proof distillate, whereas with rye in a simple pot still a final 130 to 140 proof distillate is produced.

For Scotch whiskies, the typical smokey flavor is produced at least partially from malt which has been kilned at a high temperature over a peat fire. Scotch is aged usually in sherry or partially carbonized wooden casks. Most American whiskeys are stored in oak casks. Several of the typical flavor substances such as guaiacol are leached out from the wood. The Scotch Highland whiskies are produced in simple pot stills. The Lowland whiskeys are produced in patent stills, and the malts are less heavily smoked. Most commercial Scotches are blends between Highland and Lowland Scotches. Scotches are matured for at least 3 years, bourbon and rye usually 5 years or more. Whiskey has been known in monasteries in England since the eleventh century and became commercial in the sixteenth century.

Bourbon, named after Bourbon, Kentucky, is made from grain including at least 51% but often up to 80% corn; some rye is always added, and the rest is barley malt. Grains for rye whiskey are always at least 51% rye. Canadian whiskeys usually are mainly from rye. Much of the whiskey sold constitutes blends; this may involve blending not only with other whiskeys but also with more highly purified distilled grain spirits.

Arrak is a Far Eastern liquor in which rice starch is either converted by fungi (which produce amylases) or by the amylase of the rice. Depending on the country, molasses may also be added to the mash.

Distilled alcoholic beverages

Original substrate	Name
Grape	Cognac, Armagnac, brandy
Cherry	Kirschwasser
Plum	Slivovitz and sloe gin
Agave	Tequila
Apricots	Apricot
Sugar cane	Rum
Apple	Applejack, Calvados (France)
Coconut milk	Toddy
Raspberry	Framboise

There is a law in France that "Cognac" may be applied only to the brandy made from grapes grown in the small region of Cognac. Cognac and Armagnac differ by the region (in France) where the grapes were grown and by the equipment used for distillation. Brandies are produced in many countries: United States, Spain, South Africa, Germany, Portugal. Many of the liquors in the above table are aged extensively before they reach the consumer.

Liquors obtained by addition of flavor substances to either diluted alcohol distilled from converted liquors (from grain or potatoes) or to brandy or rum include among others:

Addition	Name
Liquors	
Caraway	Aquavit, Pernod, Kümmel
No addition	Vodka
Juniper berries	Gin
Liqueurs and cordials	
Mint and sugar	Creme de menthe
Cocoa beans, sugar, and vanilla	Creme de cocoa
Eggs and sugar	Eggnog
Cherry and sugar	Cherry brandy
Coffee and sugar	Coffee liquor
Orangepeel and sugar	Grand Marnier
Honey and whiskey	Drambuie
Herbs and sugar	Chartreuse, Benedictine

The second group in the above table belongs to the group of liqueurs or cordials. Several of them contain a variety of additives. For instance, cherry brandy may contain cherry juice, distilled liquor from cherries (kirschwasser), whiskey, or rum. Among the citrus liqueurs are Curacao, Triple Sec (Cointreau), and Grand Marnier. To Chartreuse numerous herbs are added, among them anise, angelica, and coriander; the main constituents are brandy and also some honey and sweet wine. Benedictine contains brandy and honey and is made with numerous herbs, among them mint, thyme, and angelica.

Sugar and alcohol contents also vary greatly among the different liqueurs, and formulations usually have an alcohol content of 30 to 60 percent and may have total extract up to 22 percent.

Everyone is familiar with the physiologic action of ethanol, ranging from stimulation at low level to inebriation and finally narcotic effect. Alcohol is absorbed rapidly through the mucous membranes, and 30 to 90 min after consumption the top blood level is reached. It is oxidized at the rate of 7 to 11 ml/h, probably in the liver. Alcohol is a vasodilator and a central nervous system depressant. It passes rapidly through the blood-brain barrier.

Ethyl alcohol has a caloric value of 6.93/g.

Alcohol is lighter than water, and percentages may be expressed by weight or volume. With beer, weight percentages are customary; with wine, either; with liquor, volume and proof percentages.

The wastes, distillers' grain, and distillers' solubles are used in the feed industry and have valuable nutrients.

OTHER FERMENTED FOODS

In addition to alcoholic beverages, fermentation is used for bread and baked goods, vegetables, dairy products, sausage, and various oriental starchy proteinaceous foods.

BREAD AND BAKED GOODS

For bakery products fermentation is applied for texture improvement, pH, and flavor benefits. Yeast (*Saccharomyces cerevisiae*) has been used since prehistoric times to raise dough with the aid of the carbon dioxide evolved; the pH is lowered and aromatic substances are produced, giving rise to bread of improved taste. For special types of sour ryes, fermentation with lactic organisms is used in addition to yeast fermentation. The yeast ferments the mono- and disaccharides present in the dough. Additional maltose is produced from action of the amylases of the grains on the dextrins which are present. The alcohol obtained during fermentation evaporates during baking. Baking powder can replace the yeast as far as loosening and raising of the dough are concerned, but a less aromatic product results.

VEGETABLE PRODUCTS

Fermentation (resulting in low pH) together with a high salt level (with resulting increase in osmotic pressure) is used for preservation of vegetable products. Examples are sauerkraut, pickles, and olives; various lactic organisms are involved in these fermentations.

Sauerkraut is fermented, after addition of 1.5 to 2.5% sodium chloride, by the naturally adhering flora of lactic organisms and yeasts. Fermentation takes 3 to 4 weeks at 18 to 24°C. The pH is lowered to about 4.2.

Pickles are made by numerous recipes, usually calling for 2 to 8% salt and, sometimes, spices. Cucumber is the most common pickle, but sometimes other vegetables and fruits may be made into pickles. The final pH varies but may be as low as 3.5. Olives have to undergo a pretreatment with caustic to remove oleuropin which has a very bad flavor and would make the olives inedible.

DAIRY PRODUCTS OTHER THAN CHEESE

Fermentation is used to a very large extent in the dairy area. Here the purpose is souring and the development of flavor substances. Cheese making is more complex and will be discussed separately.

Sour cream and buttermilk are fermented by *Streptococcus lactis* and/or *Streptococcus cremoris* together with *Leuconostoc citrovorum*. The ratio of these organisms is important, and *L. citrovorum* produces diacetyl ($CH_3COCOCH_3$) which has a buttery flavor. Because casein is least soluble at its isoelectric point of 4.6, the acid pH effects a curdling of the casein in the milk. Butter is produced from sour cream or from sweet cream. Yogurt is made from milk by a mixed culture of *Streptococcus thermophilus* and *Lactobacillus bulgaricus*. Fermentation takes place at 42 to 46°C. Yogurt has a characteristic flavor and texture. *Acidophilus* milk is fermented with *Lactobacillus acidophilus*.

Kefir and kumys are beverages of Asia Minor and are produced through fermentation with yeast and lactobacilli. Also required in kefir fermentation is a special organism (*Bacillus caucasi*) which grows in nuggets. Kumys is made from mare's milk. Both beverages foam and contain some carbon dioxide and ethanol. Taette of Scandinavia is a milk fermented with yeast and a rope-forming strain of *S. lactis*.

CHEESE

Cheese is another fermented food which goes back to prehistoric times and probably was discovered when someone first carried milk in a calf's stomach. The stomach of young animals contains rennet, which is an enzyme complex that curdles milk. Most cheeses are made by the combined curdling action of this enzyme and a low pH brought about by the action of lactic organisms.[1] The major constituent of the curd is the predominant protein of the milk, casein. In addition to casein, cheese contains most of the minerals and fat of milk. The types of cheese manufactured and its composition depends on the starting material, whether it is skim milk, whole milk, or a milk-cream mixture, and on the method of manufacture. The liquid phase from the precipitation of the curd is the whey, containing the lactose and those proteins and minerals of milk which are soluble at low pH.

Many microorganisms are involved in cheese manufacture. The souring is produced mainly by *S. lactis* and other lactobacilli. In some cheese varieties, curdling is carried out by acidity alone without rennet. Recently, microbial proteases have been tried as a replacement for rennet. It is not yet known whether these enzymes will be sufficiently effective to replace rennet, and whether a product of equal flavor would result.

After the curdling operation the curd is separated from the whey. The fresh curd is firm and rubbery. Cheese is ripened in various ways. Texture and flavor changes and most of the enzymatic changes (proteolysis, hydrolysis) take place during ripening. A number of bacteria and molds can be involved in these secondary processes. There are hard and soft cheese and cheeses which are ripened by the action of these organisms. The texture of the finished cheese is first of all determined by the handling of the curd (how it is collected, whether and how it is cut and compressed). Other factors important for the taste are the degree of salting, fermentation temperatures, duration of aging, and, of course, the type of microflora present. Table 8-4 gives some of the characteristics of the best-known cheeses with their respective ripening procedures.

The holes in Swiss and some other cheeses are produced mainly by species of proprionibacterium; CO_2 and hydrogen are liberated and produce the holes.

The composition of a typical Cheddar cheese is as follows:

Moisture	40%
Fat	30%
Protein	25%
Ash	1–6%

There are wide variations in fat and protein content between individual

---

[1] Microbial proteases causing similar curdling are sometimes substituted.

Table 8-4 Classification of cheese

Soft cheese		
Unripened	*Ripened by bacterial action*	*Ripened by mold action*
Cottage cheese,	Limburger	Camembert
Cream cheese		Brie
Neufchatel		

Semihard cheese	
Ripened by molds	*Ripened by bacteria*
Gorgonzola	Brick
Roquefort	Muenster
Stilton	

Hard cheese ripened by bacteria	
With gas holes	*Without gas holes*
Swiss	Cheddar
Parmesan	Edam

varieties. For high-fat cheeses, cream is mixed into the curd; fat contents of 60 percent are reached and exceeded.

OTHER FERMENTED FOODS

Sausages

Some types of sausages are fermented with the organisms *Pediococcus cerevisiae* and *P. acidilactici*. A special preparation together with sugar is on the market under the names Accel and Lactacel. Important flavor contributions are ascribed to this fermentation for a variety of sausages called summer sausage.

Cocoa

Microorganisms play a role in flavor development of the cocoa bean during digestion of the outer "pulpa." Some of the metabolites pass into the inner bean (nib) and become important for aroma and taste. In coffee and tobacco, microbial fermentation, if used at all, is of lesser importance for flavor.

Oriental Fermented Foods

With canning less advanced and lack of preservation by cooling and freezing, many fermented foods have been developed in the tropical regions. A variety of substrates and microorganisms is used and appears in Table 8-5 originating from a paper by Dr. C. W. Hesseltine who has done a great deal of work in this area of fermented oriental foods.

Most of these foods employ a koji fermentation, which is a solid-phase mold fermentation. Usually, soybeans, rice, or wheat are involved and a fungus is grown on these cereals after moisture and minerals have been added. Some of the fermentations employed in oriental foods are extremely complex; Shoyu, for instance, employs four organisms, three cereals, five incubation steps, and

Table 8-5 Oriental food fermentations*

Name	Organisms used	Substrate	Nature of product	Area where article of commerce
Of fungal origin				
Tempeh	*Rhizopus*, (*R. oligosporus*)	Soybeans	Solid	Indonesia and vicinity
Sufu	Principally *Actinomucor elegans* and *Mucor* sp.	Soybeans	Solid	China, Taiwan
Ragi	*Mucor, Rhizopus*, and yeast	Rice	Solid	Indonesia, China
Tea fungus	Two yeasts and *Acetobacter* sp.	Tea extract and sucrose	Liquid	Eastern Europe, Russia, parts of the Orient
Miso	*Aspergillus oryzae, Saccharomyces rouxii*	Rice and other cereals plus soybeans	Paste	Japan, China, and some other parts of the Orient
Shoyu (soy sauce)	*Aspergillus oryzae, Lactobacillus, Hansenula*, and *Saccharomyces*	Soybeans wheat, rice	Liquid	China, Japan, Philippines, and some other parts of the Orient
Ang-kak	*Monascus purpureus*	Rice	Deep red pigmented solid	China, Indonesia, and Philippines
Of bacterial origin				
Natto	*Bacillus subtilis*	Soybeans	Solid	Japan
Nata	*Acetobacter* sp.	Various fruit	Gel	Philippines

*Reproduced by permission of Dr. C. W. Hesseltine, Northern Regional Research Laboratories.

lengthy aging. Shoyu is also called soy sauce and is imported into this country and used to flavor rice and meats.

Single-Cell Proteins

An attempt has been made to use microorganisms as direct food or a food supplement in order to supply additional protein for protein-poor diets. Brewers' yeast and torula [*Candida utilis* (syn: *Torulopsis utilis*)] have been tried to a limited extent for many years. Both yeasts suffer from deficiencies in taste, digestibility, and too high a purine concentration. Recent experiments with algae of various kinds show that these organisms have the same deficiencies; the use of algae has been considered for feeding space travelers as part of a general system to utilize all wastes in a spacecraft. A recent development has been the use of petroleum as a substrate for yeast and other microorganisms. Many petroleum companies are working in this area (for a more detailed discussion on this subject, see Chapter 13).

Most publications on microbial foods fail to mention the common edible mushroom *Agaricus campestris* var. *bisporus*. This, of course, belongs to the fungi and can be grown in its mycelial form in deep tank fermentation. In many countries other edible fungi are also collected and are articles of commerce.

GENERAL REFERENCES

Bergstrom, G. (1968): "Principles of Food Science," 2 vols. Macmillan, New York.
Cook, A. H. (ed.) (1962): "Barley and Malt," Academic, New York.
Cruess, W. V. (1947): "The Principles and Practice of Wine Making," Avi Publishing Co., Westport, Conn.
Haas, G. J. (1960): Microbial Control Methods in the Brewery, *Appl. Microbiol.*, **2**:113–162.
Hesseltine, C. W. (1965): Ammillenium of Fungi Food and Fermentation, *Mycologia*, **57**:149–197.
Lampert, L. (1947): "Milk and Dairy Products," Chemical Publishing, New York.
Roman, W. (ed.) (1957): "Yeasts," Dr. W. Junk Publishers, The Hague.
Schormueller, J. (1961): "Lehrbuch der Lebensmittelchemie," Springer-Verlag, Berlin.
——— (ed.) (1968): "Handbuch der Lebensmittelchemie," vol. VII, Alkoholische Genussmittel, Springer-Verlag, Berlin.
Underkofler, L. A., and R. J. Hickey (eds.) (1954): "Industrial Fermentations," Chemical Publishing, New York.

9
THE ROLE OF MICROBIAL GENETICS IN INDUSTRIAL MICROBIOLOGY

Richard P. Elander and Marlin A. Espenshade

Microbial genetics, more than any other branch of modern biological science, is the central theme that underlies the biochemistry of development and product synthesis by industrially important microorganisms. Therefore, in order to understand the important role that microbial genetics plays in the industrial laboratory, one must be familiar with the fundamentals and recent concepts of molecular biology. The primary aim of this chapter is to review the fundamentals of molecular and microbial genetics and to apply these concepts to representative fermentation processes.

The chapter is divided into three major sections. The first will review those fundamentals of molecular biology with special reference to genetic regulation and control; the second will describe the genetic recombinational mechanisms and life cycles of important representative industrial microorganisms; and the last will apply the concepts and methodology of molecular and microbial genetics to existing industrial processes.

CONCEPTS OF MOLECULAR BIOLOGY

The subject matter of molecular genetics revolves about two basic concepts relative to the transmission of hereditary information. One of these ideas concerns the detailed fine structure of the gene elucidated by Watson and Crick; the other is an elaborate view of gene function theorized by Monod, Jacob, and Wollman.

Although it has been reasonably clear since the pioneering work of Avery and Brachet that the genetic material contains deoxyribonucleic acid (DNA), Watson and Crick were the first to propose a detailed chemical structure which was physicochemically possible and biologically meaningful. Their proposed structure provided a scheme for the preservation, transmission, and utilization of genetic information. The collective work of Monod and his collaborators, Kornberg, Ochoa, Spiegelman, Nirenberg, Matthaei, and Khorana provided the central dogma of molecular biology. The central theme of this dogma states that the genes are composed of protein and information-containing DNA of defined, consistent, but mutable structure. These genes can be replicated by enzymatic copying to give new structures for daughter cells containing identical genetic information to that in the parent. This information, which specifies the genotype, can also be used to direct the construction within a given cell of specific proteins whose presence in the cell defines the characteristic structure and function of the cell, the phenotype. The *translation* of the information in the nucleic acids of the DNA gene is mediated by transcribing agents of ribonucleic acid (RNA). The agents are known as messenger (mRNA), transfer (tRNA), and ribosomal (rRNA). Each gene gives rise to a specific protein [the "one-gene (cistron), one-enzyme (polypeptide)" hypothesis of Beadle and Tatum] and, moreover, the production of these specific proteins is controlled according to the needs of the cell. A brief outline of the interrelations and functions of the nucleic acids is given in Fig. 9-1.

Figure 9-1 Interrelations and functions of nucleic acids.

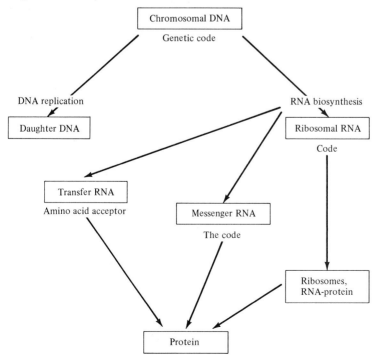

DEOXYRIBONUCLEIC ACID (DNA)

DNA consists of a linear structure of deoxyribonucleic and phosphoric acid linked by phosphodiester bonds, upon which are attached repeating pairs of nitrogenous bases. There are four chief nitrogenous bases in DNA: adenine, thymine, cytosine, and guanine. The first two bases are purines and the second

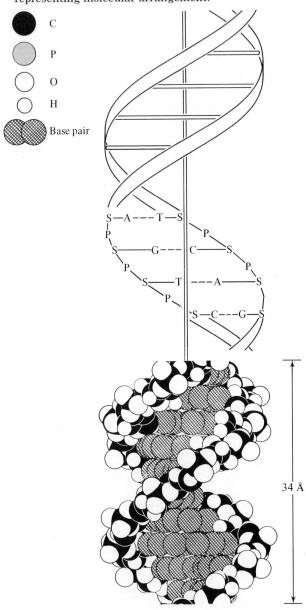

Figure 9-2 Helix of DNA with three different ways of representing molecular arrangement.

two are pyrimidines. The Watson-Crick model for DNA (Fig. 9-2) consists of two polynucleotide chains that run in opposite directions and are wrapped as a helix around each other, with the sugar-phosphate chain on the outside and the purine and pyrimidine base pairs (Fig. 9-3) on the inside. The purine and pyrimidine bases are joined to form specific pairs. Adenine is attached to thymine and guanine with cytosine. The base-to-base linkages are hydrogen-bonded and, in turn, act as horizontal supports for the axis of the double helix and hold the two chains together. The chains are antiparallel and complementary so that the arrangement of bases on one strand fixes the arrangement of the other.

The total length and molecular weight for most forms of DNA are still uncertain. One of the smallest known chromosomes is the single-stranded DNA of bacteriophage X174 which consists of approximately 5500 nucleotides. The DNA of T2 phage has a molecular weight of 1.6×10^8 and is composed of a double helix of approximately 200,000 nucleotide pairs. Autoradiographs of the *Escherichia coli* chromosome indicate that it consists of a double-stranded closed circular thread approximately 400 μm in length with a molecular weight of 1 billion.

RIBONUCLEIC ACID (RNA)

Four functionally distinct RNA's are known as transfer (tRNA), messenger (mRNA), ribosomal (rRNA), and viral RNA. All forms of RNA have in common the sugar ribose and uracil along with adenine, cytosine, and guanine. Transfer RNA has been reported to contain ribothymine. A partial chemical structure for RNA is illustrated in Fig. 9-4.

Transfer RNA has also been called soluble and acceptor RNA. Its extreme

Figure 9-3 Pairing of adenine with thymine and of guanine with cytosine by means of hydrogen bonding as in DNA. *(By permission from T. P. Bennett and E. Frieden (1967), "Modern Topics in Biochemistry," Macmillan, New York.)*

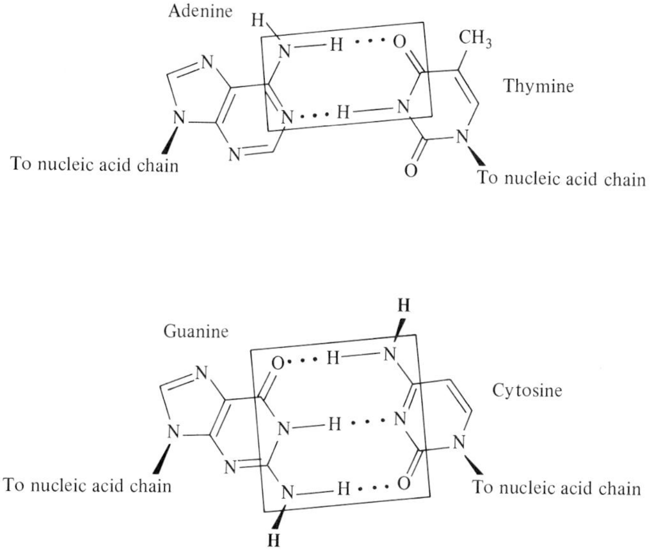

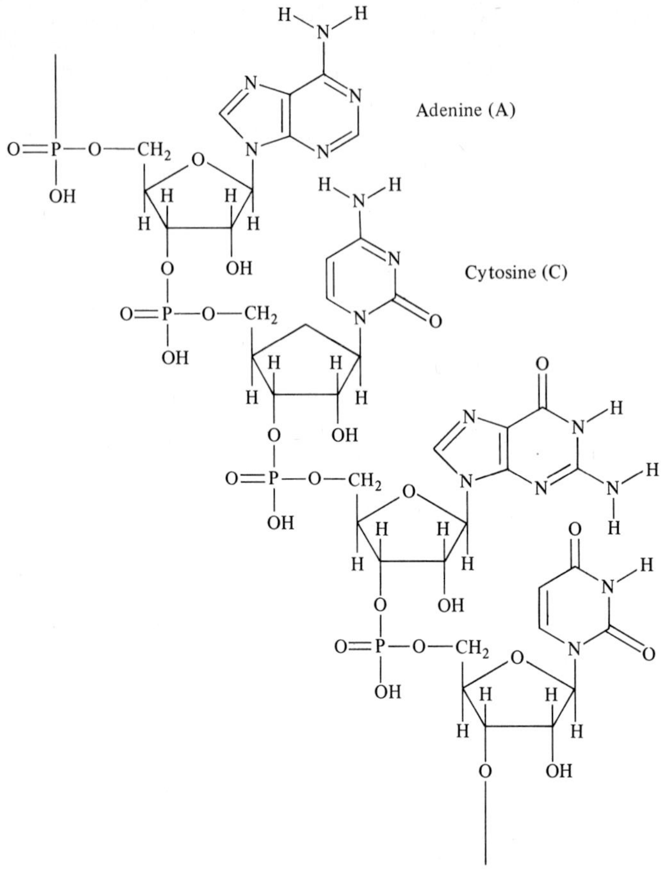

Figure 9-4 Partial chemical structure of ribonucleic acid. *(By permission from V. R. Potter, "Nucleic Acid Outlines," Burgess, Minneapolis.)*

solubility makes it unique when compared with viral and ribosomal RNA. Acceptor and transfer RNA refer to the amino acid–carrying role this species of RNA plays in protein synthesis. Unfractionated preparations of tRNA consist of a mixture of different molecular species with distinct base sequences, each capable of combining with only one of the 20 naturally occurring amino acids. All transfer RNA's have in common a terminal cytidyl-cytidyl-adenosine (-C-C-A) group. The terminal adenosine has a ribose unit with free hydroxyl groups to which an activated amino acid (acyl–amino acid) is attached (see Fig. 9-8). In recent years the nucleotide sequence has been determined for a number of tRNA's.

Template or messenger RNA (mRNA) has a primary structure complementary to a portion of a single DNA strand, i.e., a particular region is copied during the synthesis of RNA. The RNA produced has complementary bases to the DNA copied (U for A, G for C, etc.). The size of the mRNA molecule shows considerable variation and depends on the number and size of the protein it codes.

High-molecular-weight or ribosomal RNA refers to the ribonucleic acid not accounted for as tRNA or mRNA. Ribosomal RNA (rRNA) is single-stranded and contains four major (as in mRNA and tRNA) and several minor bases in a proportion that varies considerably. Two sizes of rRNA are found in all ribosomes, where they are an integral component and cannot be removed without destruction of the ribosome structure. The smaller rRNA molecule (40S), which is found in the smaller ribosomal subunit, has a molecular weight of 500,000, whereas the larger (70S) rRNA, a component of the larger ribosomal unit, has a molecular weight of 1 million.

Viral RNA includes the ribonucleic acid found in plant, animal, and bacterial viruses. In fact, RNA mycophages have been recently isolated from high-penicillin-yielding strains of *Penicillium chrysogenum*. The RNA virus-like particles also account for the interferon-inducing properties of statalon, a ribonucleic acid particle in *P. stoloniferum*. All the viral RNA's are high-molecular-weight molecules. Tobacco mosaic virus (TMV) RNA is a single strand wound into a helix. The chain contains approximately 6400 nucleotides and when fully stretched would attain a length of 5 μm. The nucleic acid is intermeshed in a sheath of protein. Together, the protein and nucleic acid form a highly organized rod-shaped particle.

BIOLOGICAL ROLES OF DNA AND RNA

DNA is an informational molecule and has two major functions. First, it provides directions for self-duplication (replication) during mitosis so that its inherent information is transmitted to progeny cells, and second, it expresses its encoded information for function and control of the complex metabolic pathways of the cell.

The central role of DNA in the propagation of the species and in the determination of inheritable characteristics has been elucidated from a number of converging experimental observations. Cytochemical studies indicate that the major portion of the DNA of a cell is located in its chromosomes. The DNA content of diploid cells is constant for tissue in a species, and haploid cells contain approximately one-half as much DNA. Mutagenic agents such as ultraviolet radiation are strongly absorbed by DNA and are more reactive with nucleic acids than with other macromolecules.

The various cellular forms of RNA appear to play important roles in the translation of genetic information inherent in DNA into protein structure. Transfer RNA serves as an adapter molecule in that its function is to adapt an amino acid to the polynucleotide template of messenger RNA. A specific enzyme attaches an amino acid to a tRNA molecule. Each molecule of tRNA contains an *anticodon* in the form of a short sequence of bases, which enables it to recognize and pair with a sequence of complementary bases, termed a *codon*, in mRNA. The tRNA base pairs with mRNA, thus bringing reactive amino acids to the template. By a series of complex reactions the amino acids lined up on the template polymerize to form polypeptides and protein molecules. Messenger or template RNA has been assigned a major role in the synthesis of proteins. The mRNA is synthesized on a DNA template in the nucleus and later becomes associated with cytoplasmic ribosomal particles where it serves as a template for the positioning of amino acyl–tRNA molecules prior to polypeptide formation. Ribosomal RNA occurs throughout the cytoplasmic membranes of a cell and is concentrated in polysomes and ribosomes. Since it is a structural component of the ribosome, a role in protein synthesis is suggested.

BIOSYNTHESIS OF MACROMOLECULES

The formation of specific and unique biological metabolites (organic acids, antibiotics, and other secondary metabolites) is accomplished by joining a number of dissimilar units through a uniform linking mechanism termed *patternization*.

All cellular macromolecules have particular surface and structural patterns which give proteins, nucleic acids, and polysaccharides their individual nature. The basis of these features is a fixed sequential arrangement of constituent subunits. For example, the biological and chemical properties of the simple tripeptide precursor molecule of benzylpenicillin is known as γ-(α-aminoadipyl)-cysteinylvaline. Such simple molecular structures, as well as those of complex macromolecules, are formed with only occasional variance.

DNA is the primary determinant of the pattern for all biosynthetic processes and their products. The DNA of the nucleus appears to be capable of determining the pattern of all new DNA as well as RNA synthesized intracellularly. The pattern written in mRNA (the code) is translated into the primary amino acid sequence of proteins.

Two general mechanisms employed by the microbial cell for patternization may be referred to as assembly-line mechanism and template mechanism. The assembly-line mechanism is important in the synthesis of small molecules and in determining larger molecular organizations such as cell walls, etc. (Fig. 9-5). The template process for regulating patternization plays an important role in the synthesis of nucleic acids and proteins.

Figure 9-5 Patternization control of macromolecular synthesis. *(By permission from T. P. Bennett and E. Frieden (1967), "Modern Topics in Biochemistry," Macmillan, New York.)*

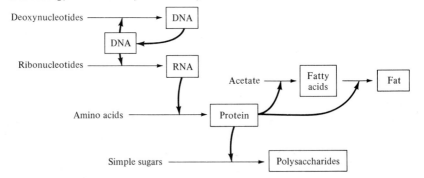

SYNTHESIS OF MACROMOLECULES

The biosynthetic relationships between DNA, RNA, and protein are outlined in Fig. 9-6. The scheme shows the flow of genetic information from DNA terminating in a new DNA molecule (replication), or from DNA to RNA (transcription), or from RNA to DNA (reverse transcription), culminating in the production of a new protein (translation). The three major groups of reactions summarized in Fig. 9-6 are replication, transcription, and translation.

In replication, free deoxyribonucleotides are assembled linearly to form an identical sequence of the original DNA structure for hereditary transmission. The first process in the overall transfer of information from DNA to protein is transcription, as the four-base-unit language (T, A, C, G) is transcribed into a similar four-symbol unit system of the various RNA's (U, A, G, C), the rRNA, mRNA, and tRNA which are complementary to the particular regions on DNA. Although the various forms of RNA are important constituents of the final translation system, it is the tRNA that carries the information specifying amino acid sequence in the final protein.

In the translation process, the coded information coming originally from DNA, contained in mRNA, programs the ribosomes for synthesis of protein. The

Figure 9-6 Biosynthetic relationship between DNA, RNA, and protein. *(By permission from T. P. Bennett and E. Frieden (1967), "Modern Topics in Biochemistry," Macmillan, New York.)*

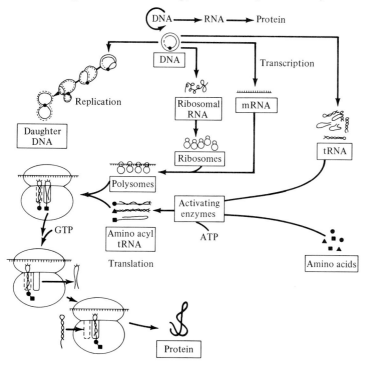

step from mRNA to protein is a complex of reactions, all involved in the 20-unit amino acid language of the amino acid sequence of proteins. The synthesis of other macromolecules is directly dependent on enzymes with specific configuration and thus indirectly related to cellular DNA.

REPLICATION OF DNA

The mechanism of DNA biosynthesis was originally worked out by Kornberg and may be summarized as follows:

$$\begin{array}{c} dATP \\ d^+CTP \\ d^+GTP \\ d^+TTP \end{array} \xrightarrow{\text{DNA template}} \begin{array}{c} DNA + P-O \sim P \\ \downarrow \\ 2P_i \end{array}$$

An enzyme called DNA polymerase catalyzes the reaction in which deoxyribonucleotides are linked together to form de novo DNA. Deoxyribonucleoside triphosphates are precursors for synthesis, and pyrophosphate is eliminated during the reaction. The characteristics of the DNA product are determined by the DNA template, and the sequential arrangement of the bases in the product is identical to that in the parent template DNA. Details of the DNA biosynthetic sequence is shown in Fig. 9-7.

The reaction between a deoxyribonucleoside triphosphate and partially completed DNA is illustrated in the preceding paragraph. The group activation, elongation, and patternization have been resolved for DNA synthesis; there is much to be learned about the replication process.

TRANSCRIPTION AND RNA SYNTHESIS

The transcription of the four-nucleotide language of DNA into the similar four-symbol system of RNA is the first process from DNA to protein. The DNA-directed RNA biosynthesis has been elucidated by Ochoa and can be summarized as follows:

$$\begin{array}{c} ATP \\ G^+TP \\ C^+TP + \\ U^+TP \end{array} \begin{array}{c} DNA \text{ template} \\ + \\ Mg^{2+} \end{array} \xrightarrow{\text{[RNA polymerase]}} \begin{array}{c} RNA + DNA \text{ template} \\ + \\ P \sim O-P \\ \downarrow \\ 2P_i \end{array}$$

The four nucleoside triphosphates, Mg^{2+}, DNA template, and RNA polymerase are needed for the synthesis. Either single- or double-stranded DNA can act as a template.

The patternization can also be determined by RNA. Cells infected with some RNA viruses produce a replicase which requires an RNA template for activity. This mechanism enables RNA in RNA-containing viruses to function as genetic material. Other RNA viruses have a reverse transcriptase enzyme which makes a DNA strand complementary to the RNA strand. Here the complementa-

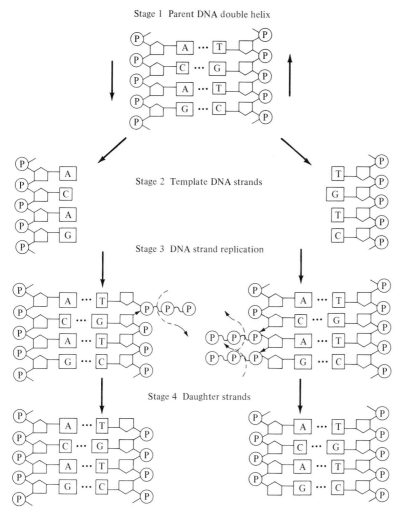

Figure 9-7 DNA replication. *(By permission from T. P. Bennett and E. Frieden (1967), "Modern Topics in Biochemistry," Macmillan, New York.)*

ry sequence device is used to pass genetic information from RNA to DNA, the opposite of the usual situation. The viruses which exhibit the transfer of genetic information from RNA to DNA are responsible for leukemia and other cancers in animals.

TRANSLATION AND PROTEIN SYNTHESIS

The translation process begins with the enzyme-mediated activation of amino acid, utilizing ATP (adenosine triphosphate) energy. The activating enzymes then attach the amino acid to tRNA to form amino acyl–tRNA which becomes associated with the ribosome and mRNA. The three nucleotide bases (anticodon) in the tRNA base pair with the nucleotide triplet (codon) of mRNA. Polypeptide chain initiation and growth occur by a series of complex reactions until the end product protein is synthesized (Fig. 9-8). Clusters of ribosomes known as polysomes are the active sites of protein synthesis. Structural analysis of the ribosomes (70S) shows they are composed of two subunits known as eosomes

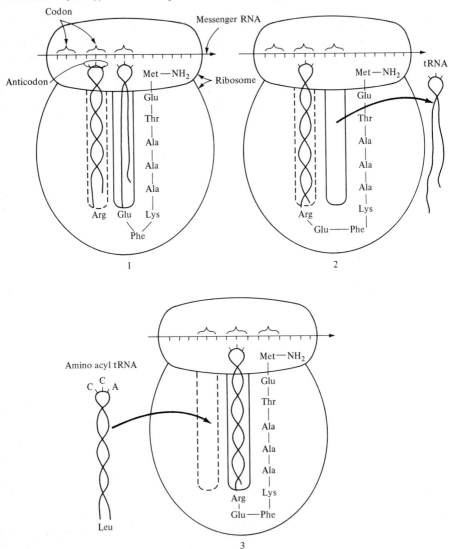

Figure 9-8 Hypothetical representation of the addition of an amino acid to the carboxyl end of a growing polypeptide chain. *(By permission from T. P. Bennett and E. Frieden (1967), "Modern Topics in Biochemistry," Macmillan, New York.)*

(30S) and neosomes (50S). There are particular binding sites on each subunit for the amino acid adapter (tRNA), genetic information (mRNA), and the growing polypeptide chain. Amino acyl–tRNA and the extending protein chain then are attached to the neosome and mRNA binds to the eosome. Ribosomes bring the components together, and here the components involved in the translational process function to synthesize protein, provided the proper tRNA, mRNA, and enzymes are present. Jacob and Monod hypothesized that messenger RNA is short-lived and that this form of RNA conveys the genetic information to the ribosome. The mRNA is formed on a DNA template and is a complementary copy of a portion of a single DNA strand. The mRNA attaches to the ribosome and links them together to form polysomes with single mRNA's functioning simultaneously on several ribosomes.

The patternization of protein molecules has been the subject of considerable speculation about how DNA determines the amino acid sequence in protein. The messenger RNA hypothesis of Jacob and Monod stimulated further speculation as well as considerable experimental work and the researches of Nirenberg and Ochoa. The patternization mechanisms in protein synthesis are well understood. In 1961, Nirenberg and Matthaei showed that synthetic polynucleotides can serve as artificial messengers. A number of synthetic mRNA's have now been synthesized by Khorana and his associates. Nirenberg and Matthaei found that polyuridylic acid (poly U) stimulated phenylalanine synthesis. This discovery, along with the genetic evidence provided by Crick, suggested a code based on triplets. Nirenberg and Matthaei postulated UUU as the triplet codon in mRNA that specified phenylalanine with the ultimate DNA codon being polyadenylic acid (AAA).

Further studies by Nirenberg and Ochoa utilized copolynucleotides to establish a codon map for all 20 amino acids. The list of mRNA codons in Fig. 9-9 reveals that the code is redundant or degenerate since a number of codons may specify the same amino acid. Alanine, leucine, and arginine have three codons. The code also appears to be ambiguous in that UUU codes for both leucine and phenylalanine. Despite the degeneracies and ambiguities noted, the code is universal and is basically the same for all organisms. A polypeptide similar to rabbit hemoglobin can be synthesized in a cell-free system derived from rabbit reticulocytes and *E. coli* cells. The DNA from polyoma and vaccinia viruses (animal viruses) is infective in competent cells of the bacterium *Bacillus subtilis*. The results support the hypothesis that there is one basic genetic code for polypeptide synthesis in present-day organisms.

GENETIC CONTROL MECHANISMS

In order for a cell to maintain metabolic balance, it must contain control mechanisms which regulate its synthetic pathways. Since a particular metabolite is often the result of a series of enzymatic reactions, its synthesis is affected if any of the preceding steps are affected. In many instances, the product itself inhibits the functioning of certain of the earlier enzymes in the pathway. Thus, when the product reaches a certain concentration within a cell, it prevents its further synthesis. This *end-product* (allosteric) inhibition of an enzyme provides immediate and sensitive control of the rate of synthesis of many metabolites.

		Second base				
		U	C	A	G	
First base	U	UUU ⎫ Phe UUC ⎭ UUA ⎫ Leu UUG ⎭	UCU ⎫ UCC ⎬ Ser UCA ⎪ UCG ⎭	UAU ⎫ Tyr UAC ⎭ UAA non UAG non	UGU ⎫ Cys UGC ⎭ UGA ? UGG Try	U C A G
	C	CUU ⎫ CUC ⎬ Leu CUA ⎪ CUG ⎭	CCU ⎫ CCC ⎬ Pro CCA ⎪ CCG ⎭	CAU ⎫ His CAC ⎭ CAA ⎫ Gln CAG ⎭	CGU ⎫ CGC ⎬ Arg CGA ⎪ CGG ⎭	U C A G
	A	AUU ⎫ AUC ⎬ Ileu AUA ⎭ AUG Met	ACU ⎫ ACC ⎬ Thr ACA ⎪ ACG ⎭	AAU ⎫ Asn AAC ⎭ AAA ⎫ Lys AAG ⎭	AGU ⎫ Ser AGC ⎭ AGA ⎫ Arg AGG ⎭	U C A G
	G	GUU ⎫ GUC ⎬ Val GUA ⎪ GUG ⎭	GCU ⎫ GCC ⎬ Ala GCA ⎪ GCG ⎭	GAU ⎫ Asp GAC ⎭ GAA ⎫ Glu GAG ⎭	GGU ⎫ GGC ⎬ Gly GGA ⎪ GGG ⎭	U C A G

Figure 9-9 In vitro mRNA codons for amino acids. (non represents possible nonsense or special function codons. Ambiguities are not indicated.) (By permission from I. H. Herskowitz (1967), "Basic Principles of Molecular Genetics," Little, Brown, Boston.)

Enzyme inhibition by the end product is one example of a *feedback mechanism*.

Another type of control involves regulation of enzyme synthesis rather than activity. The control is referred to as *repression* or *induction* of enzyme synthesis. Repression of enzyme synthesis is determined by the end product of a pathway but, unlike end-product inhibition, the synthesis, not the activity of all the enzymes of the pathway, is suppressed. Proteins that fluctuate in amounts as a function of changes in environment are called *inducible proteins* if their activity increases following the addition of a metabolite. They are called *repressible proteins* if their activity decreases after the addition of a metabolite. If the amount of a particular protein increases after removal of a specific metabolite, the phenomenon is known as *derepression*. Proteins which do not fluctuate appreciably in amounts under a variety of growth conditions are termed *constitutive*.

The polypeptide products of gene action can be controlled more directly than by end-product inhibition, at a stage prior to protein synthesis. In the *lac* region of the *E. coli* map, three separate genes are operative, and the entire cluster of genes is referred to as an operon. Two structural genes control the

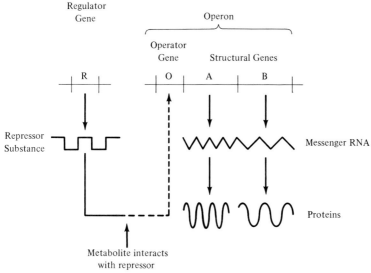

Figure 9-10 Relationships between genes regulator and operator and structural genes of the operon. *(By permission from I. H. Herskowitz (1967), "Basic Principles of Molecular Genetics," Little, Brown, Boston.)*

synthesis of two enzymes, β-galactosidase and galactoside permease. The permease is required for the uptake of lactose, and galactosidase is required for splitting the molecule into its component sugars. Both enzymes are inducible by a number of substrate and nonsubstrate inducers. In addition to the already mentioned structural genes, there is a class of regulatory genes which govern the production of a cytoplasmic substance that controls the activity of a specific structural gene. The repressor substance acts by combining with an operator gene (Fig. 9-10). The operator gene is responsible for the synthesis information contained in the group of genes whose function is repressed. The specificity of repressor action also implies an equally specific operator whose structure is genetically determined.

MUTATION

The genetic material of a microorganism does not exist by itself but, rather, exists in a constantly changing environment of living and nonliving entities. Hence, the genome and the information which it contains are subject to change. If the change is unusual and more or less permanent, this change is referred to as a *mutation* and the cell bearing the change is termed a *mutant*.

The chemical basis for mutation usually involves base changes in the nucleic acids deoxyribonucleic acid (DNA) or ribonucleic acid (RNA). Each of the bases of DNA and RNA can assume different arrangements which are termed *tautomers* (Fig. 9-11). The tautomers of uracil and adenine differ in the positions at which one hydrogen atom is attached. In both instances this change

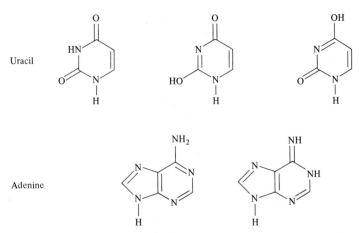

Figure 9-11 Tautomers of uracil and adenine. *(By permission from J. D. Watson (1970), "Molecular Biology of the Gene," 2d ed., W. A. Benjamin, Inc., New York.)*

is accompanied by a shift of electrons so that single bonds become double and double bonds become single. The least probable tautomers are said to be in enol (=C—OH) or imino (=NH) form.

Although the usual tautomer of adenine (in amino form) pairs with thymine, one of the less common tautomers of adenine can pair with cytosine by forming two hydrogen bonds. The imino tautomer of cytosine can pair with adenine by forming two hydrogen bonds (Fig. 9-12). Thymine and guanine can pair by forming three hydrogen bonds between guanine and thymine and a tautomer of guanine. New and incorrect purine-pyrimidine base pairing is thus made possible through tautomeric shifts in the nucleic acids which make up DNA or RNA. Tautomeric shifts are thought to play important roles in spontaneous mutations.

One of the more important chemical mutagens is nitrous acid (HNO_2). Nitrous acid causes permanent changes in the purines and pyrimidines of DNA and RNA by deamination of their respective amino groups. After deamination, adenine becomes hypoxanthine which pairs with cytosine rather than thymine; cytosine is converted to uracil, which pairs with adenine; and guanine becomes xanthine, which pairs with cytosine by only two hydrogen bonds.

The genetic consequences of the adenine to hypoxanthine mutation are called either transitions or transversions. Cytosine pairs with hypoxanthine during replication 1 and at the next division base-pairs with G (Fig. 9-13). The result of the conversion of adenine to hypoxanthine is, therefore, that the original adenine-containing strand gives rise to a second-generation strand containing guanine instead of adenine. A change from purine to purine (A⇌G) or from pyrimidine to pyrimidine (C⇌T) is called a *transition*. A net change between purine and pyrimidine (A or G⇌C or T) is called a *transversion*.

Exposure of nucleic acid to conditions of low pH may also induce base changes. In vitro low pH exposure to phage T4 results in the depurination of

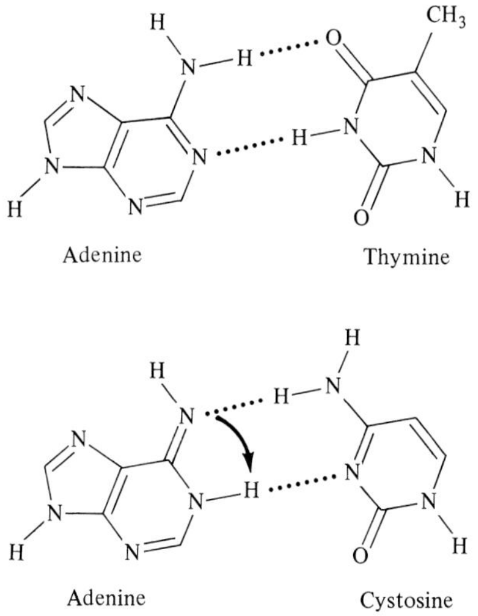

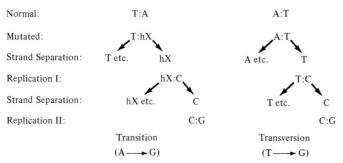

Figure 9-12 Tautomeric shift of adenine which could change its complementary base from thymine to cytosine. *(By permission from J. D. Watson (1970), "Molecular Biology of the Gene," 2d ed., W. A. Benjamin, Inc., New York.)*

phage nucleic acid, removing either guanine or adenine. When the resulting apurinic acid is returned to a higher pH, some purines are thought to rejoin. Therefore, mutations may involve either the loss of purine bases or their incorrect replacement.

Another type of mutation results when DNA base analogs are incorporated into DNA. The incorporation of 5-bromouracil (BU) may be considered an error of incorporation since bromouracil is not an ordinary component of DNA. The

Figure 9-13 Adenine → guanine transitions induced by nitrous acid and subsequent replication. (hX stands for the base hypoxanthine.)

Normal:	T:A	A:T
Mutated:	T:hX	A:T
Strand Separation:	T etc. hX	A etc. T
Replication I:	hX:C	T:C
Strand Separation:	hX etc. C	T etc. C
Replication II:	C:G	C:G
	Transition (A ⟶ G)	Transversion (T ⟶ G)

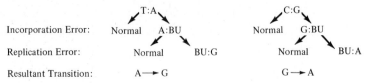

Figure 9-14 Adenine → guanine transitions involving 5-BU incorporation and subsequent replication.

incorporation of bromouracil gives rise to purine transitions of adenine ⇌ guanine as outlined in Fig. 9-14.

Ultraviolet radiation is one of the earliest known and most commonly used mutagens. The most effective wavelengths for producing mutations in microbial cells range from 253 to 265 nm of the electromagnetic spectrum. Ultraviolet radiation appears to be most effective when the treated cells are placed in an environment supporting rapid RNA synthesis. When exposed to uv radiation, pyrimidines commonly add a water molecule across the C-4, C-5 double bond of cytosine (Fig. 9-15). Although the resulting photoproduct reverts to its original form on heating, it may persist sufficiently long to weaken the H bonding with its purine complement and thus cause localized strand separation. Thymine monomers can be induced by uv light to form pyrimidine dimers (Fig. 9-16). This dimerization can result in TT, CC, UU, and mixed pyrimidine (CT) dimers.

Induced uv strand separation gives pyrimidines greater freedom of movement and thus increases the likelihood of dimerization involving bases in different strands. Interstrand dimerization results in cross linking between nucleic acid chains, thereby inhibiting strand separation and subsequent distribution to daughter cells. Ultraviolet radiation at a wavelength of 2400 Å has an opposite effect in that it promotes dedimerization, i.e., the formation of monomers from dimers (Fig. 9-16). The genetic activity of uv-inactivated DNA can be partially restored by light at 2390 Å. This phenomenon is also known as photoreactivation. Although thymine dimer formation is the major uv-lethality, hydration of the pyrimidine cytosine is most probably the major mutagenic alteration induced by ultraviolet radiation.

Figure 9-15 Effect of ultraviolet radiation on DNA pyrimidines. *(By permission from I. H. Herskowitz (1967), "Basic Principles of Molecular Genetics," Little, Brown, Boston.)*

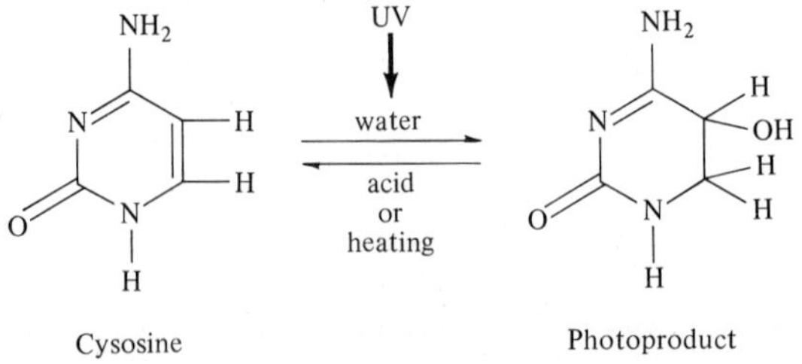

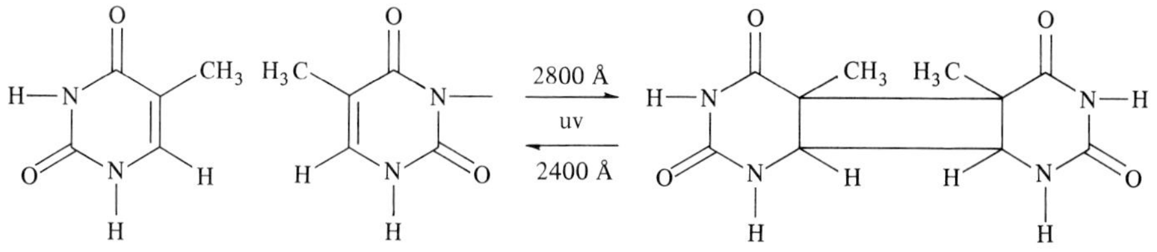

Figure 9-16 Thymine dimer formation induced by ultraviolet radiation and its reversal by photoreactivation. *(By permission from I. H. Herskowitz (1967), "Basic Principles of Molecular Genetics," Little, Brown, Boston.)*

Certain strains of the bacterium *Escherichia coli* are partially resistant to uv radiation and, after a long period, resume normal DNA synthesis. This recovery can occur in the absence of light (thus differing from photoreactivation). In the dark reaction, the dimers are enzymatically removed from the nucleotide chain. After the dimers are excised, a second protective enzyme removes an adjacent portion of the defective strand and a third repair enzyme fills in the deleted nucleotides as specified by the complementary strand. This unique repair enzyme system ensures the preservation of the original information content of the DNA.

N-methyl-N'-nitro-N-nitrosoguanidine (NG) has been widely used as a potent mutagen. It is one of the most potent mutagens known and is currently receiving wide acceptance in industrial microbial genetics laboratories. Its exact mode of action remains unresolved, and one should exercise extreme care when using this compound in the laboratory.

Aminopurine is an effective mutagen, although it is not extensively incorporated into DNA. The analog substitutes for adenine, pairing with thymine. Thymine may align with 2-aminopurine and, during replication, pairs with cytosine, resulting in a change from A-T to C-G. The alkylating agents ethyl ethanesulfonate (EES) and ethyl methanesulfonate (EMS) also react directly with DNA. They attack the N-7 position of guanine, thereby forcing misalignment of guanine to thymine which results in a change from G-C to A-T. Hydroxylamine reacts with cytosine so that it pairs with adenine, thereby resulting in a change from G-C to A-T.

Single whole-nucleotide changes which do not involve breakage of the sugar-phosphate backbone can be brought about by chemical mutagens. The acridine dyes, proflavin and acridine orange, are thought to act by binding to the outside of a double-stranded nucleic acid or by becoming inserted between successive nucleotides of a strand in a duplex. Under the latter circumstances, the strand seems to be altered lengthwise so that, if it is used as a template, an unspecified nucleotide may be inserted in the complementary strand at the position corresponding to the acridine molecule.

Purine nucleotides have been shown to decrease mutation rates, providing radiation protection. Such compounds are known as *antimutagens*. The purine nucleotides appear to be more antimutagenic than the pyrimidine nucleotides, and the ribosides are more effective than the deoxyribosides. The polyamine compounds, spermine and quinacrine, are also effective antimutagens.

MUTATION RATE

The estimation and calculation of mutation rates can be obtained for most microbial systems and are of importance because they can be used as the basis for recombinational analysis. In expressing mutation rates mathematically, it is important to define mutation rate (a) as the probability that a cell division will yield at least one mutant offspring. This can be expressed as

$$a = \frac{E}{D} \tag{9-1}$$

where E = number of mutational events
D = number of divisions

The number of mutational events (E) is not equivalent to the number of mutants (M). The number of mutants in a population increases by new mutation and by growth division of existing mutants. Because the number of cell divisions equals the increase in the number of cells, the number of divisions approximates the number of cells when the inoculum (N_0) is small.

Expressed mathematically,

$$D = N_t - N_0 = N_t \tag{9-2}$$

where N_t = number of cells at time t
N_0 = number of cells in inoculum

One can substitute the number of cells for the number of divisions in Eq. (9-1); therefore, $a = E/N_t$ and, solving for E (number of mutational events) equals the product of the mutation rate and the number of cells.

$$E = aN_t \tag{9-3}$$

GENETIC MECHANISMS AND LIFE CYCLES OF INDUSTRIALLY IMPORTANT MICROORGANISMS

In order to gain a more thorough understanding of the application of molecular and microbial genetics to industrial fermentations, it is important to consider the life cycles of a variety of representative organisms which have significance in industrial microbiology. It is important to understand life cycles, for life cycles are genetic systems, and these systems give genetic continuity and provide for variations to arise through recombination of genetic material. Both continuity and increase in numbers can be adequately provided for by asexual reproduction. Asexual reproduction occurs when a cell divides into two cells in the process of mitosis or in some other genetic process in which genetic material is divided and distributed to progeny cells. In contrast, genetic recombination occurs when the genomes of two dissimilar strains undergo an exchange of nuclear material and a hybrid genome is produced in the offspring which is different from the parental types. Sexual recombination provides the most efficient means for introducing new attributes and increased productivity into mutant strain lines. Unfortunately, most of the economically important eukaryotic organisms are asexual and, therefore, possess limited genetic diversity when compared with organisms possessing true sexual cycles.

Although the genetic systems of the great majority of industrially important organisms possess considerable similarity to the basic functions of reproduction and recombination, they vary considerably in detail. Therefore, it is necessary to understand the basic details and significance of microbial life cycles in their unique industrial environments.

VIRUSES

Viruses are particles, rather than cells, which are incapable of reproducing except in living cells and tissues. Their structure is simple, consisting largely of a protein coat and a nucleic acid core. The life cycles of only a few viruses are known in detail. The bacterial viruses or bacteriophages are best known, and among these the T even series of phages which infect the bacterium *Escherichia coli* possess life cycles which have been studied in great detail.

The T2 phage exists outside the bacterial cell in a nonreproducing state and is referred to as a "resting" phage. Upon contact with a sensitive bacterium, it combines with specific attachment pili (fimbriae) and its nucleic acid is injected into the host cell. The protein coat "ghost" is adsorbed on the external of the host cell. Within minutes, the infective foreign nucleic acid replicates at the expense of host energy and host metabolites. During this reproductive period, the phage particles are called vegetative phage. If the host cell containing the noncoated particles is disrupted prematurely, it is found that infective particles are not present. This period of nucleic acid replication is called the "eclipse" period. Shortly after phage DNA replication begins, phage protein is synthesized and during this phase of maturation the nucleic acid and protein components are combined, resulting in the formation of mature phage. The bacterial cell then lyses, and the new phage particles are released. The average burst size is approximately 300. The entire process from adsorption to lysis occurs in minutes, and upon lysis, the new phage particles are capable of reinfecting other sensitive cells. The life cycle of a typical T2 phage is outlined in Fig. 9-17.

BACTERIAL SYSTEMS

During the years 1946 to 1956, four distinct and relevant recombinational mechanisms were discovered in microorganisms. They are known as conjugation, sexduction, lysogeny and transduction, and transformation.

Conjugation

Prior to the year 1946, bacteria were considered to be entirely asexual with respect to their reproduction. In 1946, however, a brilliant young student, Joshua Lederberg, and his professor, Dr. Tatum, of Yale University reported conclusive evidence that genetic recombination occurred in *Escherichia coli* K12. Following their initial discovery, they published a series of papers concerned with a

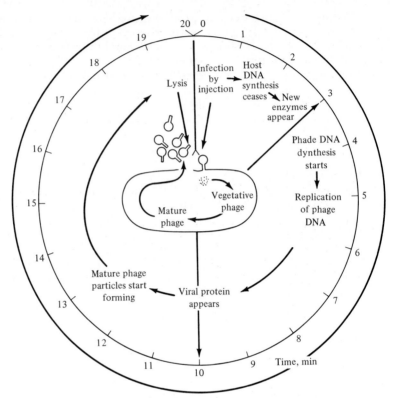

Figure 9-17 Life cycle of the *Escherichia coli* phage T2. *(By permission from R. P. Wagner and H. K. Mitchell (1964), "Genetics and Metabolism," 2d ed., Prentice-Hall, Englewood Cliffs, N.J.)*

number of alleles which were linked in a single group (indicative of a single chromosome) and which could be arrayed linearly in a definite order. In the K12 system, the frequency of recombination was extremely low; its magnitude was 1 in 10^{-6}, or one out of every million progeny cells was recombinant. Later, Hayes, Lederberg, and Cavalli recognized the existence of two different kinds of cells, namely, donor (male) and recipient cells (female). It was shown that the male or donor cells (also referred to as F^+ or Hfr) and the recipient or female cells (referred to as F^-) make physical contact by attachment to specialized sex "hairs" (also known as sex pili or sex fimbriae), through which the male chromosome passes as a thin thread to the female cell. This process is called *conjugation* and is a one-way male chromosome transfer.

Sex in *E. coli* is infectious and is determined by a cytoplasmic agent or fertility factor known as an episome. Male cells (F^+ or ♂) carry the fertility factor which they transmit readily to F^- cells (♀) by contact. Hence, a culture containing a mixture of F^+ and F^- cells may become uniformly F^+. Cells carrying F may lose the factor and are referred to as "cured." They then become F^-. The process of curing may occur spontaneously or can be induced by acridine dyes.

The F⁺ cells may become Hfr cells spontaneously at a low frequency (1 × 10^{-4} per division). The conversion involves the attachment of free-floating fertility factor F to the chromosome. The Hfr (high-frequency recombination) cells conjugate readily with F⁻ cells and transfer a portion of their chromosome with extremely high frequency. Ordinarily, when Hfr male cells are used in a mating, a recombinant cell is obtained for every 10 to 20 parent cells. The Hfr cells differ from F⁺ cells in that they do not transmit the F factor to the F⁻ cell, and they produce a high frequency of recombination after conjugation.

In a brilliant series of experiments conducted at the Pasteur Institute, Jacob, Hayes, and Wollman demonstrated that the Hfr chromosome is transferred over a period of about 30 to 40 min. They also demonstrated that the transfer is polar, i.e., the same end of the chromosome for any given Hfr strain goes in first. This point or origin, referred to as 0⟶, may be determined by the attachment of the F factor to the chromosome. In order to reconcile their data, they hypothesized that the chromosome in F⁺ cells is circular and that the attachment of the F factor causes a break at the point of attachment. The free end of the chromosome is 0⟶ and the terminal end is the region where the F factor becomes attached or integrated. The conjugation process is outlined in Fig. 9-18.

The transfer of the Hfr chromosome into the F⁻ cell can be interrupted by

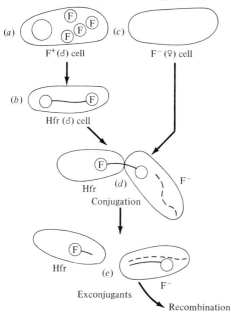

Figure 9-18 Process of bacterial conjugation in *Escherichia coli*. (*By permission from R. P. Wagner and H. K. Mitchell (1964), "Genetics and Metabolism," 2d ed., Prentice-Hall, Englewood Cliffs, N.J.*)

Table 9-1 Recombination frequency for various characters in Hfr compared with time of entrance of the same characters

	Genetic factors									
	t	l	az^f	T_1^f	lac_1	T_6	gal	λ	21	424
Recombination frequency	...	...	90	70	40	35	25	15	10	3
Times at which factors begin entering F⁻ cells, in minutes	8	8.5	9	11	18	20	24	26	35	72

SOURCE: W. Hayes (1969), "The Genetics of Bacteria and Their Viruses: Studies in Basic Genetics and Molecular Biology," 2d ed., John Wiley & Sons, Inc., New York.

separating the conjugating cells by the high-speed shearing force of a blender. Jacob and Wollman found that, when the mating is interrupted at various periods after its onset, F⁻ cells which separated from their partners before 8 min did not receive marker genes. Following 8 min of chromosome transfer, marker genes appear and in the same order as they are found to occur by standard pedigree analysis (Table 9-1). These data clearly demonstrate that the marker genes closest to the origin show the highest frequency of recombination. The same markers, because of their position with respect to 0 ⟶, also enter the F⁻ cell prior to the more distal markers (λ, 21, and 242).

The passage of the Hfr chromosome into the F⁻ cell is usually incomplete, presumably because the chromosome breaks or a portion is eliminated following penetration. The fate of the F⁻ exconjugants has been followed for several generations and it was found that they produce many different kinds of individuals. The numbers of different genetic types depends on the heterozygosity of the exconjugant zygote. Segregation may not appear for many divisions after conjugation, and then only a few lines will show evidence of segregation whereas other lines may lead to clones of the F⁻ parent genotype only. The Hfr chromosome may retain its identity over a number of generations while mating and engage in the formation of recombinant chromosomes. It is now well established that the zygote formed is a partial one (merozygote) and, therefore, differs from most of the life cycles of the fungi and other higher plants. Despite the loss of its chromosome on transfer, the Hfr exconjugants are frequently viable but, as expected, do not produce recombinant progeny. The viability of the Hfr exconjugant is explained by the fact that the chromosome replicates prior to chromosome transfer and presumably only a single chromosome actually engages in the transfer process. The lack of recombinants in the Hfr exconjugant descendants is explained on the basis that transfer is unidirectional (one-way) to the F⁻ cell.

The evidence for the presence of one chromosome per "nucleus" is based on both electron microscopy and on linkage data. There is but one linkage group for the more than 50 genetic characters that have been mapped. A typical linkage map of *E. coli* K12 is summarized in Fig. 9-19.

Sexduction

The cytoplasmic fertility factor (F) has been termed an episome to distinguish it as a separate type of cellular organelle, because it can exist free (autonomous) or

THE ROLE OF MICROBIAL GENETICS IN INDUSTRIAL MICROBIOLOGY / **215**

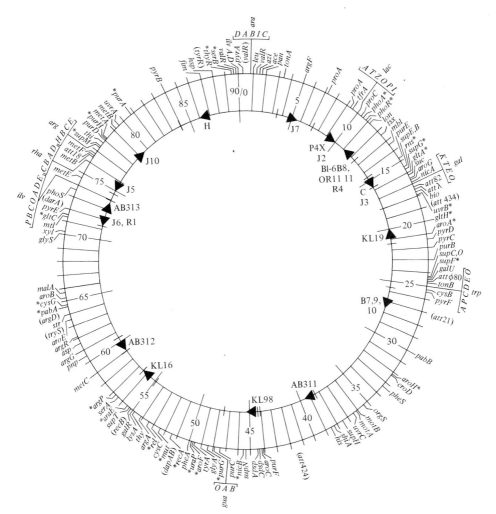

Figure 9-19 Circular linkage map of *Escherichia coli* K12. For a thorough discussion of this map and its component parts see the source, W. Hayes (1969), "The Genetics of Bacteria and Their Viruses: Studies in Basic Genetics and Molecular Biology," pp. 666–669, 2d ed., Wiley, New York. *(By permission of the publisher.)*

attached to the chromosome (integrated in Hfr). In addition to conferring the F⁺ condition on an *E. coli* cell not carrying it, the F episome may also introduce new genetic characteristics from the donor cell. The phenomenon is known as *sexduction*. The male cells have been shown to a number of episomes which carry bits of the host DNA and its genetic information. When the F⁻ cells are infected with these they not only become F⁺ but may become heterogenotes for genetic material received from the donor cells.

The F⁻ cells carrying the gene *gal*⁻ (inability to utilize galactose) may be infected with an F episome carrying gal^+ (ability to utilize galactose). The heterogenotes resulting from the episomal infection are now capable of fermenting galactose but may continue to carry gal^- on their chromosome. The cells are galactose positive only because they carry gal^+ in their episomes. They are called heterogenotes because they are heterozygous for the fragment bearing the *gal* locus. Some of the cells may cease being heterogenotes by incorporation of the gal^+ allele into their chromosomes by recombination and exchange of gal^+ for gal^-.

Sexduction is another unusual process for transferring genetic information from one cell to another. Since the material transferred by the episomes can be incorporated into the host genome, it is apparent that sexduction represents a sexual process differing from conjugation in that only a very minute amount of genetic information is transmitted by the F episome. Resistance transfer factor (RTF), conferring upon the recipient cell multiple resistance to combinations of drugs, including a number of antibiotics, is mediated by an episomal particle. Direct transmission of the multiple-drug resistance is independent of the F state of the donor or recipient cells. The resistance factor has been designated the *R factor* and may exist either in the autonomous or integrated state.

Lysogeny

Lysogenic strains of bacteria are capable of producing bacteriophages which infect and lyse sensitive bacteria. These unusual strains which act as sources of infecting phage without being lysed themselves are termed *lysogenic*. The phages they carry are termed *temperate* in contrast to the virulent phages which cause lysis of a sensitive host strain.

Temperate phages may exist in three different modes:

1 An infectious state in which the phage is free of the bacteria and capable of infecting (comparable to resting phage)
2 A vegetative state in which the phage replicates at the expense of host DNA and ultimately causes lysis of the host cell
3 The prophage state in which the genetic material of the phage is integrated to the host chromosome and is duplicated along with the bacterial genome

In contrast to the virulent phage which induces lysis, a temperate phage may enter the vegetative state and lyse the host cell or it may establish itself as a prophage and give rise to a stable clone of lysogenic bacteria. The first response is described as the *lytic cycle*, and the second type is termed *lysogenization*. The two cycles are outlined in Fig. 9-20. Once in the prophage state, the temperate phage may enter the vegetative state. About 1 out of every 10,000 cells per generation contains a prophage which enters the vegetative state and initiates the lytic cycle. The lytic cycle can also be induced by a number of physical and chemical agents.

The lysogenic state may be considered much more than a typical host-parasite relationship. In lysogeny, the phage (prophage) can actually become a part of the host's genetic apparatus and can be passed on to succeeding generations. In this sense, the prophage is actually a gene or cluster of genes. From another point of view, lysogeny may be described as an inherited ability to

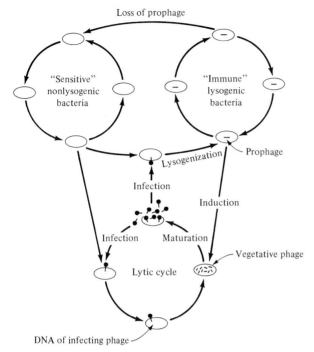

Figure 9-20 Lysis and lysogenization in enteric bacteria. *(By permission from R. P. Wagner and H. K. Mitchell (1964), "Genetics and Metabolism," 2d ed., Prentice-Hall, Englewood Cliffs, N.J.)*

pass on a specific type of virus by infection. The phenomenon is also unusual in that a bacterium containing a specific prophage is immune to infective particles of that prophage or closely related phages. However, the immunity is retained only as long as the prophage is present. The close association of prophage with the bacterial chromosome raises the question of how intregral a part of the host bacterial chromosome the prophage is. When a bacterium becomes lysogenic its prophage does not replace preexisting bacterial DNA. The prophage appears to attach to a specific locus, thus constituting an addition rather than a substitution.

Transduction

The transfer of bacterial genetic material from one cell to another by a phage acting as a vector was discovered by Zinder and Lederberg in 1952 and was termed *phage-mediated transduction*. The incorporation of the genetic material into the genome of the recipient cell completes the process.

Two types of transduction have been described:

1. Generalized or nonspecific transduction
2. Limited or restricted transduction

In generalized transduction, the phage transfer may genetically determine characteristics of the host genome. In limited transduction, a specific integrated phage known to occupy a specific locus may, when induced to enter the vegetative phase, carry with it a specific piece of bacterial-host genetic material.

The phenomenon of generalized transduction occurs with *Salmonella typhimurium* and certain of its temperate phages. If a sensitive strain of *S. typhimurium* carrying the try^+ allele (ability to synthesize tryptophan) is infected with a phage under conditions which foster the lytic cycle rather than lysogeny, a lysate can be obtained containing up to 10^{11} phage particles per milliliter. When this lysate is used to infect a sensitive host which carries a try^--containing chromosome (inability to utilize tryptophan), 1 out of every 10,000 recipients will be changed from try^- to try^+. The infecting virus carries, along with the virus genome, the try^+ allele which, in turn, is incorporated with the host genome. The descendants are stable try^+ cells which may be transduced back to the try^- genotype with phage carrying try^-. The phenomenon has been interpreted that the phage particles carry only a small fragment of bacterial genome. In generalized transduction with *Salmonella* phage P22, theoretically any genetic character should be transducible.

The galactose fermenting gal^+ or nonfermenting character gal^- in *Escherichia coli* exhibits limited transduction via phage lambda (λ). All the known gal^- mutants are closely linked to one another and to the markers λ and *succ* (succinate requirement). Any one or a group of the gal^- markers may be transduced by phage λ, which is normally carried as a prophage by the K12 strain closely linked to the gal^- loci. The ability of phage λ to transduce is limited to the λ-gal segment.

Temperate phages are complex and after the injection of their DNA into a host cell may (1) undergo rapid replication and induce lysis, (2) persist in a state of depressed replication in the host cytoplasm, or (3) associate with the bacterial chromosome and be replicated in synchrony with bacterial host cell DNA.

Since lysogenic phage may either attach to chromosomes as prophage or exist free in the cytoplasm, they show some of the characteristics of episomal mediated recombination. There appears to be only minor differences between sexduction and transduction, and one may consider both processes as part of the same general phenomenon of genetic transfer mediated by semiautonomous units or episomes.

Transformation

Bacterial transformation may be defined as the incorporation of naked DNA into the host genome. Characteristics which have been reported to be transformed in capsular types of pneumococci are penicillin (*pen*), streptomycin (*str*), and sulfanilamide (*sul*) resistance and mannitol (*man*) fermentation. The susceptibility or *competence* of the recipient cell which undergoes transformation depends on the stage of its growth cycle. During the growth cycle there are periods of greater and lesser competence. Generally, the greatest competence appears to fall immediately after fission. The generalized process of transformation is outlined in Fig. 19-21.

Although transformation may be obtained with DNA directly and sponta-

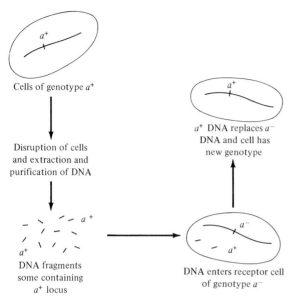

Figure 9-21 Generalized process of transformation in the bacteria. *(By permission from R. P. Wagner and H. K. Mitchell (1964), "Genetics and Metabolism," 2d ed., Prentice-Hall, Englewood Cliffs, N.J.)*

neously liberated by a donor cell, extracted and purified DNA is more commonly used. The DNA extracted from a single clone of donor bacteria contains numerous genetic factors. The detection of transformation frequency appears to be limited primarily by the sensitivity of the selective system. Genetic transformation appears to be restricted to a limited number of genera. These include *Pneumococcus, Hemophilus, Bacillus, Neisseria, Rhizobium,* and *Streptococcus*.

The principal factor limiting detection of transformation is the genetic and physiologic competence of the recipient bacteria to incorporate DNA. Competence develops only under narrow cultural conditions which must be determined empirically for each transformation system. A reasonable index of competence is the ability of a cell to take up labeled DNA.

After the DNA enters the recipient cell, it is incorporated into the recipient's genome either by actual physical exchange, which results in the ejection of a fragment of nonviable chromosome, or by influencing the replication of the chromosome so that the transforming DNA is copied rather than the host chromosome. The first mechanism is comparable to crossing over, and the second is by breakage and reunion rather than copy choice.

STREPTOMYCETE GENETICS

The formal genetics of *Streptomyces*, revealed primarily from the studies of Sermonti et al. (1956, 1963) on *Streptomyces coelicolor* (more properly classified as *S. violaceoruber*), resembles essentially that described for *Escherichia coli* (Hopwood, 1973). The major difference between the two genetic systems is that in the vegetative mycelium of streptomycetes the heterogenomic state is stabilized, whereas the relatively regular cell division process of *E. coli* establishes genetic homogeneity more rapidly. If genetic pedigrees of *E. coli* are compared with spore analyses in *S. violaceoruber*, the differences are even further minimized. A single circular linkage map for *S. coelicolor* has been reported by Hopwood (1973).

Although genetic recombination has now been described for a number of streptomycetes, it may not be correct to state that genetic recombination is widespread in the genus. Sermonti and Casciano (1963) described in *S. violaceoruber* two groups of strains referred to as R^+ and R^- (R^+ is fertile; R^- is sterile). The R^+ strain gave high frequencies of recombinants when crossed with themselves or with R^- strains, and R^- strains yielded low frequencies when mated with other R^- strains. When Sermonti and Casciano compared this system with the Hfr/F^+ and F^- fertility patterns of *E. coli*, they observed that the R^+ factor was transferred with good efficiency to R^- strains and suggested that the R^+ factor may be a nonintegrated episomal factor. A major difference is that the fertility factor was not eliminated by acridine poisons as F is in *E. coli*.

Syncytic recombination has also been demonstrated between different species of streptomycetes. Interspecific hybridization, like intraspecific hybridization, can be detected by selecting for recombinants from mixtures of nutritionally deficient parents. A simple rapid technique for detecting interspecific recombination can be used for organisms for which auxotrophic variants are not available. Wild-type strains can be mated with drug-resistant auxotrophic strains. The recombinants are then selected from mixed cultures on a minimal medium containing specific metabolic poisons. Natural markers can also be used; these include carbohydrate utilization, phage host range, and drug sensitivity. Streptomycete recombination does not appear to be as rare a phenomenon as was originally thought less than a decade ago. Of the mechanisms known, intraspecific and interspecific recombination may prove particularly valuable in that they may allow for the synthesis of strains with increased fermentation vigor or, in the case of interspecific recombination, may culminate in strains capable of producing new antibiotics with properties of related metabolites found in different species.

Actinomycete cultures exposed to phage often liberate phage on subsequent subculture. These strains are termed lysogenic, although such strains may be pseudolysogenic in that these cultures are not homogeneous but are actually mixtures of adsorbed phage on a partially resistant cell population. Pseudolysogenic cultures may be freed of phage by single spore subculture, by treatment with specific antisera, or by treatment with viricidal agents.

True lysogeny is probably widespread among actinomycetes. In fact, it has been claimed that 35 percent of the known streptomycetes are lysogenic. In order to detect temperate phages, one must test them against a wide variety of putative sensitive indicator strains. Criteria suggestive of lysogeny include spontaneous plaque formation in cultures and the presence of phage-like elements in filtrates examined by electron microscopy. It is extremely difficult to free lysogenic cultures of intracellular prophage.

Transformation has been claimed for several antibiotic-producing streptomycetes. However, the data offered in support of transformation are not sufficiently documented to be acceptable without reservation. Most of the experimentation suffers from lack of proper controls. Documentation showing a loss of transforming capacity following treatment with DNase should be mandatory for any legitimate claim of transformation. Other important factors not considered in several of the claims are short contact times between transforming DNA and recipient cells, an important factor governing competence in other bacterial systems. Conditions for DNA uptake have been established for *S. griseus* and for *S. aureofaciens*. These studies may be preludes to new and more convincing reports on transformation in streptomycetes. An interesting report of heterotransformation between two different antibiotic-producing species was given by Ramachandran et al. (1965). A transformant strain was isolated which showed simultaneous synthesis of both antibiotics.

Phage-mediated transduction in streptomycetes has been reported for *S. olivaceus* and *S. griseus*. In the *S. griseus* system, streptomycin productivity was transferred to a nonproducing mutant by actinophage. Certain of the "transactive" variants synthesized more streptomycin than nontransduced strains. It is difficult to ascertain whether the transactive variants originated through mutation and selection or by phage-mediated recombination.

YEASTS AND FILAMENTOUS FUNGI

In contrast to the viruses, bacteria, and streptomycetes, the yeasts and filamentous fungi are eukaryotic in that they possess true delimited nuclei with discernible chromosomes. Moreover, a meiotic mechanism with crossing-over and classic genetic ratios comparable to those existing in the higher plants and animals occurs in the yeasts and fungi.

True diploids are produced in the life cycles of most fungi. Unlike the "diploids" of bacteria, the homologous chromosome pairs are made up of complete chromosomes. The fungi as a group exhibit a great variety of life cycles which may be categorized into three major types: haploid, diploid, and dikaryon. The dikaryon is found as a transient phase in the sac fungi (Ascomycetes) but may represent the dormant phase in many mushroom and toadstool fungi (Basidiomycetes). The dikaryon results from delayed fusion of nuclei following fusion of two haploid cells. After cytoplasmic fusion, the binucleate dikaryotic cells divide mitotically to produce two daughter cells each with the same dikaryotic nuclei.

One of the most extensively studied genetic models is the fungus *Neurospora crassa*. The *N. crassa* is an ascomycetous fungus with a haploid vegetative phase. It is heterothallic and requires two mating types to carry out its complete life cycle which is diagramed in Fig. 9-22. The vegetative mycelium is haploid and can be propagated almost indefinitely by serial transfers of hyphal fragments or asexual spores (conidia) to fresh nutrient media. In *N. crassa*, two types of spores are produced, macroconidia and microconidia. Of the two mating types, A or a, either will produce protoperithecia on appropriate media. After the gametangia are produced, the sexual cycle is initiated by fertilization with spermatia of mycelium of the mating type opposite to that of the protoperithecium.

After the nucleus has migrated through a specialized structure known as a

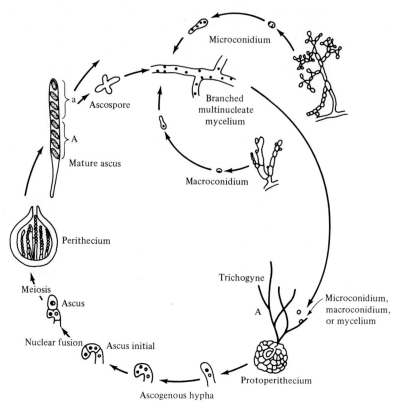

Figure 9-22 Life cycle of *Neurospora crassa*. (By permission from J. R. S. Fincham and P. R. Day (1971), "Fungal Genetics," 3d ed., Davis, Philadelphia.)

trichogyne into the protoperithecium, a dikaryon is formed which is reproduced mitotically to form a large number of dikaryotic cells which become zygotes upon fusion of pairs of the dikaryotic cells. Each fused nucleus undergoes meiosis (reduction division) to form four haploid nuclei which, in turn, undergo two successive mitotic divisions to form eight ascospores which are contained linearly in an ascus sac. A perithecium may contain 2 to 300 asci, each derived from a fusion nucleus, and when the perithecia mature, the spores are forcibly discharged from the sac. The ascospores generally require a heat shock which induces germination.

The common bakers' yeast, *Saccharomyces cerevisiae*, a heterothallic ascomycetous budding fungus, has been used for many important genetic studies. Many yeasts are thought to be degenerate sac fungi which exist vegetatively as budding unicells in which the vegetative cell also acts as the sexual cell (ascus). Saccharomyces reproduce asexually by a process called budding which is a mitotic division in which equal chromosomal but unequal division of the cytoplasm occurs. The sexual cycle of saccharomyces exists in both haploid and diploid phases and is diagramed in Fig. 9-23. As in *Neurospora* sp., the diploid zygote of a common yeast undergoes meiosis within a closed sac; hence all products of single meiotic division are kept together. Polyploidy is also common in yeast and arises as the result of a fusion of a haploid and a diploid or of diploid cells. Genetic analyses indicate that ploidy levels of up to the tetraploid (4N) condition are possible.

Another important ascomycetous fungus which displays a vast variety of genetic mechanisms is *Aspergillus nidulans*. Like *Neurospora* sp., its vegetative mechanisms convey upon it certain advantages as a tool in genetic investigations. In addition to its regular asexual and sexual cycles, *A. nidulans* possesses unique heterokaryotic and parasexual genetic mechanisms.

The normal asexual cycle consists of the germination of uninucleate conidia which produce typical haploid vegetative mycelium, a portion of which is differentiated into specialized conidial heads. The mature conidia separate and germinate, thereby repeating the cycle.

In contrast to neurosporae which are heterothallic and have two mating types, *A. nidulans* is homothallic and possesses no mating types. Hence, conidia

Figure 9-23 Haplophase and diplophase generations of *Saccharomyces cerevisiae*. (By permission from H. J. Phaff, M. W. Miller, and E. W. Mrak (1966), "The Life of Yeasts," Havard, Cambridge, Mass.)

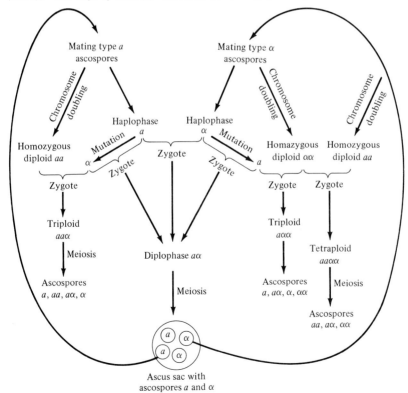

of a strain fertilize its own ascocarp initial. In contrast to neurospora, the ascocarp is a closed, thin-walled globose structure called a cleistothecium. Each cleistothecium contains numerous globose asci containing eight ascospores which are not arranged in a definite order.

Although *A. nidulans* is generally considered to be an organism with a haploid vegetative phase, it can diploidize. The ability to form rare diploid spores conveys upon it certain advantages not found in species of *Neurospora*. The unusual diploid cycle in the asexual phase of the fungus is a result of the fusion of somatic vegetative nuclei which are in the heterozygous condition. Unusual mitotic crossing-over occurs at the four-strand stage, segregation is random and mitotic, and the expected reciprocal products are found. The somatic recombination cycle has been termed *parasexuality* by Pontecorvo and his coworkers.

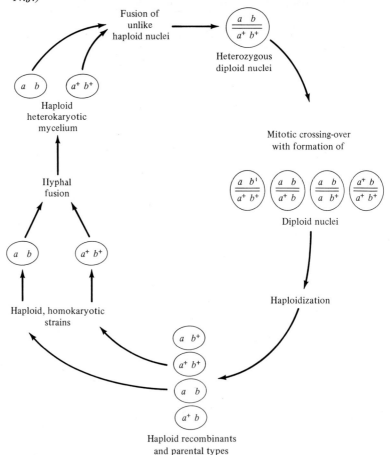

Figure 9-24 Parasexual recombination process in the filamentous fungi. *(By permission from R. P. Wagner and H. K. Mitchell (1964), "Genetics and Metabolism," 2d ed., Prentice-Hall, Englewood Cliffs, N.J.)*

Parasexuality has been discovered in a number of industrially important fungi including the penicillin fungus, *Penicillium chrysogenum*. The unusual cycle makes it possible to test the effects of genes in the heterozygous condition and has been used to map genes in *A. nidulans*. With respect to recombination efficiency, however, somatic recombination is only about $\frac{1}{500}$ as effective as sexual recombination in *A. nidulans*. An outline of the parasexual recombination mechanism is given in Fig. 9-24, and a comparison of these two important recombinational mechanisms is presented in Table 9-2.

Filamentous microorganisms also display another unusual method for genetic variation. Many fungi show hyphal anastomosis when they are actively growing in culture. Fusion results in a mixing of both the cytoplasm and the nuclei of the fused hyphae. If the hyphae represent genetically different strains (biochemical drug-resistant or conidial color mutants), the fusion may result in the production of a strain which will have nuclei of both strains. Such a strain is called a *heterokaryon*. If each parental strain has a different nutritional requirement due to gene mutation, the heterokaryon will have the wild-type allele of each and will grow on a minimal medium (prototrophic). Frequently, heterokaryotic growth is better than that obtained from the homokaryotic strains.

Heterokaryosis is of great significance in organisms in which it is found. Many fungi which are capable of forming heterokaryons are frequently heterokaryotic in their natural habitats. Heterokaryosis also is an important consideration in the cultivation and maintenance of important higher-yielding strains of industrially important microorganisms. Mutation to lower levels of productivity

Table 9-2 A comparison of the sexual and parasexual cycles in fungi

Step	Sexual cycle	Parasexual cycle
1	Hyphal anastomosis (plasmogamy)	Hyphal anastomosis (plasmogamy)
2	Nuclear exchange to form heterokaryon	Nuclear exchange to form heterokaryon
3	Nuclear fusion (karyogamy) in specialized sexual structures to yield "selfed" or "hybrid" ascocarps.	Rare nuclear fusion in vegetative cells. Heterozygotes recognized by color or nutrition. Homozygotes, if formed, are not detected.
4	*Diploidization*; zygote persists for one nuclear generation	*Diploidization*; zygote may persist through many mitotic divisions
5	Recombination at *meiosis*; crossing-over at four-strand stage, in all chromosome pairs, random assortment of each chromosome pair followed by reduction to *haploid* state	Recombination by rare accident of *mitosis*. (1) *Mitotic crossing over* at four-strand stage and usually only one exchange in a single chromosome arm; (2) *haploidization*. Independent of crossing over and random assortment of each chromosome pair
6	Products of meiosis readily recognized and isolated	Recombinants occur among vegetative cells and recognized only by use of suitable markers and selection technique

often occurs which results in "run-down" and lower production yields of valuable production strains. Such strains have to be repurified by single spore selection.

Heterokaryosis is most significant genetically since it allows for the establishment of balanced genetic systems which have selective advantage over homokaryotic strains. Heterokaryosis may be considered a process midway between syntrophism and heterozygosity in a diploid organism. Syntrophism is a relationship between two different strains of an organism resembling symbiosis, which enables both strains to survive in a particular environment whereas neither would survive alone.

The discovery of diploid nuclei in aspergilli and penicillia has resulted in a more profound understanding of recombination potential among fungal organisms. In order to detect rare heterozygous diploid nuclei, an environment which selects for a unique phenotype from an extremely large population of parental strains is necessary. A method commonly employed to demonstrate parasexual recombination is to propagate complementary biochemically deficient strains on a nonselective medium. Hyphal segments of copropagated mycelium are then plated on a selective medium, a condition which selects for heterokaryotic mycelium. Following hyphal anastomosis and fusion of the haploid mutant strains, the conidia of the heterokaryotic mycelium are then examined for the parental types which were employed for the initial mating. If parental types are found, the mycelium is heterokaryotic. Rare heterozygous conidia may also be found. Their frequency of occurrence is 1 in 10 million. They characteristically are larger than their haploid parents, they are more sensitive to ultraviolet radiation than haploid cultures, their deoxyribonucleic content is doubled, their conidial color is usually of the wild phenotype, and they are prototrophic with respect to nutritional requirements. The heterozygous diploid nuclei ultimately give rise to recombinants, either by vegetative haploidization or by mitotic crossing-over.

APPLICATIONS OF MICROBIAL GENETICS TO INDUSTRIAL FERMENTATION PROCESSES

Microbial genetics as exemplified by strain improvement forms the background for all microbiological processes. As soon as a useful organism has been isolated in a large-scale industrial screening program, it is necessary to improve the productivity of the metabolite under evaluation. The initial phases of fermentation development may consist of modifications to the medium and fermentation conditions used, but the major source of progress is to improve the performance of organisms by means of selection and genetic manipulation. At the research stage when a promising drug is under laboratory and clinical evaluation, a dependable supply of the drug is necessary. During a recent evaluation of an antitubercular antibiotic, a supply of 250 kg of pure drug was required for clinical trials alone.

MUTATION AND SELECTION OF IMPROVED MUTANT STRAINS

Large-scale programs concerned with strain selection, maintenance, and improvement begin shortly after favorable clinical reports are obtained. It is now well established that the most effective method for increasing the yield of a fermentation product has been the use of induced mutation followed by selection of improved strains. General reviews of the important aspect of applied microbial genetics have been published by Alikhanian (1962), Calam (1964), and Elander (1966, 1969). The foremost difficulty of this approach stems from the extremely low frequencies at which desired mutations occur. The desired mutant must be selected from a large nonmutant population base. The major disadvantage of the approach is the lack of a scientific rationale for selecting desirable mutants because, at this stage, usually relatively little is known of chemical structure, biosynthetic pathway, and metabolic regulation of the desired fermentation product. Therefore, the screening for the rare mutant is time-consuming and involves expensive testing procedures.

There are many methods for effective strain improvement. Initially, strain selection was dependent upon the degree of spontaneous variability encountered in natural spore populations of mold and streptomycete organisms. The concept of variation had been firmly established in microbiology long before the advent of the antibiotic era. However, by the late 1940s, a number of effective mutagenic agents were known, and certain of them were finding their way into strain development laboratories. Although mutation was generally considered to indicate loss of function, it was challenged repeatedly by the discovery of enhanced variants (gain mutants) isolated from conidial populations exposed to radiation and chemical mutagens. However, there is no evidence that increased production is not a result of decreased function of some enzyme system. As a consequence, the mutation-selection process is now undoubtedly the most important method for obtaining improved strains.

SELECTION OF STRAINS AS SPONTANEOUS VARIANTS

Fortunately, or unfortunately, antibiotic-producing microorganisms exhibit a great capacity for natural variation. Therefore, the selection of strains from natural spore populations is of great practical importance in strain development. Undoubtedly, its greatest application is in the maintenance of improved antibiotic-producing cultures. Heterokaryosis and subsequent culture rundown (Haas et al., 1956), despite the introduction of vastly improved preservation methods, are still important problems in commercial culture laboratories. The continued selection for preferred colony type is a never-ending process and one of the major factors responsible for maintenance of production at constant high levels.

Natural selection for improved strains is also important in the early stages of development programs. Most mycelial organisms are probably heterokaryotic when freshly isolated from their natural environment. Selection of single spores from heterokaryotic organisms with uninucleate conidia often leads to discovery of diverse homokaryotic colony types, many of which may represent superior antibiotic-producing entities. In the early stages of development, one discretely selects highly conidiating, stable (nonsectoring) isolates and proceeds to evaluate them for desirable characteristics: notably, antibiotic species or titers. An example of direct selection was the isolation of *Penicillium* variants which synthesized copious amounts of penicillin G in submerged fermentation (Raper, 1946).

Although advances in productivity can be obtained by direct selection techniques, the use of strains subjected to mutagenic physical and chemical agents greatly increases the probability of discovery of improved strains. Such techniques were also responsible for the selection of nonpigmented variants of *Penicillium* (Backus and Stauffer, 1955) and, indeed, for mutants capable of synthesizing antibiotics with structural modification (Ballio et al., 1960).

SELECTION OF STRAINS AS INDUCED VARIANTS

Thom and Steinberg (1939) were probably the first to apply mutation methods for the improvement of mold strains. In the middle 1940s, intensive development programs on penicillin led to the adoption of mutation and selection as a major tool in strain improvement. The pioneer work of Backus and Stauffer (1955), at the University of Wisconsin, provided the major source of high-yielding *Penicillium* cultures. Today most, if not all, strains employed for the manufacture of penicillins have their origin in one of the members of the "Wisconsin Family." Mutation methodology as applied to antibiotic-producing organisms are described in articles by Darken et al. (1960) on tetracycline; Backus and Stauffer (1955) on penicillin; Dulaney et al. (1949) on streptomycin; and Elander et al. (1961, 1974) on cephalosporin. Reviews by Nelson (1961), Alikhanian (1962), Calam (1964), and Elander (1966, 1969) offer valuable information on this subject.

ROLE OF MAJOR MUTATION IN STRAIN DEVELOPMENT

The concept and role of mutation as applied to industrially important microorganisms have two aspects. The first, major mutation, involves the selection of mutants which manifest a pronounced change in a biochemical character of practical interest. Such variants are commonly used in genetic studies and may be classified, appropriately or inappropriately, as loss mutants. Such variants are isolated routinely from populations surviving prolonged exposure of a mutagen. In contrast, minor mutants show only subtle change in a particular character. In fact, the changes are so slight that usually the variants are not morphologically distinguishable from parent strains. Such mutants are common in all our important antibiotic-producing organisms.

Examples of major mutation in strain development are numerous. The case of the important high-yielding, nonpigmented penicillia has already been mentioned. Alikhanian (1962) cites a role for major mutation in the streptomycin-producing organism *Streptomyces griseus*. The initial strain synthesized large amounts of a substance with low activity in addition to small quantities of streptomycin. The substance is mannosidostreptomycin (streptomycin B), which competes with streptomycin for biosynthetic intermediates and also interferes with efficient isolation of the antibiotic. A variant was finally isolated which produced negligible amounts of the undesired substance, thus allowing for greater synthesis and recovery of the desired moiety.

A careful study of variants exhibiting impaired antibiotic productivity may elucidate biosynthetic pathways and contribute to the identification of precursors. Studies by Miller et al. (1965) and by Hostalek and Vanek (1973) on tetracycline and Barchielli et al. (1960) on cobalamin reveal interesting data on precursor molecules involved in reaction steps prior to terminal ring closure.

Heterokaryons also offer unique systems for the study of biosynthetic pathways. Heterokaryotic strains capable of elaborating antibiotics may be synthesized from auxotrophic strains with impaired antibiotic activity. Study of the impaired homokaryotic strains may reveal precursor accumulation.

Ballio et al. (1960), using variants of *Penicillium chrysogenum* (Wis. 51-20), studied changes in the antimicrobial spectrum penicillins following incorporation of certain α-ϵ-dicarboxylic acids. Upon the addition of adipic acid, one variant was shown to elaborate a new penicillin (4-carboxy-n-butyl penicillin). This antibiotic is very similar to penicillin N (cephalosporin N), both in structure and antimicrobial activity.

In recent years, the use of major mutation has acquired particular significance and, in certain instances, has led to new and more efficacious products. The tetracycline-producing organisms appear to be particularly amenable to this approach. McCormick et al. (1957) described an interesting modification of tetracycline synthesized by a mutant strain of *Streptomyces aureofaciens*. The antibiotic was shown to be changed at the C-5a position and was almost devoid of antibiotic activity. They also reported that *S. aureofaciens* (strain S-604) synthesized 6-demethyltetracycline, a new antibiotic material not elaborated by the Duggar strain A-377. The new molecule has several advantages over the methylated form and, today, is one of the leading commercial forms of tetracycline.

The employment of mutants for the synthesis of modified antibiotic molecules appears to be a fertile area for major mutation in strain development. Mutants have been mentioned already that may accumulate precursor molecules which aid in the elucidation of pathways. An insignificant reconstruction of an antibiotic molecule could lead to new biological and therapeutic properties of a known antibiotic. Some years ago, Kelner (1949) published a paper which has great interest in this connection. He examined a series of streptomycete cultures which failed to inhibit certain test bacteria. The negative strains were then exposed to heavy doses of uv and x-ray radiation. The dose for the x-ray radiation was 300,000 R; nonirradiated strains served as controls. After examining several thousand irradiated strains, Kelner found mutant lines which exhibited antimicrobial activity. In fact, certain weak antibiotic producers then showed a significant change in spectrum. The difference in antimicrobial spectrum indicated that in certain cases this change might result from qualitative modification of an antibiotic. This has been well documented for terramycin and aureomycin.

A mutant strain of *Streptomyces fradiae*, the organism that synthesizes neomycin, was reported to produce hybrid antibiotics termed hybrimycins. Wild-type strains of *S. fradiae* normally synthesize neomycins A, B, and C, incorporating the diaminocyclitol subunit deoxystreptamine to form the neomycin moeties. Shier et al. (1969) obtained a mutant strain derived from a nitrosoguanidine-treated population which was unable to synthesize deoxystreptamine and, therefore, was dependent on outside sources for diaminocyclitols for the antibiotic synthesis sequence. When the mutant was cultured with added streptamine, a diaminocyclitol whose derivitive strepidine occurs naturally in streptomycin, two new antibiotic substances were produced which were called hybrimycin A_1 and A_2. Epistreptamine, the C-2 epimer of streptamine, gave two more antibiotics, hybrimycin B_1 and B_2. New biosynthetic analogs of butirosin, another aminoglycoside antibiotic, have been separated by Claridge et al. (1974), using mutant strains of *Bacillus circulans*.

Gorman et al. (1968) reported that modified phenylpyrroles with differing antifungal properties were synthesized by metabolism of tryptophan analogs in *Pseudomonas fluorescens*. The addition of 6-fluorotryptophan to either wild-type strains or trytophan analog-resistant mutants led to the formation of 4'-fluoropyrrolnitrin, a modified metabolite with significantly greater antifungal activity than the parent pyrrolnitrin molecule. Thus, major mutation may lead to numerous new structurally modified antibiotics. Such techniques may also be important for the screening of microorganisms which elaborate nondetectable quantities of an important antimicrobial material. Mutants may elaborate detectable amounts of the substance. This tool could serve as a valuable aid in screening for new antibiotics.

ROLE OF MINOR MUTATION IN SELECTION

Minor mutation usually plays the dominant role in strain development. By definition, these mutations affect only quantitatively the amount of product of interest synthesized. Such mutations are subtle, and variants exhibiting such features are usually similar phenotypically to the parent form. They show rapid and abundant mycelial and conidial development and produce only slightly more antibiotic than the preceding parent strain. Quantitative definition of a significant yield increase is somewhat relative and dependent on the productivity of the parent. Usually, a 10 to 15 percent increase is implied. The variants are usually selected from conidial populations exposed to small or moderate doses of a mutagen. If one repeatedly isolates "minor" (positive) variants and uses each succeeding strain for further mutation and selection, after several stages, a significant yield increase may be obtained. Such increases have also been obtained without the introduction of mutagens. Thus, the problem of strain development with respect to improvement is, in reality, the problem of increasing the concentration(s) of hereditary factor(s) responsible for productivity in the original genotype. Were it not for the improvement or productivity through minor mutation, the cephalosporins, especially cephalosporin C, its nucleus (7-ACA), and derivatives, would still be laboratory curiosities (Florey, 1955).

Minor variants exhibiting a slight change in a quantitative feature may vary also in other features owing to pleiotropic effects. Since such variation is slight, there is great dependence on efficient and accurate selection techniques. The population to be tested must be large, and the assay must be accurate and specific for the desired product. Problems of this sort are primarily statistical and are discussed in articles by Davies (1964) and Brown and Elander (1966).

Examples of gradual stepwise improvement in antibiotic production are numerous and may be seen in studies of penicillin types (Backus and Stauffer, 1955) and cephalosporin (Elander et al., 1961, 1974). The published data of Alikhanian (1962) on penicillin demonstrate well this philosophy of stepwise selection, which has been followed in the 10 to 12 percent greater penicillin productivity than the previous parent and resulted in the selection of a variant which exceeded its original parent by 64 percent. Similar examples are commonplace throughout the antibiotic industry.

Many refinements in the techniques of mutation and selection have been introduced during the past two decades. Today, there are scores of mutagens available, and for many of them a mode of action is known. An excellent review on the chemical basis of mutation is available (Orgel, 1965). Unfortunately, the phenomenon of mutation is random and not directable, so that mutation as applied to strain development is largely empirical. Little, if anything, is known of the kind of mutants desired. The limited knowledge of antibiotic biosynthesis is shown by the fact that a biosynthetic pathway is not fully elucidated for a single antibiotic. There is a paucity of solid information available about the mechanism of biosynthesis for the first important antibiotic, penicillin.

In contrast, considerable refinement in mutation and selection techniques has been demonstrated over the past two decades. Refinements in evaluation have been numerous, and these probably represent the key to a successful strain development program. The importance of evaluation is illustrated by the problem of selecting relatively few enhanced variants from a population of hundreds or thousands of individuals showing control yields or less. Many of these techniques are peculiar to organisms of commercial interest, and their publication is restricted.

Examples of sophistication in techniques are numerous. One is the work of James et al. (1956) who published an interesting method of selection with the citric acid–producing fungus *Aspergillus niger*. The preliminary test involved the propagation of mold colonies on filter paper impregnated with a suitable production medium containing an appropriate indicator dye. After a prescribed incubation period, the acidity was measured as citric acid. The best strains were then evaluated in shake flask fermentations. Many of the strains which were superior on the preliminary test synthesized greater quantities of citric acid in submerged fermentation. Alikhanian (1962) studied colony morphology as related to subsequent productivity in penicillin fermentation, and he recommended the rejection of poor-growing, poor-conidiating strains of *Penicillium*. He also suggested screening variants of *Penicillium* in the absence of phenylacetic acid on the basis that the enzymes generating penicillin nucleus (6-aminopenicillanic acid) might become rate limiting with respect to total penicillin G yield. Screening of new strains on an inferior medium was suggested, because a strain superior on such media might exhibit marked superiority under more optimal fermentation conditions. Ostroukhov and Kuznetsov (1963) described a unique plate method for the detection of superior penicillin producers on the basis of increased oxidation-reduction potential. They also mentioned that strains may exhibit more negative potential under conditions of reduced aeration.

The use of actinophage for the selection of phage-resistant strains first arose with the violent outbreak of phage infection in the streptomycin industry (Carvajal, 1952). Carvajal observed that an occasional resistant colony which survived phagolysis produced more streptomycin than the sensitive production strain. A pronounced mutagenic effect of actinophage was observed by Alikhanian (1962), who claimed that 99 percent of the colonies cultivated from a phagolysate from *Actinomyces olivaceus* was similar to uv mutants and certain of the mutant types were not found in usual uv-survivor populations. A histogram depicting the gain in the higher production of vitamin B_{12} following exposure to *A. olivaceus* phage is shown in Fig. 9-25.

A method based on a different rationale was discovered by Adelberg (1958)

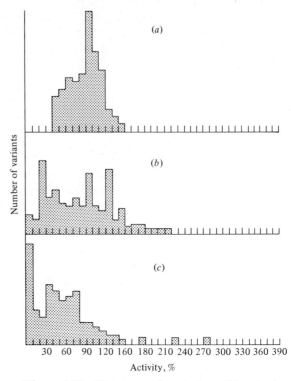

Figure 9-25 Histogram of variations of strain *Actinomyces olivaceus* N-6 in vitamin B_{12} production. *(By permission from S. I. Alikhanian (1962), Induced Mutagenesis in the Selection of Microorganisms, Adv. Appl. Microbiol., 4:1–50.)*

and later applied by Scherr and Rafelson (1962). An excellent recent application of this principle to amino acid fermentation is described by Karlstrom (1965). Adelberg reported that bacterial mutants resistant to amino acid analogs excreted small quantities of the homologous amino acid. This causes incomplete repression of enzymes synthesizing the amino acid. If it could be assumed that similar repression systems control antibiotic synthesis, and if antibiotics were essential metabolites for the organisms synthesizing them, mutants lacking repressors could be selected and thereby superior antibiotic-producing variants would result. Moreover, antibiotics are generally large complex molecules containing unusual sugar moieties, etc., which are synthesized by complex and unknown pathways. Such antimetabolites could aid in the elucidation of biosynthesis, but the design for such molecules is still a subject for the future. The technique described above has been invaluable in the selection of superior vitamin and amino acid–producing variants. It should be emphasized that, the more accurate and sophisticated the screen, the better is the probability of its success. A screen can be only as good as the design and the questions built into it.

Sophistication at the selection stage often results in an effective strain development program. In summarizing the mutation concept as applied to strain development, certain generalities appear appropriate.

1. Strains selected as obvious variants following exposure to mutagens usually are inferior in their capacity to elaborate antibiotics. Those strains with enhanced capacity for accumulation of antibiotics are extremely few in number, and selection and evaluation play extremely important roles in their detection.
2. Mutagen dose is important in strain selection methodology. The rate of mutation is a function of dose; hence, mutants sought for major mutation roles are best isolated from populations surviving prolonged doses of mutagens. Variants employed for increased productivity are generally isolated from populations surviving intermediate dose levels.
3. Strains with enhanced capacity for antibiotic synthesis generally exhibit wild-type morphology and growth habits. Strains with altered morphology, etc., may be inherently better producers but may require considerable fermentation development. Since positive variants are extremely few in number, it is better to screen a large number of variants on a few fermentation conditions rather than to screen a small number of variants under a wide variety of environmental conditions. This is especially true when one seeks only quantitative change in the variants.
4. The philosophy of stepwise selection implies small increments in increased antibiotic productivity. The range of increased productivity is generally 10 to 15 percent after the initial strains are obtained by natural selection. As productivity increases, the probability of finding superior strains decreases; hence, accurate and more sophisticated evaluation procedures play increasingly important roles. The development program is only as effective as the mutation, selection, and evaluation procedures coupled to it.
5. Variant strains often require special propagation and preservation procedures. Actual production gains depend largely on stability and reliability of performance. Maintenance through continued selection and purification plays an important role in production laboratories.
6. Although a strain may meet the numerous necessary criteria of superiority in the laboratory, there is no guarantee that enhanced productivity will occur in production fermentors. Aeration-agitation patterns, nutrient availability, etc., are often unbalanced for variant cultures. Scale-up experimentation through long-term pilot-plant studies is often necessary before any enhanced strain potential may be realized in actual production.

GENETIC CONTROL MECHANISMS IN INDUSTRIAL FERMENTATION PROCESSES

Enzyme Induction in Penicillin Fermentation

There are many industrial fermentations which clearly illustrate the genetic control phenomena of induction and catabolite repression. There is a distinct production phase (idiophase) which follows the phase of rapid cellular growth (trophophase). This phenomenon was observed in the early days of the penicillin fermentation. An analysis of the dynamics of the penicillin fermentation revealed that antibiotics synthesis became maximal when the specific rates of growth, oxygen uptake, and sugar utilization decreased. Since the idiophase follows trophophase in the penicillin fermentation, one may ask whether it is the decline in growth rate that initiates penicillin production. Pirt and Righelato (1967) reported that in chemostat culture the specific rate of penicillin production [units/(mg) (cells) (h)] is independent of growth rate until the latter falls to a critical low level at which time the penicillin rate falls to zero. The rate of decline in specific penicillin production, therefore, is inversely proportional to the original growth rate. In batch fermentations of penicillin, the best conditions for production are a rapid growth rate followed by slow idiophasic growth. Other changes must accompany the drop in growth rate in batch cultures; these changes are probably more directly involved in initiating the onset of idiophase than is the growth rate.

One of the final steps in the biosynthesis of penicillin is mediated by an enzyme known as acyltransferase which mediates the transfer of an acyl side chain to the 6-aminopenicillanic acid nucleus. A marked increase in the amount of this enzyme is correlated with the time of increase in penicillin synthesis. A decrease in activity is also noted as the penicillin synthesis rate declines. Moreover, high-yielding penicillin mutants appear to have increased levels of the enzyme. Recently, the enzyme has been shown to transfer the phenylacetyl or phenoxyacetyl group from the corresponding coenzyme A derivative to 6-aminopenicillanic acid. This enzyme appears after 48 h of fermentation (initiation of penicillin trophophase) and increases rapidly at 72 h just prior to the period of most rapid penicillin synthesis.

Feedback Regulation and Biosynthesis of Penicillin and Cephalosporin Antibiotics

Many years ago, Bonner reported the results of studies of the relationship between biochemical deficiencies and penicillin production in *Penicillium notatum* and made the significant observation that nearly one-quarter of the lysineless mutants failed to synthesize penicillin. Later, Demain (1966) reported that lysine, a well-established precursor of a penicillin intermediate, γ-(α-aminoadipyl)-cysteinylvaline, inhibited penicillin production.

Other studies indicated that d-aminoadipic acid could reverse the inhibitory effects of lysine on penicillin biosynthesis. The inhibition of penicillin synthesis by lysine is probably due to feedback inhibition and/or repression of the enzymes involved in lysine biosynthesis. This repression results in a deficiency of d-aminoadipic acid and consequent reduction in penicillin accumulation. The inhibition or repression by lysine and penicillin reduction can be overcome by additions of adipate. Excess lysine completely inhibited the accumulation of d-aminoadipic acid by a mutant strain of *Neurospora crassa*. This study adds support to the credibility of a feedback regulation of penicillin synthesis by lysine (Fig. 9-26).

The stimulation of cephalosporin C biosynthesis by methionine in strains of *Cephalosporium acremonium* appears to be indirect and may involve enzyme repression. Methionine may repress the enzymes cystathionine synthetase, cysta-

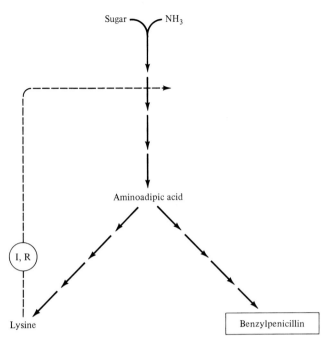

Figure 9-26 Possible mechanism of inhibition of penicillin biosynthesis by lysine. (I = inhibition; R = repression.) *(By permission from A. L. Demain (1966), Industrial Fermentations and Their Relation to Regulatory Mechanisms, Adv. Appl. Microbiol., 8:1–27.)*

thionase, and homocysteine methylase, enzymes which convert cysteine to methionine in *E. coli*. Cystathionase was reported to be identical to cysteine desulfhydrase and, in addition to reacting with cystathionine, converts cysteine to pyruvate, ammonia, and hydrogen sulfide. In this manner, methionine may repress the enzymes involved in cysteine degradation. Since the sulfur atom of the dihydrothiazine ring of cephalosporin C is derived from cysteine, the addition of methionine may spare cysteine and stimulate cephalosporin production. Cysteine desulfydrase activity has been found in *Cephalosporium*. More detailed discussions of the genetics and biochemistry of cephalosporin formation are to be found in the reviews by Lemke and Brannon (1972), Nuesch et al. (1973), and Demain (1974).

GENETIC MANIPULATION AND THE BREAKDOWN OF CONTROL MECHANISMS

Since control mechanisms are genetically controlled, mutation is an excellent means of altering regulatory mechanisms. A number of techniques are now available for obtaining (1) mutants resistant to catabolite repression, (2) constitutive mutants which are capable of producing enzymes in the presence of repressors (phosphatase production in the presence of inorganic phosphate), and (3) constitutive mutants which form enzymes without the addition of inducer substances.

A unique procedure affecting intact enzyme modification, reported first by Adelberg (1958), involves the selection of mutants resistant to a growth-inhibitory analog structurally related to the desired fermentation product. Analog-resistant mutants may overproduce the product and are resistant because the accumulation reverses the toxic effects of the growth-inhibitory analog. The phenomenon reconsidered in modern feedback concept is that the accumulation is a result, rather than the cause, of the resistance. The growth-factor analog apparently mimics the metabolite in its effects, and the majority of the sensitive wild-type population exposed to the antimetabolite is inhibited because of the resulting deficiency of end product. Rare mutants possessing modified enzymes or enzyme-forming systems which are insensitive to feedback effects of both the analog and the metabolite can be selected. Because of their insensitivity to feedback, some of the mutants overproduce and accumulate the end product.

A recent example of the use of the analog technique in the development of an antibiotic fermentation was described in Elander et al. (1971).‡ The amino acid tryptophan was found to be a precursor and stimulant of the antifungal, phenylpyrrole antibiotic, pyrrolnitrin, in strains of *Pseudomonas fluorescens* and *P. multivorans*. Since supplementation of the fermentation medium was expensive, these workers obtained mutant strains which were resistant to the analog 5-fluorotryptophan. One mutant produced three times more pyrrolnitrin than the original sensitive culture and no longer required tryptophan supplementation.

The reversion technique is also useful in manipulating control mechanisms. Many reverse mutations are pleiotropic in that the back mutations are the result of suppressor mutations and affect genetic characteristics other than the intended locus. The reversion technique has been used to obtain high-producing strains of tetracycline in *Streptomyces viridifaciens*. Revertants of auxotrophs were found which produced threefold more antibiotic than the original wild-type parental strain. The technique was also used to select mutants which no longer synthesized tetracycline followed by reversion back to production. One double-revertant strain produced nine times more tetracycline than the original producing parental strain. Other examples of end-product accumulation by analog-resistant mutants are given in Table 9-3.

RECOMBINATIONAL MECHANISMS AND THEIR APPLICATION TO FERMENTATION TECHNOLOGY
Sexual Recombination

Many eukaryotic microbes of industrial importance possess true sexual cycles involving meiosis and an alternation of haploid and diploid generations. Industrially important genera include *Saccharomyces*, *Penicillium*, *Aspergillus*, *Phycomyces*, *Claviceps*, and *Emericellopsis*. In industrially important strains of *Saccharomyces*, both the haplophase or diplophase generations can be prolonged indefinitely. Under appropriate environmental conditions, diploid cells form asci which usually contain four ascospores. The sexual spores germinate

Table 9-3 End-product accumulation by analog-resistant mutants

Analog	Product accumulated	Organism	Reference
Purines			
2,6-Diaminopurine	Adenine	*Salmonella typhimurium*	Kalle and Gots (1962)
Vitamins			
Sulfonamide	p-Aminobenzoic acid	*Staphylococcus aureus*	Oakberg and Luria (1947)
Isoniazid	Pyridoxine	*Saccharomyces microsporus*	Scherr and Rafelson (1962)
3-Acetylpyridine	Nicotinic acid	*Chlamydomonas eugametos*	Nakamura and Gowans (1964)
Amino acids			
6-Methyltryptophan	Tryptophan	*Salmonella typhimurium*	Lingens et al. (1964)
Ethionine	Methionine	*Escherichia coli*	Adelberg (1958)
p-Fluorophenylalanine	Tyrosine	*Escherichia coli*	Cohen and Adelberg (1958)
5,5,5-Trifluoro-DL-leucine	Leucine	*Salmonella typhimurium*	Calvo and Calvo (1967)
Antibiotics			
6-Fluorotryptophan	Pyrrolnitrin	*Pseudomonas fluorescens*	Elander et al. (1971)

forming homothallic or heterothallic cell populations. These cells, in turn, clump and undergo fusion and subsequent zygote formation.

The diplophase population is readily propagated and generally displays hybrid vigor. Many genetic characteristics segregate in simple mendelian fashion although certain characteristics appear to be nonchromosomal with respect to their heritability. Enebo and coworkers (1961) showed that fermentation patterns, alcohol tolerance, frothing, clumping, and lipid production are controlled by nuclear genes. Fantini (1962) reported that penicillin N production in strains of *E. glabra*, a homothallic ascomycetous fungus, is controlled by both multiple alleles and multiple loci and he studied the genetics of antibiotic formation by meiotic and mitotic recombination procedures.

Sexual recombination provides the most efficient means for introducing new attributes and increased variability into mutant strain lines. Unfortunately, most of our economically important eukaryotic organisms are primarily asexual and, therefore, possess limited genetic diversity when compared with organisms possessing true sexual cycles.

Parasexual Recombination

One of the most important advances in modern fungal genetics was the discovery of a novel recombinational mechanism known as parasexual recombination or parasexuality. This rare recombinational mechanism has now been demonstrated in a number of economically important fungi including *Aspergillpfi niger*, the koki molds, *Aspergillus sojae*, *A. oryzae*, *Penicillium chrysogenum*, *Fusarium oxysporum*, and *Ustilago maydis*.

The detection of a cryptic and rare recombinational event in a microorganism necessitates an environment which selects for a unique phenotype from an extremely large population of parental strains. A method commonly employed to demonstrate parasexual recombination is to propagate complementary biochemically deficient strains on a nonselective medium. Hyphal segments of copropagated mycelium are then plated on a selective medium, a condition which selects for heterokaryotic mycelium. The conidia of the heterokaryotic mycelium are then examined for the parental types which were employed for the initial mating. If parental types are found, the mycelium is heterokaryotic. Rare heterozygous conidia may also be found. They characteristically are larger than their haploid parents, they are more sensitive to ultraviolet radiation than haploid cultures, their deoxyribonucleic content is doubled, their conidial color is usually of the wild phenotype, and they are prototrophic with respect to nutritional requirements. The heterozygous diploid nuclei ultimately give rise to recombinants, either by vegetative haploidization or by mitotic crossing-over.

A number of strain improvement programs now use parasexual recombination for the selection of strains with improved fermentation yields.‡ After nearly a decade of parasexual research in *Penicillium*, several important factors have impeded the practical utilization of the process in industrial laboratories. The major obstacle is the marked decrease in fermentation productivity following the introduction of biochemical markers into the strains. Diploids formed following mating of mutant strains are usually inferior when compared with the productivity of the original parental strains. However, rare high-yielding diploid strains have been reported by Elander (1967). In a mating between a white-spored adenineless mutant and a yellow-spored methionineless mutant, a high-yielding diploid strain was isolated. The strain and several selections derived from it gave excellent phenoxymethyl penicillin yields. Another major obstacle is the marked instability of most diploid strains. Diploid cultures of *Penicillium* rarely breed true but typically give rise to segregant strains, most of which are of parental genotype and are inferior in their capacity to synthesize antibiotic. Diploid strains with extremely stable colony population patterns have been isolated. Diploid strains with stable homogeneous population patterns are important for production fermentations.

RECOMBINATION IN ANTIBIOTIC-PRODUCING ACTINOMYCETES

Since the discovery of syncytic recombination in the streptomycetes in 1955, many suggestions and criticisms have been offered concerning the application of recombination genetics to problems in industrial microbiology. The suggestions and criticisms have been valid in many cases; however, some have been based on observations made on one or two species and by academic investigators not always familiar with the problems and practices of industrial microbial genetics.

To date, the applications of recombinational genetics to enhanced or modified antibiotic production have not been abundant and the results obtained not overly encouraging. This is particularly evident when the results from programs of mutation and selection are compared with the results obtained from the few published reports of genetic programs. The induced mutation and selection approach is direct, practical, and usually successful.

Perhaps industrial microbial geneticists have not put sufficient effort into applying recombinational genetics to their research and development program. Although improvement in yield is the predominant activity of most applied geneticists, the potential for discovery of new antibiotic entities by techniques of interspecific hybridization, heterotransformation, and transduction has had limited application to date.

One of the major problems which have plagued streptomycete geneticists has been the problem encountered in obtaining suitable markers, a necessity for recombination studies. Alikhanian and Borisova (1961), working with *Streptomyces aureofaciens*, found only a preponderance of arginine-requiring mutants, which makes genetic studies extremely difficult. Polsinelli and Beretta (1966) also encountered similar difficulties in obtaining suitably marked strains. Fantini and Wallo (1967) isolated 388 arginineless mutants of *S. aureofaciens* and only four others with different requirements. Other difficulties frequently encountered are the poor condition following mutation induction and the weak viability of many mutant clones. Many mutants also appear to lose their ability to produce antibiotic following the introduction of genetic markers. Oftentimes, the problem of weak antibiotic activity is associated with poor vegetative growth. Although the lack of activity may have advantages in certain test systems, it is difficult to improve yields from a zero or near zero level of activity.

Despite these problems, recombination studies in a number of economically important streptomycetes have been described (Table 9-4). The studies include published reports on *S. fradiae*, *S. griseus*, *S. griseoflavus*, and *S. erythreus*. In 1961, Mindlin, Alikhanian, and Vladimirov described a breeding program with certain oxytetracycline-producing strains of *S. rimosus*. Originally, their program was designed to select a recombinant with decreased foaming characteristics on a high-carbohydrate, high-protein-containing medium. The strain finally selected, L-S-T-Hybrid, not only possessed this characteristic but also provided a significant increase in yield of oxytetracycline. Considerable research has also been undertaken with *S. aureofaciens* (Alikhanian and Borisova, 1961). They obtained recombinants with increased productivity, and certain of their arginineless mutants lost the ability to synthesize chlorotetracycline. Huang-Lo (1962) discovered evidence for recombination in the macrolide-producing organism *S. erythreus*. Certain recombinant strains synthesized excellent titers of erythromycin. Others exhibited a wide range of antibiotic yield.

The use of actinophage not only as a "mutagen" but also in transduction studies has had extensive application in certain commercial strains of streptomycetes. Phage attacks still plague antibiotic producers, and screening for phage-resistant variants is a common practice in industrial laboratories. Resistant variants isolated from lysed tank material may synthesize more antibiotic than the sensitive forms. Alikhanian and Teteryatnik (1962) reported the formation of streptomycin-producing variants from a nonproducing LS-1 strain of *Actinomyces streptomycini* through the action of actinophage. One transactive variant synthesized 5200 units/ml of antibiotic compared with 4100 units/ml for the parent strain. The author and associates recently applied phage

Table 9-4 Species of *Streptomyces* in which genetic recombination has been demonstrated

Species	Reported by	Antibiotic
Intraspecific		
S. violaceoruber	Sermonti and Spada-Sermonti (1956), Hopwood (1973), Braendle and Szybalski (1959)	Actinorhodin
S. fradiae	Braendle and Szybalski (1959)	Neomycin
S. rimosus	Alikhanian and Mindlin (1957)	Oxytetracycline
S. aureofaciens	Alikhanian and Borisova (1961)	Chlorotetracycline
S. griseoflavus	Saito (1958)	Novobiocin
S. griseus	Braendle and Szybalski (1959)	Streptomycin
S. scabies	Gregory and Huang (1964)	
S. erythreus	Huang-Lo (1962)	Erythromycin
S. sp. (8182H)	Fantini and Wallo (1967)	Unidentified antibiotic
S. antibioticus	Vladimirov (1966)	Oleandomycin
Interspecific		
S. rimosus × S. violaceoruber	Alacevic (1963)	Tetracycline and actinorhodin
S. rimosus × S. aureofaciens	Alacevic (1963)	Tetracyclines
S. aureofaciens × S. violaceoruber	Alacevic (1963), Bradley (1964)	Tetracycline and actinorhodin
S. griseus × S. viridochromogenes	Bradley (1964)	Streptomycin
S. aureofaciens × S. rimosus	Polsinelli and Beretta (1966)	Tetracyclines

techniques to certain high-producing strains of the erythromycin organism *S. erythreus*. In certain strains the phage system was more successful in obtaining mutants with desirable properties than conventional mutation-selection methods.

DNA base composition has been extensively studied in actinomycetes in recent years. Jones and Bradley (1963) and Frontali et al. (1965) studied numerous *Streptomyces* and *Nocardia* strains and reported mole percent guanosine-cytosine ratios from 74.4 to 78.5, using several sophisticated methods. Their data suggested close taxonomic relationships within the genus. More importantly, the possibility of intragenic recombination was further strengthened by such data. The discovery of polyvalent phage was also suggestive of a close relationship.

Several laboratories have recently reported interspecific recombination in *Nocardia* and *Streptomyces*. Adams (1964) reported in *Nocardia* a system closely resembling classic heterothallism. A most important paper by Alacevic (1963) describes interspecific recombination between various antibiotic-producing entities. Recombinations between *S. rimosus* and *S. coelicolor*, *S.*

rimosus and *S. aureofaciens*, and between *S. aureofaciens* and *S. coelicolor* again suggest the close relatedness of the species. No statements have been published concerning the elaboration of structurally modified antibiotics. A simple rapid technique which can be used for the detection of interspecific recombinants is diagramed in Fig. 9-27. One need scarcely belabor the importance of this concept to the antibiotic industry. Bradley (1964) reported interspecific recombination between *S. aureofaciens* (ATCC 10762) and *S. violaceoruber* (S-99). The interspecific recombinants were phenotypically similar to *S. violaceoruber*. They were unstable in culture and produced very little antibiotic.

LYSOGENIC CONVERSIONS

Virulent strains of *Corynebacterium diphtheriae* usually contain free phage, indicating that a small fraction of its population is lysogenic or pseudolysogenic. A nontoxigenic strain lysogenized with phage from a toxigenic culture may confer toxigenicity to the recipient strain. The phage-infected recipients appear to have a markedly altered metabolism which involves new macromolecular synthesis. Toxin production appears to be a result of interaction between the phage and host genomes. Examples of new macromolecular synthesis include the formation of 5-hydroxymethylcytosine (by T-even coliphages), diphtheria toxin (corynebacteria), and induced enzyme production.

Figure 9-27 Technique for the rapid selection of recombinant strains resulting from interspecific recombinations. *(By permission from R. P. Elander (1969), Applications of Microbial Genetics to Industrial Fermentations, in D. Perlman (ed.), "Fermentation Advances," pp. 89–114, Academic, New York.)*

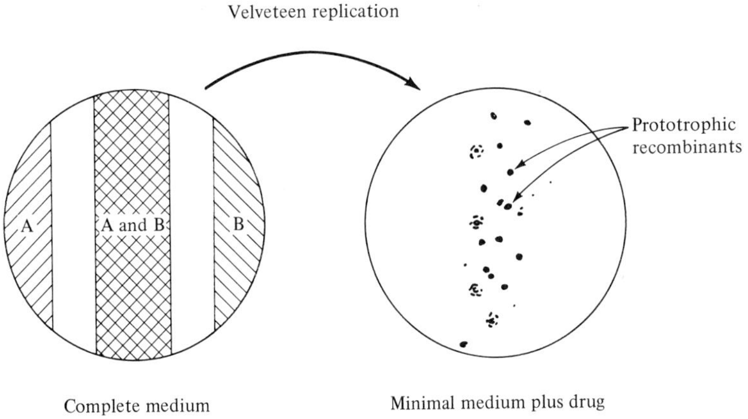

Strain A: wild-type, drug-sensitive strain, $W + D^S$
Strain B: auxotrophic, drug-resistant strain, $A^- B^- D^R$

Cross A × B: $A^+ B^+ \mid D^S \quad \times \quad A^- B^- \mid D^R$

Select recombinant: $A^+ B^+ D^R$

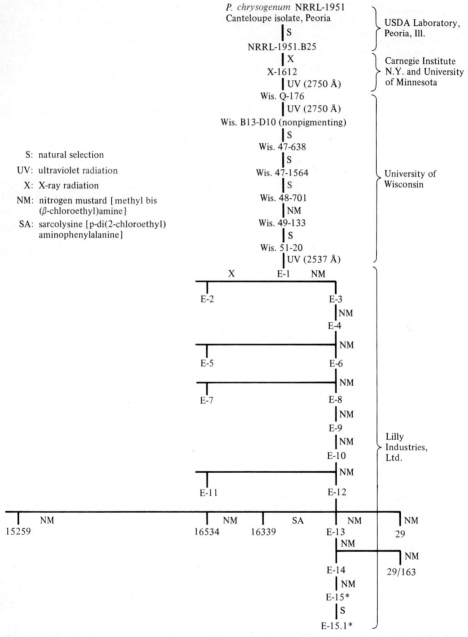

Figure 9-28 Strain lineage in the penicillin mold, *Penicillium chrysogenum*. The asterisks represent recent production cultures. *(By permission from R. P. Elander (1967), Enhanced Penicillin Production by Mutant and Recombinant Strains of Penicillium chrysogenum, Abh. Dsch. Akad. Wiss. Berlin, Kl. Med., 3:403–423.)*

Centifano (1968) described an interesting antiviral agent active against DNA viruses including vaccinia and herpes simplex. The agent is an internal phage protein, and the name phagicin has been proposed since its production is closely associated with phage replication. Phagicin is synthesized when *Escherichia coli* is infected with lambda phage. The material has not been found in uninfected *E. coli* cultures. Although internal proteins of other coliphages have been described, none has been reported to have antiviral activity.

Many actinomycete cultures are lysogenic; whether this is by chance or necessarily correlated with antibiotic production has not been established. Phage infection leading to lysis or lysogenic conversion is now known to induce new metabolites. Lysogeny and lysogenic conversion phenomena are potential sources of new chemical entities and offer potential new approaches for new antibiotic screening programs.

AN INDUSTRIAL MUTATION-SELECTION AND PARASEXUAL BREEDING PROGRAM IN *PENICILLIUM CHRYSOGENUM* ‡

Large-scale programs concerned with the induction, selection, and utilization of superior antibiotic-producing variants of *Penicillium chrysogenum* have been carried out in government, academic, and industrial laboratories for nearly 30 years. From the screens of hundreds of thousands of strains, a series of superior penicillin producers was developed, certain of which are utilized today for the commercial manufacture of penicillin. The lineage and productivity of a highly developed recent production strain of *P. chrysogenum* are illustrated in Fig. 9-28 and in Table 9-5. Most industrial microbial genetics laboratories have

Table 9-5 Antibiotic accumulation by a family of mutant strains of *Penicillium chrysogenum* in shake flask fermentation

	Improvement in productivity (fold) over			
Strain	Previous strain (Conditions of Moyer and Coghill, 1946)	Fleming strain (Conditions of Moyer and Coghill, 1946)	Previous strain (Modern laboratory conditions)	Fleming strain (Modern laboratory conditions)
Original Fleming				
NRRL-1951	0.25	0.25	0.50	0.50
NRRL-1951.B25	0.80	1.25	2.70	4.05
X-1612	0.63	2.66	0.06	4.90
Wis. Q-176	0.70	5.25	0.52	9.00
Wis. 47-1564	0.88	10.75	−0.04	8.55
Wis. 48-701	−0.14	9.00	−0.14	8.15
Wis. 49-133	0.41	13.08	0.68	12.70
Wis. 51-20	−0.37	7.83	0.73	22.60
E-15	2.73	32.00	1.05	50.00
E-15.1	−0.23	24.16	0.09	55.00

SOURCE: R. P. Elander (1967), Enhanced Penicillin Production by Mutant and Recombinant Strains of *Penicillium chrysogenum*, *Abh. Dtsch. Akad. Wiss. Berlin, Kl. Med.*, 3:403–423.

employed a strain development program consisting of natural selection, mutant selection, and parasexual breeding in their search for increased penicillin production.

A threefold program of culture selections can be established in regard to the genesis of new high-producing strains for production of penicillin. First, minor mutations are selected and screened, through a standard shake flask procedure, from natural clones or single spore clones treated through some physical agent of the electromagnetic spectrum such as the short wavelengths and high-quantum energies (as compared with visible light) of the ultraviolet spectrum of 2537 Å of shortwave or 3660 Å of long wave). Much of the ultraviolet irradiation work is performed with an ordinary Sterilamp or a Germicidal Tube Lamp and an ultraviolet intensity meter calibrated in $\mu W/cm^2$ or ergs/cm^2. Exposures of 200 to 400 $\mu W/cm^2$ may give a survival curve extended through 3 to 6 min and a 99 percent kill, depending on the particular culture being irradiated (Fig. 9-29). The ultraviolet energy source is placed one-half meter distant from the magnetically agitated spore suspension.

In order to obtain treated colonies of *Penicillium chrysogenum*, a spore suspension of a known culture is passed through a sterile filter and the spore count (using a hemacytometer) is adjusted so that the final dilution into sterile distilled water will give 200 to 400 colonies per plate when a spore dilution of 1

Figure 9-29 Frequency of morphological and biochemical variants in a survivor population of *Penicillium chrysogenum* strain E-15. [By permission from R. P. Elander (1967), Enhanced Penicillin Production by Mutant and Recombinant Strains of *Penicillium chrysogenum*, Abh. Dsch. Akad. Wiss. Berlin, Kl. Med., 3:403–423.]

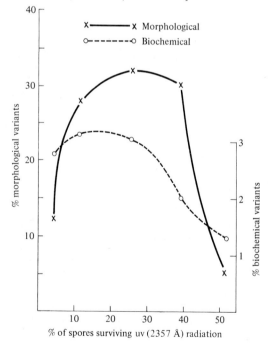

ml (of the final dilution) is flooded over the agar plate. After sampling the spore suspension for the "control set" or the "zero time" and also a 10-min warm-up for the uv source, the following exposure procedure is followed. The distance from the uv source for the experiments varies from 200 to 400 $\mu W/cm^2$ (20 to 40 ergs/cm^2) as indicated by an energy reading of the uv exposure meter. The spore suspension is treated in an uncovered petri dish on a magnetic stirrer, keeping the conidia suspended during the entire operation. The times of exposure are as follows: 0 (control), 30, 45, 60, 90, 120, 150, 180, 210, 240, 270, 300, and 360 s. After completing each exposure, the petri dish is covered with its lid over it. Protected from the ultraviolet light, 0.2-ml aliquots are taken out after each exposure and introduced to 100 ml sterile water blanks for final dilution. Then, by uncovering the dish, the exposures are continued for the desired exposure dose. The dilutions of the treated spores (and control) are added to the agar plates for germination of the conidia. One milliliter of each different time interval is spread over the surface of the replicate plates (5 or 10 plates per dilution). After the conidia germinate and the colonies grow as individual isolates at 25 to 26°C for 7 to 14 days, the plate counts are made (viable spore count), and a survival graph is made plotting time of exposure against percentage of survivors. Figure 9-29 illustrates a graph of mutation induction from a representative ultraviolet irradiation treatment.

The single-colony selections with minor mutations are made for studies of the penicillin yield by inoculating duplicate tubes from each colony near the 25 percent survival range. These are incubated for 5 to 8 days at 25°C (± 1°C), and are then ready for storage at 4°C while they are tested in the shake flask medium. After such a mutagenic treatment, selections of 100, 500, or 1000 cultures are tubed in duplicate slants and screened through a routine shake flask procedure with standard medium and proper controls (parent and/or production strain).

As a second choice of mutation and genetic studies, various chemical mutagens can be used singly, the same one in sequence several times, and also tried in conjunction with physical agents such as ultraviolet irradiation. For chemical agents of known mutagenicity for *Penicillium*, the following can be used: "nitrogen mustard" (NM) compound as mechlorethamine HCl or N-methyl bis-(2-chloroethyl)-amine HCl, nitrous acid (NA), N-methyl-N'-nitro-N-nitrosoguanidine (NG), butadiene dioxide (1,2,3,4-diepoxybutane) (DEB), or ethyl methane sulfonate (EMS). Other potential mutagens of bacteria or viruses can be tried for fungi; these include many alkylating and carcinogenic agents such as 7,12-dimethylbenz[a]anthracene (DMBA), triethylene melamine (TEM), ethylenimine (EI), and N-nitroso-N-methyl urethane.

Chemically-Induced Mutations Using N-Methyl-N'-nitro-N-nitrosoguanidine as the Mutagen (NG)

Much of the technique of NG is similar to ultraviolet irradiation procedures. Variations of the procedure are to be found in the following references: Moore (1969), Adelberg et al. (1965), and Mandell and Greenberg (1960).

The concentrations of NG used are 0.1 M or 0.01 M solutions; these concentrations may vary with time of exposure and with the strain used for the experiment. Aqueous solutions have been used and also tris-maleate buffer of pH

6.0 and 7.0 were compared, with results being of no detectable difference in the procedures used. The time of exposure may vary up to 2 h or 3 h, although platings representing intervals of 10, 20, 30, 45, 60, 120, 150, and 180 min exposure show a considerable degree of mutagenesis. The steps for plating, selecting, and testing are similar to those mentioned under the ultraviolet treatment.

As the third area of study with mold, mutagens, and mutations of strains of penicillin-producing *P. chrysogenum*, the minor mutants of the first two areas can be screened for penicillin titer and can also be used for the third choice: a study of genetic breeding and major mutations with several markers on the cultures. These can be collected by the use of "complete medium," for example, and then used as cultures for replication to minimal as well as complete medium for later use as parents in parasexual recombination studies. Cultures with hybrid vigor or "heterosis" have vast potential in changed metabolic requirements and alterations in amount of mycelium produced or products accumulated.

Auxanographic Analysis and Parasexual Recombination Techniques

Following long exposures to a variety of mutagenic agents, the treated colonies, on plates representing a 99 percent kill, either plated out directly following exposure or enriched for mutation percentage by exposure to antifungal agents, i.e., the nystatin enrichment technique of Macdonald (1968), are replicated to complete and minimal media in order to select our putative biochemically deficient strains. The formulation of the "complete medium" is as follows:

NH_4NO_3	0.3%
KH_2PO_4	0.1%
$MgSO_4 \cdot 7H_2O$	0.05%
KCl	0.05%
$FeSO_4 \cdot 7H_2O$	0.001%
Corn dextrin	3.0%
Casamino acids	0.25%
Yeast extract	0.5%
Corn-steep liquor	0.5%
Peptone	0.5%
Agar	2.0%

pH 5.5 (before sterilization)

A representative "minimal medium" is the following:

NH_4NO_3	0.3%
KH_2PO_4	0.1%
$MgSO_4 \cdot 7H_2O$	0.05%
KCl	0.05%
$FeSO_4 \cdot 7H_2O$	0.001%
Sucrose	3.0%
Ionagar (or washed agar)	1.0% (2.0%)

pH 5.5 (before sterilization)

A special type of replicator commonly used for the mass transfer of treated colonies is composed of a Teflon disc impregnated with a large number of stainless steel pins spaced according to a uniform pattern [see Clowes and Hayes (1968), for description].

Colonies which grow on a complete medium but fail to germinate on a minimal medium are termed putative auxotrophic cultures and are selected for further testing and purification.

The putative biochemical mutant is then replicated to the following media in the second phase of auxanographic analysis:

1. Complete medium
2. Minimal medium
3. Minimal medium supplemented with casein hydrolysate (vitamin-free)- selects amino acid–deficient mutants
4. Minimal medium supplemented with vitamin solution or yeast extract– selects for vitamin-requiring mutants
5. Minimal medium plus sodium ribonucleate and sodium deoxyribonucleate– selects for purine- and pyrimidine-requiring mutants

If the auxotroph makes growth only on complete (1) and minimal media supplemented with casein hydrolysate (3), the mutant is further replicated to minimal medium supplemented with the following amino acid subgroups, added either singly or in combination:

    Group A   Isoleucine
                    Leucine
                    Valine
    Group B   Threonine
                    Methionine
                    Cysteine
                    Cystine
                    Serine
    Group C   Glycine
                    Alanine
                    Aspartic Acid
                    Glutamic Acid
    Group D   Histidine
                    Tryptophan
                    Phenylalanine
                    Tyrosine
    Group E   Arginine
                    Lysine
                    Proline
                    Hydroxyproline

If growth occurs only on complete medium (1) and minimal medium supplemented with vitamin solution (4), the following individual vitamins are supplemented to minimal medium:

    Biotin
    Choline chloride
    Inositol
    Niacinamide

Nicotinic acid
Calcium pantothenate
p-Aminobenzoic acid
Pyridoxine
Riboflavin
Thiamin

If growth occurs only on complete medium (1) and minimal medium supplemented with nucleic acid salts (5), the following purine and pyrimidine bases are supplemented to the minimal medium:

Adenine
Thymine
Cytosine
Guanine
Uracil

By repeated replication on the above media and supplementations, the exact biochemical requirements for most mutant cultures may be determined.

Repeated mutagenic treatments on mutant clones result in multiple deficient cultures. Back-reversion frequencies are determined by massive platings on minimal medium. Back-reversion frequencies of less than 10^{-6} to 10^{-7} are desirable. Mutant strains which make slight growth on the various media are termed "leaky" and are normally not used for recombination studies. The mutant strains are preserved by standard lyophilization or liquid nitrogen techniques.

Other types of mutant selection techniques commonly used in recombination studies include:

1 Visual selection: conidial color (yellow, albino versus green phenotype)
2 Selection for resistance to toxic agents (antibiotics, analogs, etc.)
3 Selection by starvation: homozygous adenincless, biotinless segregants survive longer than diploids, which are homozygous for biotin independence but heterozygous for adenine requirement, in a medium lacking both biotin and adenine

Heterokaryons are synthesized by copropagation of doubly marked strains in a complete broth medium. Large numbers of spores (10^6/ml) of the complementary mutant parents are added to a complete medium and incubated for 5 to 7 days at 25°C. The resulting heterokaryotic mycelium is then washed in sterile buffer or saline solution and propagated on plates containing minimal medium. After prolonged incubation, fast-growing heterokaryotic mycelium of intermediate spore color (provided conidial color mutants are used in the mating) are observed. The heterokaryons are usually characterized by vigorous (prototrophic) growth on a minimal medium, are intermediate in color, and segregate back to the parental types on sporulation.

Rare heterozygous diploid sectors may be observed in the heterokaryotic mycelium. The diploid sectors are wild-type with respect to spore pigment, are prototrophic on minimal medium, are larger in volume when compared with haploids, contain more deoxyribonucleic acid than haploids, and are characteristically more sensitive to uv radiation. Diploid clones may also be obtained from minimal medium plates seeded with dense (10^8/ml) washed spore suspensions derived from heterokaryons.

Heterozygous diploid strains of *Penicillium chrysogenum* rarely breed true but give rise to numerous segregant sectors. The segregation frequencies can be increased by a variety of chemical and physical agents.

Segregation analyses usually reveal a high degree of parental genome segregation in *Penicillium chrysogenum*. However, new rare recombinant types are found. The genotypes of the recombinant classes can be determined by standard auxanographic analyses. Both diploid and haploid recombinant classes are routinely found as second-order segregants.

The haploid and diploid recombinant cultures are maintained according to conventional procedures. Fermentation screening of a large number of "somatic" recombinants may show rare high-yielding strains. An example of a high-yielding diploid strain and its segregants is presented in Fig. 9-30. This strain has been employed for the large-scale commercial production of benzylpenicillin.

Figure 9-30 Penicillin production by starting parent, biochemical mutants, heterokaryon, heterozygous diploid, and diploid segregants from mating combination No. 1 (y met) × (w ade). *(By permission from R. P. Elander (1967), Enhanced Penicillin Production by Mutant and Recombinant Strains of Penicillium chrysogenum, Abh. Dtsch. Akad. Wiss. Berlin, Kl. Med., 3: 403–423.)*

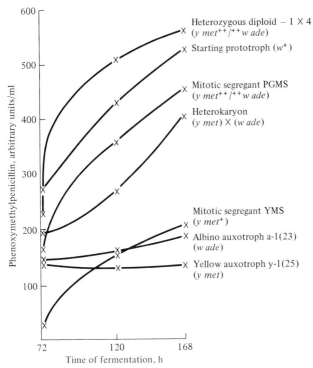

DNA (deoxyribonucleic acid) determinations indicate that the DNA content of the diploid *Penicillium* spore is approximately two times the value of the haploid parents, i.e., wild type about 0.44 to 0.58 (units $\times 10^{-7}$ µg DNA spore), haploids about 0.35 to 0.57, and diploids, 0.76 to 0.83 (Elander et al., 1973).

The shake flask evaluations for new mutant or recombinant selections are carried out in 250- or 500-ml Erlenmeyer flasks shaken at a speed of 250 r/min and a volume of 50 or 100 ml per given container. Minor deviations can be tried under the various specific conditions of different laboratories. A given uniform or standard medium should be kept basic and comparable to the production formulas used. This helps to provide data found under experimental conditions that is more meaningful in scale-up to large tank conditions in the pilot plant and production areas. A given medium would give a certain value of potency for the control strain, and the new improved cultures could be expressed as 120 percent or +20 percent over the control (see Fig. 9-31). The probabilities for improved mutants in this range of superiority are presented in Fig. 9-32.

A histogram or population distribution bar graph is made to show the skewedness of a standard curve and affords direction for the next mutagenic treatment as to culture to use as parent as well as dosage or kind of mutagen for subsequent study. By carefully examining several histograms in sequence, the path to follow for more productive mutants becomes evident from the distribution curves which show more of the slope to the plus side of 100 percent (the control average) than to the left side of the population study.

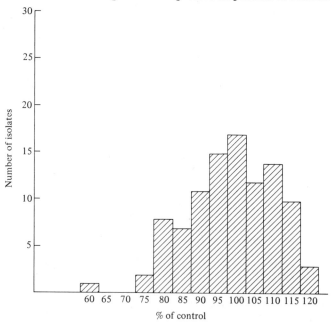

Figure 9-31 Histogram depicting a population study of 100 isolates with their potencies expressed as percent of control.

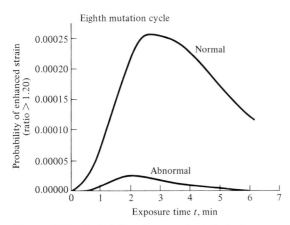

Figure 9-32 Probability that a selected strain would be truly superior to the control strain by an enrichment of 20 percent. *(By permission from W. F. Brown and R. P. Elander (1966), Some Biometric Considerations in an Applied Antibiotic AD-464 Strain Development Program, Dev. Ind. Microbiol., 8:114–123.)*

Why three pathways in a strain or culture development program? The more avenues of approach for an improvement of production yield, the better are the results and the broader the coverage for discovering methods for ameliorating the monthly production output. If a yield of 100 billion units (BU) or approximately 60 kg of penicillin G per tank can be increased to 150 billion units or approximately 90 kg by culture selection, this is the most inexpensive way to increase the total yield per day or per month. Usually the change of culture involves cost for the development work but the routine raw materials and techniques are only slightly altered. This means the expenditures for overhead, labor, materials, etc., remain the same but the yield and recovery increase for days of tank operation.

The shake flask formula for penicillin comparisons among cultures may be patterned after the following representative medium:

Corn-steep liquor	6% (wet wt/vol)
Lactose	6–8%
$CaCO_3$	0.5%
$(NH_4)_2SO_4$	0.2–0.4%
Sodium phenylacetate	0.2–0.4%
Lard oil	0.1–0.5%

pH 5.5 (before sterilization)

The fermentation medium can be dispensed to Erlenmeyer flasks (50 ml of medium in 250-ml volume, or 100 ml in 500-ml size) for rotary shaker study. The flasks are agitated at a speed of 200 to 300 r/min at an ambient temperature of 25°C. In screening selected clones by groups of 100 and saving for retest only those with a +20 percent value over the control, the highest producers after two successive evaluations are freeze-dried from the spores of a master slant. The lyophilized spores are then used for retests and the highest producer selected from a hundred or a thousand cultures is used for aerated jar tests and also for further treatment and testing.

The economic implication of a strain improvement program is that as the environment of medium and tank relationships are changed and improved, so also must the genetic potential of the culture be enhanced. As the strain is ameliorated with new selections by different mutation methods or by new and improved breeding programs of recombination, other changes in the fermentation environment are maximized by medium changes or improved engineering which result in the highest possible yield. No one condition or change will result in a large increase but all the separate and individual modifications aid in optimizing the fermentation by increased yields and, thereby, increasing both production profits and reduced consumer prices. The following statistics clearly illustrate the marked influence increased productivity has on drug costs. According to the United States Tariff Commission Report for 1945, the total United States production of penicillin was nearly 7.5 trillion units (4.5 metric tons). The total sales for this period amounted to $46.5 million. At this time penicillin sold for approximately $11 per gram. In contrast, the total amount of penicillin antibiotics manufactured in the United States alone in 1963 was 715 trillion units (480 metric tons) representing $72 million in sales at wholesale prices. Due to the tremendous gains in production, penicillin is now sold wholesale for less than 15 cents per gram. For the same year the total world production of all antibiotics was 6.7 million pounds (3050 metric tons) valued at $388 million.

THE FUTURE OF MICROBIAL GENETICS IN INDUSTRIAL MICROBIOLOGY

Applied microbial genetics will play an ever increasing role in future fermentation research. It constitutes a rationale for strain improvement programs and provides an approach for developing new and unique fermentation products. The potential of recombinant strains produced through the techniques of transformation, transduction, lysogenic conversion, and the vast potential of interspecific and intergenetic recombination offer numerous interesting avenues of application which may culminate in products now not known. The development of higher ploidy strains and strains with altered feedback control mechanisms will lead to unparalleled productivity of current and new fermentation products. Undoubtedly the greatest challenge for the applied microbial geneticist lies in the recognition of new goals. This philosophy should exert a determinative role in the development of commercial cell-free syntheses and in the directed evolution of microbes to serve both the terrestrial and extraterrestrial needs of humankind.

LITERATURE CITED

Adams, J. N. (1964): Recombination between *Nocardia canicuria* and *Nocardia erythropolis*, *J. Bacteriol.*, **88**:856–878.

Adelberg, E. A. (1958): Selection of Bacterial Mutants Which Excrete Antagonists to Antimetabolites, *J. Bacteriol.*, **76**:326–328.

———, M. Mandel, and G. C. C. Chen (1965): Optimal Conditions for Mutagenesis by N-Methyl-N'-nitro-N-nitrosoguanidine in *Escherichia coli* K12, *Biochem. Biophys. Res. Commun.*, **18**:788–795.

Alacevic, M. (1963): Interspecific Recombination in *Streptomyces*, *Nature*, **197**:1323.

Alikhanian, S. I. (1962): Induced Mutagenesis in the Selection of Microorganisms, *Adv. Appl. Microbiol.*, **4**:1–50.

——— and L. N. Borisova (1961): Recombination in *Streptomyces aureofaciens*, *Sci. Rep. 1st Super, Sanita*, **1**:470–472.

——— and S. Z. Mindlin (1957): Recombination in *Streptomyces rimosus*, *Nature*, **180**:1208–1209.

——— and A. F. Teteryatnik (1962): Formation of Streptomycin-producing Variants from *Actinomyces streptomycini* Strain LS-1 through the Action of Actinophages, *Mikrobiologiya*, **31**:54–60.

Backus, M. P., and J. F. Stauffer (1955): The Production and Selection of a Family of Strains in *Penicillium chrysogenum*, *Mycologia*, **47**:429–463.

Ballio, A., E. B. Chain, F. Dentice Di Acadia, M. F. Mastropietro-Cancellieri, G. Morpurgo, G. Serlupi-Crescenzi, and G. Sermonti (1960): Incorporation of α-ϵ-Dicarboxylic Acids as Side Chains into the Penicillin Molecule, *Nature*, **185**:97–99.

Barchielli, R., G. Baretti, A. DiMarco, P. Julita, A. Migliacci, A. Munghetti, and C. Spalla (1960): Isolation and Structure of New Factor of the Vitamin B_{12} Group. Guanosine Diphosphate Factor B, *Biochem. J.*, **74**:382–387.

Bradley, S. G. (1964): Genetic Analysis of an Unstable Mutant of *Streptomyces violaceoruber*, *Dev. Ind. Microbiol.*, **6**:296–301.

Braendle, D. H., and W. Szybalski (1959): Heterokaryotic Compatibility, Metabolic Cooperation and Genic Recombination in *Streptomyces*, *Ann. N.Y. Acad. Sci.*, **81**:824–851.

Brown, W. F., and R. P. Elander (1966): Some Biometric Considerations in an Applied Antibiotic AD-464 Strain Development Program. *Dev. Ind. Microbiol.*, **8**:114–123.

Calam, C. T. (1964): The Selection, Improvement, and Preservation of Microorganisms, *Prog. Ind. Microbiol.*, **5**:1–54.

Calvo, R. A., and J. M. Calvo (1967): Lack of End-product Inhibition and Repression of Leucine Synthesis in a Strain of *Salmonella typhimurium*, *Science*, **156**:1107–1109.

Carvajal, F. (1952): Phage Problems in the Streptomycin Fermentation, *Mycologia*, **45**:209–234.

Centifano, Y. M. (1968): Antiviral Agent from λ-Infected *Escherichia coli*, *Appl. Microbiol.*, **16**:827–834.

Claridge, C. A., J. A. Bush, M. D. DeFuria, and K. E. Price (1974): Fermentation and Mutation Studies with a Butirosin-producing Strain of *Bacillus circulans*, *Dev. Ind. Microbiol.*, **15**:101–113.

Clowes, R. C., and W. Hayes (1968): "Experiments in Microbial Genetics," Wiley, New York.

Cohen, G. N., and E. A. Adelberg (1958): Kinetics of Incorporation of p-Fluorophenylalanine by a Mutant of *E. coli* Resistant to This Analogue, *J. Bacteriol.*, **76**:328–330.

Darken, M. A., H. Berenson, R. J. Shirk, and N. O. Sjolander (1960): Production of Tetracycline by *Streptomyces aureofaciens* in Synthetic Media, *Appl. Microbiol.*, **8**:46–51.

Davies, O. L. (1964): Screening for Improved Mutants in Antibiotic Research, *Biometrics*, **20**:576–591.

Demain, A. L. (1966): Industrial Fermentations and Their Relation to Regulatory Mechanisms, *Adv. Appl. Microbiol.*, **8**:1–27.

———(1974): Biochemistry of Penicillin and Cephalosporin Fermentations, *Lloydia*, **37**:147–167.

Dulaney, E. L., M. Ruger, and C. Hlavac (1949): Observations of *Streptomyces griseus*. IV. Induced Mutation Strain Selection, *Mycologia*, **41**:388–397.

Elander, R. P. (1966): Two Decades of Strain Development in Antibiotic-producing Microorganisms, *Dev. Ind. Microbiol.*, **8**:61–73.

——— (1967): Enhanced Penicillin Production by Mutant and Recombinant Strains of *Penicillium chrysogenum*, *Abh. Dsch. Akad. Wiss. Berlin, Kl. Med.*, **3**:403–423.

——— (1969): Applications of Microbial Genetics to Industrial Fermentations, in D. Perlman (ed.), "*Fermentation Advances*," pp. 89–114, Academic, New York.

———, C. J. Corum, H. DeValeria, and R. M. Wilgus (1974): Ultraviolet Mutagenesis and Cephalosporin Synthesis in Strains of *Cephalosporium acremonium*, *Int. Symp. Genet. Ind. Microorganisms*, 2d (Abstract), Sheffield, p. 19.

———, M. A. Espenshade, S. G. Pathak, and C. H. Pan (1973): The Use of Parasexual Genetics in an Industrial Strain Improvement Program with *Penicillium chrysogenum*, in Z. Vanek, Z. Hostalek, and J. Cudlin (eds.), "Genetics of Industrial Microorganisms," vol. II, pp. 239–253, Elsevier, Amsterdam.

———, J. A. Mabe, R. L. Hamill, and M. Gorman (1971): Biosynthesis of Pyrrolnitrins by Analog-resistant Mutants of *Pseudomonas fluorescens*, *Folia Microbiol. (Prague)*, **16**:156–165.

———, J. F. Stauffer, and M. P. Backus (1961): Antibiotic Production by Various Species and Varieties of *Emericellopsis* and *Cephalosporium*, *Antimicrob. Agents Ann.* 1960: 91–102.

Enebo, L., H. Johnson, K. Nordstrom, and A. Möller (1961): Yeast Improvement by Hybridization, *J. Inst. Brew., London*, **67**:76–79.

Fantini, A. A. (1962): Genetics and Antibiotic Production by *Emericellopsis* species, *Genetics*, **47**:161–177.

———, and K. G. Wallo (1967): *Streptomyces* Genetics and Industrial Microbiology, *Trans. N.Y. Acad. Sci.*, **29**:800–809.

Florey, H. W. (1955): Antibiotic Products of a Versatile Fungus, *Ann. Internt. Med.*, **43**:480–490.

Frontali, C., L. R. Hill, and L. G. Sylvestri (1965): DNA Base Composition of *Streptomyces*; *J. Gen. Microbiol.*, **38**:243–250.

Gorman, M., R. L. Hamill, R. P. Elander, and J. Mabe (1968): Preparation of Substituted Phenyl Pyrroles Through the Metabolism of Tryptophan Analogues, *Biochem. Biophys. Res. Commun.*, **31**:294–298.

Gregory, K. F., and J. C. Huang (1964): Tyrosinase Inheritance in *Streptomyces scabies*. I. Genetic Recombination, *J. Bacteriol.*, **87**:1281–1286.

Haas, F. L., T. A. Puglisi, A. J. Moses, and J. Lein (1956): Heterokaryosis as a Cause of Culture Rundown in *Penicillium chrysogenum*, *Appl. Microbiol.*, **4**:187–195.

Hopwood, D. A. (1973): Developments in Actinomycete Genetics, in Z. Vanek, Z. Hostalek, and J. Cudlin (eds.), "Genetics of Industrial Microorganisms" vol. II, pp. 21–46, Elsevier, Amsterdam.

Hostalek, Z., and Z. Vanek (1973): Molecular Basis of Polygenic Inheritance in the Biosynthesis of Tetracycline, in Z. Vanek, Z. Hostalek, and J. Cudlin (eds.), "Genetics of Industrial Microorganisms," vol. II, pp. 353–371, Elsevier, Amsterdam.

Huang-Lo, L. (1962): Hybridization in *Actinomyces erythreus*, *Mikrobiologiya*, **31**:61–65.

James, L. V., S. D. Rubbo, and J. F. Gardener (1956): Isolation of High Acid-yielding Mutants of *Aspergillus niger* Using a Paper Culture Selection Technique, *J. Gen. Microbiol.*, **14**:223–227.

Jones, L. A., and S. G. Bradley (1963): Thermal Denaturation of Deoxyribonucleic Acid from Actinomycetes and Actinophages, *Bacteriol. Proc.*, p. 148.

Kalle, G. P., and J. S. Gots (1962): Feedback Inhibition of Purine Biosynthesis in Adenine-excreting Mutants of *Salmonella typhimurium*, *Proc. Soc. Exp. Biol. Med.*, **109**:281–284.

Karlstrom, O. (1965): Methods for the Production of Mutants as Amino Acid Fermenting Organisms, *Biotechnol. Bioeng.*, **7**:245–268.

Kelner, A. (1949): Studies on the Genetics of Antibiotic Formation. I. The Introduction of Antibiotic-forming Mutants in Actinomycetes, *J. Bacteriol.*, **57**:73–92.

Lemke, P. A., and D. R. Brannon (1972): Microbial Synthesis of Cephalosporin and Penicillin Compounds, in E. H. Flynn (ed.), "Cephalosporin and Penicillins: Chemistry and Biology," pp. 370–437, Academic, New York.

Lingens, F., H. Kraus, and S. Lingens (1964): Regulation of Tryptophan Biosynthesis, *Z. Physiol. Chem.*, **339**:1–8.

Macdonald, K. D. (1968): The Selection of Auxotrophs of *Penicillium chrysogenum* with nystatin, *Genet. Res. Comb.*, **11**:327–330.

Mandell, J. D., and J. Greenberg (1960): A New Chemical Mutagen for Bacteria, 1-Methyl-3-nitro-1-nitrosoguanidine, *Biochem. Biophys. Res. Commun.*, **3**:575–577.

McCormick, J. R. D., N. O. Sjolander, U. Hirsch, E. Jensen, and A. P. Doershuk (1957): A New Family of Antibiotics: The Demethyl Tetracyclines, *J. Am. Chem. Soc.*, **79**:4561–4563.

Miller, P. A., J. H. Hash, M. Lincks, and N. Bohonos (1965): Biosynthesis of 5-Hydroxytetracycline, *Biochem. Biophys. Res. Commun.* **18**:325–331.

Moore, D. (1969): The Mutagenic Action of N-Methyl-N'-nitro-N-nitrosoguanidine on *Coprinus lagopus*, *J. Gen. Microbiol.*, **55**:121–125.

Nakamura, K., and C. S. Gowans (1964): Nicotinic acid–excreting Mutants of *Chlamydomonas*, *Nature*, **202**:826–827.

Nelson, T. C. (1961): Screening Methods in Microbiology. Mutation and Plant Breeding, *N. A. S. N. R. C. Publ.* 891, pp. 311–335.

Nuesch, J., H. J. Treichler, and M. Liersch (1973): The Biosynthesis of Cephalosporin C, in Z. Vanek, Z. Hostalek, and J. Cudlin (eds.), "Genetics of Industrial Microorganisms," vol. II, pp. 309–334, Elsevier, Amsterdam.

Oakberg, E. F., and S. E. Luria (1947): Mutation to Sulfonamide Resistance in *Staphylococcus aureus*, *Genetics*, **32**:249–261.

Orgel, L. E. (1965): The Chemical Basis of Mutation, *Adv. Enzymol.*, **27**:289–346.

Ostroukhov, A. A., and V. D. Kuznetsov (1963): An Accelerated Method of Selecting Active Variants of Penicillin Producers by Means of rH_2- Indicators, *Antibiotiki (Moscow)*, **8**:33–35.

Pirt, S. J., and R. C. Righelato (1967): Effect of Growth Rate on the Synthesis of Penicillin by *Penicillium chrysogenum* in Batch and Chemostat Cultures, *Appl. Microbiol.*, **15**:1284–1290.

Polsinelli, M., and M. Beretta (1966): Genetic Recombination in Crosses Between *Streptomyces aureofaciens* and *Streptomyces rimosus*, *J. Bacteriol.*, **91**:63.

Ramachandran, S., R. S. Sukapure, and M. J. Thirumalachar (1965): Bio-transformation of Thiolutin Production from *Streptomyces pimprina* to *Streptomyces aureofaciens*, *Hind. Antibiot. Bull.*, **7**:197–202.

Raper, K. B. (1946): The Development of Improved Penicillin-producing Molds, *Ann. N.Y. Acad. Sci.*, **48**:41–57.

Saito, H. (1958): Heterokaryosis and Genetic Recombination in *Streptomyces griseoflavus*, *Can. J. Microbiol.*, **9**:571–580.

Scherr, G. A., and M. E. Rafelson, Jr. (1962): The Directed Isolation of Mutants Producing Increased Amounts of Metabolites, *J. Appl. Bacteriol.*, **25**:187–194.

Sermonti, G., and S. Casciano (1963): Sexual Polarity in *Streptomyces coelicolor*, *J. Gen. Microbiol.*, **33**:293–301.

———and I. Spada-Sermonti (1956): Gene Recombination in *Streptomyces coelicolor*, *J. Gen. Microbiol.*, **15**:609–616.

Shier, W. T., K. L. Rinehart, and D. Gottlieb (1969): Preparation of Four New Antibiotics from a Mutant of *Streptomyces fradiae*, *Proc. Nat. Acad. Sci. U.S.A.*, **63**:198–204.

Thom, C., and R. A. Steinberg (1939): The Chemical Induction of Genetic Changes in Fungi, *Proc. Nat. Acad. Sci. U.S.A.*, **25**:329–335.

Vladimirov, A. V. (1966): Genetic Recombination in *Actinomyces antibioticus* Producing Oleandomycin, *Antibiotiki (Moscow)*, **11**:117–122.

GENERAL REFERENCES

Bennett, T. P., and E. Frieden (1967): "Modern Topics in Biochemistry," Macmillan, New York.
Bradley, S. G. (1966): Industrial Microbial Genetics, *Soc. Ind. Microbiol., Spec. Publ. 3*, Arlington, Va.
Clowes, R. C., and W. Hayes (1968): "Experiments in Microbial Genetics," Wiley, New York.
Collins, C. H. (1967): "Progress in Microbiological Techniques," Plenum, New York.
Curtis, H. (1968): "Biology," Worth Publishers, New York.
Esser, K., and R. Kuenen (1967): "Genetics of Fungi," Springer-Verlag, New York.
Fincham, J. R. S., and P. R. Day (1971): "Fungal Genetics," 3d ed., Davis, Philadelphia.
Hayes, W. (1969): "The Genetics of Bacteria and Their Viruses: Studies in Basic Genetics and Molecular Biology," 2d ed., Wiley, New York.
Herskowitz, I. H. (1967): "Basic Principles of Molecular Genetics," Little, Brown, Boston.
Hopwood, D. A. (1973): Developments in Actinomycete Genetics, in Z. Vanek, Z. Hostalek, and J. Cudlin (eds.), "Genetics of Industrial Microorganisms," vol. II, pp. 21–46, Elsevier, Amsterdam.
Hostalek, Z., and Z. Vanek (1973): Molecular Basis of Polygenic Inheritance in the Biosynthesis of Tetracycline, in Z. Vanek, Z. Hostalek, and J. Cudlin (eds.), "Genetics of Industrial Microorganisms," vol. II, pp. 353–371, Elsevier, Amsterdam.
Lehninger, A. L. (1970): "Biochemistry: The Molecular Basis of Cell Structure and Function," Worth Publishers, New York.
Levene, R. P. (1968): "Genetics," 2d ed., Holt, New York.
Phaff, H. J., M. W. Miller, and E. M. Mrak (1966): "The Life of Yeasts," Harvard, Cambridge, Mass.
Potter, V. R. (1960): "Nucleic Acid Outlines," Burgess, Minneapolis.
Sermonti, G. (1969): "Genetics of Antibiotic Producing Microorganisms," Wiley, New York.
Stent, G. S. (1963): "Molecular Biology of Bacterial Viruses," Freeman, San Francisco.
Swanson, C. P. (1964): "The Cell," 2d ed., Prentice-Hall, Englewood, N.J.
Terui, G. (ed.) (1973): "Fermentation Technology Today," Society for Fermentation Technology, Japan.
Vanek, Z., Z. Hostalek, and J. Cudlin (eds.) (1973): "Genetics of Industrial Microorganisms," vols. I (Bacteria) and II (Actinomycetes and Fungi), Elsevier, Amsterdam.
Wagner, R. P., and H. K. Mitchell (1964): "Genetics and Metabolism," 2d ed., Wiley, New York.
Watson, J. D. (1970): "Molecular Biology of the Gene." 2d ed., W. A. Benjamin, Inc., New York.

10
FOOD MICROBIOLOGY
John H. Litchfield

In this chapter we will present the microbiological aspects of food processing and preservation from the standpoint of characteristics of microorganisms involved in food spoilage and in foodborne diseases. Plant diseases that affect market quality of fruits and vegetables will not be discussed in this chapter, since this subject properly belongs in the area of plant pathology.

HISTORICAL DEVELOPMENT

In 1810 Nicholas Appert was awarded a prize of 12,000 francs by Napoleon for the development of a method for preserving foods by heating in sealed glass containers. However, the role of microorganisms in food spoilage remained a controversial subject until 1866 when Louis Pasteur announced his book "Études sur le vin" in which wild yeasts and bacteria were shown to be the causative agents of wine spoilage. Pasteur pointed out that heating wine at 50 to 60°C ("pasteurization") solved the spoilage problem.

William Thompson Sedgwick at the Massachusetts Institute of Technology in 1892 demonstrated the relationship of the bacterial count of milk to its sanitary quality. Subsequently, Theobald Smith in 1899 determined the thermal death time of *Mycobacterium tuberculosis* in milk to be 60°C for 15 min.

Pioneering work in the 1890s by Prescott and Underwood at the Massachusetts Institute of Technology and by H. L. Russell at the University of Wisconsin identified certain spore-forming bacteria as the causative agents of canned food spoilage. These investigators pointed out the need for adequate thermal process times and temperatures to destroy heat-resistant spores of spoilage bacteria. However, it remained for Bigelow and Ball in the 1920s to determine the actual temperatures in hermetically sealed containers of food during heating and cooling. By relating this information to the thermal resistance of bacterial spores, they were able to develop mathematical methods for determining safe thermal process times for canned foods.

Every serious student of food microbiology should read the original papers written by these pioneering investigators. These publications have been conveniently reprinted in an anthology (Goldblith et al., 1961).

FOOD PRESERVATION
GENERAL CONSIDERATIONS

Microorganisms require a minimum of a carbon and energy source, a nitrogen source, inorganic nutrients, and water for growth. Many organisms also require certain amino acids and vitamins. Since foods provide all these nutrients to some degree, methods for controlling the growth of microorganisms in foods or for preserving foods depend upon the following:

1 Killing microorganisms by heating (canning), mechanical disruption, or ionizing (gamma irradiation, electron beam) or nonionizing (ultraviolet, microwave) radiations
2 Inhibiting growth of microorganisms by drying (dehydration, freeze-drying), chemical preservatives, refrigeration or freezing, high osmotic pressures (sugar syrups and salt brines), low pH (acidulants), or low oxidation-reduction potential (anaerobic conditions)
3 Removing microorganisms by filtration or centrifugation, in the case of liquid foods such as fruit juices
4 Preventing contamination of foods by maintaining aseptic conditions

Numerous combinations of these methods are used in the food industry; for example, ham and bacon are prepared by curing these cuts of pork with salt, sugar, sodium nitrate, and sodium nitrite and then by thermal processing at a sufficient temperature during smoking to reduce the numbers of food poisoning bacteria, such as staphylococci, to minimal levels.

Smoke constituents usually provide only a mild bacteriostatic effect at the surface of the meat. These products are then packaged and must be held under refrigerated conditions to prevent outgrowth of the microorganisms that survive processing. However, the storage life of these products is limited since some surviving organisms will grow out eventually even under refrigeration temperatures. In addition, cured hams may be packed in metal cans and processed at a "pasteurizing" time and temperature sufficient to reduce the numbers of vegetative cells of aerobic spore-forming spoilage microorganisms such as *Bacillus megaterium* to minimal levels. However, this heat treatment will not destroy either the spores of this aerobe or those of *Clostridium botulinum* which is the causative organism of botulism. Such a product must be refrigerated at all times before use. In this type of product, the brine concentration [salt × 100/(moisture + salt)] and refrigerated storage are sufficient to inhibit the growth of any surviving microorganisms.

To carry this example one step further, cured hams may also be packed in metal cans and processed at a sufficient temperature-time combination to destroy heat-resistant spores of spoilage bacteria and of food poisoning organisms such as *Clostridium perfringens* and *C. botulinum.* Such a product is termed "commercially sterile." We will point out subsequently that the destruction of microorganisms by heat follows a logarithmic relationship. Thus, there is always a probability that a spore will survive the process. Commercial sterility indicates that conditions in the product after processing are unfavorable for the outgrowth of any surviving spores.

Every student of elementary microbiology is introduced to the nature of microbial growth and death curves. To refresh the reader's memory, the microbial growth cycle consists of a lag phase, a logarithmic (or exponential) growth phase, a stationary phase, a phase of death or decline, and a survival phase (Monod, 1949).

In food microbiology, the lag and logarithmic phases are of greatest importance, since foods are usually spoiled by the time that microbial populations reach the stationary phase. Ideally, foods should be held under conditions that maintain microbial populations in the lag phase.

The shape of the bacterial growth curve, i.e., length of lag, logarithmic, stationary, decline, and survival phases, is determined by such factors as the following:

1. The initial number of organisms present: Usually, the lag phase of growth will be shorter and the growth rate more rapid as the number of bacteria present initially increases.
2. The physiologic state of the bacterial cell: The lag phase of growth generally increases with increasing age of the bacterial cells present. Also, transferring bacteria frequently to new growth media maintains them in a rapid growth phase.
3. The particular species of organism present: A pure culture under optimum growth conditions may exhibit more rapid growth than mixed cultures of bacteria competing for the same nutrient supply. Also, in mixed cultures, some of the microorganisms present may produce metabolic products that are toxic to others.
4. The environmental conditions present: Temperature, pH, oxygen supply or lack of it, water activity, and composition of the growth medium, i.e., concentrations of carbon and energy source, nitrogen source, salt and macro and micro element contents, and vitamin content, all markedly influence growth and survival of microorganisms in foods.

GROWTH

During the logarithmic growth phase, the increase in numbers of bacteria by binary fission as a function of time is proportional to the numbers present and can be described by the equation

$$\frac{dN}{dt} = KN \tag{10-1}$$

and after integration:

$$N = N_0 e^{Kt} \tag{10-2}$$

where N_0 = initial number of bacteria present
N = number present after observation time
t = observation time
e = base of natural logarithms
K = velocity constant of growth

Equation (10-2) can be solved for K and expressed in terms of logarithms of cell numbers (to the base 10):

$$K = \frac{2.303}{t} \log \frac{N}{N_0} \tag{10-3}$$

For yeasts which grow by budding, the increase in cell mass as a function of time is a more accurate measure of growth. Consequently, cell mass should be substituted for cell numbers in the previous equations. A logarithmic plot of cell numbers or cell mass during the logarithmic growth phase on the ordinate as a function of time (linear scale) on the abscissa yields a straight line.

During the logarithmic growth phase the doubling of cell numbers (or cell mass) can be expressed by

$$N = N_0 2^n \tag{10-4}$$

where n is the number of generations. The number of generations n can also be expressed as

$$n = \frac{t}{g} \tag{10-5}$$

where g is the generation time.
Consequently,

$$N = N_0^{t/g} \tag{10-6}$$

or

$$\log \frac{N}{N_0} = \frac{t}{g} \log 2 \tag{10-7}$$

Solving for g,

$$g = t \frac{\log 2}{\log N/N_0} \tag{10-8}$$

By combining Eqs. (10-3) and (10-7)

$$K = \frac{\log 2}{2.303 g} \tag{10-9}$$

These equations allow the computation of generation time or of number of organisms or cell masses at any subsequent time in the logarithmic growth phase.

In the case of filamentous organisms such as molds or actinomycetes, growth occurs by terminal extension of hypae and not by binary fission. With these organisms, the logarithm of hyphal length can be plotted as a function of time to obtain a growth curve.

Let us consider, as an example, a sample of sliced ham having a *Staphylococcus aureus* count of 3000 per gram. If this product is held outdoors at a picnic at summer temperatures (32 to 35°C) for 8 h, what will be the *S. aureus* count, assuming an exponential growth rate and a generation time of 40 min?

Using Eq. (10-5)

$$n = \frac{t}{g} = \frac{480}{40} = 12$$

Using Eq. (10-4)

$$N = N_0 2^n = (3000)(2^{12})$$
$$N = 12 \times 10^6 \text{ organisms per gram}$$

This is a dramatic indication of the rapidity at which food poisoning bacteria can grow in a contaminated product. A similar statement applies to food spoilage bacteria.

Effects of Temperature

Growth rates of microorganisms vary with temperature. It should be emphasized that the temperature for optimum growth rate may be different from the temperature for optimum yields of cells. For the purposes of this discussion, temperature ranges affecting microbial growth are defined as follows:

1. Optimum: The temperature at which the maximum growth rate is observed
2. Minimum: The lowest temperature below the optimum at which growth will first occur
3. Maximum: The highest temperature above the optimum at which growth will occur

The velocity constant of growth, K, is at a maximum at temperature 1 and approaches 0 at temperatures 2 and 3. These temperatures for a given organism will vary under different environmental conditions.

Based on optimum, maximum, and minimum growth temperatures, microorganisms have been classified into the following groups:

Psychrophiles: growth at 0°C regardless of optimum or maximum growth temperature (Ingraham, 1958; Ingraham and Stokes, 1959)
Mesophiles: optimum temperatures between 20 and 45°C
Thermophiles: optimum temperatures above 45°C, usually 55°C and above

Psychrophiles such as bacteria of the genera *Pseudomonas, Achromobacter, Alcaligenes,* and *Flavobacterium,* some yeasts, and many molds are involved in the spoilage of refrigerated meats, poultry, dairy products, and fish, and of frozen foods during prolonged thawing (Farrell and Rose, 1967; Witter, 1961). They may survive in foods stored at temperatures as low as −7°C, depending upon the freezing point of the solute and water in the food.

Mesophiles include some of the causative organisms of foodborne diseases, such as *Clostridium botulinum, C. perfringens, Staphylococcus aureus, Salmonella* serotypes and other members of the Enterobacteriaceae, as well as many spoilage bacteria in the genera *Micrococcus, Lactobacillus, Bacillus,* and *Clostridium* and most yeasts and molds.

Thermophiles include members of the genera *Bacillus* (e.g., *B. stearothermophilus*) and *Clostridium* (e.g., *C. thermosaccharolyticum*) that are involved in the spoilage of thermally processed foods. Thermophiles may be facultative or obligate. Thermophilic actinomycetes and fungi are widespread in nature but have not been reported as causative agents of food spoilage (Cooney and Emerson, 1964).

Thermoduric bacteria include a variety of species that survive pasteurization temperatures (161°F or 71.7°C). The more important non-spore-forming thermodurics are *Streptococcus thermophilus, Lactobacillus bulgaricus,* and *Microbacterium* and *Micrococcus* species. Spore-forming thermodurics include *Bacillus cereus, B. coagulans, B. subtilis,* and *Clostridium butyricum.*

The effect of temperature on growth rate is often expressed in terms of the temperature coefficient, which is the ratio of the growth rate constant at a given temperature (K_2) to that at a temperature 10°C lower (K_1).

$$Q_{10} = \frac{K_2}{K_1} \tag{10-10}$$

or, for a temperature interval other than 10°C,

$$Q_{10} = \frac{K_2}{K_1} \frac{10}{\Delta T} \tag{10-11}$$

A more general method for expressing the effects of temperatures on reaction rates is the Arrhenius equation:

$$K = Ae^{-\mu/RT} \tag{10-12}$$

where K = rate constant
    A = constant of integration
    e = base of natural logarithms
    μ = temperature characteristic or energy of activation of reaction, cal/g mol
    R = gas constant, 1.987 cal/mol-deg
    T = absolute temperature, K

The usual van't Hoff–Arrhenius form of Eq. (10–12) is expressed after integration as

$$\mu = 2.303 R \log \frac{K_2}{T_1} \frac{T_2 T_1}{T_2 - T_1} \qquad (10\text{-}13)$$

or, where T_1 and T_2 differ by 10°C,

$$\mu = 2.303 R \log Q_{10} \frac{T_2 T_1}{T_2 - T_1} \qquad (10\text{-}14)$$

For most chemical reactions an Arrhenius plot of log K on the ordinate against $1/T$ on the abscissa yields a straight line. For biological systems, deviations from linearity are observed. Typical values of μ obtained from Arrhenius plots range from 9000 for psychrophiles such as *Pseudomonas* sp. to 14,000 for *Escherichia coli*, a mesophile, and 19,400 cal for a thermophilic strain of *Bacillus circulans* (Farrell and Rose, 1967).

For chemical reactions, Q_{10} values range from 2 to 4 for each 10°C temperature increase. Since numerous chemical reactions are involved in bacterial cell growth and metabolism, Q_{10} values do not follow this rule except in the middle of the active temperature range for growth. For example, for a strain of *Klebsiella (Aerobacter) aerogenes*, the Q_{10} values were 1.6 at 20 to 30°C, 3.1 at 10 to 20°C, and 9.1 at 0 to 10°C (Edwards and Ewing, 1972).

The effect of temperature on the length of the lag phase is of considerable interest in food microbiology. In general, the shortest lag period is observed at the optimum temperature for growth. The lag phase increases with decreasing temperature and approaches infinity near the minimum growth temperature. This minimum temperature cannot be determined precisely, but studies in a temperature-gradient-type incubator indicate minimum growth temperatures of 14.3°C for *Staphylococcus aureus* and 7.0°C for *Salmonella typhimurium*, for example (Michener and Elliott, 1964).

Water Activity

Life on earth requires an aqueous environment, and microorganisms are no exception to this general rule. Drying has been an important method of food preservation since ancient times. Although improved methods for dehydration have been developed in recent years (vacuum-drying, freeze-drying), the major objective is still that of reducing the water content to levels where microorganisms will not grow.

Since foods vary widely in chemical composition and in particle size, moisture contents as determined gravimetrically are of little help in judging the susceptibility of a food to microbiological spoilage. Wheat is susceptible to mold attack at 16 percent moisture but is resistant at 14 percent moisture. On the other hand, unrefined cane sugar having a 2 percent moisture content may spoil because of a thin film of syrup around each crystal in which the moisture content is substantially greater than that in the bulk (Scott, 1957). Consequently, it is more useful to determine the moisture requirements for microbial growth in terms of *water activity* (a_w), which is a fundamental property of aqueous solutions:

$$a_w = \frac{P}{P_0} \qquad (10\text{-}15)$$

where P and P_0 are the vapor pressures of the solution and the pure solvent (water), respectively. When water vapor is in equilibrium with an aqueous solution, a_w is equal to the corresponding relative humidity. In real systems, as opposed to ideal solutions, a_w varies slightly with temperature over the range favorable for microbial growth, and these variations are greater in concentrated solution than in very dilute solutions.

Water activity may be controlled experimentally by (1) placing a sample in water vapor equilibrium with a solution that maintains a known relative humidity above it, (2) determining water sorption isotherms, or (3) adding solutes. The first method is widely used, especially for determining mold growth on thin layers of foods, but varying periods of time ranging from hours to weeks may be required to reach equilibrium. According to Scott (1957), equilibrium water contents of many foods range from 10 percent dry weight basis at $0.50a_w$ to 15 to 50 percent at $0.80a_w$. Table 10-1 shows typical a_w values for microorganisms of importance in foods.

Organisms that tolerate low water activity are termed *xerophilic*; those that tolerate high osmotic pressure, *osmophilic*; and those that tolerate high salt concentrations, *halophilic*.

Oxygen Requirements

Food spoilage and food poisoning microorganisms can be classified from the standpoint of their oxygen requirements. Obligate aerobes require molecular oxygen for growth. *Pseudomonas* spp., many of which are responsible for psychrophilic spoilage of dairy products, meat, fish, and poultry, some yeasts, and many molds fall into this class. They are usually found on surfaces of those food products where maximum exposure to oxygen occurs.

A wide variety of food spoilage and food poisoning microorganisms fall into the facultative anaerobe class. Typical examples are aerobic spore-forming bacilli, such as the *B. subtilus-mycoides* group which may be responsible for spoilage of a canned cured meat containing nitrate. These products receive only a mild heat treatment during processing. In this case, nitrate is reduced by acting as a hydrogen acceptor, hence making molecular oxygen available for bacterial growth. The aerobic bacilli such as *Bacillus coagulans* var. *thermoacidurans* involved in the spoilage of tomato products; *B. macerans* and *B. polymyxa*, in fruit spoilage; *B. stearothermophilus*, in flat sour spoilage of low-acid-foods; fermentative yeasts; and food poisoning organisms such as *Staphylococcus aureus* and *Salmonella* serotypes are also typical examples of facultative anaerobe types.

In general, in the presence of an adequate oxygen supply for growth, or in foods having a high oxidation-reduction (OR) potential, facultative organisms convert carbohydrates in foods such as simple sugars to bacterial cell substance, carbon dioxide, and water during growth. However, when oxygen supplies are deficient or the food has a low OR potential, facultative organisms convert carbohydrates to intermediary metabolic products such as alcohols and organic acids.

Microaerophilic organisms such as *Lactobacillus* spp. grow under greatly reduced oxygen tension as compared with facultative anaerobes. These organ-

Table 10-1 Minimum water activity requirements for growth of selected microorganisms

Organism	Minimum water activity a_w	Reference
Bacteria		
Achromobacter sp	0.96	Scott (1957)
Klebsiella (Aerobacter) aerogenes	0.945	Wodzinski and Frazier (1961a)
Bacillus subtilis	0.950	Burcik (1950)
Clostridium botulinum		Baird–Parker and Freame (1967)
Type A	0.96	Baird–Parker and Freame (1967)
Type B	0.96	Baird–Parker and Freame (1967)
Type E	0.96	Baird–Parker and Freame (1967)
Clostridium perfringens	0.960	Strong et al. (1970)
Escherichia coli	0.995	Ware et al. (1955)
Lactobacillus viridescens	0.935	Wodzinski and Frazier (1961b)
Micrococcus roseus	0.905	Burcik (1950)
Pseudomonas fluorescens	0.965	Wodzinski and Frazier (1960)
Salmonella newport	0.95	Christian and Scott (1953)
S. oranienburg		
Glucose mineral salts medium	0.97	Christian (1955)
Glucose and amino acids and vitamins	0.95	Christian (1955)
Staphylococcus aureus	0.86	Scott (1957)
Streptococcus faecalis	0.982	Ware et al. (1955)
Vibrio metschnikovii	0.97	Scott (1957)
Yeasts		
Candida utilis	0.93	Burcik (1950)
Saccharomyces rouxii	0.62–0.65	Von Schelhorn (1951)
Schizosaccharomyces spp.	0.94	Burcik (1950)
Trichosporon pullalans	0.894	Burcik (1950)
Molds		
Aspergillus niger	0.88	Heintzeler (1939)
A. glaucus	0.71	Heintzeler (1939)
Penicillium spp.	0.80–0.90	Galloway (1935)
Sporodonema sebi	0.983	Vaisey (1954)
Xeromyces bisporus	0.62–0.65	Scott (1957)

isms are often found as spoilage agents in wine and fermented food products. The enterococci such as *Streptococcus faecalis* and *S. faecium* are also microaerophilic organisms.

Obligate anaerobes constitute an important group of food spoilage and food poisoning microorganisms. Food poisoning strains, in particular, are discussed in greater detail subsequently in this chapter. These organisms grow in the absence of free oxygen and require a low oxidation-reduction potential in the medium. These organisms include both mesophilic species such as *Clostridium botulinum*, *C. sporogenes*, *C. butyricum*, *C. histolyticum*, *C. pasteurianum*, and *C. perfringens*, and thermophilic species such as the saccharolytic *C. thermosaccharolyticum* which does not produce hydrogen sulfide and *C. nigrificans* which does produce hydrogen sulfide.

Reducing substances in foods such as ascorbic acid (vitamin C) and sulfhydryl (SH) compounds such as cysteine maintain low oxidation-reduction potentials. Under these circumstances, diffusion of oxygen from the surface of a food into the interior is slight and there is a favorable environment for the growth of anaerobes.

In addition to converting carbohydrates in foods to organic acids, anaerobes also convert amino acids to amines by decarboxylation which give putrefactive spoilage odors. Hydrogen sulfide may also be produced from the breakdown of sulfur-containing amino acids.

Temperature and oxygen requirements for growth are often interrelated. The temperature for maximum growth rates of an organism in batch culture may be several degrees higher than the temperature giving the largest cell yields. Sinclair and Stokes (1963) explain this effect on the basis of the greater solubility and availability of oxygen at low temperatures, since equivalent cell yields are obtained at higher temperatures on vigorous aeration of the culture.

pH

The hydrogen-ion concentration as measured by pH is an important determinant of the types of microorganisms that will grow in a food. The proteins, organic acids such as acetic and citric, and phosphates in foods act as buffers and resist changes in pH as a result of microbiological growth, depending on the relative concentrations of these substances present. We should emphasize at this point that the titratable acidity (the amount of a standard solution of an alkali such as sodium hydroxide required to bring the food to a given pH end point) of different foods having the same pH varies widely because of the presence of buffers.

Most aerobic and facultative food spoilage and food poisoning bacteria, such as the pseudomonads, grow best in the neutral pH range (6.5 to 7.5). Acid-forming microaerophilic bacteria such as the lactobacilli tolerate lower pH values down to pH 3.5 in the case of some strains. On the other hand, anaerobes, particularly the proteolytic types involved in the spoilage of meat and animal products, grow in the alkaline range pH 7.5 to 8.5. The group D *Streptococci* such as *S. faecalis* grow at pH values as high as 9.6. Yeasts and molds involved in food spoilage grow over a wide range of pH values and tend to be more tolerant of acid pH values than most food spoilage bacteria. Many aerobic and facultative yeasts are active in the range pH 4.0 to 6.0, and molds tolerate pH values as low as 2.0.

Effect of Food Composition on Growth

Foods contain varying amounts of carbohydrates, lipids, proteins, minerals, and vitamins that can serve as substrates for the growth of food spoilage and food poisoning microorganisms, provided that the environmental conditions such as

Table 10-2 Classification of selected organisms of importance in food microbiology according to the Cold Spring Harbor scheme

Energy source: Chemoorganotrophic, energy provided by exogenous organic substances

Nutrition

Autotrophic: Essential metabolites are synthesized.
  1. Autotrophic, in the strictest sense: Oxidized inorganic nutrients can be reduced; *Pseudomonas fluorescens*.
  2. Mesotrophic: Requires one or more reduced inorganic nutrients; *Escherichia coli*, *Salmonella typhimurium* (nonexacting types).

Heterotrophic: Unable to synthesize all essential metabolites (amino acids, vitamins, growth factors); *Clostridium botulinum, C. perfringens, Salmonella typhimurium* (amino acid-requiring strains), *Staphylococcus aureus, Streptococcus faecalis*.

Hypotrophic: Multiply by reorganization of complex structures of the host cell; bacteriophages, viruses.

SOURCE: *Cold Spring Harbor Symp. Quant. Biol.* (1946), vol. 11:302.

temperature, pH, OR potential, and water activity are in a suitable range. Thus, the organisms of interest in food microbiology can be classed as chemoorganotrophs from an energy source standpoint and as either autotrophs or heterotrophs from a nutritional standpoint, using the Cold Spring Harbor Classification System. Table 10-2 shows some examples of the classification of foodborne bacteria according to this scheme.

Some examples of utilization of food constituents as substrates for growth by microorganisms are of interest at this point. Simple sugars in fruit juices can be fermented by yeasts to yield ethanol and carbon dioxide. The ethanol can be oxidized to acetic acid by *Acetobacter* species. Glucose can be oxidized to acetic acid by *Acetobacter* species. Glucose can be oxidized to gluconic acid by pseudomonads and molds. Heavy sugar syrups and honey may serve as substrates for the growth of osmophilic yeasts that tolerate the high osmotic pressures of these products.

Fats can be hydrolyzed by bacteria, yeasts, and mold lipases to free fatty acids and glycerol, resulting in hydrolytic rancidity. Unsaturated fatty acids can be oxidized by microbial oxidative enzymes and by plant lipoxidases in addition to chemical autooxidation, resulting in oxidative rancidity. Other lipid constituents in foods such as phospholipids and sterols can be degraded to simpler molecules by microbial enzymes.

Proteins in foods, particularly in meat, fish, and poultry, can be hydrolyzed to peptides and amino acids by microbial proteinases and proteases. The amino acids in turn can be deaminated, decarboxylated, or a combination of both through oxidative, hydrolytic, or reductive reactions by food spoilage bacteria.

Other constituents of foods including pectins, higher polysaccharides, organic acids, and alcohols can be converted by food spoilage microorganisms to a variety of end products. In fish, trimethylamine oxide in the muscle can be converted by bacteria of the genera *Pseudomonas, Achromobacter*, and *Flavobacterium* to trimethylamine which produces the "fishy" odor in spoiled fish.

Table 10-3 presents a list of typical food spoilage microorganisms and the foods that they affect adversely. Note the wide variety of bacteria, yeasts, and

Table 10-3 Typical microorganisms involved in the spoilage of selected foods

Food	Type of spoilage	Organism
Cereal products		
Bread	Ropiness	*Bacillus subtilis* var. *mesentericus*
	Bread mold	*Aspergillus niger*
		Neurospora sitophila
		Mucor sp.
		Pencillium expansum
		Rhizopus nigricans
	Red (bloody) bread	*Serratia marcescens*
		Oidium aurantiacum
	Chalky bread	*Endomycopsis fibuliger*
		Trichosporon variable
Flour	Acid flour paste	*Bacillus* spp.
Pasta products (Macaroni)	Swelling	*Enterobacter cloacae*
Confectionery	Exploded fondant	*Clostridium* sp.
		Yeasts
Syrups		
Sucrose	Yeast spoilage	*Candida* sp.
		Rhodotorula sp.
		Zygosaccharomyces sp. (syn. for *Saccharomyces*)
		Bacillus sp.
		Leuconostoc sp.
Molasses	Gas production	*Zygosaccharomyces* sp. (syn. for *Saccharomyces*)
		Clostridium butyricum
Maple syrup	Ropy syrup	*Klebsiella* (*Aerobacter*) *aerogenes*
		Leuconostoc spp.
	Green syrup	*Pseudomonas fluorescens*
	Cloudy syrup	*Achromobacter* sp.
		Flavobacterium sp.
	Red syrup	*Micrococcus roseus*
	Yeasty syrup	*Saccharomyces* sp. (*Zygosaccharomyces*)
Dairy products		
Milk		
Raw	Lactic acid souring	*Streptococcus lactis*
		S. cremoris
		Lactobacillus bulgaricus
	Proteolytic	*Achromobacter* spp.
		Flavobacterium spp.
		Micrococcus spp.
		Pseudomonas spp.
		Streptococcus faecalis
	Gas formation	*Klebsiella* (*Aerobacter*) *aerogenes*
		Bacillus polymyxa
		Clostridium spp.

Table 10-3 Typical microorganisms involved in the spoilage of selected foods (*Continued*)

Food	Type of spoilage	Organism
Milk (raw)	Ropiness	*Alcaligenes viscolactis*
		Enterobacter spp.
Pasteurized	Lactic acid souring	*Lactobacillus thermophilus*
		Microbacterium lacticum
		Streptococcus thermophilus
	Butyric acid	*Clostridium butyricum*
	Proteolysis	*Bacillus* spp.
		Clostridium spp.
		Micrococcus spp.
	Gas formation	*Bacillus* spp.
		Clostridium spp.
	Ropiness	*Alcaligenes viscolactis*
		Micrococcus spp.
Cream	Blue fluorescence	*Pseudomonas fragi*
Evaporated milk	Coagulation	*Bacillus coagulans* var. *thermoacidurans*
Meat		
Fresh meats (primal cuts)	Psychrophilic slime formation	*Achromobacter* spp.
		Flavobacterium spp.
		Micrococcus spp.
		Pseudomonas spp.
		Yeasts
	Whiskers	*Mucor mucedo*
		Thamnidium elegans
Ground fresh beef	Sourness	*Lactobacillus* spp.
		Leuconostoc spp.
		Micrococcus spp.
		Streptococcus spp.
Cured and smoked ham	Souring	*Achromobacter* spp.
		Bacillus megaterium
		Clostridium putrefaciens
		C. sporogenes
		Lactobacillus spp.
		Micrococcus spp.
		Pseudomonas spp.
	Mushiness	*Clostridium histolyticum*
Bacon	Greening	*Lactobacillus* spp.
		Leuconostoc spp.
	Molding	*Aspergillus* spp.
		Alternaria spp.
		Mucor racemosus
		Rhizopus (nigricans) stolonifer
Sausage	Greening	*Lactobacillus* spp.
		Leuconostoc spp.
	Slime	*Micrococcus* spp.
	Moldiness	Molds and yeasts
Dried beef	Springiness	*Bacillus* spp.

Table 10-3 Typical microorganisms involved in the spoilage of selected foods (*Continued*)

Food	Type of spoilage	Organism
Dried beef	Discoloration	*Halobacterium cutirubrum*
		Pseudomonas spp.
		Rhodotorula spp.
Poultry and eggs		
Eggs	Green rot	*Pseudomonas fluorescens*
	Black rot	*Proteus melanovogenes*
	Colorless rot	*Achromobacter* spp.
		Pseudomonas spp.
	Moldiness	*Penicillium* spp.
		Cladosporium spp.
		Sporotrichum spp.
	Off-flavors	*Achromobacter perolens*
		Pseudomonas graveolens
		P. mucidolens
		Streptomyces spp.
Poultry	Sliminess	*Achromobacter* spp.
		Pseudomonas spp.
		Alcaligenes spp.
		Flavobacterium spp.
	Sour slime	*Alcaligenes* spp.
		Pseudomonas spp.
Fish		
Fresh fish, ice-chilled	Fishiness	*Achromobacter* spp.
		Flavobacterium spp.
		Micrococcus spp.
		Pseudomonas spp.
		Serratia spp.
	Discoloration	*Pseudomonas fluorescens*
		Micrococcus spp.
		Sarcina spp.
		Molds
		Yeasts
Shellfish		
Shrimp, (ice-chilled)	Off-flavors	*Achromobacter* spp.
Crabmeat, oysters,	Sourness	*Pseudomonas* spp.
Lobsters		*Flavobacterium* spp.
		Micrococcus spp.
		Serratia spp.
		Proteus spp.
Canned foods		
Tomato juice	Flat sours	*Bacillus coagulans* var. *thermoacidurans*
Peas, corn	Flat sours	*B. stearothermophilus*
Peas, corn	Sulfide	*Clostridium nigrificans*
Peas, corn, meat, fish, poultry	Putrefaction	*C. sporogenes*
Vegetables, meats	TA (thermophilic anaerobic spoilage)	*C. thermosaccharolyticum*

Table 10-3 Typical microorganisms involved in the spoilage of selected foods (*Continued*)

Food	Type of spoilage	Organism
Canned foods		
Fruits	Butyric spoilage	*C. pasteurianum*
	Moldiness	*Byssochlamys fulva*
Soft drinks (bottled, carbonated, or noncarbonated)	Cloudiness	*Candida* spp.
		Torulopsis spp.
	Slime	*Leuconostoc* spp.
		Bacillus spp.
		Achromobacter spp.
Jellies, jams	Mold	*Aspergillus* spp.
		Penicillium spp.
Salad dressings	Gas	*Lactobacillus brevis*
		Saccharomyces spp.

molds involved. To summarize, food spoilage organisms include aerobic microaerophilic and anaerobic types from the standpoint of oxygen relationships; psychrophilic, mesophilic, and thermophilic types from the standpoint of temperature tolerance; acid, neutral, and alkaline pH tolerant types; and xerophilic, osmophilic, and halophilic types from the standpoint of low water activity, high osmotic pressure, and high salt tolerance, respectively.

MICROBIAL INHIBITORS IN FOODS

In addition to the nutrients in foods that promote growth of microorganisms, natural inhibitors which limit growth may be present. Unsaturated fatty acids, such as oleic, linoleic, and linolenic acids, inhibit the outgrowth of bacterial spores. This inhibition has been attributed to the formation of inhibitory free radicals from the unsaturated fatty acids since sodium nitrite, a known quenching agent for peroxide free radicals, reverses this effect and allows normal vegetative cell outgrowth. A similar mechanism is apparently involved in the antibacterial activity of chlorophyll after it has been illuminated.

Many natural anthocyanins, leucoanthocyanins, and phenolic acids that are present in fruits and fruit juices are mildly inhibitory toward *E. coli*, *S. aureus*, and *Lactobacillus casei*. Cranberries contain benzoic acid, a well-known preservative. Other natural inhibitors of microbial growth in foods include protamines in fish (Brackkan and Boge, 1964) and lactoperoxidase and agglutinins in raw milk used for cheese making (Auclair, 1964). During cooking, inhibitory substances may be formed from sugars undergoing nonenzymatic browning reactions (Mälkki, 1964).

In addition to inhibitors naturally present in foods, microorganisms may produce metabolic products during growth that are inhibitory to other microorganisms. For example, such common foodborne bacteria as *Achromobacter* sp., *Klebsiella (Aerobacter) aerogenes*, *Bacillus cereus*, *E. coli*, *Proteus vulgaris*, *Pseudomonas* sp., and *Serratia marcescens* may inhibit the growth of enterotoxin-forming strains of *S. aureus* (De Giacinto and Frazier, 1966; Seminiano and Frazier, 1966; Troller and Frazier, 1963b; and Twedt and Novelli,

1971). Such antagonisms against food poisoning microorganisms are discussed subsequently in this chapter.

DESTRUCTION OF MICROORGANISMS IN FOODS

The preceding discussion of inhibitors in foods leads us to consider factors affecting the death of foodborne microorganisms. Thermal processing, freezing, irradiation with ionizing or nonionizing radiation, and the use of chemical agents are means for destroying microorganisms in food. The following discussion gives the general principles involved in the destruction of foodborne microorganisms by heat, irradiation, or chemical agents. The reader is referred to more specialized references cited at the end of this chapter for additional information.

Death

In general, the death of microorganisms as a result of exposure to heat, chemicals, or ultraviolet or ionizing radiation also follows a logarithmic relationship in that a plot of logarithms of the numbers of survivors as a function of time usually yields a straight line. There may be marked deviations from a logarithmic order of death because of clumping of cells, variations in the resistance among individual cells or spores in the population, or variations in cellular permeability barriers to chemical agents.

In the ideal case of logarithmic order of death, the decrease in numbers of organisms with time can be determined in a manner similar to that for growth and $dn/dt = -Kt$ or, after integration,

$$K = \frac{2.303}{t} \log \frac{N_0}{N} \tag{10-16}$$

where K is the velocity constant of death.

In food processing it is conventional to use the term decimal reduction time (D) or D value, which refers to the time interval required to reduce the number of organisms by 90 percent or the time period for the survivor curve to traverse one logarithmic cycle. In this instance $N = 0.1 N_0$ and, on substitution in Eq. (10-16),

$$K = \frac{1}{t_{10\%}} = \frac{2.303}{D} \tag{10-17}$$

Heat

Microorganisms vary widely in their resistance to destruction by heating. Vegetative cells of bacteria, yeasts, and molds are less heat-resistant than their spores, and bacterial spores are more heat-resistant than those of yeasts and molds. In addition, heat resistance may vary considerably within a given population of spores or vegetative cells of a given culture.

The following factors influence the heat resistance of vegetative cells and spores, according to Pflug and Schmidt (1968).

1 Inherent resistance: Different strains of the same species may have a different heat resistance under the same growth and environmental conditions.
2 Environmental factors active during growth: Age, composition of the medium, and incubation temperature may influence heat resistance.
3 Environmental factors active during heating:
   a Initial number of cells present: The higher the number, the greater is the time-temperature treatment required for destruction.
   b pH: Heat resistance is usually greatest in a medium at neutral pH.
   c Composition of the medium: Presence of carbohydrates, fats, proteins, salt, and mineral contents, presence or absence of growth factors or inhibitors influence heat resistance.

In general, microorganisms are less resistant to destruction by moist heat (flowing steam under pressure) than by dry heat. This difference has been attributed to destruction of a key protein in the case of moist heat as compared with death by oxidation in the case of dry heat. However, this proposed explanation appears oversimplified in view of the wide range of heat resistance among vegetative cells and spores and the complex pattern of nutritional requirements and sensitivity to inhibitors of survivors of heat treatment (Pflug and Schmidt, 1968). Thermal processing of foods is concerned with the destruction of vegetative cells and spores by moist heat, where the water activity is 1.00 (100 percent relative humidity).

Since heat resistance of microorganisms and types of microbiological spoilage occurring in foods are closely related to pH, foods have been classified according to their pH range as follows:

Low-acid foods	pH above 4.5
Acid foods	pH 4.0–4.5
High-acid foods	pH below 4.0

In general, few spore-forming bacteria grow at a pH of 4.0 or below. In the few cases where this occurs, such as in flat sour spoilage by *Bacillus coagulans* var. *thermoacidurans*, a heavy initial inoculum is present.

The term *thermal death time* is used to describe the heat resistance of vegetative cells or spores. It is defined as the time required at a given temperature to kill a specified number of vegetative cells or spores. Table 10-4 shows thermal death time values for some selected organisms of interest in the food industry.

The heat treatments used in food processing include pasteurization, which refers to the destruction of a portion of the microorganisms in a food usually at temperatures below 100°C, and sterilization or destruction of all viable cells or spores, which involves thermal processing with steam under pressure to achieve a temperature of 121°C or higher. High-temperature short-time (HTST) pasteurization or sterilization processes in current use enable the destruction of microorganisms to be acccomplished in much shorter time periods with less damage to the flavor, aroma, and color of foods than in the long-time holding processes used in the past. This is particularly true in the case of milk and dairy products where an HTST pasteurization process of 71.7°C (161°F) for 15 to 17 s has largely

Table 10-4 Moist heat resistance of selected microorganisms of importance in foods

Organism	Medium	Thermal death time				Thermal resistance				Reference
		t, °C	F value, min	z_F °C	z_F °F	t, °C	D value, min	z_D °C	z_D °F	
Bacterial spores:										
Bacillus coagulans	Distilled H$_2$O	...	...	...	...	110	0.344	...	...	Youland and Stumbo (1953)
	0.05 M phosphate pH 7.0					100	270	10	18	Murrell and Warth (1965)
	Evaporated milk					110	0.359	...	...	Frank (1955)
	Tomato juice					110	0.219	...	...	Frank (1955)
B. stearothermophilus	0.05 M phosphate pH 7.0					100	714	10	18	Murrell and Warth (1965)
	0.067 M phosphate pH 7.0	120	28	...	...	...	...	...	...	Williams and Hennessee (1956)
Clostridium botulinum Types A and B	0.067 M phosphate pH 7.0	121	2.45	9.8	17.6	121	0.20	10	18	Schantz (1964)
Type E	Trypticase peptone glucose broth	...	...	...	...	80	2.3	5.3	15	Schantz (1964)
C. perfringens	Distilled H$_2$O	...	...	...	...	100	17	10.5	19	Roberts (1968)
C. sporogenes (PA 3679)	0.067 M phosphate pH 7.0	121	14.0	10.7	19.2	121	2.033	10.5	18.9	Esselen and Pflug (1956)
	Green beans	121	6.0	10.9	19.6	121	0.932	9.5	17.1	Esselen and Pflug (1956)
	Beef	121	13.9	...	...	121	2.23	11.8	21.2	Kaplan et al. (1954)

Organism	Substrate								Reference
Non-sporeforming bacteria:									
Micrococcus radiodurans	Beef	...	...	...	60	0.75	10.65	19.17	Duggan et al. (1963b)
Salmonella manhattan	Custard	19.00	7.64	13.75	60	2.44	8.00	14.40	Angelotti et al. (1961)
Salmonella manhattan	Chicken a la king	3.00	5.17	9.30	60	0.40	4.97	8.95	Angelotti et al. (1961)
S. meleagridis	Liquid whole egg	9.00	4.4	7.9	60	1.6	4.3	7.8	Anellis et al. (1954)
S. montevideo	Liquid whole egg	8.9	4.6	8.2	60	1.4	4.6	8.2	Anellis et al. (1954)
S. oranienburg	Liquid whole egg	7.6	4.5	8.1	60	1.3	4.5	8.1	Anellis et al. (1954)
S. pullorum	Liquid whole egg	2.9	4.1	7.4	60	0.4	4.2	7.5	Anellis et al. (1954)
S. senftenberg	Custard	78.00	6.56	11.80	60	11.32	6.71	12.5	Angelotti et al. (1961)
S. senftenberg	Chicken a la king	81.50	6.37	11.45	60	9.61	6.58	11.83	Angelotti et al. (1961)
S. typhimurium	Liquid whole egg	15.1	4.6	8.3	60	2.2	4.3	7.8	Anellis et al. (1954)
S. worthington	Liquid whole egg	6.9	4.3	7.7	60	1.1	4.3	7.7	Anellis et al. (1954)
Staphylococcus aureus	Custard	59.00	5.83	10.50	60	7.82	5.78	10.40	Angelotti et al. (1961)
Staphylococcus aureus	Chicken a la king	40.00	5.17	9.30	60	5.17	5.41	9.73	Angelotti et al. (1961)
Streptococcus faecalis	Chicken a la king	...	...	...	65.5	1.9	6.8	12.3	Ott et al. (1961)
Molds:									
Aspergillus chevalieri	Plum extract	...	...	...	80	3.3	12.8	23	Pitt and Christian (1970)
Byssochlamys fulva	Grape juice	...	...	...	87.8	10	6.1	11	King et al. (1969)
Xeromyces bisporus	Sucrose-malt extract	...	...	...	80	2.7–3.6	12.2	22	Dallyn and Everton (1969)

replaced the long-time holding process of 62.8°C (143 to 145°F) for 30 min with equivalent results.

For low-acid foods, the thermal death time of *Clostridium botulinum* spores is the basis for determination of the required thermal process for these products. Esty and Meyer (1922) established the thermal resistance of spores of *C. botulinum* in phosphate buffer as 2.78 min at 250°F (121.1°C). This was subsequently corrected by Townsend et al. (1938) to 2.45 min at 250°F. When the logarithm of the thermal death times of *C. botulinum* spores at various temperatures is plotted as a function of temperature, a straight-line relationship is observed. The slope of this curve expressed as the number of degrees Fahrenheit required for the curve to pass through one logarithmic cycle (a tenfold reduction) is termed z. Townsend et al. (1938) determined a corrected z value of 17.6°F for *C. botulinum*. Most of the important heat-resistant spore-forming bacteria have z values of approximately 18°F.

The term F or F *value* refers to the time in minutes required in a given medium at 250°F (121.1°C) to destroy the organism in question. The term F_0 is the F value when $z = 18$. Thermal process times are calculated values [see Stumbo (1972) for details]. The decimal reduction time D, defined previously, can be used to express the heat resistance of spores in terms of the time required in minutes to bring about a 90 percent reduction in the numbers of an organism during heating at a specified temperature. Table 10-4 compares F and D values for some microorganisms of importance in the food industry.

As in bacterial growth, the Q_{10} value can be determined from the ratio of velocity constants of death at 10°C intervals by use of Eq. (10-10).

It can be shown that Q_{10} and z are related as follows:

$$z = \frac{18}{\log Q_{10}} \tag{10-18}$$

The "activation" energy for the destruction of bacterial spores can be determined from the Arrhenius equation [Eq. (10-14)]. By substituting Eq. (10-18) in Eq. (10-14),

$$\mu = 2.303 R \frac{T_2 T_1}{(T_2 - T_1)} \frac{18}{z} \tag{10-19}$$

For $z = 18$, and temperatures in the range of 111 and 121°C, respectively, (384 and 394 K) $\mu = 69{,}295$ cal.

Spores of some anaerobes, notably *Clostridium sporogenes*, are more resistant to heat than the spores of *C. botulinum*. The thermal process times for low-acid foods recommended by the National Canners Association are based on a $12D$ dose, that is, the heat treatment required to reduce a population of 10^{12} spores to 1 surviving spore for $z = 18$.

Chemical Preservatives

Chemical preservatives are added to foods to prevent microbiological spoilage. They are classified as food additives under the Federal Food, Drug and Cosmetic Act and must be approved by the Food and Drug Administration as safe from the human health standpoint as well as effective for their intended use.

Chemical preservatives either kill microorganisms or inhibit their growth by inhibiting enzyme activity in the cells or by interfering with the cell membrane or the genetic mechanisms involved in reproduction.

Both inorganic and organic compounds are used as chemical preservatives in foods. Salt (sodium chloride) is the oldest inorganic preservative in use, having been used since ancient times. In low concentration, salt stimulates microbial growth but in high concentrations inhibits it. Some organisms are halophilic and tolerate relatively high concentrations of salt, in the 10 to 25 percent range. The enzymes of obligate halophilic organisms (those grown only at high salt concentrations) actually require salt for their activity. In general, salt exerts its preservative action through the osmotic withdrawal of water from the microbial cell, resulting in dehydration and plasmolysis. The salt concentration in a food is related to its water activity, depending upon the amount of dissolved solute (primarily salt in salted foods) and water in the food. We mentioned previously that salt content of cured and salted foods is usually expressed as brine concentration: (salt × 100)/(moisture + salt).

The antimicrobial effectiveness of salt is less at refrigeration temperatures than at room temperature. At low pH values, less salt is required than at neutrality to prevent microbial growth. Salt also has a synergistic effect with preservatives such as benzoic acid; lower concentrations are required for equivalent antimicrobial effect when salt is present than in its absence. [See Ingram and Kitchell (1967) for a further discussion of salt as a food preservative.]

Salt exercises a selective antimicrobial effect in some food products. In pickle fermentation, salt favors the development of the proper sequence of lactic acid bacteria and inhibits undesirable types such as coliform organisms.

Sugars such as glucose, maltose, sucrose, and lactose also perform an important role as preservatives in foods when used in sufficiently high concentrations to produce unfavorable osmotic pressures for the growth of food spoilage microorganisms. Jellies, jams, and preserves, honey, sugar syrups, fruit juice concentrates, confectionery products, and sweetened condensed milk are examples of foods where high sugar concentrations exert a preservative effect. In general, sugar concentrations must be greater than 40 percent to be effective.

Even at the high sugar concentrations of 70 to 80 percent present in honey, osmophilic yeasts such as *Saccharomyces (Zygosaccharomyces) rouxii* and related species may ferment the sugars to ethyl alcohol or polyhydric alcohols, organic acids, and carbon dioxide. Off-flavors and colors and crystallization of the honey usually result from osmophilic yeast spoilage. Similar reactions occur in syrups and confectionery products infected by osmophilic yeasts [see Onishi (1963) for a review of this subject].

To prevent osmophilic spoilage, tanks and pipelines for conveying sugar syrups at food processing plants must be cleaned and sanitized at frequent intervals, especially at the end of each working shift or after every production stoppage. All tanks for holding syrups should be covered. Ultraviolet lamps may be helpful in reducing airborne yeast contamination.

The more important inorganic and organic compounds that are used as antimicrobial agents in foods and as food preservatives are listed in Table 10-5. In addition, gaseous sterilizing agents such as ethylene oxide and propylene oxide are used for fumigation of foods and, in particular, disinfection of natural products such as spices. Gaseous sterilization is discussed in Chap. 14.

Table 10-5 Selected antimicrobial chemicals used as preservatives

Chemical	Use	Level of use
Acetic acid	Acidulant in pasteurized cheese, cottage cheese, syrups, sherbets, beverages, baking products, confections	0.005–1.0%
Dehydroacetic acid	Preservative in cut squash	65 ppm
Sodium diacetate	Mold and rope inhibitor in bakery products	0.06–0.32%
Benzoic acid (also sodium benzoate)	Preservative in bottled soft drinks, fruit juices, jellies, preserves, confections, oleomargerine	0.05–0.1%
p-Hydroxybenzoic acid (methyl, ethyl, propyl esters)	Preservative in acid foods	0.1%
Carbon dioxide (also carbonic acid)	Preservative in fruit juices, soft drinks, beer, controlled gas storage of fruits, vegetables	
Citric acid	Acidulant	Sufficient quantity to achieve desired pH or acidity
Hydrogen peroxide	Pasteurization of milk for cheese manufacture	
Lactic acid	Acidulant, preservative in carbonated beverages, cottage cheese	Sufficient quantity to achieve desired pH or acidity

The organic acids listed in Table 10-5 have their greatest antimicrobial effectiveness when in their undissociated form. This means that those compounds are most effective at low pH values, particularly below pH 4 to 5, where the fraction of the acid present in undissociated form increases and that present as the ionized form decreases.

At the present time the Food and Drug Administration is reviewing all food additives including preservatives for safety and efficacy. The allowable levels of some of the substances listed in Table 10-5, notably nitrites and nitrates, may be reduced considerably in the future as a result of Food and Drug Administration regulatory action.

Chemical agents are also used to destroy or reduce the numbers of microorganisms on food processing equipment or food utensils, or in treating water to be used in food processing. For the purpose of this discussion, a *disinfectant* is defined as a chemical agent that destroys microorganisms associated with spoilage and foodborne diseases (but not necessarily bacterial spores) particularly on inanimate surfaces. A *sanitizer* is a chemical agent that reduces the numbers of microorganisms on food contact surfaces to safe levels based on public health requirements. The term *sterilization* refers to the total destruction of all living organisms. As in the case of heat sterilization, the death

Table 10-5 Selected antimicrobial chemicals used as preservatives (*Continued*)

Chemical	Use	Level of use
Calcium propionate	Mold and rope inhibitor in bakery products, processed cheese	0.125–0.30%
Sodium propionate	Mold and rope inhibitor in bakery products, processed cheese	0.0001–0.25%
Phosphoric acid	Aciditizing agent in carbonated beverages	Sufficient quantity to achieve desired pH or acidity
Sodium nitrate	Curing of meat and fish	500 ppm*
Sodium nitrite	Curing of meat and fish	200 ppm*
Sorbic acid (also potassium sorbate)	Fungistat in syrups, fresh fruit, cheese wrappers, cakes, salads	0.05–0.2%
Sulfur dioxide	Preservative, antioxidant, anti-browning agent in fruit juices, wine, dried fruits, syrups	0.0015–0.02%
Sodium sulfite	Preservative, antioxidant, anti-browning agent in fruit juices, wine, dried fruits, syrups	
Sodium bisulfite (also potassium bisulfite)	Preservative, antioxidant, anti-browning agent (see sodium sulfite)	Up to 0.0025% (as SO_2)
Sodium metabisulfite (also potassium metabisulfite)	Preservative, antioxidant, anti-browning agent (see sodium sulfite)	Up to 0.0025% (as SO_2)

*Total of nitrate plus nitrite must not exceed 200 ppm in finished product.

of microorganisms resulting from exposure to chemical agents also follows an exponential relationship. There is always a probability that an organism will survive chemical sterilization, particularly if the initial population present is large.

Table 10-6 presents some antimicrobial compounds used in food plant sanitation. Chlorine and its compounds are the most widely used since sterilization can be achieved when sufficient concentrations are used. Chlorine reacts with water to form hypochlorous acid and hydrochloric acid [Eq. (10-20)]. Hypochlorous acid also ionizes to hypochlorite ion [Eq. (10-21)] which, in turn, can undergo hydrolysis to hypochlorous acid [Eq. (10-22)].

$$Cl_2 + H_2O \rightleftharpoons HOCl + HCl \tag{10-20}$$
$$HOCl \rightleftharpoons H^+ + OCl^- \tag{10-21}$$
$$OCl^- + H_2O \rightleftharpoons HOCl + OH^- \tag{10-22}$$

Below pH 2.0, molecular chlorine (Cl_2) is present in aqueous solution; from pH 2.0 to 4.0, hypochlorous acid (HOCl) is present in undissociated form; from pH 4.0 to 7.5 hypochlorous acid predominates; from pH 7.5 to 9.5, hypochlorite ion (OCl^-) predominates; and at pH 10 and above, almost all the chlorine is present

Table 10-6 Some antimicrobial chemicals used in food sanitation

Chemical	Chemical formula	Commercial form
Chlorine	Cl_2	Liquefied in 100–150-lb cylinders
Chlorine dioxide	ClO_2	Generated from sodium chlorite ($NaClO_2$)
Chlorinated lime	$Ca(OCl)_2\text{-}CaCl_2$	Powder, 35% available chlorine
Calcium hypochlorite	$Ca(OCl)_2$	Powder, 70% available chlorine
Sodium hypochlorite	$NaOCl$	Aqueous solutions containing 2–15% available chlorine
Chloramine T	$p\text{-}CH_3C_6H_5SO_2NNaCl\cdot 3H_2O$	Powder, 25% available aqueous chlorine
1, 3-Dichloro-5, 5-dimethyl hydantoin (halane)	$C_5H_7ON_2Cl_2$	Powder, 66% available chlorine
p-Sulfondichloramido benzoic acid (halazone)	$COOHC_6H_5SO_2NCl_2$	Tablets prepared with NaCl anhydrous Na_2CO_3 and borax, 48–52% available chlorine
Dichloroisocyanuric acid (also sodium salt)	$C_3HO_3N_3Cl_2$	Powder, 70–72% available chlorine
Trichloroisocyanuric acid	$C_3O_3N_3Cl_3$	Powder, 88–90% available chlorine
N-chlorosuccinimide (succinchlorimide)	$C_4H_2O_2NCl$	Powder, 50–54% available chlorine
Trichlormelamine (TCM)	$C_3H_3N_6Cl_3$	Powder, 70–129% available chlorine
Iodophors (Wescodyne, Iobac, etc.)		Aqueous solutions of nonionic detergents, iodine complexes 0.96–1.75% available iodine
Quaternary ammonium compounds (benzalkonium chloride, USP, etc.)	Alkyl (C_8-C_{18}) dimethyl benzylammonium chloride	Aqueous solution, 200–250 ppm of active quaternary ammonium compound

as hypochlorite ion. Hypochlorous acid is considered to be the form of chlorine that is effective against microorganisms. Hence, pH conditions during sanitation operations should be adjusted to maximize the quantity of this chemical species (Parker and Litchfield, 1962).

Chlorine and hypochlorites are strong oxidizing agents and probably kill microorganisms through reaction with oxidizable groups, particularly sulfhydryl groups in key enzymes of the microbial cell.

Hypochlorous acid, formed from the addition of chlorine or hypochlorites to water, reacts with nitrogenous compounds to produce chloramines.

$$HOCl + NH_3 \longrightarrow NH_2Cl + H_2O \quad (10\text{-}23)$$
$$2HOCl + NH_3 \longrightarrow NHCl_2 + 2H_2O \quad (10\text{-}24)$$

Chloramines persist in water for a longer period of time than chlorine but have a lower bactericidal action.

The term *available chlorine* is usually used to indicate the quantity of chlorine present in hypochlorites and represents the quantity of chlorine equivalent to the iodine released in the following typical reaction:

$$Ca(OCl)_2 + 4KI + 4HCl \longrightarrow 2I_2 + CaCl_2 + 4KCl + 2H_2O \qquad (10\text{-}25)$$

Thus, "available chlorine" is not a true measure of antimicrobial effectiveness; however, hypochlorites continue to be sold on this basis. For example, commercial calcium hypochlorite usually contains 70 percent available chlorine, whereas sodium hypochlorite solutions usually contain 2 to 15 percent available chlorine.

The antimicrobial effectiveness of sodium and calcium hypochlorite solutions is affected markedly by changes in temperature, pH, and the presence of organic matter. The U.S. Public Health Service Model Milk Ordinance and Code of 1965 specifies the use of a 50-ppm available chlorine solution of hypochlorities for 1 min at 75°F in sanitizing dairy utensils after the removal of all organic matter by thorough cleaning.

Chlorine dioxide is being used to an increasing extent in food plant sanitation, particularly for in-plant chlorination and water reuse systems, because of its strong oxidizing action in the presence of organic matter, its persistence in water, and its lack of reaction with nitrogenous compounds. It is generated by the reaction of chlorine and sodium chlorite in the pH range 3.0 to 3.5.

$$2NaClO_2 + Cl_2 \longrightarrow 2ClO_2 + 2NaCl \qquad (10\text{-}26)$$

Chlorine dioxide residuals in water reuse systems in vegetable canning operations have generally been in the range of 5 to 10 ppm.

Organic chlorine compounds such as sodium *p*-toluene sulfonchloramide (chloramine T), *p*-sulfondichloramido benzoic acid (halazone), 1,3-dichloro-5,5′-dimethyl hydantoin, dichloroisocyanuric acid, trichloroisocyanuric acid, and trichlormelamine (TCM) release hypochlorous acid in aqueous solution. These compounds may be used in food utensil sanitation both in food processing plants and in food service operations.

Iodine, iodine-releasing compounds, and iodine formulated into nonionic surface active agents known as *iodophors* are also used in food and restaurant sanitation for disinfecting food utensils.

Iodine and iodide ion react in aqueous solutions to form triiodide ion.

$$I_2 + I^- \longrightarrow I_3^- \qquad (10\text{-}27)$$

Diatomic iodine has a greater antimicrobial activity than triiodide ion whereas iodide ion has no significant antimicrobial activity. In dilute aqueous solutions of iodine (20 ppm) at pH values below 7.0, essentially all the titratable iodine in solution is in diatomic form. Iodine is less reactive than chlorine. Its antimicrobial activity is not reduced in the presence of organic matter to the same extent as with chlorine. However, organic matter should be removed from all surfaces to achieve maximum antimicrobial effectiveness from iodine compounds. In general, concentrations of iodine used in sanitizing food utensils range from 50 to 200 ppm for best results.

Quaternary ammonium compounds are also used in food industry and food service sanitation operations. These compounds are cationic surface active agents and have the following structure:

$$R_1-\overset{\overset{R_2}{|}}{\underset{\underset{R_4}{|}}{N^+}}-R_3 \; X^-$$

where R_1 is a long-chain (C_8 to C_{18}) alkyl or alkylaryl group, R_2, R_3, and R_4 are hydrogen, alkyl, aryl, or heterocyclic groups, and X is an inorganic anion, usually chloride or bromide ion. These compounds, although surface active, are poor detergents but are good bactericides in contrast to surface active anionic compounds such as the alkyl benzyl sulfonates which are good detergents but have little antimicrobial activity.

Quaternary ammonium compounds are effective against both gram-positive and gram-negative bacteria and are stable to heat, colorless and odorless, readily soluble in water, and noncorrosive. Their highest antimicrobial activity is exhibited above pH 6.0. High concentrations of calcium and magnesium ions, as in hard water, and ferrous and ferric ions reduce the antimicrobial effectiveness of quaternary ammonium compounds. Also, they are incompatible with soaps, anionic detergents, and inorganic polyphosphates (sodium hexametaphosphate, sodium tetraphosphate, and sodium tripolyphosphate). They are compatible with trisodium phosphate and tetrasodium pyrophosphate.

The U. S. Public Health Service Milk Ordinance and Code specifies a $1/2$-min time exposure at 75°F or higher for a 200-ppm solution of quaternary ammonium compounds at pH 5.0 or higher in the presence of 500 ppm water hardness and 0.2 percent tetrasodium pyrophosphate in sanitizing dairy utensils.

Radiation

Both nonionizing radiations such as ultraviolet or radiofrequency radiation and ionizing radiations either in the form of gamma radiation from cobalt 60 (^{60}Co) or high-energy electron beams from linear accelerators have been applied to the destruction of microorganisms of significance in food processing. Only ultraviolet radiation is used in commercial food processing at the present time, primarily in the disinfection of air in processing and storage areas; in controlling microorganisms on bread slicing equipment in bakeries; in preventing wild-yeast growth in the production of fermented products including pickles, sauerkraut, vinegar, and beer; and in margerine processing and packaging areas, among others.

Nonionizing Radiation

ULTRAVIOLET

Ultraviolet irradiation is effective against microorganisms over the range 2400 to 2800 Å, but commercial quartz–mercury-vapor lamps emit 90 percent or more of this radiation at 2537 Å. Only small amounts of ozone are produced by commercial ultraviolet lamps, and its bactericidal effect is only 1 percent of that of ultraviolet radiation at 2537 Å. However, ozone may induce oxidative rancidity in foods containing unsaturated fats.

In general, the death of bacteria exposed to ultraviolet radiation follows logarithmic survivor curves similar to those obtained in the presence of heat or chemical agents. Marked deviations from a logarithmic relationship are observed frequently. Factors affecting the bactericidal effectiveness of ultraviolet radiation include clumping of cells and absorption of ultraviolet radiation by dust, thin films of grease, opaque or turbid solutions, or by photoreactive molecules in the material being irradiated that are more sensitive than the bacterial cells being irradiated. For example, ascorbic acid may exert a protective effect against the destruction of bacterial spores by ultraviolet (Parker and Litchfield, 1962).

The bactericidal action of ultraviolet radiation is a function of time of exposure and intensity (power of the lamp and distance from the source). It is attributed to the absorption of wavelengths in the 2400 to 2800 Å lethal range by purine and pyrimidine bases in bacterial nucleic acids and by aromatic amino acids such as tryptophan and tyrosine in key bacterial enzymes. Gram-negative non-spore-forming bacteria such as the coliforms are very sensitive to ultraviolet radiation; staphylococci are more resistant, requiring 5 to 10 times as much exposure; and bacterial and mold spores are most resistant, requiring 10 to 50 times as much exposure as gram-negative organisms. Viruses are also extremely resistant to ultraviolet. It should be mentioned at this point that bacteria, particularly *Escherichia coli*, are subject to photoreactivation by visible light after presumably being killed by ultraviolet (Hollaender, 1955; Kelner, 1951).

RADIOFREQUENCY

Radiofrequency heating of foods such as dielectric, induction, and diathermal heating of foods has been used commercially only to a limited extent. Microwave radiation in the frequency range of 2450 MHz is used in household ranges, in restaurant and institutional ranges, and even in some industrial-scale processing applications such as cooking of poultry. Infrared radiation either from electric lamps or from gas-fired units with porcelain reflectors also can be used in processing of foods, particularly with meat products such as frankfurters. All these methods provide pasteurizing rather than sterilizing doses. [See Goldblith (1966) for a further discussion of microwave applications.]

Ionizing Radiation

Ionizing radiations that are of interest in food processing include x-ray, gamma radiation from isotopes such as cobalt 60, and high-energy electrons such as beta rays, cathode rays, and electron beams from high-energy linear accelerators. X-rays are costly to generate and have been used only to a limited extent in experimental studies. With the increasing availability of radiostopes since World War II, cobalt 60 gamma irradiation has been studied extensively for its usefulness in pasteurization and sterilization of foods as well as for inhibition of sprouting by potatoes and onions during storage and for disinfestation of grain. The use of high-energy electron beams from linear accelerators has been studied in a similar manner.

Gamma radiation is attenuated exponentially with increasing thickness of absorbing material and thus is never absorbed completely. On the other hand, electrons, regardless of energy, have a finite range, and the thickness of absorbing material required for complete attenuation is directly proportional to the energy of the radiation and inversely proportional to the density of the material.

For irradiation of materials other than air, the term *rad* is used to represent the absorption of 100 ergs per gram of material. The term *roentgen* is defined as the quantity of x or gamma radiation that produces one electrostatic unit of charge of either sign in one cubic centimeter of air at standard conditions.

Microorganisms exposed to ionizing radiations also follow a logarithmic order of death. Two theories have been advanced to explain the death of microorganisms in foods exposed to ionizing radiations. The target theory holds that death results from ionization in or near the cell, causing the destruction of an enzyme or key molecule concerned with cellular reproduction. The indirect action theory states that ionizing radiations produce reactive free radicals such as hydrogen and hydroxyl free radicals from the irradiation of water that in turn destroy microorganisms. It is probable that both these mechanisms operate during the exposure of food to ionizing radiations (Silverman and Sinskey, 1968).

Table 10-7 shows the relative resistance to ionizing radiations of microorganisms of importance in food processing. Note that radiation resistances of a given species may vary significantly, depending upon the composition of the medium in which the organisms are suspended during irradiation. Bacterial spores, particularly those of *Clostridium* species, are more resistant to ionizing radiation than vegetative cells, with the exception of *Micrococcus radiodurans* which is sensitive to heat but highly resistant to ionizing radiation (Duggan et al., 1963a). Viruses, such as foot-and-mouth disease virus, are even more resistant than bacterial spores. According to Massa (1966), a dose of 2 Mrad is necessary to sterilize carcasses of animals infected with foot-and-mouth disease virus.

In recent years, several new terms have been proposed to define treatment of foods by ionizing radiation (Goresline et al., 1964). *Radappertization* refers to the radiation process that produces a commercially sterile product analogous to the $12D$ dose used in the heat processing of foods for destroying spores of *Clostridium botulinum*. In canned meats, doses of 4.8×10^6 to 5.0×10^6 rad are required to achieve this level of inactivation of *C. botulinum* spores. *Radication* refers to the reduction of populations of microorganisms of public health significance in foods, such as the *Salmonella* or other members of the Enterobacteriaceae to safe levels, analogous to heat pasteurization. *Radurization* refers to the application of a dose of radiation sufficient to extend the shelf life of

Table 10-7 Resistance to ionizing radiation of selected microorganisms of importance in foods

Organism	Medium	D_{10}, Mrad	Reference
Bacteria*			
Clostridium botulinum			
Type A	Water	0.12–0.14	Roberts and Ingram (1965)
	Phosphate buffer	0.33	Grecz (1965)
	Canned chicken	0.31	Grecz (1965)
	Canned bacon	0.22	Grecz (1965)
Type B	Water	0.11	Roberts and Ingram (1965)
	Phosphate buffer	0.32–0.33	Grecz (1965)
	Canned chicken	0.27–0.37	Grecz (1965)
	Canned bacon	0.16–0.21	Grecz (1965)
Type D	Water	0.22	Roberts and Ingram (1965)
Type E	Water	0.08–0.16	Roberts and Ingram (1965)
Type F	Water	0.25	Roberts and Ingram (1965)
Clostridium perfringens			
Type A	Water	0.12	Roberts and Ingram (1965)
Clostridium sporogenes			
(PA 3679)	Water	0.16–0.22	Roberts and Ingram (1965)
Salmonella binza	Fish meal	0.10	Mossel (1966)
Salmonella enteritidis	Brain heart infusion	0.048–0.06	Previte et al. (1971)
Salmonella gallinarum	Frozen whole egg	0.057	Ley et al. (1963)
Salmonella oranienburg	Fish meal	0.09	Mossel (1966)
Salmonella senftenberg			
(775W)	Frozen whole egg	0.047	Ley et al. (1963)
Salmonella thompson	Brain heart infusion	0.084	Previte et al. (1971)
Salmonella typhimurium	Frozen whole egg	0.068	Ley et al. (1963)
Escherichia coli	Phosphate buffer	0.009	Bellamy and Lawton (1955)
Staphylococcus aureus	Phosphate buffer	0.02	Bellamy and Lawton (1955)
Streptococcus faecium	Dry	0.28	Christensen and Holm (1964)
Micrococcus radiodurans	Phosphate buffer	0.22	Duggan et al. (1963a)
Yeasts and molds			
Aspergillus niger	Saline + 0.5% gelatin	0.047	Lawrence et al. (1953)
Penicillium notatum	Saline +0.5% gelatin	0.02	Lawrence et al. (1953)
Saccharomyces cerevisiae	Saline +0.5% gelatin	0.05	Lawrence et al. (1953)
Viruses			
Foot-and-mouth disease	Liquid	0.481	Massa (1966)
	Dried	0.626	Massa (1966)

*All data on clostridia are resistances of spores.

foods without affecting sensory quality (flavor, aroma, color) adversely. In this case, doses must be less than 5×10^5 rad to avoid adverse changes in sensory quality. Doses in this range will destroy yeasts, molds, and common non-spore-forming spoilage bacteria, such as *Pseudomonas* spp., but will not destroy fecal streptococci, spores of clostridia, or viruses.

In general, the commercialization of ionizing radiation methods for food processing will depend upon demonstration of safety and wholesomeness from microbiological, nutritional, and toxicological standpoints to the satisfaction of regulatory agencies.

Other Physical Methods

Microorganisms may be removed from liquid foods by filtration or centrifugation. Beer and fruit drinks are examples of products that have been "pasteurized" by passage through filters having a sufficiently small pore size to exclude most vegetative microorganisms but not spores. Centrifugation, particularly with fruit juices, removes suspended solids and a portion of the microbial population present. However, it is not as effective as filtration for removing sufficient numbers of microorganisms to prevent spoilage.

Several processes involving the use of mechanical pressure for destroying microorganisms have been studied for their applicability in foods, particularly in Europe, but none of these is of commercial significance at present.

Sound waves have also been investigated for the destruction of microorganisms in foods. Ultrasonic vibrations (sound waves above audible frequencies) at frequencies greater than 20,000 Hz destroy microorganisms by the heat produced and by cavitation effects at the cell membrane. Ultrasonic treatment has been used experimentally for pasteurization of milk. However, this method may also destroy vitamins and causes undesirable physical changes in foods. Accordingly, ultrasonic treatment of foods has not been attractive commercially.

FOODBORNE DISEASES OF MICROBIAL ETIOLOGY

Foodborne diseases in which microorganisms are the causative agents can be classified in two groups as follows:

1 True food poisoning or intoxication resulting from ingestion of toxins produced by microorganisms in food
2 Food infections resulting from the consumption of foods containing pathogenic organisms

Examples of the first group include botulism, caused by toxins produced by various strains of *Clostridium botulinum*, and food poisoning from ingestion of toxins produced by *Clostridium perfringens*, *Staphylococcus aureus*, various molds and higher fungi (mycotoxins), and algae or planktonic organisms.

The second group includes those pathogens that may be carried by but will not grow in foods including *Mycobacterium tuberculosis* (tuberculosis); *Brucella abortus*, *B. melitensis*, *B. suis* (brucellosis); *Entamoeba histolytica* (amoebic dysentery); *Vibrio cholerae* (cholera); *Rickettsia*; *Coxiella burnetii* (Q fever), and viruses such as infectious hepatitis; and those pathogens that can utilize foods as substrates for growth, including the Enterobacteriaceae (salmonellae, etc.,) the streptococci, and the genus *Bacillus*.

BOTULISM

Botulism is a neuroparalytic disease caused by toxins produced by various types of *Clostridium botulinum*. Generally, the central nervous system is affected, resulting in nausea, vomiting, difficulty in swallowing and speaking, muscular weakness, and ultimately in death from respiratory failure [See Dack (1956), Riemann (1969), and Lewis and Cassel (1964) for further discussion]. The incubation period after consuming food containing the toxin is usually from 12 to 36 h but it may be shorter or longer. In fatal cases, death occurs in 3 to 6 days. Mortality is high and in the United States approaches 65 percent among those consuming foods containing the toxins. The only known treatment is to administer type specific antitoxin before irreversible damage to the central nervous system has taken place.

Clostridium botulinum is a gram-positive anaerobic spore-forming rod, 0.3 to 0.8 by 3.0 to 8.6 μm. The spores are ovoid and may be central, subterminal, or terminal. Cells bearing spores may be swollen. The cells are motile with peritrichous flagella. The natural habitat of *C. botulinum* is probably the soil (Buchanan and Gibbons, 1974).

Clostridium botulinum types A, B, E, and F are involved in human cases of botulism; types C and D affect fowl and domestic animals, respectively. Types A and B have been involved in outbreaks resulting from the consumption of underprocessed home-canned low-acid vegetables and from commercial products such as canned liver paste. Types A and B generally produce a distinctive foul putrefactive odor, but some strains of type B do not.

Type E strains have become increasingly important in recent years in outbreaks associated with canned or packaged fishery products. Vacuum-packaged smoked whitefish and canned tuna have been involved in two major outbreaks of type E botulism.

Table 10-8 shows some of the important characteristics of the various types of *C. botulinum*. In general, a neutral pH is most favorable for growth and toxin production. It is generally agreed that toxin is not produced in most foods at pH 4.5 or lower. The temperature range for growth and toxin formation is somewhat higher for types A and B than for type E. The thermal resistance, radiation resistance, and salt tolerance of type A and B spores are greater than the corresponding values for type E spores.

Clostridium botulinum produces the most poisonous toxins known (Lamanna, 1959). The toxins are proteins and have been purified and crystallized. The toxins are heat labile as compared with the spores and are readily destroyed by boiling a suspect food for 15 min. Type E toxin can be activated 10- to 100-fold by treatment with trypsin.

Neither toxins nor the spores are destroyed by freezing. Low-acid frozen foods, particularly precooked formulated frozen foods, should be kept frozen until cooked.

Clostridium botulinum is an obligate anaerobe. Consequently, vacuum-packed nonsterile products can be troublesome, as is evidenced by the outbreaks

Table 10-8 Summary of characteristics of *Clostridium botulinum* types A, B, D, E, F.

Characteristics	Comments
Food poisoning types	A, B, E, F (D rarely)
Temperature range (growth and toxin production)	A, B, D : 10–48°C E : 5–45°C } (Schmidt, 1964)
Optimum temperatures	A, B, D : 34–35°C E : 25–28°C } (Schmidt, 1964)
pH Range (growth)	A, B : 4.8–8.3 (Schmidt, 1964) E : 5.0–8.3
Optimum pH	A, B, D, E : 7.4–7.6
Salt tolerance	A, B : 8.5–10.5% NaCl (spores and vegetative cells) E : 5.0% NaCl (spore outgrowth) 5.8% NaCl (vegetative cell growth)
Sucrose tolerance	A, B : 30% sucrose (spores and vegetative cells) E : 38.5% sucrose (spore outgrowth)
Water activity (a_w) (minimum)	A : 0.96 Vegetative cell B : 0.96 Outgrowth, pH 7.0 } (Baird-Parker, and Freame, 1967) E : 0.98 30°C in NaCl
Thermal resistance (pH 7.0)	
Spores	A, B : $D_{250°F} = 0.20$ min (Schmidt, 1964) E : $D_{212°F} = 0.003–0.017$ min (Schmidt, 1964) $D_{176°F} = 0.6–3.3$ min (Roberts and Ingram; Schmidt, 1964)
Toxin	A : 80°C, 5–6 min B : 90°C, 15 min } (Lamanna, 1959; Schantz, 1964)
Radiation resistance (gamma irradiation)	A : $D = 0.251–0.309$ Mrad B : $D = 0.172–0.283$ Mrad } (Roberts and Ingram, 1965; Schmidt, 1964) E : $D = 0.125–0.138$ Mrad
Toxin	
Molecular weight	A, B, D : 900,000 (Lamanna, 1959; Schantz, 1964) E : 18,600 (Dolman, 1964)
Detection	*Presumptive:* Intraperitoneal injection of food extract or culture filtrate into mice; observation for death in 72 h (Lewis and Angelotti, 1964; Sharf, 1966; Thatcher and Clark, 1968) *Confirmatory:* Passive immunization of mice with type specific antitoxin; observation for death within 72 h after injection of presumptive toxic material (Lewis and Angelotti, 1964; Sharf, 1966; Thatcher and Clark, 1968)
Specific toxicity (Mice) LD_{50} units/mg N	A, B : $2.4–2.6 \times 10^8$ (Lamanna, 1959; Schantz, 1964) E : 7.7×10^4 (activated by trypsin) (Dolman, 1964)
Media for isolation from and detection in foods	*Enrichment:* Fluid thioglycollate; cooked meat + 0.2% soluble starch; trypticase - peptone - glucose - yeast extract - trypsin (Lewis and Angelotti, 1964; Sharf, 1966; Thatcher and Clark, 1968) *Plating* (anaerobic under nitrogen or hydrogen): blood agar, brain heart agar (Lewis and Angelotti, 1964; Sharf, 1966; Thatcher and Clark, 1968)

of type E botulism resulting from the consumption of vacuum-packed whitefish mentioned previously. In most foods packaged in hermetically sealed containers such as metal cans or glass jars, the oxidation-reduction potential is sufficiently low for outgrowth of any surviving *C. botulinum* spores.

The best means for prevention of botulism are the following:

1. Clean all raw vegetables before processing to remove soil that may contain *C. botulinum* spores.
2. Use National Canners' Association approved thermal processes and process controls.
3. Maintain food plant sanitation practices during all stages of handling of raw materials prior to processing.
4. Inspect warehoused processed products to detect and remove all swollen cans and to check the contents of suspect cans for presence of spores or toxin.
5. Do not taste any processed food having a questionable odor.

Recently the U.S. Food and Drug Administration (1973) adopted stringent regulations for in-plant controls over thermal process conditions in the canning industry. These regulations also provide for certification of supervision of retort operations in canning plants. The $12D$ dose mentioned previously will provide a sufficient margin of safety against botulism to the consumer if all low-acid canned foods are processed to this extent.

CLOSTRIDIUM PERFRINGENS FOOD POISONING

Clostridium perfringens food poisoning is caused by ingestion of food containing large numbers of cells of this organism. The cells produce an enterotoxin in the intestinal tract which is the causative factor. Symptoms include nausea, diarrhea, gas, abdominal distention and pain, but vomiting is rare. The incubation period after consuming foods containing this organism is 8 to 12 h. The illness lasts only a few hours to a day. Mortality is very rare, but a few fatalities have been reported.

Clostridium perfringens is a gram-positive nonmotile anaerobic spore-forming rod, 0.9 to 1.3 by 3.0 to 9.0 μm, and occurs singly and in pairs but infrequently in chains. The oviod spores are central and do not swell the cells (Buchanan and Gibbons, 1974).

Clostridium perfringens strains require the amino acids arginine, aspartic acid, cystine, glutamic acid, histidine, isoleucine, leucine, methionine, phenylalanine, threonine, tryptophan, tyrosine, and valine (Smith, 1972). Alanine and serine are also required by some strains. The vitamin and growth factor requirements are biotin, nicotinamide, pantothenate, pyridoxal, and adenine. Some strains also require riboflavin and uracil (Smith, 1972).

Heat-resistant type A strains are generally considered to be the causative organisms of human food poisoning although a few cases have been attributed to types C and F. Insufficiently cooked meats that are allowed to cool slowly and then are eaten a day or more later, either cold or reheated, are frequent sources of *C. perfringens* food poisoning. Such meats and meat dishes as cold chicken, reheated beef stew, and reheated roast beef and gravy are typical examples.

Table 10-9 Summary of characteristics of *Clostridium perfringens*

Characteristics	Comments
Food poisoning types	A (perhaps D and F)
Temperature range	15–55°C ⎫ (Smith, 1972)
Temperature optimum	43–47°C ⎭
pH Range	4.9–8.3 ⎫ (Smith, 1972)
pH Optimum	7.0–7.2 ⎭
Salt tolerance	Growth: 5% NaCl ⎫ (Smith, 1972)
	Survival: 22% NaCl ⎭
Water activity (a_w)	
(minimum)	0.960, Vegetative cells, pH 6.5, 37°C in glucose (Strong et al., 1970)
Thermal resistance	
Spores	$D_{90°C}$ = 0.015–8.71 min (Nakamura and Converse, 1967; Roberts, 1968)
Toxin	Inactivated, 60°C, 10 min (Hauschild, 1971)
Toxin (type A enterotoxin)	
Molecular weight	36,000–40,000 (Hauschild, 1971)
Isolectric point	pH 4.3
Specific toxicity (mice)	
LD_{50} units/mg N	2000
Detection	Skin erythema without necrosis in guinea pigs and rabbits
Media for isolation and detection in foods	Enrichment: Fluid thioglycollate
	Cooked meat-dextrose broth
	Plating: McClung-Toabe egg yolk agar, sulfite-polymyxin-sulfadiazine agar (Thatcher and Clark, 1968)

Table 10-9 summarizes some of the important characteristics of *C. perfringens*. Infective doses of cells are usually in the range of 5×10^8 to 5×10^9 per gram (Smith, 1972).

Clostridium perfringens does not grow to any significant extent below 15°C or below pH 5.0. Best growth takes place at neutral pH and 43 to 47°C. Some spores of heat-resistant strains have survived at 100°C for 5 h and in frozen meat at −5°C and −20°C for 6 months (Barnes et al., 1963). The spores will also survive indefinitely in curing brines (20 to 25% salt).

Considerable information has been obtained on the characteristics of *C. perfringens* type A enterotoxin in recent years. It is relatively heat labile and is sensitive to some proteolytic enzymes such as *Bacillus subtilus* protease but not to trypsin, chymotrypsin, or papain. This toxin probably exerts its effect through causing increased capillary permeability and vasodilation in the intestine followed by diarrhea, resulting from excessive fluid accumulation in the intestinal lumen (Hauschild, 1971).

Outbreaks of *C. perfringens* food poisoning can be prevented by cooling stews and other meats properly after cooking. Meats should not be rewarmed for extended periods prior to serving or repeatedly cooled slowly and rewarmed.

STAPHYLOCOCCUS FOOD POISONING

Staphylococcus food poisoning results from ingestion of food containing enterotoxin produced by enterotoxigenic strains of *Staphylococcus aureus*. In recent years, staphylococcus food poisoning has been the most frequently reported foodborne disease. The symptoms of this disease include salivation, nausea, vomiting, retching, severe abdominal cramps, and diarrhea. There may be marked prostration and either fever or subnormal temperatures in severe cases (Dack, 1956). Symptoms occur in 1 to 6 h (usually 3 h) after ingestion of food containing the toxin. The illness lasts from 1 to 3 days. Fatalities are very rare.

Staphylococcus aureus is a gram-positive nonmotile, facultative coccus. The cells are spheres, 0.8 to 1.0 μm in diameter and occur singly, in pairs, in short chains, or in clumps that appear like a bunch of grapes (Buchanan and Gibbons, 1974). *Staphylococcus aureus* colonies usually produce an orange pigment although some strains may produce dull dirty-white colonies.

Most enterotoxigenic strains of *S. aureus* ferment mannitol anaerobically and are coagulase-positive, but a few exceptions have been reported. An organic nitrogen source containing amino acids such as arginine, aspartic acid, cystine, glycine, proline, and valine and the vitamins thiamine and niacin are required for growth and enterotoxin production in a medium containing glucose as the carbon and energy source (Minor and Marth, 1971).

Table 10-10 summarizes the important characteristics of *S. aureus*. Six chemically and serologically different enterotoxins have been described and are designated types A, B, C, C_2, D, and E. These toxins are polypeptides. The molecular weights of types A and B are estimated to be 28,366 and 34,486, respectively (Minor and Marth, 1972a).

Five bacteriophage types are recognized (I to IV and miscellaneous). Food poisoning strains are usually in group III. Some strains are type specific for one bacteriophage whereas others show a pattern of lysis by several bacteriophages.

Growth and enterotoxin production can take place over a wide range of temperatures. However, at temperatures below 20°C or above 46°C, enterotoxin production is extremely slow or inhibited completely in some strains.

Water activity markedly affects toxin production. A decrease of a_w from 0.99 to 0.97 results in a significant decrease in enterotoxin B production by large populations of *S. aureus*.

There is a significant interaction among pH, NaCl concentration, and temperature as regards their effect on growth and enterotoxin production. Enterotoxin A is produced at NaCl concentrations as high as 10 percent. Enterotoxin B is produced in broth containing 10% NaCl at pH 6.9 or 4% NaCl at pH 5.1 but is inhibited in 4 or 8% NaCl at temperatures below 35°C or in 12% NaCl in the temperature range 4 to 35°C. Enterotoxin C is produced in broths containing 0, 4, 8, and 10% NaCl in the pH ranges 4.0 to 9.83, 4.40 to 9.43, 4.50 to 8.55, and 5.45 to 7.30. Neither enterotoxin A or B production or growth of *S. aureus* is affected by sodium nitrate ($NaNO_3$) or sodium nitrite ($NaNO_2$) at concentrations of 1000 ppm or 200 ppm, respectively (Minor and Marth, 1972a).

Staphylococcus aureus cells are relatively sensitive to heat in that they are less resistant than bacterial endospores but are more resistant than many gram-negative food spoilage organisms. However, the enterotoxins are quite resistant, and boiling at 100°C for 2 to 60 min may be required for inactivation.

Table 10-10 Summary of characteristics of *Staphylococcus aureus*

Characteristics	Comments
Enterotoxin types	A, B, C, C_2, D, E
Temperature range (growth and toxin production)	4–46°C (Minor and Marth, 1971)
Optimum temperature	37°C (Minor and Marth, 1971)
pH Range (growth and toxin production)	4.8–8.0 (Minor and Marth, 1971, 1972d)
Optimum pH	7.4 (Minor and Marth, 1971)
Salt tolerance	15–20% NaCl (in broth media) (Minor and Marth, 1971)
Sucrose tolerance	50–60% Sucrose (Minor and Marth, 1971)
Water activity (a_w) (minimum)	0.86 (aerobic) 0.90 (anaerobic) } (Minor and Marth, 1971)
Thermal resistance Cells	$D_{60°C}$: 5.2–7.8 min (depending upon food) Thermal inactivation at 66°C, 12 min } (Angelotti et al., 1961; Minor and Marth, 1971)
Toxins (Enterotoxin B in milk)	$D_{210°F}$: 68.5 min (Read and Bradshaw, 1966a) Thermal inactivation: 274 min. 210°F (Read and Bradshaw, 1966a)
Toxins	Detection: single or double gel diffusion with type specific antisera Production of symptoms by feeding culture filtrate to cats, human volunteers (Dack, 1956; Minor and Marth, 1972a)
Media for isolation and detection	Enrichment: cooked meat + 10% NaCl Plating: mannitol-salt agar, Staphylococcus 110 agar, Tellurite-glycine agar, Egg-tellurite-glycine-pyruvate agar (Crisley et al., 1965; Sharf, 1966)

Enterotoxin B in purified form is more stable to heat than enterotoxin A (Minor and Marth, 1972a; Read and Bradshaw, 1966a, 1966b).

A wide range of foods may be involved in staphylococcus food poisoning, including ham, turkey, chicken and chicken salad, baked products, especially filled pastries, table-ready meats (sausage, etc.) precooked frozen foods, and dairy products (Minor and Marth, 1972b, 1972c).

There are a number of organisms found in food that may either inhibit or stimulate *S. aureus* (Graves and Frazier, 1963). Inhibitory bacteria include *Achromobacter* sp., *Pseudomonas* sp., *Klebsiella aerogenes*, *Bacillus cereus*, *Escherichia coli*, and *Proteus vulgaris* (DeGiacinto and Frazier, 1966; Seminiano and Frazier, 1966; Troller and Frazier, 1963a, 1963b). Streptococci, lactobacilli, gram-positive rods, gram-negative gas formers, and yeasts may either inhibit or stimulate growth of *S. aureus*, depending upon the specific strain. Recent studies have shown that lactic acid starter cultures such as *Streptococcus lactis* used in cottage cheese making may inhibit *S. aureus* (Minor and Marth, 1972b).

Staphylococcus aureus and its enterotoxins may survive in a wide range of foods. In frozen foods the cells may survive at −10°C. In the presence of egg white and corn syrup, *S. aureus* survived for more than 70 days at −30°C because of the protective effects of these substances.

Staphylococcus aureus may survive in cured meats although $NaNO_3$ and $NaNO_2$ are inhibitory in broth media, as mentioned previously. Staphylococci grew in hams containing 10% NaCl, 0.135% $NaNO_2$, and 0.0075% $NaNO_3$, but all strains were destroyed to a greater extent than 99.9 percent after heating for 60 min at 58°C (Silliker et al., 1962).

Staphylococci may survive in foods during freeze-drying and subsequent storage and then may grow out and produce enterotoxin during rehydration unless hot water (80°C or higher) is used (May and Kelly, 1965). Chicken and shrimp are excellent examples of foods that are freeze-dried and are susceptible to contamination with *S. aureus* prior to processing.

In general, the survival of *S. aureus* is best in foods that contain high concentrations of sugars, eggs, and buffering components such as phosphates and proteins. Salt concentrations below 9.5 percent, temperatures above 20°C, and a pH in the range 6 to 8 are favorable for growth and enterotoxin formation.

Staphylococcal food poisoning can be prevented by avoiding contamination of food with *S. aureus*. Exudates from skin lesions (pimples, boils) and nasal discharges of food handlers are a rich source of staphylococci. Susceptible foods should be cooked rapidly to lethal temperatures for *S. aureus* cells (65 to 70°C, 12 to 15 min) and then cooled promptly to below 5°C. Prolonged storage at room temperature of filled pastries, meat salads, and similar products that receive only a minimal heat treatment should be avoided. Cream-filled baked products such as eclairs should be heated to 218.3°C (425°F) for 20 min, followed by cooling to 10°C (50°F) or below within 1 h after baking.

SALMONELLOSIS

The salmonellae constitute the major group of bacteria of the Enterobacteriaceae that are causative agents of foodborne infections. The literature on this group of organisms is voluminous, and the reader is referred to the text by Edwards and Ewing (1972), the report of the Committee on Salmonella (1969) of the National Research Council, and a recent review (Litchfield, 1973) for additional information beyond that presented here. In this discussion, we will emphasize salmonella serotypes involved in foodborne illness rather than *Salmonella typhi*, the causative agent of typhoid fever in human beings.

In salmonella food poisoning, the onset of symptoms may be sudden and violent, with headache, chills, nausea, vomiting, abdominal pain, and diarrhea. Symptoms ordinarily occur within 12 to 24 h after ingestion of food contaminated with salmonellae. Recovery is usually complete within 2 to 4 days, and the mortality is less than 1 percent in foodborne outbreaks.

Salmonellae are gram-negative motile or nonmotile facultative non-spore-forming rods that do not ferment lactose, sucrose, adonitol, or salicin, that do not peptonize milk, produce indole or acetylmethyl carbinol, liquefy gelatin, or hydrolyze urea (Buchanan and Gibbons, 1974). Biochemical reactions are not sufficient to identify salmonella serotypes. The Kauffman-White scheme enables us to classify members of this genus that have similar biochemical characteristics, on the basis of somatic (O) and flagellar (H) antigens. Some 1400 serotypes have been identified, although only about 10 account for the majority of food poisoning outbreaks. *Salmonella typhimurium* and *S. enteritidis* are still the most frequently encountered of all serotypes.

Table 10-11 presents a summary of the characteristics of salmonellae. Some strains of a given serotype will grow in a simple medium containing glucose, an inorganic nitrogen source such as ammonium salts, and the usual mineral salts, whereas other strains of the same serotype may require specific amino acids or vitamins or both for growth.

The effects of pH on growth of salmonella serotypes in foods depend upon a complex interaction among temperature, nutrients, salt or sugar concentration, and the presence or absence of inhibitors such as $NaNO_2$.

Other members of the Enterobacteriaceae may produce antagonistic substances against salmonella serotypes. These antagonists include colicins produced mainly by strains of *E. coli* and noncolicin-type inhibitors produced by such organisms as *Bacillus polymyxa*. Competitive effects among the natural flora in foods may exert a strong effect on the growth of salmonellae.

The infective dose range of salmonella serotypes for causing food poisoning symptoms has been studied, using human volunteers. Generally, large numbers

Table 10-11 Summary of characteristics of foodborne salmonellae

Characteristics	Comments
Most common food poisoning serotypes	*Salmonella typhimurium, S. heidelberg, S. enteritidis*
Temperature range	5-47°C (Committee on Salmonella, 1969)
Optimum temperature	35-37°C (Committee on Salmonella, 1969)
pH Range	4.1-9.0 (Committee on Salmonella, 1969)
Optimum pH range	6.5-7.5 (Committee on Salmonella, 1969)
Salt tolerance	14-24% NaCl in broth (Committee on Salmonella, 1969)
Sucrose tolerance	10% (Committee on Salmonella, 1969)
Water activity (a_w) (minimum)	0.94-0.97 (depending upon serotype and medium) (Christian, 1955; Christian and Scott, 1953)
Thermal resistance	$D_{60°C}$ = 0.06-11.32 min (depending upon food and serotype) (Anellis et al., 1954; Angelotti et al., 1961; Committee on Salmonella, 1969)
Radiation resistance	D_{10} = 0.05-0.10 Mrd (depending upon food and serotype) (Ley et al., 1963; Mossel, 1966; Previte et al., 1971)
Media for isolation and detection	Selective enrichment in selenite-cystine and/or tetrathionate broth, selective isolation on differential agar such as brilliant green, bismuth sulfite, deoxycholate-citrate, confirmation on triple sugar iron and urea agars, and by agglutination of polyvalent and group specific antisera (Edwards and Ewing, 1972; U.S. Food and Drug Administration, 1972; Lewis and Cassel, 1964; Litchfield, 1973; Sharf, 1966; Thatcher and Clark, 1968)

of most serotypes, typically 10^6 to 10^9 organisms, must be ingested to cause illness (McCullough and Eisele, 1951a, 1951c).

A wide range of foods including chicken, eggs and egg products, turkey, meat pies, cured meats (ham, sausage), milk and dairy products, and confectionery have been implicated in salmonella food poisoning outbreaks. Animals and human beings are natural reservoirs of salmonellae. In particular, poultry (chickens and turkeys), cattle, and swine may be infected with salmonella serotypes. These organisms may then be carried over into the meat of these animals. Rats, mice, and flies also bear salmonellae, and unprotected foods contaminated by them may become a source of salmonella infection. Household pets such as cats and dogs may also be a problem. Human carriers also may serve as a source of salmonella infection, particularly if they are employed as food handlers.

Under the provisions of the U.S. Food, Drug and Cosmetic Act, foods containing salmonellae are considered adulterated and subject to seizure if shipped in interstate commerce. Egg products now must be pasteurized before sale in interstate commerce. This requirement has eliminated eggs as a major source of salmonella contamination in foods.

Salmonella serotypes may survive for extended periods in frozen, dehydrated, and baked foods. *Salmonella typhimurium* survived over a year in frozen fish at $-17.8°C$ and over 3 months in frozen comminuted beef at $-20°C$ (Georgala and Hurst, 1963). The survival of salmonellae in poultry and egg products is well known. *Salmonella typhimurium* was protected from destruction when it was frozen in 4 percent (volume basis) egg white at $-11°C$, and at $-30°C$ it was protected from the effects of freezing in both egg white and 4 percent (volume basis) corn syrup media (Woodburn and Strong, 1960). The results of other studies have demonstrated that salmonellae may survive in egg powder for as long as 65 weeks at $3°C$ (Wilson, 1948). *Salmonella typhimurium*, when freeze-dried, has a greater survival in a nitrogen atmosphere than in air (Sinskey et al., 1967).

The control of foodborne salmonella infections requires the following:

1 Preventing food contamination by human carriers, especially food handlers.
2 Avoiding the use of animal products from domestic livestock that are grossly infected with salmonellae.
3 Avoiding the use of food ingredients that contain salmonellae.
4 Processing all foods susceptible to salmonella contamination at time-temperature schedules sufficient to destroy these organisms.
5 Refrigerating all foods susceptible to salmonella contamination and avoiding prolonged holding of these foods at room temperature. Heating foods so that all portions reach $66°C$ for 12 to 15 min will assure destruction of even the most resistant salmonella serotypes.

OTHER ENTEROBACTERIACEAE ASSOCIATED WITH FOODBORNE ILLNESS

Shigella

Shigella is sometimes involved in food poisoning outbreaks. Symptoms of shigella food infections include nausea, abdominal cramps, diarrhea, vomiting, and elevated temperatures. The onset of symptoms occurs within 24 h after ingestion. The mortality associated with *Shigella dysenteriae* infections is about 20 percent, but it is much lower with the other species, *S. flexneri, S boydii*, and *S. sonnei*.

The important characteristics of shigellae are summarized in Table 10-12. These organisms are differentiated from salmonellae on the basis of biochemical reactions (acid but no gas from numerous carbohydrates, positive methyl red test, negative Voges-Proskauer, urea, salicin, adonitol, inositol, and citrate tests), nonmotility and serological tests (Edwards and Ewing, 1972). The infective dosage in human beings is in the range 10×10^9 to 50×10^9 cells (Shaughnessy et al., 1946).

Foods that have been implicated in food poisoning outbreaks involving shigella include milk and dairy products, fresh and cured meats, fish, and potato salad, for example. *Shigella* may survive for extended periods in foods. Taylor and Nakamura (1964) reported that *S. flexneri* survived at 25°C as follows: flour, 170 days; milk, 190 days; eggs and egg white, 55 and 10 days, respectively; oysters, 40 days; shrimp, 55 days; clams, 75 days; cooking oil, 12 days; orange juice, 10 days; and tomato juice, 15 days.

The control of shigella foodborne infections is similar to that of salmonellae; avoiding contamination of foods by animal or human carriers or their excrement, thorough cooking, and prompt cooling.

Escherichia coli

Enteropathogenic strains of *Escherichia coli* have been implicated as the causative organisms in food poisoning outbreaks. Symptoms include acute gastroenteritis, diarrhea, vomiting, headache, and elevated temperature. The incubation period after ingestion of food is 1 to 5 or 7 h. Recovery usually takes place in 18 to 24 h but may take 2 to 3 days.

Escherichia coli is differentiated from other members of the Enterobacteriaceae on the basis of biochemical and serological reactions, especially acid and gas production from glucose and lactose, indole formation, nitrate reduction, positive methyl red reaction, and negative Voges-Proskauer gelatin liquefaction,

Table 10-12 Summary of characteristics of shigellae

Characteristic	Comments
Food-poisoning types	*Shigella dysenteriae, S. flexneri, S. boydii, S. sonnei*
Temperature range	10–40°C
Optimum temperature	37°C
pH Range	7.0
Optimum pH	7.0
Salt tolerance	5–6 percent
Thermal destruction	72–82°C, 1–5 min (depending upon food)
Media for isolation from and detection in foods	Enrichment in GN or tetrathionate broth, isolation on SS, xylose-lysine-deoxycholate or deoxycholate-agars, serological confirmation (Edwards and Ewing, 1972; Thatcher and Clark, 1968)

hydrogen sulfide, urea, and citrate tests. Serotypes that have been associated with food poisoning outbreaks include 026 B_6, 055 B_5, 086 B_7, 0111 B_4, and 0124 B_{17}. [See Edwards and Ewing (1972) for a further discussion of serological typing.] The infective dose as determined in human volunteers is 1.7×10^9 cells or greater (Dack, 1956).

Escherichia coli grows well over a wide range of temperatures, 20 to 40°C, with a minimum growth temperature at 10°C and an optimum at 37°C. Heating at 65°C for 15 to 20 min is lethal (Dack, 1956).

The pH range for growth is 4.0 to 8.5, with an optimum in the range pH 7.0 to 7.5. *E. coli* will grow in the presence of 5 percent salt at 37°C but 10 percent is inhibitory.

A large number of media have been developed for the isolation and enumeration of *E. coli* in foods. A most-probable-number procedure using multiple tubes of lactose broth, or direct plating on violet red bile agar, Endo agar, and eosin methylene blue agar, is usually used (Lewis and Angelotti, 1964; Sharf, 1966; Thatcher and Clark, 1968).

Escherichia coli food poisoning outbreaks have been attributed to consumption of milk, cheese, ice cream, meats, fish, and macaroni, which are contaminated with various serotypes. *E. coli* is relatively sensitive to destruction by drying or freezing but some survivors may exist for extended periods.

Prevention of *E. coli* foodborne infections depends upon avoiding the use of food or food ingredients that have high coliform counts, using adequate cooking procedures for destruction, and prompt refrigeration.

Other Enterobacteriaceae

A few food poisoning outbreaks have been attributed to organisms in the *Citrobacter*, *Arizona*, *Proteus*, and *Klebsiella* groups of the Enterobacteriaceae. Methods for the isolation and identification of these organisms are similar to those used for other members of the Enterobacteriaceae (Edwards and Ewing, 1972). Further information is needed on the involvement of these organisms in foodborne infections and on factors affecting their incidence, growth, and survival in foods.

STREPTOCOCCUS INFECTIONS

Members of *Streptococcus* have been implicated in several food poisoning outbreaks. There is some question about the validity of these implications. The symptoms reported for streptococcal food poisoning include nausea, vomiting, colic-like pains. The incubation period is claimed to be 4 to 18 h with a recovery period of 6 to 24 h (Dack, 1956).

The organism implicated in these outbreaks was *Streptococcus faecalis*. In several reported outbreaks, *S. faecalis* was present at counts of 10^7 to 10^8 per gram of food. Several human volunteer studies have given inconclusive results. In one case no ill effects were obtained when doses of 50×10^9 *S. faecalis* cells were consumed in sterilized hams and as a 48-h culture growth in milk (Deibel and Silliker, 1963); in another study, typical food poisoning symptoms were observed on ingestion of this organism but not a culture filtrate (Dack, 1956).

Further evidence is needed before *S. faecalis* can be considered a causative organism of human food poisoning. However, it is only prudent to avoid using foods and food ingredients having high counts of this organism.

MISCELLANEOUS BACTERIAL FOODBORNE DISEASES
Vibrio parahaemolyticus

Vibrio parahaemolyticus, a pathogenic halophile, has received increasing attention as an etiologic agent of food poisoning in the Far East, particularly in Japan. This organism has been isolated from Pacific oysters along the coast of the State of Washington (Baross and Liston, 1968, 1970), from shrimp in the Gulf of Mexico (Ward, 1968), and from Chesapeake Bay blue crabs (Krantz et al., 1969). Possibly four outbreaks of this disease have occurred in the United States.

Symptoms include diarrhea and abdominal pain, sometimes accompanied with nausea, headache, and vomiting. The incubation period is approximately 13 h, and recovery takes place in 4 to 5 days. The mortality rate is very low.

Morphologically, *V. parahaemolyticus* appears as short pleomorphic gram-negative rods, including curved, straight, coccus, and swollen forms. It is motile with a single polar flagellum (Twedt et al., 1969). It grows well in broth media containing 7% NaCl, and many strains grow in the presence of 10% NaCl (Covert and Woodburn, 1972). Hemolytic activity is variable. This organism grows in the range 22 to 42°C with an optimum at 35 to 37°C. The pH range for growth is 5.0 to 11.0 (Vanderzant and Nickelson, 1972a). Isolation from seafoods can be accomplished by enrichment in Trypticase soy broth containing 7% NaCl followed by isolation on agar media such as a peptone-yeast extract. A 7% NaCl agar containing corn starch is used since starch hydrolysis is a distinguishing characteristic (Twedt and Novelli, 1971; Vanderzant and Nickelson, 1972a). Fluorescent antibody techniques can be used for confirmation.

When *V. parahaemolyticus* was inoculated into whole shrimp, survivors were present after 8 days at −18°C. There were no survivors present in inoculated shrimp homogenates heated to 100°C for 1 min (Vanderzant and Nickelson, 1972b).

Bacillus cereus

Bacillus cereus is not a common cause of food poisoning, but several outbreaks have been reported in the United States and even more in Europe.

Symptoms are mild, namely a gastroenteritis, but young children may be affected more severely. Nausea, abdominal pain, tenesmus, and some vomiting may occur. The symptoms occur 8 to 16 h after consumption of contaminated food; recovery is complete in 6 to 12 h (Dack, 1956).

Bacillus cereus is a gram-positive spore-forming aerobic rod. It is ubiquitous in nature. The temperature range for growth is 10 to 49°C with an optimum of 30°C. The temperature range for the first spore germination stage is −1 to +59°C, and for the second stage, 10 to 44°C. The pH range for growth is 4.9 to 5.3 (Goepfert et al., 1972).

The spores of *B. cereus* are not highly heat-resistant. Typical $D_{100°C}$ values are 8, 2.7 to 3.1, and 5 min in phosphate buffers, pH 7.0, skim milk, and low acid foods, respectively (Goepfert et al., 1972).

Foods associated with outbreaks of *B. cereus* food poisoning include turkey, sausages, roast pork, and pudding desserts. From food poisoning outbreak studies, the infective dose for human beings appears to be in the range of 10^6 to 10^8 organisms (Dack, 1956). *Bacillus cereus* produces a phospholipase which liberates phosphoryl choline from lecithin. This compound may produce food-poisoning-like symptoms.

Bacillus cereus can be isolated on an egg yolk–polymyxin medium, which enhances spore production, followed by identification by means of serological procedures (Goepfert et al., 1972).

Considerably more information than is available at present needs to be obtained on the rule of *B. cereus* in human food poisoning. It is probably widely distributed in foods, but leads to food poisoning outbreaks only under conditions that have yet to be determined.

Mycotoxins

Mycotoxins have given considerable cause for concern because they are produced by a variety of molds in staple foods such as corn, wheat, rice, and peanuts. These toxins produce illness or death in animals and in many cases are carcinogenic.

Aflatoxin, produced principally by *Aspergillus flavus* but also by other molds, is a cause of hepatoma in South African Bantus who consume "moldy" foods (Wogan, 1965).

Aflatoxins B, B_2, G, and G_2 are structurally related coumarin derivatives that are probably formed from aromatic amino acid precursors such as phenylalanine, tryptophan, and tyrosine (Adye and Mateles, 1964). They can be isolated by solvent extraction with hexane or ethyl ether followed by separation, using thin-layer chromatography. They produce a characteristic fluorescence on excitation with ultraviolet light at a wavelength of 365 nm. Detection of pathologic lesions in the livers of ducklings is a common bioassay procedure (Wogan, 1965).

Other diseases produced by mycotoxins distinct from aflatoxins include moldy corn toxicosis, affecting swine and horses, which is caused by *A. flavus* and *Penicillium rubrum* toxins; alimentary toxic aleukia in human beings from consumption of grain allowed to stand outside over the winter which is contaminated by *Fusarium sporotrichiella* and other fungi; and ergot poisoning in human beings and animals from consumption of rye or bread made from rye infected with *Claviceps purpurea* (Wogan, 1965).

At the present time, aflatoxins are of the greatest concern to public health authorities. The manufacturers of peanut butter now routinely examine peanuts as well as the finished product for any evidence of aflatoxin content.

ALGAL OR PLANKTONIC FISH POISONINGS

Fish poisoning can result from the ingestion of fish or shellfish that have fed upon algae toxic to human beings. Paralytic shellfish poisoning is caused by ingestion of shellfish such as scallops, Alaska butter clams, and California mussels which have consumed toxic dinoflagellates such as *Gymnodinium brevis* (Florida), *Gonyaulax catenella* (Pacific Coast and Gulf of Mexico), and *Gonyaulax tamarensis* (Canadian Atlantic and New England Coasts).

Symptoms appear within 10 min after ingestion and include gastrointestinal distress, paresthesia of the lips, tongue, and fingertips, followed by ataxia, muscular uncoordination, and ascending paralysis. Death may occur within 2 to 12 h from cardiovascular collapse or respiratory failure.

The human lethal dose of *G. catenella* toxin is considered to be 3 to 4 mg or 20,000 mouse units (McFarren et al., 1960). All lots of clams harvested commercially along the Pacific Coast of Canada are now routinely assayed for presence of paralytic shellfish toxin.

VIRUSES AND RICKETTSIAE IN FOODS

A large number of foodborne disease outbreaks each year are of unknown etiology. Failure to find bacterial or other agents of food intoxication or infection in the food leads one to suspect that viruses or rickettsiae may be involved.

Viruses can survive only in foods containing living host cells necessary for their propagation. Investigators in this field have been handicapped because of the lack of acceptable techniques for propagating some viruses in tissue cultures. Usually, animal cell tissue cultures such as rhesus monkey kidney cells are used as a host system.

The evidence for transmission of viruses and rickettsiae in foods can be classified as follows (McCrea and Horan, 1962):

I   Direct isolations from the food, supported by laboratory confirmation
II   Isolation from patients with epidemiologic evidence of viral infection
III Presumptive clinical and epidemiologic evidence of viral infection in food
IV Viruses present in food for which there is no evidence for the initiation of disease in human beings

Virus and rickettsiae may enter foods in two ways: (1) from primary sources in which the animal itself may be infected and the meat or other products derived from it then become infected or (2) from secondary sources in which food is accidentally contaminated by human beings, animals, or insects.

Table 10-13 presents a classification of virus and rickettsiae diseases by contamination source and class as developed by McCrea and Horan (1962).

Foot-and-mouth disease is of great economic importance in countries, such as Argentina, that produce beef cattle from which the meat is ultimately exported. Considerable research on the nature and control of this disease has been conducted by the U.S. Department of Agriculture at its Plum Island station in New York.

A number of serious outbreaks of infectious hepatitis have occurred from consumption of shellfish, such as oysters and clams, infected with these viruses. Infectious hepatitis has also resulted from consumption of contaminated milk, corn beef, and custard.

Q fever is a rickettsial disease caused by *Rickettsia* or *Coxiella burnetii*. It is spread by ticks that parasitize wild rodents and domestic livestock, especially cattle. The rickettsiae are shed in the milk of an infected animal; this constitutes a route of infection in human beings.

This disease has an incubation period of 11 to 18 days. Symptoms are similar to those of a severe common cold; i.e., fever, headache, cough, nausea,

Table 10-13 Classification of viral and rickettsial diseases by source and class

Contamination source*	Class*			
	I	II	III	IV
Primary	Q fever (*Coxiella burnetii*) Russian tickborne complex Foot-and-mouth disease Rabies		Cowpox Cancer Lymphoma Leukemia	New Castle disease (experimental)
Secondary	Poliomyelitis	Infectious hepatitis	Epidemic viral diarrhea Lymphocytic choriomeningitis Smallpox Vaccinia	

*Refer to text for definitions of sources and classes.
SOURCE: J. F. McCrea and R. R. Horan (1962), Literature Survey of Viruses and Rickettsiae in Foods, Final Report, Contract DA-19-129-QM-1810, Quartermaster Food and Container Institute for the Armed Forces, Chicago.

and vomiting. It may last 2 to 3 weeks. Mortality is usually less than 1 percent. Serological procedures are used in detection.

In studies at the University of California, Davis, Enright et al. (1957) established that *C. burnetii* may survive at pasteurization of milk at 61.8°C for 30 min. Increasing this temperature to 62.7°C (145°F) for 30 min is sufficient to destroy the rickettsia. The usual high-temperature short-time (HTST) pasteurization procedure, 71.7°C (161°F) for 15 to 17 s, is adequate for eliminating *C. burnetii* from whole raw milk.

Enteroviruses such as Coxsackie, ECHO, and poliomyelitis have all been associated with foodborne outbreaks although unequivocal evidence for the occurrence of these viruses in foods has been difficult to obtain. In general, animal cell tissue culture techniques are used to propagate these viruses.

Symptoms of these viral infections range from those typical of bacterial food poisoning, including fever, nausea, abdominal cramps, and diarrhea, to central nervous system involvement including paralysis in the case of poliomyelitis.

Foods that have been associated with enterovirus infection include raw cow's milk, ice cream, and poultry products.

According to Cliver et al. (1970), enteroviruses, particularly poliovirus, survived in low-moisture foods, such as cheese sandwiches, bacon squares, beef cubes, and dry banana pudding, for greater than 2 weeks at room temperature and greater than 2 months at 5°C. Virus stability was greatest at pH 7 or higher; at pH 5.5, stability was a complex function of the interaction between protein and salt content.

There is a need for considerable improvement in methods for detecting viruses in foods. Improved methods will lead to a better understanding of the roles of viruses in human foodborne disease.

SUMMARY

Microorganisms present in food are an important consideration in the food processing industry from the standpoint of prevention of food spoilage and foodborne disease of microbial etiology. Control of microorganisms in foods depends upon the following:

1. Using sound raw materials having low microbial populations
2. Avoiding microbial contamination from human carriers of pathogenic organisms, from rodents and domestic animals, and from insects
3. Preventing growth of microorganisms in foods by drying, refrigeration, or freezing
4. Destroying microorganisms by suitable thermal, radiation, or other physical processes
5. Proper packaging and storage of foods in channels of distribution to prevent postprocessing contamination
6. Handling, cooking, and refrigerating foods properly in the home or in institutions such as hotels, restaurants, schools, and hospitals

LITERATURE CITED

Adye, J., and R. I. Mateles (1964): Incorporation of Labelled Compounds into Aflatoxins, *Biochim. Biophys. Acta,* **86**:418–420.

Anellis, A., J. Lubas, and M. M. Rayman (1954): Heat Resistance in Liquid Eggs of Some Strains of the Genus *Salmonella, Food Res.,* **19**:377–395.

Angelotti, R., M. J. Foter, and K. H. Lewis (1961): Time-Temperature Effects on Salmonellae and Staphylococci in Foods, *Appl. Microbiol.,* **9**:308–315.

Auclair, J (1964): Les substances antibactériennes du lait cru et leur rôleen technologie laitière, in N. Molin and A. Erichsen (eds.), "Microbial Inhibitors in Food," pp. 281–296, Almquist & Wiksell, Stockholm.

Baird-Parker, A. C., and B. Freame (1967): Combined Effect of Water Activity, pH and Temperature on the Growth of *Clostridium botulinum* from Spore and Vegetable Cell Inocula, *J. Appl. Bacteriol.,* **30**:420–429.

Barnes, E. M., J. E. Despaul, and M. Ingram (1963): The Behaviour of a Food Poisoning Strain of *Clostridium welchii* in Beef, *J. Appl. Bacteriol.,* **26**:415–427.

Baross, J., and J. Liston (1968): Isolation of *Vibrio parahaemolyticus* from the Northwest Pacific, *Nature,* **217**:1163.

――― and ――― (1970): Occurrence of *Vibrio parahaemolyticus* and Related Hemolytic Vibrios in Marine Environments of Washington State, *Appl. Microbiol.,* **20**:179–186.

Bellamy, W. D., and E. J. Lawton (1955): Studies on Factors Affecting the Sensitivity of Bacteria to High Velocity Electrons, *Ann. N.Y. Acad. Sci.,* **59**:595–603.

Brackkan, O. R., and G. Boge (1964): Protamines from Fishes as Inhibitors on the Growth of Microorganisms, in N. Molin and A. Erichsen (eds.), "Microbial Inhibitors in Food," pp. 271–279, Almquist & Wiksell, Stockholm.

Buchanan, R. E., and N. G. Gibbons (eds.) (1974): "Bergey's Manual of Determinative Bacteriology," 8th ed., Williams & Wilkins, Baltimore.

Burcik, E. (1950): Über die Beziehungen zwischen Hydratur und Wachstum bei Bakterien und Hefen, *Arch. Mikrobiol.,* **15**:205–235.

Christian, J. H. B. (1955): The Influence of Nutrition on the Water Relations of *Salmonella oranienburg, Aust. J. Biol. Sci.,* **8**:75–82.

――― and W. J. Scott (1953): Water Relations of Salmonellae at 30°C, *Aust. J. Biol. Sci.,* **6**:565–573.

Christensen, E. A., and N. W. Holm (1964): Inactivation of Dried Bacteria and Bacterial Spores by Means of Ionizing Radiation, *Acta Pathol. Microbiol. Scand.,* **63**:281.

Cliver, D. O., K. D. Kostenbader, Jr., and M. R. Vallenas (1970): Stability of Viruses in Low Moisture Foods, *J. Milk Food Technol.,* **33**:484–491.

Committee on Salmonella (1969): An Evaluation of the Salmonella Problem, *Nat. Acad. Sci. Pub.* 1683, Washington, D.C.

Cooney, D. G., and R. Emerson (1964): "Thermophilic Fungi," Freeman, San Francisco.

Covert, D., and M. Woodburn (1972): Relationships of Temperature and Sodium Chloride Concentration to the Survival of *Vibrio parahaemolyticus* in Broth and Fish Homogenate, *Appl. Microbiol.,* **23**:321–325.

Crisley, F. D., J. T. Peeler, and R. Angelotti (1965): Comparative Evaluation of Five Selective and Differential Media for the Detection and Enumeration of Coagulase-positive Staphylococci in Foods, *Appl. Microbiol.,* **13**:140–156.

Dack, G. M. (1956): "Food Poisoning," 3d ed., University of Chicago Press, Chicago.

Dallyn, H., and J. R. Everton (1969): The Xerophilic Mould, *Xeromyces bisporus* as a Spoilage Organism, *J. Food Technol.,* **4**:399–403.

Deibel, R. H., and J. H. Silliker (1963): Food Poisoning Potential of the Enterococci, *J. Bacteriol.,* **85**:827–832.

Dolman, C. E. (1964): Growth and Metabolic Activities of *C. botulinum* types, in K. H. Lewis and K. Cassel, Jr. (eds.), Botulism, Proceedings of a Symposium, pp. 43–68, U.S. Department of Health, Education, and Welfare, Public Health Service Publ. 999-FP-1.

DeGiacinto, J. V., and W. C. Frazier (1966): Effect of Coliform and *Proteus* Bacteria on Growth of *Staphylococcus aureus, Appl. Microbiol.,* **14**:124–129.

Duggan, D. E., A. W. Anderson, and P. R. Elliker (1963a): Inactivation of the Radiation-Resistant Spoilage Bacterium, *Micrococcus radiodurans.* I. Radiation Inactivation Roles in Three Meat Substrates and in Buffer, *Appl. Microbiol.,* **11**:398–403.

——, ——, and —— (1963b): Inactivation Rate Studies on a Radiation-Resistant Spoilage Microorganism. III. Thermal Inactivation Rates in Beef, *J. Food Sci.,* **28**:130–134.

Edwards, P. R., and W. H. Ewing (1972): "Identification of Enterobacteriaceae," 3d ed., Burgess, Minneapolis.

Enright, J. B., W. W. Sadler, and R. C. Thomas (1957): Pasteurization of Milk Containing the Organism of Q Fever, *Am. J. Public Health,* **47**:695–700.

Esselen, W. B., and I. J. Pflug (1956): Thermal Resistance of Putrefactive Anaerobe No. 3679 Spores in Vegetables in the Temperature Range of 250–290°F, *Food Technol.,* **10**:557–560.

Esty, J. R., and K. F. Meyer (1922): The Heat Resistance of the Spores of *B. botulinus* and Allied Anaerobes, XI, *J. Infect. Dis.,* **31**:650–653.

Farrell, J., and A. H. Rose (1965): Low-temperature Microbiology, *Adv. Appl. Microbiol.,* **7**:335–378.

—— and —— (1967): Temperature Effects on Microorganisms, in A. H. Rose (ed.), "Thermobiology," pp. 147–218, Academic, New York.

Frank, H. A. (1955): The Influence of Cationic Environments on the Thermal Resistance of *Bacillus coagulans, Food Res.,* **20**:315–321.

Galloway, L. D. (1935): The Moisture Requirements of Mold Fungi with Special Reference to Mildew in Textiles, *J. Text. Inst.,* **26**:T123–129.

Georgala, D. L., and A. Hurst (1963): The Survival of Food Poisoning Bacteria in Frozen Foods, *J. Appl. Bacteriol.,* **26**:346–358.

Goepfert, J. M., W. M. Spira, and H. U. Kim (1972): *Bacillus cereus*: Food Poisoning Organism. A Review, *J. Milk Food Technol.,* **35**:213–227.

Goldblith, S. A. (1966): Basic Principles of Microwaves and Recent Developments, *Adv. Food Res.,* **15**:277–301.

———, M. A. Joslyn, and J. T. R. Nickerson (1961): "An Anthology of Food Science," vol. 1, Introduction to Thermal Processing of Food, Avi Publishing Co., Westport, Conn.
Goresline, H. E., M. Ingram, P. Macuch, G. Mocquot, D. A. A. Mossel, C. F. Niven, and F. S. Thatcher (1964): Tentative Classification of Food Irradiation Processes with Microbiological Objectives, *Nature,* **204**:237.
Graves, R. R., and W. C. Frazier (1963): Food Microorganisms Influencing the Growth of *Staphylococcus aureus, Appl. Microbiol.,* **11**:513–516.
Grecz, N. (1965): Biophysical Aspects of Clostridia, *J. Appl. Bacteriol.,* **28**:17–35.
Hauschild, A. H. W. (1971): *Clostridium perfringens* Enterotoxin, *J. Milk Food Technol.,* **34**:596–599.
Heintzeler, I. (1939): Das Wachstum der Schimmelpilze in Abhangigkeit von der Hydraturverhältnissen unter verschiedenen Aussenbedingungen, *Arch. Mikrobiol.,* **10**:92–132.
Hollaender, A. (ed.) (1955): "Radiation Biology," vol. II, Ultraviolet and Related Radiations, McGraw-Hill, New York.
Ingraham, J. L. (1958): Growth of Psychrophilic Bacteria, *J. Bacteriol.,* **76**:75–80.
——— and J. L. Stokes (1959): Psychrophilic Bacteria, *Bacteriol. Rev.,* **23**:97–108.
Ingram, M., and A. G. Kitchell (1967): Salt as a Preservative for Foods, *J. Food Technol.,* **2**:1–15.
Kaplan, A. M., H. Reynolds, and H. Lichtenstein (1954): Significance of Variations in Observed Slopes of Thermal Death Time Curves for Putrefactive Anaerobes, *Food Res.,* **19**:173–181.
Kelner, A. (1951): Action Spectra for Photoreaction of Ultraviolet Irradiated *E. coli* and *Streptomyces griseus, J. Gen. Physiol.,* **34**:835–852.
King, A. D., Jr., H. D. Michener, and K. A. Ito (1969): Control of *Byssochlamys* and Related Heat-resistant Fungi in Grape Products, *Appl. Microbiol.,* **18**:166–173.
Krantz, G. E., R. R. Colwell, and E. Lovelace (1969): *Vibrio parahaemolyticus* from the blue crab *Callinectes sapidus* in Chesapeake Bay, *Science,* **164**:1286–1287.
Lamanna, C. (1959): The Most Poisonous Poison, *Science,* **130**:763–772.
Lawrence, C. A., L. E. Brownell, and J. T. Graikoski (1953): Effect of Cobalt 60 Gamma Radiation on Microorganisms, *Nucleonics,* **11**:9.
Lewis, K. H., and R. Angelotti (eds.) (1964): Examination of Foods for Enteropathogenic and Indicator Bacteria, U.S. Department of Health, Education and Welfare, Public Health Service Publ. 1142.
——— and K. Cassel, Jr. (eds.) (1964): Botulism—Proceedings of a Symposium, U.S. Department of Health, Education and Welfare, Public Health Service Publ. 999-FP-1.
Ley, F. J., B. M. Freeman, and B. C. Hobbs (1963): The Use of Gamma Radiation for the Elimination of Salmonellae from Various Foods, *J. Hygiene,* **61**:515–529.
Litchfield, J. H. (1973): Salmonella and the Food Industry—Methods for Isolation, Identification and Enumeration, *CRC Crit. Rev. Food Technol.,* **3**:415–456.
Mälkki, Y. (1964): The Effect of Heated Sugar Solutions on the Growth of the *Clostridium* species P. A. 3679, in N. Molin and A. Erichsen (eds.), "Microbial Inhibitors in Food," pp. 347–352, Almquist & Wiksell, Stockholm.
Massa, D. (1966): Radiation Inactivation of Foot and Mouth Disease Virus in the Blood, Lymphatic Glands and Bone Marrow of the Carcasses of Infected Animals, *Food Irradiat.,* pp. 329–341.
May, K. N., and L. E. Kelly (1965): Fate of Bacteria in Chicken Meat During Freeze Dehydration, Rehydration, and Storage, *Appl. Microbiol.,* **13**:340–344.
McCrea, J. F., and R. R. Horan (1962): Literature Survey of Viruses and Rickettsiae in Foods, Final Report, Contract DA19-129-QM-1810, Quartermaster Food and Container Institute for the Armed Forces, Chicago.
McCullough, N. B., and C. W. Eisele (1951a): Experimental Human Salmonellosis. I.

Pathogenicity of Strains of *Salmonella meleagridis* and *Salmonella anatum* Obtained from Spray Dried Whole Egg, *J. Infect. Dis.,* **88**:278–279.
―――― and ―――― (1951b): Experimental Human Salmonellosis. III. Pathogenicity of Strains of *Salmonella newport, Salmonella derby* and *Salmonella bareily* Obtained from Spray Dried Whole Egg, *J. Infect. Dis.,* **89**:209–213.
―――― and ―――― (1951c): Experimental Human Salmonellosis. IV. Pathogenicity of Strains of *Salmonella pullorum* Obtained from Spray Dried Whole Egg, *J. Infect. Dis.,* **89**:259–265.
McFarren, E. F., M. L. Schafer, J. E. Campbell, K. H. Lewis, E. T. Jensen, and E. J. Schantz (1960): Public Health Significance of Paralytic Shellfish Poison, *Adv. Food Res.,* **10**:136–179.
Michener, H. D., and R. P. Elliott (1964): Minimum Growth Temperatures for Food Poisoning, Fecal Indicator, and Psychrophilic Microorganisms, *Adv. Food Res.,* **13**:349–396.
Minor, T. E., and E. H. Marth (1971): *Staphylococcus aureus* and Staphyloccal Food Intoxications. A Review. I. The Staphylococci: Characteristics, Isolation and Behavior in Artificial Media, *J. Milk Food Technol.,* **34**:557–564.
―――― and ―――― (1972a): *Staphylococcus aureus* and Staphylococcal Food Intoxications. A Review. II. Enterotoxins and Epidemiology, *J. Milk Food Technol.,* **35**:21–29.
―――― and ―――― (1972b): *Staphylococcus aureus* and Staphylococcal Food Intoxications. A Review. III. Staphylococci in Dairy Foods, *J. Milk Food Technol.,* **35**:77–82.
―――― and ―――― (1972c): *Staphylococcus aureus* and Staphylococcal Food Intoxications. A Review. IV. Staphylococci in Meat, Bakery Products and Other Foods, *J. Milk Food Technol.,* **35**:228–241.
―――― and ―――― (1972d): Loss of Viability by *Staphylococcus aureus* in Acidified Media. I. Inactivation by Several Acids, Mixtures of Acids and Salts of Acids, *J. Milk Food Technol.,* **35**:191–196.
Monod, J. (1949): The Growth of Bacterial Cultures, *Ann. Rev. Microbiol.,* **3**:371–394.
Mossel, D. A. A. (1966): Perspectives for the Use of Ionizing Radiation in the Decontamination (Salmonella Radicidation) of Some Frozen Proteinaceous Foods and Dry Mixed Feed Ingredients, *Food Irradiat.,* pp. 365–380.
Murrell, W. G., and A. D. Warth (1965): Composition and Heat Resistance of Bacterial Spores, in L. L. Campbell and H. O. Halvorson (eds.), "Spores," vol. III, pp. 1–24, American Society for Microbiology, Washington, D.C.
Nakamura, M., and J. D. Converse (1967): Heat Resistance of Spores of *Clostridium welchii, J. Hyg.,* **65**:359–365.
Onishi, H. (1963): Osmophilic Yeasts, *Adv. Food Res.,* **12**:53–94.
Ott, T. M., H. M. El Bisi, and W. B. Esselen (1961): Thermal Destruction of *Streptococcus faecalis* in Prepared Frozen Foods, *J. Food Sci.,* **26**:1–10.
Parker, M. E., and J. H. Litchfield (1962): Food Plant Sanitation, Reinhold, New York.
Pflug, I. J., and C. F. Schmidt (1968): Thermal Destruction of Microorganisms, in C. A. Lawrence and S. S. Block (eds.), "Disinfection, Sterilization and Preservation," pp. 63–105, Lea & Febiger, Philadelphia.
Pitt, J. I., and J. H. B. Christian (1970): Heat Resistance of Xerophilic Fungi Based on Microscopical Assessment of Spore Survival, *Appl. Microbiol.,* **20**:682–686.
Previte, J. J., Y. Chang, and H. M. El Bisi (1971): Effects of Radiation Pasteurization on *Salmonella.* III. Radiation Lethality and the Frequency of Mutation to Antibiotic Resistance, *Can. J. Microbiol.,* **17**:385–389.
Read, R. B., Jr., and J. G. Bradshaw (1966a): Staphylococcal Enterotoxin B Thermal Inactivation in Milk, *J. Dairy Sci.,* **49**:202–203.
―――― and ―――― (1966b): Thermal Inactivation of Staphylococcal Enterotoxin B in Veronal Buffer, *Appl. Microbiol.,* **14**:130–132.
Riemann, H. (ed.) (1969): "Food-borne Infections and Intoxications," Academic, New York.

Roberts, T. A. (1968): Heat and Radiation Resistance and Activation of Spores of *Clostridium welchii, J. Appl. Bacteriol.,* **31**:133-144.
——— and M. Ingram (1965): The Resistance of Spores of *Clostridium botulinum* Type E to Heat and Radiation, *J. Appl. Bacteriol.,* **28**:125-141.
Schantz, E. J. (1964): Purification and Characterization of *C. botulinum* Toxins, in K. H. Lewis and K. Cassel, Jr. (eds.), Botulism, Proceedings of a Symposium, pp. 91-104, U.S. Department of Health, Education, and Welfare, Public Health Service Publ. 999-FP-1.
Schmidt, C. F. (1964): Spores of *C. botulinum*: Formation, Resistance, Germination, in K. H. Lewis and K. Cassel, Jr., (eds.), Botulism, Proceedings of a Symposium, pp. 69-82, U.S. Department of Health, Education, and Welfare, Public Health Service Publ. 999-FP-1.
Scott, W. J. (1957): Water Relations of Food Spoilage Microorganisms, *Adv. Food Res.,* **7**:84, 127.
Seminiano, E. N., and W. C. Frazier (1966): Effect of Pseudomonads and Achromobacteriaceae on Growth of *Staphylococcus aureus, J. Milk Food Technol.,* **29**:161-164.
Sharf, J. M. (ed.) (1966): "Recommended Methods for the Microbiological Examination of Foods," 2d ed., American Public Health Association, Washington, D.C.
Shaughnessy, H. J., R. C. Olsson, K. Bass, F. Friewer, and S. O. Levinson (1946): Experimental Human Bacillary Dysentery, *J. Am. Med. Assoc.,* **132**:362-368.
Silliker, J. H., C. E. Jansen, M. M. Voegeli, and N. W. Chmura (1962): Studies on the Fate of Staphylococci during the Processing of Hams, *J. Food Sci.,* **27**:50-56.
Silverman, G. J., and T. J. Sinskey (1968): The Destruction of Microorganisms by Ionizing Irradiation, in C. A. Lawrence and S. S. Block (eds.), "Disinfection, Sterilization and Preservation," pp. 741-760, Lea & Febiger, Philadelphia.
Sinclair, N. A., and J. L. Stokes (1963): Role of Oxygen in the High Cell Yields of Psychrophiles and Mesophiles at Low Temperatures, *J. Bacteriol.,* **85**:164-167.
Sinskey, T. J., G. J. Silverman, and S. A. Goldblith (1967): Influence of Platen Temperatures and Relative Humidity during Storage on the Survival of Freeze-dried *Salmonella typhimurium, Appl. Microbiol.,* **15**:22-30.
Smith, L. D. S. (1972): Factors Involved in the Isolation of *Clostridium perfringens, J. Milk Food Technol.,* **35**:71-76.
Strong, D. H., E. F. Foster, and C. L. Duncan (1970): Influence of Water Activity on the Growth of *Clostridium perfringens, Appl. Microbiol.,* **19**:980-987.
Stumbo, C. K. (1973): "Thermobacteriology in Food Processing," 2d ed., Academic, New York.
Taylor, B. C., and M. Nakamura (1964): Survival of Shigellae in Food, *J. Hyg.,* **62**:303-311.
Thatcher, F. S., and D. S. Clark (eds.) (1968): "Microorganisms in Foods: Their Significance and Methods of Enumeration," University of Toronto Press, Toronto.
Townsend, C. T., J. R. Esty, and F. C. Baselt (1938): Heat-resistance Studies on Spores of Putrefactive Anaerobes in Relation to Determination of Safe Processes for Canned Foods, *Food Res.,* **3**:323-346.
Troller, J. A., and W. C. Frazier (1963a): Repression of *Staphylococcus aureus* by Food Bacteria. I. Effect of Environmental Factors on Inhibition, *Appl. Microbiol.,* **11**:11-14.
——— and ——— (1963b): Repression of *Staphylococcus aureus* by Food Bacteria. II. Causes of Inhibition, *Appl. Microbiol.,* **11**:163-165.
Twedt, R. M., and R. E. Novelli (1971): Modified Selective and Differential Isolation Medium for *Vibrio parahaemolyticus, Appl. Microbiol.,* **22**:593-599.
———, P. L. Spaulding, and H. E. Hale (1969): Morphological, Cultural, Biochemical, and Serological Comparison of Japanese Strains of *Vibrio parahaemolyticus* with Related Cultures Isolated in the United States, *J. Bacteriol.,* **95**:511-518.

U.S. Food and Drug Administration (1972): Bacteriological Analytical Manual, Washington, D.C.
——— (1973): Thermally Processed Low-acid Foods Packaged in Hermetically Sealed Containers, *Fed. Regist.*, **38**:2398–2410.
Vaisey, E. B. (1954): Osmophilism of *Sporendonema epizoum*, *J. Fish. Res. Board Can.*, **11**:901–903.
Vanderzant, C., and R. Nickelson (1972a): Procedure for Isolation and Enumeration of *Vibrio parahaemolyticus*, *Appl. Microbiol.*, **23**:26–33.
——— and ——— (1972b): Survival of *Vibrio parahaemolyticus* in Shrimp Tissue under Various Environmental Conditions, *Appl. Microbiol.*, **23**:34–37.
Von Schelhorn, M. (1951): Control of Microorganisms Causing Spoilage in Fruit and Vegetables, *Adv. Food Res.*, **3**:429–482.
Ward, B. Q. (1968): Isolation of Organisms Related to *Vibrio parahaemolyticus* from American Estuarine Sediments, *Appl. Microbiol.*, **16**:543–546.
Ware, G. C., E. Childs, and H. M. Smith (1955): The Effect of Salt Concentration on the Growth of Bacteria in Dilute Nutrient Solutions, *J. Appl. Bacteriol.*, **18**:446–448.
Williams, O. B., and A. D. Hennessee (1956): Studies on Heat Resistance. VII. The Effect of Phosphate on the Apparent Heat Resistance of Spores of *Bacillus stearothermophilus*, *Food Res.*, **21**:112–116.
Wilson, M. E. (1948): Survival of Salmonella Organisms in Stored Egg Powder, *Food Ind.*, **20**:873–874, 971–972.
Witter, L. D. (1961): Psychrophilic Bacteria—a Review, *J. Dairy Sci.*, **44**:938–1015.
Wodzinski, R. J., and W. C. Frazier (1960): Moisture Requirements of Bacteria. I. Influence of Temperature and pH on Requirements of *Pseudomonas fluorescens*, *J. Bacteriol.*, **79**:572–578.
——— and ——— (1961a): Moisture Requirements of Bacteria. II. Influence of Temperature, pH, and Malate Concentration on Requirements of *Aerobacter aerogenes*, *J. Bacteriol.*, **81**:353–358.
——— and ——— (1961b): Moisture Requirements of Bacteria. III. Influence of Temperature, pH, and Malate and Thiamine Concentration on Requirements of *Lactobacillus viridescens*, *J. Bacteriol.*, **81**:359–365.
Wogan, G. N. (ed) (1965): "Mycotoxins in Foodstuffs," M.I.T., Cambridge, Mass.
Woodburn, M. J., and D. H. Strong (1960): Survival of *Salmonella typhimurium*, *Staphylococcus aureus*, and *Streptococcus faecalis* Frozen in Simplified Food Substrates, *Appl. Microbiol.*, **8**:109–113.
Youland, G. C., and C. R. Stumbo (1953): Resistance Values Reflecting the Order of Death of Spores of *Bacillus coagulans* Subjected to Moist Heat, *Food Technol.*, **7**:286–291.

GENERAL REFERENCES

Ayres, J. C., A. A. Kraft, H. E. Snyder, and H. W. Walker (eds.) (1962): "Chemical and Biological Hazards in Food," Iowa State University Press, Ames.
Ball, C. O., and F. C. W. Olson (1957): "Sterilization in Food Technology," McGraw-Hill, New York.
Buchanan, R. E., and N. E. Gibbons (eds.) (1974): "Bergey's Manual of Determinative Bacteriology," 8th ed., Williams & Wilkins, Baltimore.
Charm, S. E. (1971): "Fundamentals of Food Engineering," 2d ed., Avi Publishing Co., Westport, Conn.
Dack, G. M. (1956): "Food Poisoning," 3d ed., University of Chicago Press, Chicago.
Edwards, P. R., and W. H. Ewing (1972): "Identification of Enterobacteriaceae," 3d ed., Burgess, Minneapolis.
Elek, S. D. (1959): "*Staphylococcus pyogenes*," E. and S. Livingstone, Edinburgh.

Elliott, R. P., and H. D. Michener (1965): Factors Affecting the Growth of Psychrophilic Microorganisms in Foods, *U.S. Dep. Agri. Tech. Bull.* 1320.
Frazier, W. C. (1967): "Food Microbiology," 2d ed., McGraw-Hill, New York.
Goldblatt, L. A. (ed.) (1969): "Aflatoxin," Academic, New York.
Guthrie, R. K. (ed.) (1972): "Food Sanitation," Avi Publishing Co., Westport, Conn.
Housler, W. J. (ed.) (1972): "Standard Methods for the Examination of Dairy Products," 13th ed., American Public Health Association, Washington, D.C.
Jay, J. M. (1970): "Modern Food Microbiology," Van Nostrand-Reinhold, New York.
Lawrence, C. A., and S. S. Block (eds.) (1968): "Disinfection, Sterilization, and Preservation," Lea & Febiger, Philadelphia.
Lewis, K. H., and K. Cassel, Jr. (eds.) (1964): Botulism, Proceedings of a Symposium, U.S. Department of Health, Education, and Welfare, Public Health Service Publ. 999-FP-1.
Litsky, B. Y. (1973): "Food Service Sanitation," Modern Hospital Press, Chicago.
Molin, N., and A. Erichsen (1964): "Microbial Inhibitors in Food," Almqvist & Wiksell, Stockholm.
Nickerson, J. T. R., and A. J. Sinskey (1972): "Microbiology of Foods and Food Processing," American Elsevier, New York.
Parker, M. E., and J. H. Litchfield (1962): "Food Plant Sanitation," Reinhold, New York.
Riemann, H. (ed.) (1969): "Food-borne Infections and Intoxications," Academic, New York.
Sharf, J. M. (ed.) (1966): "Recommended Methods for the Microbiological Examination of Foods," 2d ed., American Public Health Association, Washington, D.C.
Slanetz, L. W., C. O. Chichester, A. R. Gaufin, and Z. J. Ordal (eds.) (1963): "Microbiological Quality of Foods," Academic, New York.
Stumbo, C. R. (1973): "Thermobacteriology in Food Processing," 2d ed., Academic, New York.
Weiser, H. H., G. J. Mountney, and W. A. Gould (1971): "Practical Food Microbiology and Technology," 2d ed., Avi Publishing Co., Westport, Conn.
Wogan, G. N. (ed.) (1965): "Mycotoxins in Food Stuffs," M.I.T., Cambridge, Mass.

11
MICROBIOLOGICAL DETERIORATION OF PULPWOOD, PAPER, AND PAINT

Richard T. Ross
and
C. George Hollis

The economic importance of microorganisms is nowhere more dramatically seen than in the consideration of their role in the deterioration of materials. It has been estimated that biological deterioration of all materials excluding foodstuffs exceeds $5 billion a year. As a form of deterioration, that produced by microorganisms is exceeded only by the corrosion of metals. The figure presented above would be ten times larger were it not for the microorganism control practices carried out by many industries. To appreciate the technology as well as the economic importance of industrial microorganism control, this chapter reviews the microbiological problems encountered in two important industries and their solutions, where they exist.

BIODETERIORATION IN THE PULP AND PAPER INDUSTRY

The production of pulp and paper is the fifth largest industry in the United States. In 1963 about 45.5 million cords of wood was used to produce the paper and paperboard requirements for the United States. This compares with the utilization of 8.7 million cords in 1936 and 26.5 million cords during 1952. It is expected that in 1975 pulpwood requirements will have increased to about 65 million cords, based on the current rate of growth.

The United States consumes more paper per capita than any other nation in the world, utilizing in excess of 500 lb per person per year. During 1966, more than 47 million tons of paper and paperboard were produced. The industry continues to grow at a rapid rate (Fig. 11-1). Paper and paperboard sales in 1967 amounted to more than $8 billion.

Losses to the pulp and paper industry through the activity of microorganisms are substantial. The degradation process begins with the first stage of biological infestation of the wood when it is cut, extends through the entire

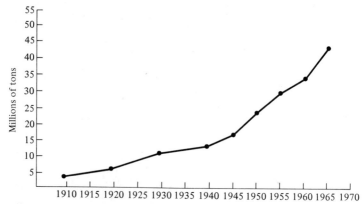

Figure 11-1 Growth of paper and paperboard production in the United States.

papermaking process, and ends when the paper product no longer functions for its intended purpose. The following discussion deals with the more important areas of deterioration and losses caused by microorganisms in the pulp and paper industry.

THE PAPERMAKING PROCESS

The papermaking process is a complex series of microbiological ecosystems. Figure 11-2 shows a typical, fully integrated pulp and paper mill system for the production of bleached Kraft paper.

WOOD HANDLING

Large corporations may own thousands of acres of carefully managed timberlands to ensure an adequate supply of pulpwood to satisfy the demands of an ever-increasing market. In addition, pulpwood is bought from local timber owners. Sawmill and plywood wastes are utilized as a fiber supply in many papermaking operations. Sawdust from sawmills is not used extensively as yet. However, it has been shown that sawdust is an economical and useful fiber source for some grades of paper and paperboard. That portion of the log not utilized for commercial lumber production can be processed into chips and transported to the paper mills for use in papermaking. However, most of the fiber for making paper originates as pulpwood.

Pulpwood is purchased by the cord. A cord is equivalent to a stack of wood 4 ft wide, 8 ft high, and 8 ft long. The pulpwood is debarked in barking drums and chipped by high-speed chipping knives; the chips go directly to the digesters for cooking or to the chip storage piles to be used later.

THE PULP MILL

Papermaking fibers are produced either by chemical or mechanical means. Mechanical pulps are produced by grinding bolts of wood on rapidly revolving stones. Such pulps are referred to as stone groundwood. Some modern installa-

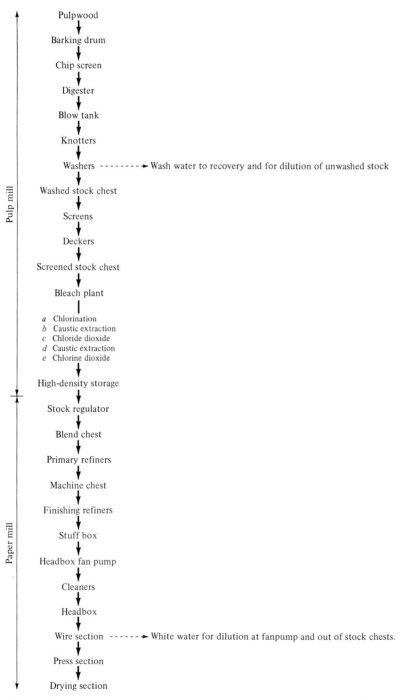

Figure 11-2 Schematic flow diagram of a typical pulp and paper mill producing bleached paper.

tions produce mechanical pulp by separation of fibers from chips by forcing the chips between enclosed, rapidly revolving, motor-driven, steel discs, with flat cutting and shearing surfaces. This type of mechanical pulp is called refiner groundwood from the name of the equipment which produces it. Mechanical pulps are used principally for newsprint, directory, and similar grades. Magazine paper may also contain a large portion of groundwood fiber. Paper produced from all groundwood would have little strength; hence, most grades of paper, including newsprint, have varying proportions of chemically produced "long fiber" to obtain the necessary strength of the final sheet.

Chemical pulps are so called because the cellulose fibers are separated by chemical means, generally at elevated temperatures and pressures. The chips are fed into large vessels called digesters, not unlike an autoclave. Cooking chemicals are added, steam is applied either directly into the digester or indirectly by steam lines encircling the digester, and the mass of chips and cooking liquor is heated. When all the air is exhausted from the digester, it is closed and the contents of the digester are raised to the desired temperatures and pressure and allowed to cook until sufficient solubilization of the lignins, pectins, and other wood adhesives occurs. Pressure, temperature, time, strength, and ratio of cooking chemicals to wood may be varied to obtain the degree of cooking necessary to satisfy the requirements for particular grades of paper. For example, pulp to be bleached to go into fine writing paper is cooked more than that which goes into a brown unbleached paper bags.

In the production of chemical pulps, the cooking chemicals may be acid in nature, as in the acid sulfite process, or they may be alkaline, as in the sulfate (Kraft) process.

When cooking is completed, the digester is opened at the bottom at full temperature and pressure. This sudden release of pressure when the digester is "blown" literally explodes the chips, separating the loosely bound cellulose fibers. In addition to cellulose as the primary product from the digesting process, valuable by-products are obtained. Sulfate turpentine is obtained as the volatile turpentine ingredients are recovered during the cooking process. These solubilized wood by-products present in the cooking liquor yield such useful materials as tall oil, dimethylsulfoxide, and vanillin. Sulfite liquor is processed and used as a nutrient material for the growth of yeast as a supplement to animal feeds. In years to come, it is certain that other valuable by-products can be recovered from the spent cooking liquor, which will not only provide increased economical advantages to processing pulpwood into fiber for papermaking but also result in less deleterious materials which have to be discharged into streams and become a source of pollution.

The remainder of the time the pulp spends in the pulp mill is for final preparation of the cellulose fiber to send to the paper mill. The pulp from the digesters and from the grinding devices is screened to removed uncooked chips, knots, and slivers. Chemical pulps are washed several times to remove as much of the cooking chemicals as feasible. In the Kraft process, the liquor containing the cooking chemical is condensed to a more concentrated form and then finally burned for the recovery of cooking chemicals which go back into makeup liquor for addition at the digester.

For the production of unbleached paper and board, the fibers move on to the paper mill. If the pulp is to be bleached for the production of writing, printing, and publication papers, the cellulose fibers go through a series of bleaching stages, for which chlorine compounds such as gaseous chlorine, hypochlorite, and chlorine dioxide are usually used, until the desired degree of brightness of the pulp is obtained.

Following bleaching, or if the pulp is moving to the paper mill in an unbleached state, slurry of cellulose fibers in water moves into storage, where it is stored for use in the paper mill at consistencies of from 5 to 15 percent fiber and water.

THE PAPER MILL

It is the function of the paper mill to take the raw cellulose fibers from storage in the pulp mill and process the fibers into the finished sheet of paper. As the pulp is moved from storage in the pulp mill, it is diluted to a consistency of 3.0 to 4.0 percent fiber in water. For chemically produced, long-fibered pulps, the first stage of processing in the paper mill is to pass the slurried fiber through refiners. The refiners may be of several types but consist essentially of rapidly rotating enclosed discs or cones inside a steel shell through which the fibers are forced. The shearing and cutting action of the refiners brushes and cuts the fiber surfaces to the proper degree of fibrillation so that later as the fibers move on to the paper machine they have an ability to interlock and form a uniform sheet of paper with the proper amount of strength. Following the refining process, various types of chemicals are added, with the variety and the amount depending on the final use of the paper or paperboard. Tables 11-1 and 11-2 show a

Table 11-1 General furnish components of a paper machine producing unbleached board for boxes

Component	Amount
Fiber:	
Pine Kraft	70%
Hardwood Kraft	30%
Broke*	20%
Additives:	
Rosin size	2–8 lb/ton
Alum	20 lb/ton
Gums†	2–4 lb/ton
Defoamer‡	0.5–3.0 lb/ton
Microbicide§	0.2–1.0 lb/ton
Dispersant	0.5–1.0 lb/ton
Wire life extender¶	0.2–0.8 lb/ton

*Broke is redispersed fiber from culled paper originally made on the paper machine and therefore would have the same or nearly the same ratio of fiber sources as the sheet.

†Strength additive such as guar, locust bean, and mannogalactan gums.

‡May be fatty acid derivatives, alcohols, or hydrocarbons, etc.

§Depends on type, concentration, etc.

¶For corrosion control and wear reduction of copper-containing wires.

Table 11-2 General furnish components of a paper machine producing coated magazine paper

Component	Amount
Fiber:	
Bleached softwood Kraft	40%
Bleached hardwood Kraft	20%
Groundwood	40%
Coated broke	15%
Uncoated broke	10%
Additives (amounts vary across the industry):	
Rosin size	0.0–4.0 lb/ton
Alum or sodium aluminate	0.0–3.0 lb/ton
Clay	0.0–6.0 lb/ton
Starch	0.15–2.5% of furnish
Dyes	0.0–0.0075 lb/ton
Retention aid	0.25–2.5 lb/ton
Defoamer	0.025–0.25 lb/ton
Microbicide	0.025–0.25 lb/ton
Dispersant	0.025–0.25 lb/ton
Pitch control agent	0.025–0.25 lb/ton
Wire life extender	0.025–0.25 lb/ton
Coating:	
Clay	20.0–45.0 wt %
Binder (starch, protein, or synthetic)	15.0–20.0 wt %
Titanium dioxide	0.5–5.0 wt %
Defoamer	10–50 ppm
Dispersant	0.2–2.0 wt %
Lubricant	1.0–3.0 wt %
Microbicide	100–1000 ppm

Note: Footnotes to Table 11-1 apply here also.

typical list of chemical additives which are used for the production of unbleached Kraft paper and for bleached magazine paper, respectively.

The largest and most complicated piece of equipment in the paper mill is the paper machine itself. Its function is to form the mass of fiber into a sheet of uniform quality. A modern paper machine may operate at speeds up to 3000 ft/min and may be more than 30 ft wide. Such a machine may produce more than 500 tons of standard newsprint per day, or at lower speeds may produce as much as 1200 tons/day of unbleached Kraft board for use in boxes.

To form a uniform sheet on a high-speed modern papermaking machine, the consistency of the fiber slurry is usually less than 1 percent. It moves into a complex system of flow spreaders and on into a pressurized headbox where the fibers are maintained in a uniform suspension. The headbox is equipped with a flow spreader which ejects the dilute slurry of fibers onto a rapidly moving paper machine wire. This is a fine-mesh endless belt, which revolves at the speed of the paper machine. Fibers are laid onto the surface of the wire, the sheet is formed, and water is removed. The sheet continues to dry as it moves down the wire. The sheet of partially dried paper passes from the wire into the press section of the paper machine. The press section is equipped with a series of two or three woolen, or combinations of woolen and synthetic, felts. The sheet is pressed between heavy rolls and the felt, removing more water from the sheet. As the water is removed from the sheet, the felt then passes over water-removing devices so that the felt maintains an absorbent character as it contacts the sheet again. As the sheet leaves the press section, it passes into a series of steam-heated drying drums, where more water is evaporated from the sheet, and the finished paper is wound into large rolls at the end of the machine.

The roll of paper is then cut into standard roll sizes to fit customer requirements, or it may be cut into standard sheet sizes. The paper is then packaged and shipped to customers of the papermaking organization to be converted into forms of multiple use that we see every day.

BIODETERIORATION OF PULPWOOD

It is conservatively estimated that 10 percent of all the pulpwood cut is lost through the activity of decay organisms (Shema, 1955). For the 65 million cords to produce the paper and paperboard requirement of the United States in 1975, nearly 72 million cords had to be harvested.

Temperature and moisture, together with an adequate supply of oxygen, are the chief environmental factors which influence the rate of wood decay. Adequate nutrients are present in the substrate. Generally, the best growth of the wood-destroying fungi occurs at temperatures of 15 to 33°C. There are some exceptions. Most of the wood-decay fungi are cold-tolerant and exist for long periods of times at freezing temperatures or below. At the other extreme, the wood-decay fungi generally are unable to live at temperatures in excess of 65°C.

It is generally recognized that decay fungi rarely grow in wood with a moisture content less than 25 percent. The optimum moisture for growth varies with individual species. Spore germination of most species occurs where a free film of water exists. Some wood-staining fungi which cause discloration of wood and pulp can grow at moisture levels as low as 10 percent, but optimum growth occurs at higher levels.

Since the wood-decay fungi are aerobic, the presence of an adequate supply of oxygen is a requisite. Normally, the oxygen supply to support growth is ample. The fact that the wood-decay fungi are aerobic has prompted a number of paper mills to store wood under water, either totally submerged or sprayed continually, greatly reducing losses due to microbiological degradation.

With the temperature requirements of the decay fungi, it would be expected that pulpwood losses would be greater in the southern climates than in the colder regions. However, this does not seem to be the case. Of greater importance than regional temperature differences are the length of storage time, the season during which the wood is cut, the size of the stored pulpwood, and the method of piling. In the southern United States, for example, rarely is pulpwood stored in excess of 30 days in the mill yards, so that the time between cutting and utilization is comparatively short. This is possible because a favorable cutting environment is present the year round and timber resources are close to the mills. In the far northern United States and in Canada where the cutting season is shorter, storage times may be as long as 9 months or even a year. Therefore, in actual practice, loss of usable fiber by decay in the southern United States is less than in those areas where the environment is too severe for a year-round harvesting of wood pulp. Wood stored for 6 months during the warm months from spring until fall loses two to three times more wood substance than for similar storage times during the colder season. Table 11-3 shows a partial list of the fungi isolated from decayed wood (Shema, 1955).

WOOD ROTS

Wood degradation by microorganisms is usually classified by the gross effect it has on the characteristics of the wood. *White rots* are characterized by the

Table 11-3 A partial list of decay fungi identified from pulpwood

Coniophora cerebella	*Pleurotus ostreatus*
Coniophora puteana	*Polyporus abietinus*
Corticium evolvans	*Polyporus asseus*
Corticium galactinum	*Polyporus balsameus*
Echinodontium tinctorium	*Polyporus betulina*
Flammula connisans	*Polyporus chioneus*
Fomes annosus	*Polyporus circinatus*
Fomes ignarius	*Polyporus hirsutus*
Fomes laricis	*Polyporus paragamenus*
Fomes nigricans	*Polyporus schweinitzii*
Fomes nigrolimitatus	*Polyporus zonatus*
Fomes pini	*Poria asiatica*
Fomes pinicola	*Poria cocos*
Fomes roseus	*Poria monticola*
Ganoderma appalanatum	*Poria subacida*
Hypoxylon cohaerans	*Poria tsugina*
Irpex fusco-violaceus	*Poria vaporia*
Lentinus lepideus	*Schizophyllum commune*
Lenzites saepiaria	*Stereum abierinum*
Lenzites trabea	*Stereum chailletii*
Merulius himantioides	*Stereum hirsutum*
Odontia bicolor	*Stereum purpureum*
Omphalia campanella	*Stereum rugosiusculum*
Peniophora abietinus	*Stereum sanguinolentum*
Peniophora gigantea	*Stereum sulcatum*
Peniophora luma	*Trametes serialis*
Peniophora septrionalis	*Trechispora brinkmannii*
Peniophora velutina	*Trechispora raduloides*
Pholiota adiposa	

degradation of the brownish lignin, leaving a white, spongy cellulosic mass, or white pockets or streaks in the wood. The whitish areas show a predominance of cellulose to lignin. Although it may be that the lignin is attacked preferentially, cellulose is also destroyed by the white rot fungi. Some of the common white rot fungi include:

Polyporus abietinus
Polyporus paragamenus
Polyporus ancepts
Polyporus circinatus
Fomes pini
Polystictus versicolor
Stereum frustulosum
Stereum subpileatum

The white rots may be further classified by gross appearance as white pocket rot, white mottled rot, white spongy rot, and white stringy and flaky rot.

Brown rots are the result of preferential microbiological degradation of the cellulose, leaving behind a brown, punky mass with a predominance of lignin. The fungi which attack the cellulose generally show little preference for the lignin content. Some of the common brown rot fungi are:

Lenzites trabea
Lenzites saepiaria
Polyporus balsameus
Polyporus schweinitzii
Echinodontium tinctorium
Coniophora cerebella
Poria monticola

In recent years, studies have shown that significant degradation occurs by the softening of the surfaces of stored moist wood, generally by other than the basidiomycetes which cause the white and brown rots. These are the *soft rots* caused by members of the ascomycetes and some Fungi Imperfecti.

Histologically, in the wood with brown rot, the fungal hyphae ramify through the lumen of the cells and penetrate the walls at right angles, forming holes, the bores of which are larger than the hyphae. In the white rots, the cell walls become thin due to general degradation, and the cell walls are penetrated by the hyphae, showing both lignin and cellulose degradation. In the soft rots, distinct elongated cavities occur within the middle layer of the secondary wall of fibers, tracheids, and vessels. Both lignin and cellulose are attacked by soft rot organisms. Generally, the rate of soft rot decay is slower than for brown or white rot. Some of the fungi known to cause soft rot are:

Ascomycetes
  Ceratocystis pilifera
  Chaetomium globosum
  Chaetomium elatum
  Chaetomium fumicola
  Ophiostoma caerulescens
  Ophiostoma picae
  Ophiostoma pini

Fungi Imperfecti
Alternaria humicola
Bispora effusca
Camarosporium ambiens
Coniothyrium fuckelii
Biscula pinocola
Helicosporium aureum
Phialophora richardsiae
Trichosporium heteromorphum
Trichurus terrophilus

The increased tendency toward underwater or spray storage of pulpwood could result in a greater tendency for soft rots. The wood cooling towers used in the paper industry for cooling fresh and process water provide an ideal habitat for the growth of soft rot fungi.

In addition to the loss of fiber caused by decay fungi, other problems are encountered during the use of fiber from wood infested with microorganisms. These include (1) increased amounts of chemicals required to satisfactorily pulp decayed wood by the Kraft process, (2) increased amounts of bleaching chemicals required to obtain equivalent pulp brightness when compared with nondecayed wood, (3) the lower bursting, tensile, and tearing strength and generally lower folding endurance of paper made from decayed wood, and (4) reduced operating efficiency of the paper machine.

Among the early problems found with pulpwood and chip storage was wood staining. Figure 11-3 shows freshly cut chips together with chips stored outside for 2 weeks. The stored chips are infected with an *Ophiostoma* sp. Staining of wood for use in bleached papers increases the amount of bleaching chemicals needed to obtain the required degree of brightness. Fungal staining of pulpwood for use in unbleached grades of paper and board is of little consequence.

To alleviate problems of fungal attack on wood, the following practices are recommended for the most efficient use and storage of pulpwood in roundwood or bolt form:

1. The wood should be used as soon as possible after cutting.
2. The wood should be stored in such a manner as to obtain good water drainage and air circulation through the pile.
3. The woodyard should be kept clean and free of rubbish, bark, and other debris. At periodic intervals, the woodyard should be sprayed with an effective fungicide.
4. The woodyard should be located in a well-drained area.
5. Wood should be used on a first-in, first-out basis so that excessive storage times are avoided.
6. Old and new wood should not be mixed in the same pile.
7. If practical, bolts should be peeled before storage.

In recent years, the storage of wood in chip form has gained wide acceptance throughout the paper industry. Chip storage has several advantages over storage of roundwood bolts. With chippers installed at most lumber and veneer operations, much of the wood waste can be reduced to chip form, providing an increased supply of fiber to the paper industry. In addition, when

Figure 11-3 Freshly cut and stored pine chips used to produce pulp.

wood is chipped in the green or wet stage less fiber loss occurs than when wood is chipped from dry roundwood. A cleaner woodyard, more uniform flow of wood to the woodyard, and savings in the cost of woodyard manpower are other advantages.

Losses caused by microorganisms during chip storage amount to about 6 percent of cellulose after storage of 6 months. Where chips are to be stored for extended periods of time, it is generally more practical to treat chips chemically with an effective microbicide than it is to treat roundwood.

One of the chief disadvantages of outside chip storage is the increased loss of wood by-products, such as turpentine and tall oil, which are valuable by-products of the sulfate (Kraft) digestion process. Turpentine is usually lost by volatilization and tall oil by enzymatic oxidation. Whether microorganisms have any effect on degradation of lipid and resin acids during storage has not been determined. Since it is known that spray storage of roundwood reduces by-product loss, perhaps work being done by spray storage of chips will alleviate this problem. In spray storage, the roundwood or chip pile is equipped with a sprinkling system, and the wood piles are continually wetted with water. By maintaining the chips and wood in a wet state, growth of troublesome fungi is retarded by the lack of oxygen. However, it may be that soft rots will become increasingly important, and some bacterial attack will occur on prolonged storage in this state.

BIODETERIORATION OF PULP

Pulp is nothing more than wood which has been disintegrated by chemical or mechanical means. Chemically produced pulps provide less nutrients for microbiological growth than do mechanically produced pulps because the chemicals used in pulping solubilize much of the wood substance, leaving cellulose and some lignin. Most of the solubilized substances are subsequently removed from the chemical pulps by repeated washing. On the other hand, mechanically produced pulps, such as stone groundwood, contain all the nutrients present in the original wood. For this reason the possibilities for biodeterioration is greater during storage of mechanical pulp than it is during storage of chemical pulps, other factors being equal.

The principal problems associated with the use of microbiologically infested pulp are (1) loss of usable cellulose fiber, (2) loss of strength of paper made from degraded pulp, (3) discoloration of pulp, (4) spots, specks, and holes in paper made from deteriorated pulp, (5) lost production and reduced operating efficiency of papermaking machines, and (6) odor in the finished product. The severity of each of these problems depends on the way the pulp is manufactured, shipping and storage conditions, and the nature of the microflora present as contaminants.

Loss of Cellulose

Although the intermediate cellulolytic pathways of microbial degradation of cellulose remain obscure, it is apparent that the traditional pathway from cellulose to cellobiose to glucose through the activity of cellulase and cellobiase is one skeletal system. There is no reason to feel that either cellobiose or glucose has to be an end product, and so other pathways are probable. Whatever the mechanism, the end product or products will be soluble carbohydrates or carbohydrate derivatives, and some indication of the amount of cellulose degradation can be obtained by solubility values of pulp. For example, Bray and Andrews (1923) showed that solubility of decayed pulp in hot water and in 1 percent sodium hydroxide was greatly increased over that for sound pulp.

Generally, cellulose degradation, whether by direct chemical or by biological means, is measured by viscosity changes of cellulose in suitable solvents. Gadd (1962) utilized a water-soluble cellulose derivative, carboxymethyl cellulose (CMC), and, by measuring viscosity at 1-min intervals, followed the decomposition rate. He found that a 3 percent solution of CMC was completely degraded to glucose in the presence of 10 mg/l of a commercial cellulase preparation. In the pulp and paper industry the degree of polymerization (DP) of the cellulosic components is measured by viscosity of the cellulose in cuprammonium hydroxide (Schweizer's reagent) or cupriethylenediamine. Both these reagents are used to determine the number of times a glucose unit appears in the cellulose preparation.

Structurally, cellulose is composed of repeating glucose units held together by β-glucosidic linkages. This may be represented by the formula

Therefore it is readily seen that upon complete hydrolysis cellulose yields glucose units. Incomplete hydrolysis can yield mixtures of polysaccharides such as hydrocellulose. As hydrolysis proceeds, cellodextrins, oligosaccharides, and glucose are formed. The appearance and amounts of the degradation products may be measured by the use of Fehling's solution (alkaline copper tartrate) which measures the reducing ability of the carbohydrates. Purified cellulose has only a slight reducing action whereas glucose is a reducing compound. Two of the oligosaccharides formed during the hydrolysis of cellulose are cellobiose and cellotriose composed of two and three glucose units, respectively.

It must be emphasized that work on purified cellulose and cellulose preparations provides basic information which is valuable academically to explain basic processes. However, fiber from the cell walls of wood for use in papermaking contains cellulose as described and, in addition, contains other carbohydrates and some lignin. Holocellulose consists of cellulose with its repeating glucose units whereas hemicelluloses contain units other than glucose. Xylose and mannose are the principal sugar units, and small amounts of glucose, galactose, arabinose, and sometimes rhamnose units are found. The hemicelluloses also contain uronic acid groups which are typically linked to xylose units and are called xylans or pentosans. Other groups composed of glucose and mannose are called glucomannans or simply mannans. Therefore, it can be readily seen the microbial degradation of cellulosic products such as wood fiber for use in papermaking is far more complex than hydrolysis of cellulose, in the stricter sense.

A large number of bacteria and fungi have been isolated which show cellulolytic properties (Siu, 1951) on cotton fabrics. The number of microorganisms isolated from wood pulp are fewer in number. Russell (1961) lists the decay fungi associated with degradation of moist groundwood pulp as follows:

Corticium evolvans
Flammula spumosa
Fomes annosus
Hymenochaete corrugata
Lentinus lepideus
Lenzites saepiaria
Paxillus panuoides
Peniophora gigantea
Polystictus abietinus
Stereum sanguinolentum
Trametes serialis

It is noteworthy that these are higher fungi belonging to the Agaricales, chiefly in the families Polyporaceae, Agaricaceae, and Thelephoraceae. Siu (1951) provides an impressive list of fungi which have been shown to degrade fabrics. For the most part this list is devoid of higher fungi. However, many of the fungi responsible for staining and discoloration of pulp have cellulolytic properties. The following genera of organisms are listed by Russell (1961) as common pulp-staining microorganisms: *Alternaria, Aspergillus, Cladosporium, Geotrichum, Gliocladium, Margarinomyces, Oidiodendron, Papulospora, Penicillium, Phialophora, Phoma, Pullularia, Stemphylium, Torulopsis, Trichoderma, Trichosporium, Verticillium.*

Most of these genera have species which are cellulolytic (Siu, 1951); therefore, although most of the cellulose degradation of moist pulp can be attributed to the higher fungi, the cellulolytic activity of members of the Phycomycetes, Ascomycetes, and Fungi Imperfecti is well known and must contribute to the degradation of the cellulosic content of stored pulp. This has been shown to be the case in some stored pulp badly infested with staining organisms and other molds but devoid of the usual wood-rotting fungi.

Loss of wood substances, including cellulose and pentosans, can amount to as much as 75 percent under storage conditions approaching ideal for the growth of microorganisms (Acree, 1919). Barker (1961) showed a fiber loss of 6.0 percent after storing a slurry of groundwood pulp for 5 months. Storage of pulp under the same conditions with 0.1 percent copper sulfate (based on bone-dry fiber) showed a fiber loss of 1 to 2 percent.

It should be noted that fiber loss can take two forms, from a practical standpoint. First, that amount of wood substance which is actually enzymatically hydrolyzed is no longer available to produce a commercial sheet of paper. Second, the reduction of the fiber length which can be attributable to microbiological activity results in a considerable amount of cellulose "fines" much of which is lost to the sewer during the manufacturing process and cannot be recovered. Such materials present in the effluent from pulp and paper mills contribute to pollution of the receiving streams.

Loss of Strength

The strength of paper made from degraded wood fiber has been the subject of several investigations. Kress et al. (1925) noted a decrease in bursting and tensile strength in paper made from decayed pulp, substantiating earlier work by Acree (1919), who found that paper made from infected pulp retained only 25 percent of the original strength of the paper made from sound pulp (Table 11-4). Barker (1961) reported losses in tensile strength of 8.7, 30.3, and 44.6 percent of groundwood pulp stored in slurry form over periods of 30, 60, and 150 days, respectively. Pulp treated with copper sulfate lost 0.1, 2.9, and 3.1 percent during the same storage periods.

From a practical standpoint, it is difficult to detect small losses in strength attributable directly to microbiological action because of other variables in the papermaking process. However, at the moment pulp moves into storage in the presence of cellulolytic bacteria and fungi, the wood substances are attacked and degradation occurs to some degree. The magnitude of loss of strength and loss of fiber substance then depends on the storage time, storage conditions, and amount of contamination.

Discoloration

There are many spurious fungi and bacteria which can cause localized discoloration of stored pulp. The isolation and identification of these are of academic interest for the most part because they do not cause measurable problems in the case of the pulp. It is recognizable that any microorganisms which in themselves produce colored pigments in the hyphae or spores can cause disfigurement. There are fungi which are capable of producing colored substances by the chemical alteration of their substrate. This has been mentioned previously in relation to the early decomposition of cellulose and lignin by certain wood-rotting fungi. Gadd (1949) described the stains produced by certain of the wood-rotting fungi as follows:

Coniophora cerebella   light-brown
Polyporus abietinus   dark-brown

Table 11-4 Strength properties of a laboratory-formed waterleaf sheet of paper made from spruce groundwood pulp stored for 6 months and 12 months

Description	Average fiber length, mm	Weight of 500-sheet ream (24 × 35 in), lb	Points per pound	Average breaking length, m
Freshly ground	1.5	47.0	0.30	2780
After 6 months	0.9	49.8	0.27	2340
After 12 months	1.0	46.0	0.22	2175

Lentinus lepideus   yellow-white
Lenzites saepiaria   yellow-white
Fomes annosus   brown
Stereum sanguinolentum   brown

Therefore, where pulp is extensively attacked by wood-rotting fungi, discoloration can be a problem and result in the loss of brightness of the pulp.

Russell (1957) and Gadd (1949) characterized some of the typical stains produced by fungi other than those associated with wood decay. *Trichoderma viridae* produces a yellow to pale green discoloration. *Pullularia pullulans* causes bluing of wood and wood pulp and macroscopically is difficult to separate from the blue to black discoloration caused by species of *Cladosporium*, *Alternaria*, *Stemphylium*, and *Aspergillus* or by *Margarinomyces fasciculatis* and *Trichosporium heteromorphum*. Purplish stains can be produced by *Phoma lignicola* and by *Penicillium roqueforti*. *Penicillium rubrum* produces brown stains. Species of yeast, such as those of *Torula*, produce pink stains.

It is only when the typical staining fungi grow extensively on the surface and proliferate to the interior of the pulp that losses in brightness occur during the use of the pulp. Kress et al. (1925) compared the brightness of groundwood pulp made from spruce after 6 and 12 months storage, utilizing an Ives tint photometer. Their results are shown in Table 11-5. The increase in yellowing of the pulp is measured by the increase in the red plus green and corresponds to the decrease in white.

The darkening of pulp during storage makes it necessary to use more bleaching chemicals or brightening agents to bring the finished sheet up to brightness standards.

Loss in Finished Production

The use of pulp which is badly infested with fungi or bacteria can result in poor quality of the sheet and in lost production time. The strength of the sheet is impaired in some cases where decay fungi are present. In addition, it is difficult to completely redisperse fiber contaminated by fungi. If the fiber cannot be completely dispersed, small knots or masses of fiber and fungal hyphae will cause spots, specks, and holes in the sheet, lowering its quality. In work with fungi causing discoloration, Gadd (1962) said that if 0.1 percent of the pulp is discolored it is to be expected that knots will appear during sheet formation. Kress et al. (1925) said that decayed pulp had a tendency to stick to some of the

Table 11-5 Comparison of color of paper made from spruce groundwood after 6 and 12 months storage

Description	Color (tint photometer)		
	White, %	Black, %	Red plus green, %
Freshly ground	68.5	20.0	11.5
After 6 months	67.5	19.8	12.7
After 12 months	64.6	22.8	12.6

rolls during manufacture, causing lost production time. In addition, some 20 times as many specks per unit area of the sheet were noted when using decayed pulp. Troublesome foaming of water and pulp has been found when using decayed pulp (Kress et al., 1925).

The use of pulp badly contaminated with bacteria and fungi will introduce large numbers of microorganisms into the papermaking system. Many of these organisms can develop and grow into slime deposits in the paper mill and cause problems.

BIODETERIORATION OF PAPER

The growth of microorganisms on the surfaces of lines, chests, and tanks in the paper mill results in a decrease in quality and quantity of production of paper. The term *slime* describes those deposits consisting of bacteria and/or fungi, fiber, and debris. The slime masses may be uniform over surfaces or they may occur in patches. Depending on the predominant microflora together with the accumulated nonliving material in the system, the deposits may be slippery, gelatinous, fluffy, stringy, pasty, rubbery, mealy, tapioca-like, or hard. In a system as complex as a paper mill it is often difficult to decide if a deposit is primarily microbiological because pitch, scale, and other chemical materials have a tendency to become an integral part of deposits. Therefore, both microbiological and chemical analyses are necessary to determine the nature of paper mill deposits.

LOSSES CAUSED BY SLIME

Buckman and Kirchen (1958) estimated that losses caused by the uncontrolled growths of microorganisms in the pulp and paper mill amount to $1.00 to $5.00 per ton of paper produced. They detail the following types of losses:

I   Lost production time
  1 Washups and squirtups not required by order changes
  2 Wire cleaning, removal of slime spots, etc.
  3 Felt washing, felt repairs, and felt changes to the degree that plugging and deterioration by microorganisms require washing, repair, or change during normal production time
  4 Breaks of the sheet during production, including time to remove paper from the dryers and calender stacks
  5 Miscellaneous items resulting from slime, such as cleaning screens, suction boxes, presses, and press doctors; sticking of sheet to the wire, couch, and presses

II Decreased equipment performance
  1 Decreased average speed of operation of machines resulting from slime breaks and the loss of efficiency of press felts plugged with slime and slime-bound particles
  2 Increased steam required to dry paper received from wires and felts operated at decreased effectiveness
  3 Increased steam required to overdry paper to obtain uniformity of moisture content and compliance with customer's specifications

III Loss of heat, chemicals, filler, fiber, and water
1. Loss of heat, chemicals, filler, and fiber in white water lost to the sewer
2. Increased use of fresh water to replace white water discharged to the sewer to reduce the difficulties caused by slime

IV Loss of finished product and customer acceptance
1. Loss of profit, cost of segregating, and cost of repulping paper culled because of slime spots, slime holes, and nonuniformity caused in different ways by the growth of microorganisms for the culled paper reutilized at the original grade
2. Additional loss in grade or complete loss of that portion of the total culled paper which cannot be reprocessed at the original grade
3. Loss involved in the settlement of customer claims resulting from defects in the finished product as the result of the growth of microorganisms
4. Loss in weight and in some cases loss in paper quality as a result of overdrying paper to obtain uniformity of moisture content and compliance with customer's specifications
5. Decreased customer acceptance of the paper mill's products as a result of gross defects caused by the growth of microorganisms in the pulp and paper mill system

In addition to the above losses, equipment can be damaged by microbiologically induced corrosion. The production of acid in the microenvironment beneath and in slime, scale, or other deposits will cause pitting of metallic surfaces and decomposition of concrete and some tile adhesives. The production of hydrogen sulfide by sulfate-reducing bacteria, such as *Desulfovibrio* sp., will result in corrosion of ferrous metals.

Production time on a large modern expensive paper machine is a critical factor. Therefore, the efficiency of a mill operation is a function of the actual time the paper machine is producing saleable paper at maximum operating speeds. Interruptions in production or production of substandard paper decreases efficiency. *Breaks* is a term used to describe the separation, tearing, or "breaking" of the continuous sheet of paper during periods of operation of the machine. The period of the break may be short, e.g., 10 min or less, or it may be longer. Slime deposits which slough off the lines, tanks, or chests can cause the sheet to stick to the machine components from the wire section through part of the dryer section, and since the sheet will not transfer stresses this causes it to tear. Smaller masses cause spots and specks or holes in the sheet. (Fig. 11-4). Large masses cause the sheet to split or tear off, completely interrupting production.

The value of production time varies, depending on the grade of paper or board produced together with the potential production of a particular machine. Some of the new board machines are approaching a production rate of 1 ton/min with rates of 30 to 40 tons/h commonplace. Newer newsprint machines are capable of producing 20 to 25 tons/h. Therefore it is readily seen that economic values can be placed on each minute of production time for many machines.

Figures 11-5 and 11-6 show the results of the initiation of improved programs for slime control on paper machines.

Figure 11-4 Hole in a sheet of bleached paperboard caused by slime.

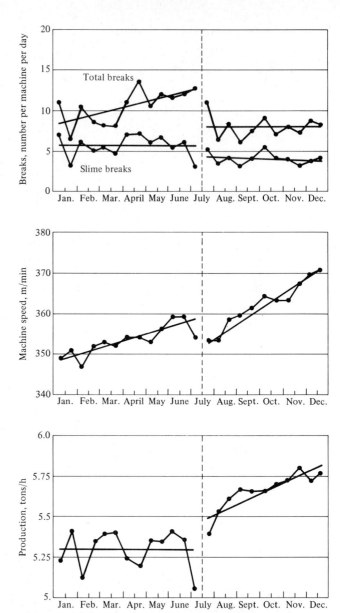

Figure 11-5 Results of the initiation of a slime control program in the middle of July in a Swedish newsprint mill. [N. Lundgren (1949), N. Slemkontroll i ett Tidnigspapersbruk, *Sven. Papperstidn.*, **52**(9):221–226.]

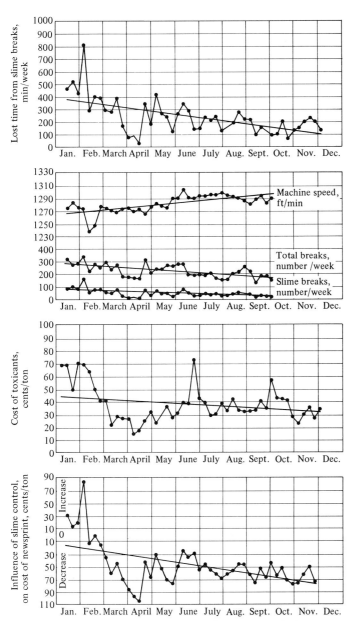

Figure 11-6 Results of the initiation of an improved slime control program in a Canadian newsprint mill. [E. L. Neal and S. F. Jennings (1948): Slime Control at Anglo-Canadian Pulp and Paper Mills, Ltd., *Pulp Pap. Mag. Can.*, **48**(3):137–142.]

MICROORGANISMS ASSOCIATED WITH SLIME

Since slime is caused by the deposition of microorganisms and the subsequent entanglement of fiber, fines, and other debris from the water, fiber, and other components of the papermaking medium, it is logical to assume that those organisms that either are capsulated or are in filamentous form would be common components of slime masses. Therefore it is not surprising to see a large variety of both bacteria and fungi in mill systems. Among the bacteria, *Enterobacter (Aerobacter) aerogenes* and *Bacillus* sp. are quite common. On those paper machines operating above pH 5.5 the filamentous iron bacteria, such as *Sphaerotilus natans*, can usually be found as a predominant part of the slime mass.

Fungi identified as causes of slime deposits include species of *Mucor*, *Penicillium*, *Trichoderma*, and *Fusarium*. Yeasts are fairly common, and species of *Torula* and *Rhodotorula* are easily isolated.

Pink slime is particularly troublesome in some mills producing bleached paper. *Alcaligenes viscosus* var. *dissimilis* has been identified in pink slime from a number of mills (Appling et al., 1959).

Desulfovibrio sp. are common anaerobic bacteria found where oxygen is absent or minimal.

It is possible to isolate tremendous numbers of different kinds of microorganisms including bacteria, fungi, protozoa, and even nematodes and sometimes even algae from paper mill systems. It does not mean that all such organisms cause problems of slime formation. Although academically impressive lists of microorganisms can be obtained, specific identification provides information of limited usefulness as far as practical control is concerned. It is important to recognize whether the problem is primarily bacterial or fungal and whatever other identification is necessary to trace the source of contamination.

CONTROL OF SLIME

In industrial processes, such as that involved with paper production, it is not necessary to sterilize a system to achieve desired benefits. The primary goal is to provide that amount of control consistent with economical production. Depending on the grade of paper or paperboard being produced, some machines require cleaner systems than others to achieve this goal.

In a natural environment microorganisms are limited or controlled by such factors as the availability of nutrients, moisture, and oxygen, together with the range of temperature and pH of the microhabitat. A paper machine system provides an adequate supply of nutrients, oxygen, and moisture to support the growth of a wide variety of microorganisms. These nutrients may be derived from the pulp and from the water, or they may be supplied by the addition of materials such as starch, casein, gums, phosphates, and innumerable others to the pulp. The practice of reusing much of the water removed from the sheet of paper during manufacture tends to concentrate nutrients, thereby providing for increased growth. Pumping, screening, sluicing, and otherwise moving the pulp and water slurry maintain adequate oxygen to satisfy all requirements.

Paper and paperboard are made in the pH range from 4.0 to 5.0. Unbleached grades for bags, wrapping, and boxes are made between pH 5.5 and 6.5. Corrugating medium for use in boxes and some other specialty grades are made on the alkaline side at pH 7.5 to 9.0. There is a tendency for more of the grades now being produced on the acid side to be manufactured closer to neutrality or on the alkaline side. Therefore, from a practical standpoint pH is not a limiting factor in slime formation when considering the diversity of microorganisms capable of causing slime deposits. As noted previously, *Sphaerotilus natans* will not grow and do well at less than pH 5.0. However, bacterial counts in excess of 10 million per gram are commonly obtained on paper machine furnishes operating at pH 4.2 to 4.5. Similar population numbers are found on machines operating with alkaline furnishes.

Paper machines may operate with headbox furnish temperatures from about 80 to 150°F. However, individual machine temperatures vary within a narrow range. The majority of paper is made on machines operating at temperatures between 90 and 135°F. As can be seen, temperature can be a limiting factor in the growth of a number of different kinds of microorganisms, particularly on those machines operating in the higher temperature ranges. The majority of bacteria usually have an optimum growth temperature between 90 and 110°F. Generally, expenditures for slime control chemicals are less on those machines operating in excess of 115°F. However, deposit problems and corrosion can occur even at higher temperatures by the growth and activity of thermophilic bacteria and fungi. Where excess steam is available for heating paper machine systems it can be used to advantage as part of a slime control program. However, the operating temperature of a machine must be consistent with all facets of papermaking because increased temperature increases the rate of corrosion, increases drainage of the sheet up to a point, and can affect other operating conditions.

Slime control consists of a number of interrelated factors and cannot be generalized for the industry as a whole. Individual slime control programs for each machine have to be designed according to the factors peculiar to that machine. Some of these factors include the grade and weight of the paper being produced; temperature, pH, and degree of water reuse in the system; operating time between cleanups and thoroughness of cleaning during cleanup; the nature of slime deposits; and paper machine idiosyncrasies.

The first item on any slime control program should be a consideration of good housekeeping. Thorough cleaning during those periods when the paper machine is down for repairs or equipment changes is a prerequisite for inexpensive slime control. Ideally, washing and high-pressure hosing of all internal and external surfaces are desired. At periodic intervals hot "boilout" solutions may be made up with caustic soda (NaOH) at about 60 lb per 1000 gal of water, heated to about 160°F, and circulated stepwise through the system for at least 2 h. Other detergents of demonstrated value, such as sodium metasilicate, may be used instead of caustic soda. It is often desirable to add dispersants and microbicides to the boilout solution, particularly if contact time is limited or high temperatures cannot be maintained. A paper machine system which is thoroughly cleaned at regular intervals will be easier to keep clean during operation by the use of microbicides.

A large number of chemical formulations and chemical products are available for use as microbicides in the paper industry. Many of these fall under the following categories:

    Chlorine and chlorine compounds
    Phenolic compounds
    Organosulfur compounds
    Organobromine compounds
    Heavy-metal compounds
    Mixtures of microbicides

Gaseous chlorine provides its greatest value to slime control in the paper mill by its addition to the incoming fresh water and is similar in this respect to treatment of municipal potable water. Chlorine will control most of the vegetative forms of bacteria and reduce such bacterial contamination. However, in the amounts utilizable chlorine will not seriously affect bacterial spores or many of the fungi. Chlorine is corrosive to some paper mill equipment and will degrade the wool in paper machine felts. Therefore, the amount of chlorine added at the water treatment plant should be just sufficient to carry a trace residual into critical areas of the paper mill. Chlorine is inactivated quickly when it contacts the furnish on most paper machines; for this reason chlorine is not widely used in the paper mill itself. It does find some application inside the mill for treatment of restricted volumes of paper machine white water and as a supplement to other microbicides. Excessive use of chlorine or chlorine-donor compounds around the paper machines can cause a buildup of chlorides in the system and, in addition, react adversely with paper dyes.

A number of chlorine-donor compounds have been introduced. These include sodium chlorite, chloroamines, and chlorine dioxide. The purpose of these compounds is to provide a slow but effective concentration of chlorine throughout the paper machine system in an effort to avoid some of the immediate inactivation by the organic material in the furnish. However, these materials have shown limited effectiveness but are used to supplement other microbicides in some instances.

Chlorine and chlorine compounds derive their microbicidal activity from the fact that when added to water hypochlorous acid is formed which decomposes to form HCl and nascent oxygen, as shown in the following reaction:

$$Cl_2 + H_2O \longrightarrow HCl + HClO$$
$$HClO \longrightarrow HCl + O$$

The oxygen released is a strong oxidizing agent which combines with the cellular constituents of microorganisms, thereby killing them. Chlorine itself can poison cells by direct chlorination of some cellular substances.

Phenolic compounds normally used in paper mill systems include the sodium or potassium salts of tri- and pentachlorophenols. The corresponding phenols have limited solubility in aqueous media, hence the use of the salts. Products such as sodium or potassium 2,4,5-trichlorophenate, 2,4,6-trichlorophenate, and pentachlorophenate or mixtures of the chlorinated phenolic salts are in common use. Orthophenylphenol is useful where the odor of chlorinated phenols is objectionable. The widest use of the phenolic compounds is as fungicides, usually in conjunction with other more powerful bactericides. Chlorinated phenols are often mixed with other microbicides such as organosulfur and organomercury compounds to provide a product with a broader spectrum of microbicidal activity.

The mode of action of the phenolic compounds seems to depend on their ability to denature proteins or damage cell membranes. Many of these compounds reduce surface tension which in turn allows for easier destruction of microorganisms.

The organosulfur compounds are a diverse group of sulfur-containing molecules usually represented by a thio group within their structure. The dithiocarbamates have been used to control plant pathogens for a number of years. The dithiocarbamates are derivatives of the reaction between carbon disulfide, CS_2, and amines to produce the general formula RNHCSSH, and commercial products are the result of further reactions. Examples of the type of products useful in the paper industry are (1) potassium N-methyldithiocarbamate and (2) sodium dimethyldithiocarbamate.

$$(1) \quad \begin{array}{c} H-N-C-SK \\ | \quad \| \\ CH_3 \quad S \end{array} \qquad (2) \quad \begin{array}{c} H_3C-N-C-SNa \\ | \quad \| \\ CH_3 \quad S \end{array}$$

Another sulfur compound which has found considerable use in the paper industry is 3,5-dimethyl-1,3,5,2H-tetrahydrothiadiazine-2-thione, sometimes simply called 35D. It appears from existing evidence that one of the products of the solubilization of 35D in basic materials such as amines or caustic soda (NaOH) is N-methyldithiocarbamate. This reaction may be represented as follows:

$$\begin{array}{c} S \\ H_2-C \diagup \diagdown C=S \\ | | \\ CH_3C-N \diagdown \diagup N-CH_3 \\ C \\ | \\ H_2 \end{array} \xrightarrow[\text{amine or NaOH}]{\text{base}} \begin{array}{c} -S H \\ \diagdown | \\ C-N-CH_3 \\ \| \\ S \end{array}$$

The dithiocarbamates contain a sulfhydryl (SH) configuration in the molecule, and evidence suggests that the mode of action of these compounds is related to the inactivation of enzymes also containing the SH grouping, thereby interfering with metabolic processes of the cell.

Other organic sulfur compounds of demonstrated effectiveness for the control of paper mill slimes are disodium cyanodithioimidocarbonate and methylene bis-isothiocyanate. Although the latter product is widely used alone or in combination with other microbicides, it is reported that under some conditions it will slowly decompose, releasing cyanide.

The organic bromine compounds used in the paper industry are represented by two principal compounds. These are bromohydroxyacetophenone and bis-1,4,-bromoacetoxy-2-butene.

The mode of action of the organic bromine compounds is not completely understood. However, it appears that their action is not based on the presence of the halogen alone.

Products containing heavy metals have been known for their microbicidal activity for many years. Chief among those used in the paper industry are based on mercury and include organic mercury compounds such as phenylmercuric acetate, ethyl mercuric phosphate, and pyridylmercuric acetate. However, phenylmercuric acetate, alone and in combination with other compounds, has found the widest application. Silver fluoride and copper sulfate are sometimes found to be useful but neither product has gained widespread acceptance because they are corrosive to paper mill equipment.

The mercurials have lost dominance in the paper industry in recent years because they are sustentative to fiber to some degree and therefore not allowed by the U.S. Food and Drug Administration for use in paper for food packaging. Since many of the paper mills produce some grades which may be intentionally or unintentionally used for food packaging, one of the criteria for the use of microbicides in paper mills is that the product must be allowable by the U.S. Food and Drug Administration. Among the last major users of phenylmercuric acetate were the Scandinavian countries where it was used as a wet pulp preservative. In recent years it has been noted that fish in the waters near the pulp mills showed increasing concentrations of mercury in their flesh. This, in addition to the fact that less wet pulp is being sold on the world market, has greatly reduced the use of the mercurials.

The heavy-metal compounds derive their activity from their ability to inactivate enzymes containing the sulfhydryl configuration. In addition, in higher concentrations the heavy-metal compounds act as protein precipitants. Enzyme inactivation may be figuratively shown as follows:

$$\text{Enzyme} \begin{array}{c} \diagup \text{SH} \\ \diagdown \text{SH} \end{array} + \text{Hg}^{2+} \longrightarrow \text{enzyme} \begin{array}{c} \diagup | \text{S} \\ \diagdown | \text{S} \end{array} \text{Hg} + 2\text{H}^+$$

It is desirable in the paper industry to have products which are not only effective in controlling the troublesome microflora but are also convenient to use. For this reason the majority of the most effective and useful products in common use are mixtures of two or more microbicides. For example, it is well known that some of the fungi, notably *Penicillium roqueforti*, is relatively unaffected by usual concentrations of phenylmercuric acetate. Therefore, the use of phenylmercuric acetate alone is not recommended where fungi are a problem. It is an excellent bactericide. The combination of phenylmercuric acetate and chlorinated phenols provides a product with a wide spectrum of activity against troublesome microorganisms. The most effective preservation of wet pulp has been obtained by the combined use of phenylmercuric acetate and 8-hydroxyquinoline. Other mixtures include organic sulfur and chlorinated phenols, different salts of chlorinated phenols, and some including quaternary ammonium compounds. The synergistic action obtained by mixtures of compounds of limited activity when used singly provides the maximum way to take advantage of resistant microorganisms and provide several different modes of action against a mixed population.

Measuring the Efficiency of Microbicides ǂ

With the large number of potentially effective microorganism control products on the market and the continuing research to find better ones, it is desirable to be able to determine the relative effectiveness of products in the laboratory prior to applying them to the paper mill system. Although this is desirable, no satisfactory method has been found that is exclusively used or is satisfactory throughout the industry. Therefore, screening procedures are fraught with disappointment. The main reason for the inability to satisfactorily screen products in the laboratory for use on a paper machine is the complex nature of the papermaking process itself. At best, laboratory procedures are static in nature and the papermaking process is dynamic. In another sense the mere killing or inhibiting of microorganisms from a mixed population in the laboratory does not mean the controlling of the slime-forming potential in actual process. There is a difference between killing viable cells and preventing the mechanical process of forming slime deposits. A number of methods have been proposed and used to some extent for evaluating microbicides for use in the paper industry; they may be generally summarized as follows:

1. Tests using pure laboratory-maintained cultures in an artificial substrate
2. Tests using pure laboratory-maintained cultures and a natural substrate
3. Tests using fresh mixed populations of microorganisms from the actual papermaking process in an artificial medium
4. Tests using fresh mixed populations of microorganisms from the actual papermaking process and a natural substrate

One would expect that more correlation could be obtained by the use of the last type of testing procedure. Of those methods in use it appears to be the most reliable. However, since the physical conditions of the paper mill cannot be duplicated in the laboratory, even this procedure produces unpredictable results.

Whether a chemical material is selected for use based on initial laboratory screening procedures or whether it is selected by some other process, the final evaluation must take place in the actual operation. This is true whether the microbicide is used to prevent slime formation, preserve pulp, or preserve any type of material capable of being degraded by microorganisms.

Since the end result of any program of control in the paper mill is directed toward increased and improved production, it is logical to measure the ability of a microbicide based on these factors. This type of evaluation is demonstrated in Figs. 11-5 and 11-6. As is noted, such evaluations must cover a statistically valid period of time. Little useful information can be obtained in less than 60 days. Longer periods are preferred. To offset this drawback, there are techniques to help determine the amount of control being obtained during the course of the use of the microbicide. Inspection of critical components of the paper machine at every opportunity when it is not operating will provide information on the amount and nature of the deposits occurring. During operation selected portions of the machine system may be examined to determine any unusual pattern of deposit formation. However, it is scientifically desirable to have quantitative data to demonstrate the degree of deposit control being obtained.

One method which has been used quite extensively is the measurement of the amount of deposits occurring on a specified area in the system or on a surface suspended in the stock and water system of a paper machine. Figure 11-7 shows one apparatus designed to measure the amount of deposited materials. This "slime measuring unit" is installed in the system in an area where a flow of dilute stock and water can pass over the surface of the board. Scrapings from the board are measured at regular intervals. Once a sufficient amount of such collected data can be correlated to the operating conditions of the paper machine, the measurements can be used as a guide in determining whether the machine is getting dangerously close to operating difficulties caused by deposits.

Total population counts have been widely used in the paper industry to quantitatively determine slime-forming potential in the systems. Total counts would be helpful if one could be assured that the microorganisms being counted were the ones causing deposit problems. To assume that a definite proportion of a

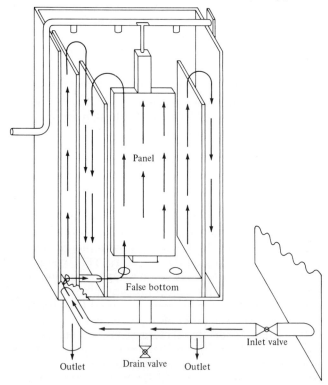

Figure 11-7 Schematic layout of a slime measuring unit used to measure the amount of slime deposited from an aqueous system on a wood panel over a period of time. [C. W. Dean (1953): Slime Measuring Unit, Patent 2,660,884 (assigned to Buckman Laboratories, Inc.).]

mixed free-flowing population of organisms in an aqueous system is directly related to the number that will form deposits is not sound. A machine system could contain hundreds of millions of microorganisms but unless there are sufficient numbers and kinds of bacteria and fungi capable of depositing and entraining debris then deposits problems are unlikely. On the other hand, a few heavily capsulated or filamentous forms can deposit, grow, and become troublesome. Therefore, there is no population level at which one could say that problems are potentially close at hand or that a machine is free of problems.

As indicated previously, slime control is the prevention of deposits caused by microorganisms and is not necessarily the same as controlling numbers of microorganisms. Conceivably, slime control could be obtained by a product which would prevent microorganisms from sticking to surfaces in the system and in this sense would not necessarily have to be part of the aqueous system at all. For total counts to be of any value they must be done at frequent intervals under standard conditions and with culture media that will provide for the growth of the variety of potential slime-forming bacteria and fungi. The medium should be adjusted to the pH of the system, and incubation should be at the temperature of the system. In addition, the counts should be done immediately after collection of the samples. Such counts may provide a guideline for determining the lethal properties of the microbicide being used against those microorganisms which will grow on the medium used to recover them. The most useful information obtained from counts is to determine areas of potential contamination to the system. For example, if by total counts one finds proportionately large numbers of microorganisms in one or two components of the furnish to the paper machine, such as fresh water, or white water, or one fiber source, the information will suggest a point in the system to introduce a portion of the chemical treatment.

As can be seen, both subjective and objective measurements are used to obtain information on the effectiveness of a program of slime control. However, success of the program is gauged by its contribution to the operating efficiency of the paper machine.

BIODETERIORATION OF PAPERMAKING ADDITIVES

The paper industry uses a large number of chemicals in the process of making paper. These chemicals may be added to the fiber and water prior to forming the sheet, i.e., internal additives, or they may be added to the surface of the sheet after sheet formation, such as surface sizes or coatings. Many of the additives or coatings are subject to degradation by microorganisms, or they may support growth to the extent that their use introduces microbial contamination to the paper machine system. In other cases the growth of microorganisms may cause deposits or slime in the additive system, resulting in spots and specks in the sheet or inconsistent flows of the additive through the automatic feeding devices going to the papermaking system. For these reasons many of the commonly used papermaking additives require preservation.

The need for preservation is dependent on the storage or holding time of the additive; sanitation practices, including housekeeping, around the additive system; and whether the additive passes through once or is cycled in the system.

Not all mills require preservation of all additives, but most preserve those that provide for the most luxuriant growth of microorganisms. Many of the additives are made up in solution or slurries in water and therefore the microbiological nature of the water is important. Those mills utilizing freshwater with a large amount of microbial contamination typically require preservation of most of their additives. Some of the additives such as starches and gums are solubilized (cooked) prior to use and may retain their sterility sufficiently long to preclude further preservation. The following materials frequently require preservation:

- Rosin size (gum, tall oil, and wood)
- Gums (guar, Locust bean, mannogalactan)
- Starch slurries (tapioca, potato, corn)
- Cooked starch
- Glue (Saveall)
- Papermaker's alum (dilute)
- Clay (slurried with phosphate dispersants)
- Titanium dioxide (slurried)
- Special sizes (emulsified with starch)
- Retention aids (polyacrylamide and polyamide types)
- Dyestuffs (most commercial blues, violets, yellows, etc.)
- Starch surface sizes
- Paper coatings (starch, protein, or mixtures of binders)

Many of the commercially available microbicides are used for additive preservation. Preliminary checks for compatibility with the additive, together with its preserving ability, can be determined with small batches of the additive.

Preservatives for sizes and coatings require closer scrutiny because of U.S. Food and Drug Administration regulations for those products which come in contact with food. Since the addition of materials after sheet formation will likely remain with the finished sheet it is important that they be nonmigratory from the sheet to food materials. Different federal regulations cover components of paper intended for dry food packaging and for wet or oily food packaging. Table 11-6 provides information on the degree of preservation of a casein-based paper coating with varying amounts of a modified barium metaborate.

Paper coatings consist of pigments such as clay and titanium dioxide and binders such as starch, protein, or synthetic materials. The starch and protein are readily degraded by microorganisms. Hydrolytic enzymes degrade starch, producing soluble carbohydrates and organic acids which change the pH and viscosity of the coating. Proteolytic enzymes degrade protein-based coatings, producing amino acids and putrefactive end products with offensive odors. Viscosity changes upset the flow and spread of the coating on the paper during the coating process. Extensive degradation causes lumpy coatings and unevenness on the sheet. Degradation of the binders in the coating makes it lose its cohesive characteristics and no longer functional as a coating.

The finished paper or paperboard is subject to degradation by microorganisms in much the same way as pulp or pulpwood. Cellulolytic enzymes can destroy the cellulosic fibers. There are times when it is necessary to produce paper or paperboard which is mold- (fungus-) resistant. Modified barium metaborate, phenylmercuric acetate, and chlorinated phenol compounds have been used for this purpose. Such items as soap wrap, paper sand bags, and structural paper (gypsum board liner, aluminum foil backing for air ducts, ground cover,

Table 11-6 Degree of preservation in a coating formulation (casein adhesive system, 60% solids) containing various amounts of modified barium metaborate added to the adhesive during makeup, inoculated with spoiled coating, and incubated at 32°C for 28 days

	Degree of preservation after number of days as measured by changes in										
	pH			Viscosity, KU*			Color-odor[†]			Microbial growth[‡]		
Amount of preservative	0	14	28	0	14	28	0	14	28	0	14	28
0 ppm $BaB_2O_4 \cdot H_2O$	9.2	6.5	6.5	62	40	42	0	4	4	0	++++	++++
500 ppm $BaB_2O_4 \cdot H_2O$	9.2	8.4	7.8	65	58	50	0	2	3	0	+++	+++
1000 ppm $BaB_2O_4 \cdot H_2O$	9.2	8.4	8.0	65	62	59	0	0	1	0	0	++
2000 ppm $BaB_2O_4 \cdot H_2O$	9.2	9.2	9.1	65	64	67	0	0	0	0	0	0
4000 ppm $BaB_2O_4 \cdot H_2O$	9.2	9.2	9.1	65	65	66	0	0	0	0	0	0

*Kreb units, read at 32°C on a Stormer viscometer, average of three readings.
[†] Rating 0 indicates no change and 4 indicates maximum change.
[‡] Rating 0 indicates no growth and ++++ indicates maximum growth.
SOURCE: R. W. Lutey (1967), Paper Coating Improvements, *Chem. Eng. Prog.*, **63**(3): 44–51.

etc.) are commonly impregnated with antifungal as well as anti-insect agents to prolong their usefulness. Figure 11-8 shows the amount of growth on coated paper samples inoculated with a mixture of *Aspergillus terreus*, *Chaetomium globosum*, and *Penicillium chrysogenum* and containing various amounts of modified barium metaborate. In coated paper and paperboard the mold resis-

Figure 11-8 Results of using various amounts of modified barium metaborate (Busan 11-M1) in laboratory-coated paper inoculated with a mixture of fungi and incubated at 28°C for 30 days.

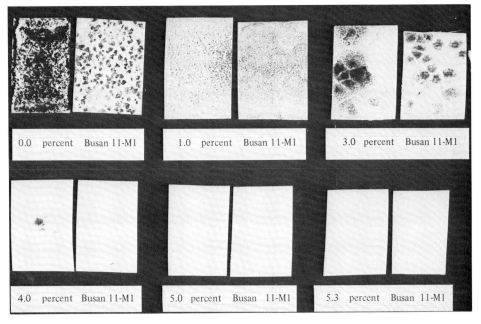

tance may be obtained by adding the fungicide to the coating formulation. In uncoated paper the chemical is usually added in a water solution after drying the paper.

BIODETERIORATION OF PAINT AND PAINT FILMS

Paints and paint films, like most organic substrates, are subject to attack by both microorganisms and macroorganisms. The degree to which they are attacked is governed by the chemical structure of the organic materials included in the formulation and the influence of the physical and chemical environment on both the substrate and the challenging microorganism.

Paints, simply defined, are opaque and ancillary pigments suspended in an organic polymer or mixture of polymers which, when exposed as a thin layer to oxygen or heat, polymerize or are heat converted to "dry" films. Paints, both in the liquid state or as dry films, are subject to attack, although the predominant microflora is different from one physical state to the other.

To properly classify the biological problems of paints and paint films, it is desirable to define briefly the types of paints involved and the organisms encountered. Paints are generally classified as house paints, maintenance paints, and industrial finishes. Typical examples of these would be interior wall paints for homes, ship bottom paints, and automotive finishes, respectively. Although industrial finishes can be attacked by certain fungi, the infrequency of such attacks makes the problem of no industrial significance. Maintenance paints, including not only the example given but paints used on steel storage tanks, concrete bridges, etc., are susceptible to biological attack. However, from the standpoint of industrial importance, marine paints and maintenance coatings used in industrial environments of high humidity, such as those found in breweries, bakeries, etc., are of greatest concern. From the standpoint of biodeterioration, the latter may be considered along with house paints, since their formulation and microbiological problems are similar.

House paints are generally of two basic types: (1) water-thinned paints and (2) solvent-thinned paints. The water-thinned paints most frequently used are actually emulsions of film-forming resins or oil in water. Solvent-thinned paints employ drying oils or other oleoresinous materials as the film former, and they are thinned with mineral spirits, turpentine, or other hydrocarbon solvents.

Although painting practices are undergoing evolutionary changes, it is safe to say that solvent-thinned paints currently make up the bulk of paint products used on exterior surfaces and that water-thinned paint is the type most frequently used on interior surfaces.

Water-thinned paints are subject to attack by microorganisms in the container. Such paints contain methylated or ethylated cellulosic materials, proteins, and numerous synthetic additives which provide a nutritional environment suitable for the growth of a number of microorganisms. The pH and storage temperatures of water-thinned paints are in a range favorable for microbial growth.

The types of spoilage encountered in water-thinned emulsion paints depend upon the types of emulsion stabilizers employed and include putrefaction, fermentation, gassing, and viscosity changes.

A number of organisms have been isolated from "spoiled" water-thinned paints; those most frequently isolated are the Pseudomonads, particularly *Pseudomonas aeruginosa*. However, *P. aeruginosa* is easily controlled by a number of paint preservatives, and the attack of water-thinned paints in the container is an industrial problem only when inadequate levels of these preservatives are incorporated in the paint and raw materials used in paint.

Occasionally, pigment pastes, tinting colors, wetting agents, and other additives used in formulating water-thinned emulsion paints are contaminated with bacteria when received by the paint manufacturer, since producers of these raw materials rarely incorporate a microbicide in their products. Where the cell count is sufficiently high, some free bacterial cellulase is transferred to the paint when the additive is used. In paints employing methyl cellulose or carboxymethyl cellulose as viscosity stabilizers, the free cellulase is sufficient to decompose these materials by hydrolyzing the glucosidic linkages. Thus the paint viscosity may be drastically altered even in a paint containing no viable bacteria. As little as 0.1 ppm cellulase can produce a 20 percent decrease in viscosity of paints stabilized with cellulosic thickeners.

Most of the more frequently employed paint microbicides are ineffective in inhibiting enzyme activity at concentrations normally used. The best means of preventing enzymatic deterioration would be for all suppliers of water-wet raw materials to adequately preserve them prior to shipment to the paint manufacturer. Accompanying this should be a consistent program of "good housekeeping" within the paint factory.

The nonaqueous nature of solvent-thinned paints prevents their being altered in the container by microorganisms. Fungi have occasionally been found in solvent-thinned paints but generally in a static condition.

The conversion of liquid paints to paint films provides a radical environmental change in both water-thinned and solvent-thinned paints as substrates for the growth of fungi and bacteria. In addition to the chemical changes of the paints themselves, microbial growth is sharply influenced by the surface being painted and its general environment. The growth of fungi occurs more frequently and is more luxuriant on paint films applied over wooden surfaces than on paint films applied over masonry or metal substrates. Paint films on exterior surfaces present the greatest number of microbiological problems, yet the severest disfigurement of painted surfaces by fungi occurs on the interior walls and ceilings of breweries, dairies, food processing plants, etc., where conditions of temperature and humidity are ideal for maximum growth.

MICROFLORA OF PAINT FILMS

The most complete surveys of the types of microorganisms associated with paint films were made by Rothwell (1958), who examined exterior solvent-thinned paint films, and by Drescher (1958), who examined exterior water-thinned paint films. A list of microorganisms isolated from paint films is presented in Tables 11-7 and 11-8. Although a large number of microorganisms were observed or isolated, most of them were isolated infrequently and were undoubtedly chance inhabitants. At least four fungi and one bacterium were isolated in sufficient numbers, however, to be considered major contributors to the disfigurement or

Table 11-7 Microorganisms isolated from oil paint films exposed on test fences at six different locations in the United States

Fungi	Bacteria
Alternaria dianthicola	Alcaligenes recti
Aspergillus flavus	Alcaligenes sp.
Botryodiplodia malorum	Bacillus cereus
Botrytis cinerea	Bacillus mycoides
Cladosporium sphaerospermum	Bacillus sphericus
Cladosporium sp.	Bacillus sp.
Cephalosporium carpogenum	Flavobacterium invisibile
Fusarium flocciferum	Flavobacterium marinum
Helminthosporium spiciferum	Micrococcus albus
Paecilomyces varioti	Micrococcus candidus
Penicillium oxalicum	Micrococcus ureae
Phoma glomerata (Peyronellaea)	Sarcina flava
Pullularia pullulans	
Stemphylium consortiale	

deterioration of paint films. These include *Pullularia pullulans, Phoma glomerata, Cladosporium* sp., *Alternaria* sp., and *Flavobacterium marinum.*

Various geographical locations tend to support more extensive growth of certain of the fungi. Examination of exterior paint films exposed at West Coastal areas indicated that *Phoma glomerata* accounts for a major part of the fungal disfigurement in that area. Conversely, this particular fungus was rarely found in other areas of the United States. *Cladosporium* sp. was isolated from disfigured paint films exposed along the Eastern seaboard, with a far greater frequency than from paint films exposed in other geographical areas. *Pullularia pullulans* was isolated in every geographical location and is the single greatest offender. It has been repeatedly isolated and reported as the fungus most

Table 11-8 Microorganisms isolated from emulsion paint films exposed on test fences at six different locations in the United States

Fungi	Bacteria
Alternaria dianthicola	Bacillus mycoides
Aspergillus flavus	Flavobacterium marinum
Cladosporium sphaerospermum	Sarcina flava
Cladosporium sp.	
Cephalosporium carpogenum	
Helminthosporium spiciferum	
Penicillium oxalicum	
Phoma glomerata (Peyronellaea)	
Pullularia pullulans	
Stemphylium consortiale	
Torula nigra	
Unknown yeasts	

Figure 11-9 Photograph showing a house paint severely discolored by *Pullularia pullulans*.

frequently associated with exterior paint films (Goll and Coffey, 1948; Klens and Lang, 1956). A house paint severely disfigured by *Pullularia pullulans* is shown in Fig. 11-9.

The microflora of interior paint films in breweries, dairies, canneries, etc., was reported by Krumperman (1958) to include extensive growth of many fungi rarely found on exterior surfaces. Prominent among these are *Aspergillus* species and *Penicillium* species. His investigation again indicated the frequent occurrence of *F. marinum*.

In general, regardless of the type of exposures or geographical locations, the following observations concerning the growth of fungi were consistent.

1 Fungal growth was most profuse when associated with fissures in the paint films.
2 The presence of chemical preservatives in paint films had a decided effect on the gross appearance of fungi. Whereas unpreserved paint films generally supported mycelial growth forms, the fungal growth on preserved paint films was more often in the form of minute spore clusters, indicative of fungistatic or at least unfavorable growth conditions.
3 The presence of fungi on paint films increases the retention of dirt, particularly when present in the vegetative state. Dirt particles become firmly entrained by the mycelium. The association of fungi and dirt particles is an extremely important one since nutrients, in addition to that provided by the paint itself, are available to the fungi.

MICROBIOLOGICAL DEGRADATION OF PAINT BINDERS

The presence of fungi as the predominant disfiguring agent of paint films dictated the use of antimicrobial agents in finish coats as far back as 1935. Until 1958, paint primers rarely if ever contained antimicrobial chemicals since it was believed that the microbial problem was strictly a surface phenomenon. However, during the course of isolating fungi from the surface of paint films, Rothwell (1958) isolated bacteria from the primer. These bacteria when inoculated on polymerized oil or oil-alkyd binders were observed by Ross (1958) to drastically alter both the physical and chemical characteristics of the polymerized films. His investigations revealed that the principal offending organism was *Flavobacterium marinum*. The initial attack of oil or alkyd resins by *F. marinum* is at the ester linkage. Cleavage of this linkage is followed by a beta oxidation of the free fatty acids and conversion of glycerol to glyceraldehyde. The diagram presented in Fig. 11-10 outlines the proposed biodegradation of linseed

Figure 11-10 Proposed biodegradation of linseed oil films.

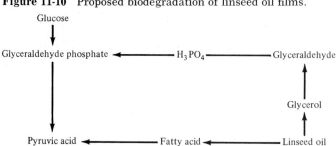

oil films. The ultimate utilization of oleoresinous materials by *F. marinum* was found to be a continued aerobic oxidation with an eventual accumulation of oxalic acids. *Pullularia pullulans*, on the other hand, converts linseed oil to oxalic, acetic, and acetoacetic acids.

Further work by Ross and Wienert (1962) described the biodegradation of oil or alkyd primers by bacteria and the importance of this as related to adhesion of paint films. The most consistent form of paint failure is that generally encompassed by the term "peeling." Peeling of paint is the result of adhesion loss, either between paint films or, as more frequently encountered, at the paint-substrate interface.

The nature of the substrate over which the paint is applied, the degree to which moisture accumulates behind and within paint films, and the impermeability and stress created by several layers of paint all contribute to adhesion loss. The large populations of bacteria consistently isolated at the paint-substrate interface when peeling occurs suggested that biodeterioration of the resinous binders of paint films plays perhaps the most important role in bringing about loss of adhesion. The importance of ensuring that paint primers are adequately protected against microbial attack is indicated by studies which revealed that the bacterial population at paint-wood interfaces of unprotected paint was between 1000 to 10,000 times as great as that found at the interface of adequately preserved paints. The appearance of primers supporting large populations of *Flavobacterium marinum* is shown in Fig. 11-11. Physically, the primer is converted to a powdery residue following deterioration of the resinous binder.

Figure 11-11 Paint chips removed from houses on which the paint film was peeling from the wood. The photograph shows the deteriorated prime coat which was next to the wood.

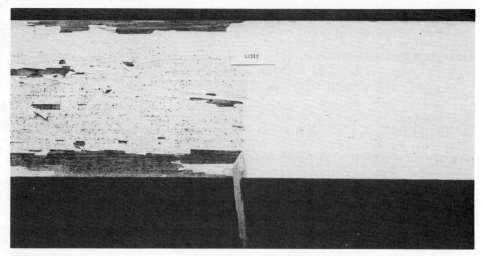

Figure 11-12 Superior film durability resulting from the use of a preservative in the primer. The right half of each panel contains two coats of paint with barium metaborate. The same paint was used as the topcoat on the left half but was applied over an unpreserved primer.

That the durability of paint films can be significantly increased by adequate preservation of paint primers is demonstrated in Fig. 11-12. The decrease in cracking and peeling of paint systems when the primer is preserved has been confirmed on numerous substrates in geographical areas as diverse as southern United States and Sweden.

MICROORGANISM CONTROL OF HOUSE PAINTS AND INDUSTRIAL MAINTENANCE PAINTS

The total resistance of paint films to the growth of microorganisms depends not only on the inclusion of chemical antimicrobial agents but also on the types of raw materials employed and their volumetric ratio to one another.

Selection of Paint Raw Materials

Paint binders are the film-forming ingredients of paints and consist of both naturally occurring drying oils and resins and synthetic polymers. Oil and alkyd binders are gellike structures which, when wetted, absorb varying amounts of moisture. Since the amount of moisture which is retained within a paint film is a critical determinant of the degree to which it is attacked by microorganisms, the greater the water resistance of paint binders, the less favorable is the environment for microbial growth. The higher viscosity oils and alkyd resins generally provide more mold- and bacterial-resistant paint films because of their increased water resistance. Synthetic vinyl-type paint binders, such as polyvinyl acetate copolymers and acrylic resins, provide increased resistance to microorganisms not only because of their increased water resistance but because of their chemical structure as well. However, in the preparation of emulsion paints using this type of binder, the inclusion of emulsifiers, surfactants, and other additives counteracts the improved resistance of the vinyl resins.

The total amount of binder used in relation to the pigment of a paint film also influences the paint film's overall resistance to microorganisms. Any binder not directly in contact with the pigment is more easily hydrated and more readily accessible to microbial attack.

The choice of nonfungicidal pigments and the pigment volume concentration affect the mold and bacterial resistance of a paint film. Lead pigments tend to decrease the water absorptivity of a paint film. Diatomaceous silica and other flatting pigments tend to increase water contact with the binder.

Formulating Techniques

In paint terminology *pigment volume concentration* refers to the ratio of the volume of pigment to the sum of the volume of nonvolatile binder and the volume of pigment in a given paint. The *critical pigment volume concentration* refers to that pigment volume concentration at which there is just enough binder in the dry film to fill the voids between the pigment particles. Increasing the volume of pigment in a given paint closer to the critical pigment volume concentration reduces the amount of *free binder* and thus reduces the amount of moisture that can be trapped behind or held within a paint film.

Other than the use of chemical preservatives, the most frequently used technique of paint chemists to formulate paints which will not become disfigured by fungi is to choose hiding pigments which will chalk or slough off surface pigment layers after a short exposure period. "Chalking," or sloughing off of unbound pigment at the surface of paint films, obviously has little effect on the bacterial population concentrated at the paint-substrate interface. It does, however, remove much of the fungal growth on the paint film surface by physical removal of spores and dirt particles. It is extremely important that "chalking" not be excessive so that chalk wash or rapid erosion results.

Chemical Inhibitors

The only effective means of preventing microbial attack of liquid paints and paint films is through the use of chemical inhibitors. Although many microbicides and microbistatic agents have been available, only a few have been found useful in paints. The unique physiochemical characteristics required for a chemical to be an effective paint microbicide on the one hand while not interfering with desirable paint characteristics on the other have ruled out many products.

The choice of paint preservatives is dictated by a number of considerations. They must be effective in controlling microorganisms, yet present no hazard to either the workmen preparing the paint, the painter applying the paint, or to the people living in the painted environment. Paint preservatives must have both a low solubility in water and a low leach rate to ensure protection of paint films over a number of years. Regardless of its effectiveness, a paint preservative must be physically and chemically compatible with other ingredients in a paint. Thus it should not react with the resinous binder in a way that would harden the paint film. It must not precipitate or flocculate pigments to a degree that it reduces the opacity of a paint film. Paint preservatives ideally should be white or colorless and remain stable in a paint so as to not discolor.

No interference in the conversion of the liquid paint to the solid paint film can be allowed. Finally, paint preservatives must be effective at a cost level that will not increase excessively the overall cost of the finished paint product.

The antimicrobial agents used by the paint industry may be classified in two categories. The first includes the microbicidal or microbistatic inorganic pigments which are used in relatively high concentrations and fulfill a measurable percentage of the total pigment demand. As such, they cannot be added to a finished paint but rather must be incorporated at the time the paint is made. Barium metaborate, zinc oxide, and cuprous oxide are examples of this type of inhibitor. Cuprous oxide must generally be used in primers or colored paints because of the off color it produces. Zinc oxide is a reactive pigment with oleoresinous paint binders and, although it has been used extensively in the paint industry for fungus control, it is well known that paints containing zinc oxide harden prematurely and generally fail by cracking and peeling. Its use has been confined primarily to oil house paints. Barium metaborate is a slightly soluble alkaline pigment which has been found effective in controlling fungus in paints formulated to have maximum chalk resistance. Its slight solubility has prevented its incorporation into certain water paints containing polyvinyl alcohol because of the gelation that occurs between polyvinyl alcohol and salts of boric acid.

The second category of paint preservatives is considerably larger than the first and may be classified as organic additives. They are used in relatively low concentrations, rarely exceeding a level of 2 percent based on the total weight of the paint.

The most frequently employed organic additive preservatives have been the organometallics and more specifically various phenylmercury compounds. These compounds are broad-spectrum microbicides and hence have been used as bactericide for preservation of liquid paints and at higher concentrations for inhibition of fungi and algae. Almost all the commercially used products have the general structure RHgX in which R is generally a phenyl radical and X is either an organic or inorganic group such as acids, amides, and phenol. The latter group has a dissociable hydrogen ion bound to mercury. It has been generally accepted that the mercury metal content of these type of compounds determine their lethal characteristics. The anion has little influence on toxicity but does influence the solubility, volatility, and stability of the compound. There are distinct differences in the degree to which phenylmercury preservatives influence pigment flocculation; this has been attributed to the nature of the anion.

Investigations by Hoffmann et al., (1964, 1963, 1960, 1966) indicated that, regardless of their chemical structure, all phenylmercury preservatives are leached from paint films prematurely. The degree of leaching depends upon the exposure environment, the fastest leaching occurring in hot, humid areas.

Organo-tin compounds have a high degree of activity against fungi but have nevertheless provided inconsistent results when used as fungicides in house paints. Their usefulness in marine paints, however, has increased significantly during recent years.

Chlorinated and substituted phenols are among the oldest antimicrobial agents known and have been used extensively in the paint industry. The most frequently employed chlorinated phenols have been tetrachlorophenol and pentachlorophenol. These products have some history of discoloring paints when used at high concentrations and, like the phenylmercury compounds, are reduced to insufficient levels after 1 to 2 years exposure. Sodium and potassium salts of orthophenylphenol have also been used as paint microbicides either alone or in combination with phenylmercury compounds.

The relatively high mammalian toxicity of phenylmercury compounds has received increased attention in recent years, and many paint manufacturers have or are actively seeking non-mercury-containing substitutes. A number of new compounds have been developed to meet this need. It is somewhat early to judge their overall utility as paint microbicides, but many are known to have a high degree of activity against bacteria or fungi or both. These newer products range from quaternary compounds to halogenated sulfonylpyridines. A list of the better known microbicides as well as newer ones is presented in Table 11-9.

Table 11-9 Antimicrobial agents used in paints and paint films

I Inorganic pigments
   a Barium metaborate
   b Zinc oxide
   c Cuprous oxide

II Organic additives
   a Phenylmercury compounds
      1 Phenylmercuric acetate
      2 Phenylmercuric oleate
      3 Phenylmercuric borate
      4 Phenylmercuric propionate
      5 Phenylmercuric mercaptide
      6 Di(phenylmercury)dodecenylsuccinate
   b Organo-tin compounds
      1 Bis-tributyl tin oxide
      2 Tributyl tin linoleate
   c Chlorinated phenols and substituted phenols
      1 2,3,4,6,-Tetrachlorophenol
      2 Pentachlorophenol
      3 Orthophenylphenol
      4 *p*-Phenylphenol
   d Quaternary compounds
      1 Cetyldimethylethyl ammonium bromide
      2 *N*-(3-chloroallyl)heximinium chloride
   e Halogenated sulfonylpyridines
      1 Trichloro-4(methylsulfonyl)pyridine
      2 Tetrachloro-4(methylsulfonyl)pyridine
   f Captans
      1 *N*(trichloromethylthio)phthalimide
      2 *N*(trichloromethylthio)4-cyclohexene-1,2-dicarboximide
   g Triazines
      1 2,4,6-Dichloro-6(*O*-chloroanilino)-*S*-triazine
      2 Hexahydro 1,3,5-triethyl-*S*-triazine

None are without limitations and there is certainly a continuing need for new and improved paint microbicides.

EVALUATION OF PAINT MICROBICIDES

It was mentioned previously that one of the major problems in developing paint microbicides has been the difficulty in accurately evaluating microbial resistance of a given paint formulation. Indeed, no other paint property is more difficult to predict. Although accelerated tests methods have been used, the reliability of the results obtained has been poor.

The most widely used laboratory method for evaluating microbicides in paint films has been the simple zone-of-inhibition test. This test, often referred to as the "agar plate test," was originally developed for the evaluation of disinfectants, antibiotics, and other antimicrobial agents that were employed in aqueous systems. The test consists essentially in applying the test paint to filter paper, placing the coated paper on the surface of Tryptone glucose agar (or equivalent) and inoculating the paint film and surrounding agar surfaces with spore-mycelial mixtures of one or more designated fungi. Some test procedures call for leaching or other special conditioning of the paint film prior to its inoculation in the test. The inoculated plates are incubated for 7 days. The effectiveness of the preservative is then determined by the absence of fungal growth on the surface of the paint film and the zone of inhibition of fungal growth on the agar surface surrounding the paint film. Popular interpretation of the results has been that the greater the zone of inhibition, the greater is the fungal resistance of the paint film and, consequently, the effectiveness of the included microbicide.

Many paint technologists, as well as microbiologists, have frequently observed that the test provides not simply misleading results but results directly the reverse of those obtained when the paint was applied under normal use conditions. The inadequacy of the test is attributed to its artificiality. The microorganisms employed in the agar plate test are obtained from pure cultures of designated fungi which have been maintained for varying periods of time on a synthetic medium complete with all the nutritional requisites for maximum growth. This is opposed to the natural inoculation that occurs when fungal spores, adjusted to an exterior environment, come in contact with a painted surface. The most commonly encountered fungus on exterior exposures is *Pullularia pullulans*. The growth pattern of this fungus on exterior paint films is different from its growth pattern on an agar medium. On exterior paint films, its growth pattern is initially a long period of spore reproduction resulting in spore clusters with a minimum of vegetative growth. The yeastlike growth observed on Tryptone glucose agar, used in the agar plate test, is such that *Pullularia pullulans* is not well suited for use as the inoculating organism for the agar plate test, and thus it has been necessary to use other fungi as inocula.

In addition to high concentrations of free moisture available to microorganisms, the agar plate test provides an abundance and a variety of food materials that are rarely, if ever, available in the paint film on practical exposures. For those organisms commonly employed in the agar test, the presence of an organic nitrogen source (peptones, polypeptides, amino acids) ensures a rapid proliferation of mycelium. This abundant vegetative growth seldom occurs on exterior paint films where the principal nitrogen source is inorganic nitrates in soil particles falling on the painted surface. The rapid proliferation of mycelium can often provide misleading results in the agar plate test by growing over the paint film without actually deriving any nutrient from that film.

Perhaps the most important artificiality of the agar plate test is that it favors characteristics of a microbicide which, although desirable in controlling fungus under conditions of the test, are highly undesirable in providing long-term fungus resistance under exterior-use conditions. These characteristics are those of migration and miscibility in water. The greater the migration of the microbicide from the paint film, the more effective it will be in producing a large zone of inhibition in the agar plate test. This migration of a microbicide from a paint film is highly undesirable under use conditions since it represents a means of losing the microbicide from the paint film. This solubility or miscibility of some microbicides explains their remarkable effectiveness as measured by the agar plate test and their marginal long-term effectiveness in coatings for exterior use. In view of the artificial conditions of the agar plate test, it is not surprising that there are differences between results obtained by this accelerated method and results obtained by actual exposure in use environments.

The Biodeterioration Subcommittee of the American Society for Testing and Materials has recently developed a test which is reported to provide more accurate results for measuring the effectiveness of paint microbicides. The test is restricted to evaluating the fungus resistance of interior paint films and simulates interior environments and a natural inoculation as closely as possible. The test employs a chamber in which the temperature and humidity can be carefully controlled. Test paints containing the microbicide under test are applied on duplicate panels constructed of the substrate (wood, wallboard, concrete) on which the paint will be used commercially.

The important step in the test is the continuous natural inoculation of the test paint film. This is achieved by employing a soil bed with a 16-mesh screen floor above a water bath and placed below the panels inside the cabinet. A small circulating low-speed fan placed between the soil bed and the panel provides sufficient air movement to maintain a constant inoculation of fungal spores on the paint surface.

It is too early to know if the ASTM test will be broadly adopted by the paint industry. In any event, additional tests are needed to evaluate the fungus resistance of exterior paints.

MARINE PAINTS

A discussion of biodeterioration of paints and paint films would not be complete without a review of the specific problems associated with marine paints. Marine fouling provides the biologist with one of the most interesting ecological studies in an industrial environment. No other paint film supports at one time or another such a broad spectrum of micro- and macroorganisms. Marine fouling generally begins with the attachment and growth of slime-forming bacteria, particularly species of the genera *Enterobacter* and *Flavobacterium*. In freshwater fouling, the initial growth may be algal. Protozoa, diatoms, and fungi are "secondary invaders," becoming physically trapped within the bacterial or algal slime. If no antimicrobial agent is present in marine paints, the invasion of protozoa, diatoms, and fungi often occurs almost as the slime deposit is formed by the rapidly proliferating bacteria.

The paint film may or may not serve as a source of nutrition for these microorganisms. Frequently it serves only as an inert growth substrate. The bacterial slime layer, however, serves as an excellent food source for the attachment of the fouling macroorganisms such as the anneleds, barnacles, bryozoa, and molluscs.

Zobell (1939) proposed two additional ways in which bacterial slime contributes to the attachment of the higher biological forms other than simply as a nutrition source. He suggested that the slime barrier would also act as a barrier between the cyprid larvae of barnacles and any biocide incorporated in the film. He also pointed out that the slime would probably have a more alkaline pH than most paint films and hence favor the deposition of the calcareous cements secreted by the cyprid barnacle. The latter has indeed been found to be an important contribution to barnacle growth. It is well known that the surfaces accumulating the heaviest bacterial slime will become the most fouled. The total encrustation of barnacles on painted ship bottoms often reaches several inches thick.

The economic importance of marine fouling is very great. Encrustation of the organisms previously mentioned markedly increases weight and drag on ships, with a subsequent increased demand in power and fuel consumption. It has been estimated that the weight increase resulting from marine fouling is equivalent to 1 ton for each 1/4-in thickness of encrustation on large ocean-going vessels. Of perhaps even greater importance is the frequent interruption and destruction of anticorrosive paint primers by marine fouling, resulting in corrosion undercutting.

A number of other factors determine the degree and rate at which marine fouling occurs. Ship speed is important, since fouling generally does not occur at speeds above 4 knots. Most fouling occurs when water temperatures are above 48 to 50°F; hence it creates the greatest problem in offshore waters of West Africa, South America, Hawaii, Ceylon, Phillipines, and the Far East. Since slime-forming bacteria, particularly species of *Enterobacter*, *Flavobacterium*, and *Pseudomonas* are present in water very near the shore, fouling likewise occurs most often when ships are stationary near land bodies.

To combat biodeterioration of marine paints, the paint formulator has relied primarily on the incorporation of biocides. Because of the diversity of the microflora, it has frequently been necessary to employ more than one chemical inhibitor in the paint formulation. Copper compounds and more specifically cuprous oxide (Cu_2O) have historically been the most frequently employed inhibitor. The cuprous ion is effective against bacteria, algae, and fungi and hence inhibits the formation of the slime base upon which the cyprid larvae attach themselves.

Saroyan (1969) stated that, for an antifouling paint based on cuprous oxide to be effective, the cuprous ion must have a steady-state leaching rate equal to 100 µg of Cu^{2+} per square centimeter per day. This steady-rate leaching rate is achieved by incorporating cuprous oxide into a paint containing a polymeric binder that is slowly degraded in seawater. Binders based on glycerolesters of rosin in combination with vinyl chloride are frequently used in this type of paint.

Obviously, it is necessary to employ rather high concentrations of cuprous oxide to provide a steady leach rate over any period of time. It is not unusual to incorporate levels of cuprous oxide equal to 55 percent of the total weight of an antifouling paint. Moreover, the use of such levels of cuprous oxide create corrosion problems when ships are anchored in water where the sulfur content is high. The formation of cuprous sulfide causes a depolarization at the cathode, increasing the loss of ferrous ions at the anode of the electrolytic cells, almost always present on metal ship bottoms.

Hence, many other biocides have been evaluated to either partially or completely replace cuprous oxide. These have included organometallics including organic tin compounds such as triphenyl tin, chloride, phenylmercury salts, and organo-lead compounds such as tributyl lead acetate and organic coppers, principally copper 8-hydroxyquinolinate. These compounds have provided supplementary benefits when used with cuprous oxide but generally have not been sufficiently inhibiting to be used alone in antifouling paints. One reason for this has been their leaching rate from paint films which, despite the low water solubility, has been relatively high. Many a biocide that looked extremely effective at the end of 6 months was found to be totally inadequate 1 year later.

More recent attempts to prevent barnacle fouling has been to seek a compound which would interfere mechanically with the adhesive substance secreted at the paint surface. It has also been demonstrated that incorporation of radioactive thallium in antifouling paints is effective against marine fouling but only when the radiation intensity is too high for practical use. It is obvious that effective biocides are still needed for marine paints.

LITERATURE CITED

Acree, S. F. (1919): Destruction of Wood and Pulp by Fungi and Bacteria, *Pulp Pap. Mag. Can.,* **17**:569–571.

Appling, J. W., N. J. Ridenour, and S. J. Buckman (1959): Pink Slime in Paper Mills. Control and Identification of Certain Causative Organisms, *Tappi,* **34**(8):347–352.

Barker, E. F. (ed.) (1961): How Austrian Mill Overcame Difficulties in Long Term Storage of Groundwood Pulp, *Pap. Trade J.,* May 15, pp. 20–23.

Bray, M. W., and T. M. Andrews (1923): Chemical Changes of Groundwood during Decay, *Ind. Eng. Chem.,* **15**:137–139.

Buckman, S. J., and C. P. Kirchen (1958): Economics of Microorganisms Control in the Pulp and Paper Industry, *Tappi,* **41**(7):144A–149A.

Dean, C. W. (1953): Slime Measuring Unit, Patent 2,660,884 (assigned to Buckman Laboratories, Inc.).

Drescher, R. F. (1958): Microbiology of Paint Films. IV. Isolation and Identification of the Microflora on Exterior Emulsion Paints, *Am. Paint J.,* **42**(38):80–102.

Gadd, G. O. (1949): Fungal Damages in Groundwood Pulp, *Pulp Pap. Mag. Can.,* **50**(11):98–99.

––––– (1962): Enzyme Cellulase and Its Influence on Pulp, *Papper och Tra,* **8**:405–412.

Goll, M., and G. Coffey (1948): *Paint, Oil Chem. Rev.,* pp. 14, 16, 17, 30, Aug. 5.

Hoffmann, E. (1964): Phenylmercury Compounds as Fungicides, Part III, *J. Oil Colour Chem. Assoc.,* **47**(11):871–877.

––––– and B. Bursztyn (1963): Phenylmercury Compounds as Fungicides, Part II, *J. Oil Colour Chem. Assoc.,* **46**(6):460–466.

———— and O. Georgoussis (1960): Phenylmercury Compounds as Fungicides, *J. Oil Colour Chem. Assoc.,* **43**(6):779–786.

————, A. Saracz, and J. R. Barned (1966): Phenylmercury Compounds as Fungicides, Part IV, *J. Oil Colour Chem. Assoc.,* **49**(8):631–638.

Klens, P. F., and J. F. Lang (1956): *J. Oil Colour Chem. Assoc.,* **39**(12):887–899.

Kress, O., C. J. Humphry, C. A. Richards, M. W. Bray, and J. A. Staidl (1925): Control of Decay in Pulp and Pulpwood, *U.S. Dep. Agric. Bull.* 1298.

Krumperman, P. H. (1958): Microbiology of Paint Films. V. Microorganisms Found on the Interior Paint Films of Food Processing Plants, *Am. Paint J.,* **42**(38):72–84.

Lundgren, N. (1949): N. Slemkontroll i ett Tidnigspapersbruk, *Sven. Papperstidn.,* **52**(9):221–226.

Lutey, R. W. (1967): Paper Coating Improvements, *Chem. Eng. Prog.,* **63**(3):44–51.

Neal, E. L., and S. F. Jennings (1948): Slime Control at Anglo-Canadian Pulp and Paper Mills, Ltd., *Pulp Pap. Mag. Can.,* **48**(3):137–142.

Ross, R. T. (1958): Microbiology of Paint Films. III. Attack of Linseed Oil and Linseed Oil Films by Microorganisms, *Off. Dig. Fed. Paint Varn. Prod. Clubs,* **30**(399):377–391.

———— and L. A. Wienert (1962): Microbiology of Paint Films. VIII. The Roll of Microorganisms in Peeling of Exterior Paint Films, unpublished.

Rothwell, F. P. (1958): Microbiology of Paint Films. II. Isolation and Identification of Microflora on Exterior Oil Paints, *Off. Dig. Fed. Paint Varn. Prod. Clubs,* **30**(399):368–376.

Russell, P. (1957): Problems Concerning the Infection of Groundwood Pulp, *Proc. Tech. Sec. Br. Pap. Board Maker's Assoc.,* **38**(2):241–254.

———— (1961): Microbiological Studies in Relation to Moist Groundwood Pulp, *Chem. Ind. (London),* **20**:642–649.

Saroyan, J. R. (1969): Marine Biology in Antifouling Paints, *J. Paint Technol.,* **41**(531):285–303.

Shema, B. F. (1955): The Microbiology of Pulpwood, *TAPPI Monog. Ser.,* **15**:28–54.

Siu, R. G. H. (1951): "Microbial Decomposition of Cellulose," Reinhold, New York.

Zobell, C. E. (1939): Fouling of Submerged Surfaces and Possible Preventive Procedures, *Paint, Oil Chem. Rev.,* **101**:74.

12
POLLUTANTS AND AQUATIC ECOSYSTEMS: BIOLOGICAL ASPECTS OF WATER QUALITY PROBLEMS

Robert A. Coler
and
Warren Litsky

SOCIOLOGICAL ASPECTS OF PROBLEM

It is not the purpose of this chapter to outline the development of pollution control measures, or to trace the webs of influence which continue to inhibit the passage of the stringent legislation necessary to markedly decrease present pollution. Rather, we intend to describe some of the more significant environmental *effects* of unregulated pollution which have been occurring in recent years, and which have increased to the degree where they are encountered in all aspects of our studies.

Historically, administration of pollution control measures has fallen to public health agencies, whereas their implementation has been carried out by sanitary engineers, who often work for the private sector. This is something of an anomaly, since corrective legislation has been precipitated by pressure from private industrial interests in the United States and England, rather than being a result of the widespread effects of pollution on the public sector in the form of cholera or typhoid epidemics (U.S. Department of the Interior, 1968). This means that, although control of pollution falls to the "general public," the heterogeneity of this group makes the definition of its clear-cut "interest" nearly impossible; yet it is the public which is the primary victim of pollution abuses. It follows that private industry, smaller and more articulate, and with its interests well in mind, has had far greater access to the agencies which oversee this area of national activity. Since the "interests" of the public at large and of private industry are not always identical, the kinds of pollution noticed by public agencies have often in the past been those that industry *wanted* them to notice, whereas other—sometimes more profitable—abuses have continued unregulated, right up to the present moment. This may explain why early legislation originated not in the public sector but in the private one, and why effective enforcement of standards has been so difficult to institute.

In the past, the states have had little power to prevent expanding industries and population centers from exploiting the waterways. With the evolution of a water carriage sewerage system and chlorination, authority of the public health sector, never significant to begin with, became further diminished. Hampered by poorly defined departmental responsiblities, administrative overlap, and a dependence on common law, these agencies had only boards of health regulations and quasi-legal rules to serve as their source of authority to restrict abuses.

All this is beginning to change. By the mid-1950s a growing grass-roots concern with the deteriorating environment (launched by Rachel Carson) provided the impetus for effective federal and state legislation, with the delegation of clear-cut authority to specific administrative agencies.

In short, pollution has proceeded to the extent where it is now basically and fundamentally changing the *substance* of the biological sciences, as well as the nature of practical problems encountered by engineers on a daily basis. We feel that it is necessary to enlarge the horizons of pollution studies from the more rigid and limited perspectives which have prevailed in the past. This is because we are coming to understand, theoretically and practically, that the ecological perspective is essential for water classification, the determination of biologically meaningful pollutant levels, and the evaluation of water quality itself. For these reasons, the chapter is presented with strong ecological overtones.

BIOLOGICAL ASPECTS OF PROBLEM

When evaluating the impact of a pollutant on the aquatic biota, it is essential to realize that the physiologic, morphologic, and behavioral adaptations of the resident populations are the total expression of their favorable interaction with selective environmental demands. Environmental stress is imposed on organisms by chemical or biological components which exert restrictive tensions on populations. Pollutants, however, rarely act through the disruption of a single homeostatic mechanism or on a single species, nor are their effects always so drastic as death or extinction. Rather, there is more often a *shift* in population densities and in the distribution of the species which constitute the community. This indicates that most pollutants have a global or total effect on communities into which they are introduced: The resident species change, the remaining ones play different roles, and they occupy vital "spaces" of rearranged size.

Some organisms which have adapted to the rigors of a marginal habitat assume a dominant role within the various levels of the community, because their "marginality" enables them to respond to a greater range of environmental conditions. Such conditions may include a decrease in the oxygen supply and an increased percentage of organic matter in the water. These saprobic conditions select favorably for the marginal organism, whose rising numbers give a fairly precise indication of the extent of pollution. Organisms are therefore considered biological, as opposed to possible chemical, indices of pollution and are associated with various levels of the saprobic condition. Microscopic organisms are among the earliest and most sensitive indicators of pollution because they almost immediately reflect the condition of biological mechanisms.

When thus selected for, such organisms embody the attributes of "pioneer" communities in that they have an explosive capacity to exploit the habitat at low energy efficiencies. Through short, telescoping life cycles, high reproductive rates, and the ability to free a niche in an otherwise hostile habitat, they are preadapted to fill an ecological void. Thus, their population explosion results in a decline in species diversity and in the successful adaptation of the few; the complex becomes reduced to the simple, and man's role as pollutor emerges as the motive force in a kind of reverse evolution. The simplified ecosystem has fewer responses to pollutants than does a complex one: It simply extends its simplicity to more (that is, fewer) levels. This new simplicity is embodied in the genetic characteristics of the remaining populations which now lack former phenotype variation to draw upon.

Therefore, it is essential to recognize the operative physiologic mechanisms exerted by pollutants, and the adaptations for which they select, if we are to interpret accurately the total biological response. The problem is to understand a unity of interrelated responses by organisms to pollution. The authors have developed this chapter with a biological overview in order to show the effects of pollutants on populations of microbes and the effectiveness of using these populations as indices of pollution.

Following will be a review of the more critical environmental restrictions militated by the adoption of the aquatic mode of life. We will attempt to catalog briefly the specific effects of selective environmental demands often imposed by industrial pollutants, and the common biological responses evoked by the toxicity of those stress factors.

Our discussion of the biological realities is sometimes followed by a series of recommendations concerning the quantity and composition of effluents and other waste products which can safely be introduced into aquatic communities by cities or industry. Although we recognize the urgent requirements of many modern industrial processes, it is hoped that those who work on their implementation will appreciate the ecological problems attendant upon environmental imperatives.

THE ECOSYSTEM

BIOLOGICAL EVALUATION OF A TOXIC SUBSTANCE

Toxicity studies require considerable time and competent interpretation of results. Although a large body of information has been published regarding the toxicity of specific ions to the aquatic biota, the complexity of environmental variables precludes a standard definition; "toxic" varies with different environments and different pollutants. Thus, biological assay methods have had to evolve *inductive* approaches to discover ranges of permissible toxicant levels.

Most of the available toxicity data on aquatic life are reported as the median tolerance limit (TL_m), the concentration that kills 50 percent of the test organisms within a specified time period, usually 96 h or less. This system of reporting has been misapplied by some who have erroneously inferred that a TL_m value is a safe value, whereas it is actually the level at which *half* the test organisms are killed. In most cases the discrepancy between environmental conditions that are sufficient to permit growth or reproduction and TL_m values is several orders of magnitude; that is, when half the organisms are killed, very few can reproduce.

Most TL_m values apply to a specific situation. The wide variability of species, water quality, and the nature of the test substance make generalizations extremely precarious. Although there are many types of assays, two are in general use: the *static bioassay* in which the test solution is not changed during the period of exposure and the *constant-flow* (flow-through) bioassay in which fresh test solution is continually circulated through the system.

With static tests, the initial concentration rapidly drops to subclinical levels through absorption and adsorption, making TL_m extrapolations irrelevant, and unwise for concentrations below 1 mg/l. However, short-term static tests have been shown to be very effective when testing for concentrations of highly toxic waste materials as long as the data are not used as absolute values. In general, such tests have more validity when both initial and final concentrations are known.

It should be understood that mildly toxic compounds in particular, and all compounds in general, can be assayed with much more accuracy by the constant-flow bioassay. The chemical nature of the substances to be tested is of prime consideration since substances that are volatile, unstable, or relatively insoluble may not be accurately assayed. The techniques for maintaining a supportive milieu, selecting the test animals, and applying the toxicant in static tests are outlined in detail in "Standard Methods for the Examination of Water and Wastewater" (American Public Health Association, 1971).

The constant-flow bioassay, although lacking in convenience and ease of operation, offers many advantages over the static assay: Concentrations of test material can be maintained; environmental prerequisites can be controlled; metabolites and waste products can be removed (i.e., the animal can be fed); absolute rather than relative TL_m values can be obtained; and volatile, unstable, and sparingly soluble materials can be tested. The constant renewal test can be more effectively monitored on either a continuous or intermittent basis. In addition, it is thus far the only suitable method for chronic exposure studies.

A major breakthrough has recently been made with the application of the Japanese rice fish (*Oryzias latipus*) to chronic toxicity studies. This aquarium species can undergo a complete generation from egg to egg in less than 3 months, providing the biologist with an ecological perspective in evaluating a toxicant. Thus, the use of *O. latipus* affords some of the advantages of microorganisms in similar studies, because short generations enable us to observe long-range as well as immediate change.

However, the difficulty with this, as with all such approaches, is the tendency to overemphasize the role of a single factor in evaluating changes in environments which are highly complex and interrelated by nature. Thus, to substitute conclusions based on a one-factor analysis for the complexity of factors which *is* the environment is to adversely simplify our thinking in much the same way that pollutants adversely simplify ecosystems. Each factor can be understood and interpreted only in its environmental context.

METABOLISM

With one thirty-thousandth the concentration in water that it has in air, perhaps the most metabolically significant aquatic environmental factor is dissolved oxygen. Normally recharged at a greater rate than it is respired by photosynthesis and atmospheric diffusion, oxygen concentrations are rapidly dissipated by domestic sewage. The introduction of sewage, then, places a priority on those adaptations permitting survival and reproduction in oxygen-poor media; the community changes from predator-dominated to saprobic. These changes are characterized by an abundance of aquatic invertebrates that are dependent on atmospheric oxygen (sewage flies, pulmonary snails), have a respiratory pigment (midges, sewage worms), have large absorptive area per unit volume (sewage worm), or have a low oxygen requirement. In general, the abandonment of the aggressive mode for the sedentary life-style is reflected in lower metabolic rates for individual organisms.

It is important to recognize that the concern here is not with lethal limits of anoxia, but with incipient limiting levels, i.e., those levels at which the organism goes into respiratory dependence. Because oxygen tensions limit the rate of oxygen intake and utilization for aquatic organisms, reduced dissolved oxygen concentrations can restrict the efficacy of a species in predation, escaping predation, and resistance to disease. These effects are debilitating for the *total* aquatic environment, because the restrictions of vital processes imposed on the organism result in the eventual displacement of a species from its ecological niche and alter the structure of the niche itself. The total impact is as drastic, though not as dramatic, as acute toxicity.

This evolution, from a predator-dominated community to a saprobic one, has been observed in a variety of cases where domestic sewage, agricultural runoff, fermentation, dairy or canning effluents, pulp and paper effluents, and a host of other readily degraded waste products are introduced to aquatic communities in large quantities. In addition to the more immediate influences exerted on community composition, these by-products contribute to the eutrophication of the receiving waters through the release of *phosphates, nitrates,* and *carbonaceous matter*. The definition of standards, then, must take into account these three fertilizing components, as well as the dissolved oxygen (DO) concentrations.

The parameters most frequently employed to gauge organic enrichment and consequent enhanced productivity are:

1. *Biological*: coliform count, biological oxygen demand (BOD), community composition
2. *Chemical*: CO_2, dissolved oxygen (DO), PO_4, NO_3, chemical oxygen demand (COD)
3. *Physical*: turbidity

A brief expansion of their significance and range of occurrence follows.

Dissolved Oxygen

For most water uses, an ample amount of dissolved oxygen is desirable. Low concentrations of dissolved oxygen contribute to a marginal existence for aquatic life, and the absence of dissolved oxygen creates the obnoxious odors associated with anaerobic decomposition. The latter conditions prevail because of the presence of oxygen-demanding organic substances, inorganic chemical oxygen demand, and increased water temperature.

RECOMMENDED LEVELS[1]

Domestic water supplies The presence of dissolved oxygen increases the palatability of water; however, there are no recommended limits stipulated.

Industrial water supplies The absence of dissolved oxygen is desirable for many industries (particularly in cooling waters), as a means of inhibiting corrosion. For boiler feed waters, the limiting dissolved oxygen concentration has been given as 20 mg/l for boiler pressure up to 250 lb/in^2, whereas 0.0 mg/l is recommended for pressures exceeding 250 lb/in^2 (American Water Works Association, 1950).

Aquatic life For optimum diversity, DO concentration of a eutrophic environment should be above 5 mg/l. During periods of winter-summer stagnation and summer nights, however, concentrations may extend temporarily below 5 and 4 mg/l without impairing carrying capacity. In stratified lakes, the DO requirements do not apply to the hypolimnion.

These stipulations should pertain to all waters except legally specified mixing zones. In *lakes*, such zones must be restricted to limit the effect on the biota, and in *streams*, adequate and safe passageways for migrating forms are required. These must be sufficiently extensive to permit the majority of plankton and other drifting organisms safe passage.

For the *cold-water oligotrophic biota*, DO concentrations should be at or near saturation levels, particularly in spawning areas where DO levels must be maintained above 7 mg/l.

Trout-salmon habitats should not have DO concentrations below 7 ppm, except for short periods when they may range between 6 and 5 ppm, provided the water quality is otherwise favorable. In general, for partially stratified streams that serve principally as *migratory routes*, DO levels may range between 4 and 5 mg/l for periods up to 6 h but must never be below 4 mg/l. Similarly, the hypolimnion of *oligotrophic lakes* should not be lowered below 6 mg/l at any time as a consequence of oxygen-demanding waste.

BIOCHEMICAL OXYGEN DEMAND (BOD)

Biochemical oxygen demand is the measurement of the amount of oxygen utilized by bacteria for the oxidation of organic substrates. These data do not reveal the concentration of a specific pollutant but, rather, the effect of a combination of substances and conditions.

CHEMICAL OXYGEN DEMAND (COD)

The chemical oxygen demand indicates the amount of oxygen required for combination with oxidizable materials in the water. Its application arises from the fact that, though the COD and ultimate BOD of pure organic substances are theoretically the same, in practice they are not. Some organic compounds (acetic

---

[1] Unless noted otherwise, recommended levels are those cited by McKee, J. E., and H. W. Wolf (eds.) (1963), "Water Quality Criteria," 2d ed., Publication No. 3A, California State Water Quality Control Board, Sacramento Calif.

acid) are refractory to chemical oxidation, whereas others (cellulose and elemental carbon) are biologically inert but do exert a COD.

Carbon Dioxide (CO_2)

Free CO_2 in water is the product of aerobic and anaerobic respiration and is closely bound in the complex carbonate equilibria.

RECOMMENDED LEVELS

Domestic water supplies Although free CO_2 affects taste, it appears to have no gross physiologic effect in normal concentrations.

Industrial water supplies Requirements of pH, acidity, and alkalinity indirectly set restrictions on the content of free CO_2 in water used for industrial processes. For example, the manufacture of various types of paper limits free CO_2 to 10 mg/l. The presence of free CO_2 in water increases the solubility of lead (plumbosolvency) and other heavy metals, accelerates corrosion of iron and steel, and promotes the solvent action of calcium carbonate in cement and concrete.

Aquatic life It is recommended that the free CO_2 concentration should not exceed 25 mg/l, because higher concentrations interfere with oxygen uptake in blood as a result of the acid-base balance. Likewise, the acid environment generated by the increased CO_2 level causes solubilization of mulluscan shells.

TURBIDITY

Turbidity, the degree to which light intensity is reduced by suspended or agitated matter, is an expression of the degree to which light is scattered and absorbed rather than transmitted or reflected. Its limnological significance is outlined in the section on solids.

RECOMMENDED LEVELS

Domestic water supplies The 1962 Drinking Water Standards specify that turbidity shall not exceed 5 units (U.S. Department of Health, Education and Welfare, 1962).

Industrial water supplies Turbidity is undesirable in industrial waters and should not exceed 25 units except in the manufacturing of pulp and paper where the range is between 5 and 100 units, depending on the grade produced.

Aquatic life Turbidity in the receiving waters due to the discharge of wastes ought not to exceed 50 Jackson units in warm-water streams and 10 Jackson units in cold-water streams. There should be no discharge to warm-water and cold-water lakes which would cause turbidities exceeding 25 and 10 Jackson units, respectively.

Recreational waters For primarily contact recreation water, clarity should be such that a Secci disc remains visible at a minimum depth of 4 ft. In "learn to swim" areas, however, it is recommended that water clarity permit discernment of a Secci disc on the bottom.

NUTRIENTS AND NUISANCE ORGANISMS

A burgeoning human population and an expanding economy act in concert to exploit our waterways as disposal systems. As a consequence of the uncontrolled nutrient and thermal input of industrial and domestic wastes, the aging process of our lakes and ponds has been greatly accelerated. This acceleration is commonly referred to as eutrophication.

Within certain limits, the more abundant the nutrient and the greater the temperature, the greater is the productivity. Such growth may be bacteria, aquatic fungi, phytoplankton, filamentous algae, or submerged and floating plants—littoral plant life. When these become so dense that they either interfere with the balance of the aquatic habitat or with human utilization of the water resource, they are defined as *nuisance growths*.

A case in point is the sheath-forming bacteria, the primary nuisance growth in rivers, lakes, and ponds. The most common offenders belong to the genus *Sphaerotilus* which prevails in waters receiving raw domestic sewage, improperly stabilized paper pulp effluents, and effluents containing simple sugars. These organisms foul fish lines, clog nets, and create unsightly conditions. Their metabolic demands and decomposition after death impose a high BOD load on the stream and severely deplete it of dissolved oxygen. Large populations of *Sphaerotilus* may render the habitat noxious and hence actively exclude desirable fish and invertebrates.

Another example of imbalanced population increase may be seen in the proliferation of algae. Limited concentrations of algae are essential to the food web, but the effects of overproduction may be serious. Some blue-green algae, green algae, and diatoms produce odors and scums that detract from the recreational value of water bodies. By increasing turbidity, dense growths of such planktonic algae may limit photosynthetic activity to a layer only a few inches beneath the surface of the water. Under certain conditions, the algae die, and their decomposition depletes dissolved oxygen, precipitating massive fish kills and further increasing the amount of dead organic matter in the water. Eutrophication is the result.

Many forms of plankton and filamentous algae clog sand filters in water treatment plants, produce undesirable tastes and odors in drinking water, and secrete oily substances that interfere with domestic use and the industrial processes whose wastes constituted part of the initial growth stimuli. Some algae cause foaming during heating, metal corrosion, and clogging of screens, filters, and piping; or they may coat cooling towers and condensers.

Although most algae require some simple organics such as amino acids, vitamins, and trace elements such as manganese and copper, these elements are not always necessary. Nitrogen and phosphorus, however, are often decisive in governing the abundance and distribution of organisms, because they are in short supply. Studies to date indicate that different phosphorus requirements range between 0.01 and 0.05 mg/l. At these minute levels, without the interference of other limitations, algal blooms, with their attendant problems, may be expected.

The nitrogen-phosphorus ratio may also be of importance. This ratio is dependent upon season, temperature, and geological formation and may range from 1 or 2:1 to 100:1. In natural waters, it is often very near 10:1, and this appears to be an indication of normal conditions. The major sources of nitrogen entering fresh waters are atmospheric [approximately 5 lb/(acre)(year) (Hutchinson, 1962)], domestic sewage effluents, animal and plant processing wastes, animal manure, fertilizer, chemical manufacturing spillage, various types of industrial effluents, and agricultural runoff. Freshwater sources of phosphorus are domestic sewage effluents, animal and plant processing wastes, fertilizer, and chemical manufacturing spillage.

Benthic forms also respond to increased nutrition with adverse secondary results. Cladophora (sewage blanket) becomes abundant in lakes and rivers where nutrients are increased beyond base levels. In these cases, cladophora replaces the original diverse benthic flora.

RECOMMENDED LEVELS

Domestic water supplies The 1962 Drinking Water Standards state that the maximum limit for nitrates is 10 mg/l as N. No limit has been established for phosphates.

Industrial water supplies Nitrates are harmful to fermentation processes and cause a disagreeable taste in beer (LeClerc and Dujanquier, 1946). Nitrates and free nitric acid at concentrations of 15 to 30 mg/l render water deleterious for various industrial processes.

Phosphates are an asset to industrial process or cooling waters because of their weakly dissociated acid reaction and their solubilizing effects upon scale formation. They are harmful, however, where they encourage biological aftergrowth.

Aquatic life In order to limit nuisance growths, the addition of all organic wastes such as sewage, food processing, and cannery or industrial wastes, all of which contain nutrients, vitamins, trace elements, and growth stimulants, should be carefully controlled. Furthermore, because of colloidal dynamics, the addition of sulfates or manganese oxide to a lake should be limited if iron is present in the hypolimnion, as they may increase the quantity of available phosphorus.

The naturally occurring ratios and amounts of nitrogen, as NO_3 or NH_4, to total phosphorus should not be radically changed by the addition of foreign matter. As a guideline, the concentration of total phosphorus should not be increased to levels exceeding 100 µg/l in flowing streams, or 50 µg/l where streams enter lakes or reservoirs.

Since present knowledge of conditions promoting nuisance growth and eutrophication is limited, a biological monitoring program should be instituted and funded to determine the effectiveness of control measures. A monitoring program can also detect the early development of undesirable change in population levels and varieties of rooted aquatics and algal growth. With periodic monitoring, such undesirable developments can be detected at relatively early stages and corrected by more stringent regulations and control measures.

TEMPERATURE

Temperature changes result from seasonal and diurnal cycles and from the introduction of industrial wastes (i.e., from distilling effluents) or the discharges of cooling waters (i.e., from atomic-powered installations). The monthly average temperature of any aquatic body fluctuates within limits prescribed by other environmental influences, and as such is the natural context in which the organisms within that body have developed homeostatically. Thus, to alter this proportion or to raise the mean average temperature is to disequilibrate the balance of living systems within the aquatic environment.

The effects of thermal pollution on aquatic life are diverse, for unlike specific toxicants, thermal pollution elicits a complex of responses whose disruptive effects are still imperfectly understood. However, we may point to three primary ways in which artificial alterations of water temperatures have been shown to have toxic effects. First, an increase in temperature may kill certain species outright.

Second, temperature increases decrease oxygen solubility and mixing; although such an increase may fall within the tolerance zone defined by the upper and lower incipient lethal limits, it will nonetheless decrease the physiologic range of spawning or aggressive behavior in some species by increasing their metabolic rates and hence their need for oxygen. For any species, temperature increases cause a peak in the active metabolism, and a steady rise in the resting metabolism. Since the scope of an organism's activity is the difference between these two O_2 consumption rates, it is clear that their increased convergence is a direct threat to life. The organism is trapped between its need for more oxygen and its increased metabolism; between its rate of oxygen use when active and its rate of use when resting. The active-resting differential increases rapidly at the optimum temperature and decreases rapidly above and below this point, so that, although consumption may increase with temperature, the amount available for active pursuits decreases rapidly. A species might then survive the immediate stress of increased temperature but be decimated by the chronic debilitating effects of decreased efficiency in predation, reproduction, and resistance to disease. It should also be noted that, for the same reasons, the responses of species to toxic substances are increased.

Third, increased water temperature speeds biological degradation processes, which also accelerate the demand on the dissolved oxygen reserves. The resultant change in water chemistry precipitates profound limnological changes that militate against desirable fish life. Putrefaction and scavengers are favored, and the number and variety of flora are decreased.

RECOMMENDED LEVELS

Domestic water supplies For drinking purposes, a temperature of 50°F is usually satisfactory, whereas 60°F or higher may be objectionable. The survival time of microbial pathogens, cysts, and ova of parasitic worms is generally reduced with an increase of water temperature.

Industrial water supplies The water temperature may be of great importance for various industrial processes; i.e., warm water accelerates corrosion in pipelines and cooling systems.

Aquatic life; warm waters To maintain a diversified aquatic population, it is necessary to limit temperature extremes to conform as closely as possible with the natural levels of the original proportion. During any month of the year, heat should not be added to a stream in excess of the amount that would raise the temperature of the water (at the expected minimum daily flow for that month) more than 5°F. In lakes, the temperature of the epilimnion ought not to be raised more than 3°F. The increase should be based on the monthly average of the maximum daily temperature. Unless a special study shows that a discharge of a heated effluent into the hypolimnion will be desirable, such practice is not recommended; water for cooling should not be pumped from the hypolimnion to be discharged to the same body of water. The normal daily and seasonal temperature variations that were present before the addition of heat must be maintained. The recommended maximum temperatures that are not to be exceeded for various species of warm-water fish are given in Table 12-1.

Aquatic life; cold waters Because of the large number of trout and salmon habitats which have been made marginal, nonproductive, or destroyed, the remaining trout and salmon waters must be protected if these species are to be preserved.

The temperatures of inland trout streams, headwaters of salmon streams, trout and salmon lakes, and the hypolimnion of lakes and reservoirs containing salmonids and other cold-water forms should not be warmed or used for cooling. Further, no heated effluents should be discharged in the vicinity of spawning areas.

For other types and reaches of cold-water streams, reservoirs, and lakes, the principles discussed with reference to warm waters should be applied. The following restrictions are recommended: During any month of the year, heat should not be added to a stream in excess of the amount that would raise the temperature of the water more than 5°F (based on the minimum flow for that month). In lakes, the temperature of the epilimnion should not be raised more than 3°F. Again, it is essential that the normal daily and seasonal temperature fluctuations that existed before the addition of heat be maintained.

BUFFERING

The potential of a community to exploit its environment is a function of the capacity of resident populations to adapt to the rigors imposed by the physical-

Table 12-1 Provisional maximum temperatures recommended as compatible with the well-being of various species of fish and their associated biota

93°F:	Growth of catfish, gar, white or yellow bass, spotted bass, buffalo, carpsucker, threadfin shad, and gizzard shad
90°F:	Growth of largemouth bass, drum, bluegill, and crappie.
84°F:	Growth of pike, perch, walleyes, smallmouth bass, and sauger
80°F:	Spawning and egg development of catfish, buffalo, threadfin shad, and gizzard shad
75°F:	Spawning and egg development of largemouth bass, white and yellow bass, and spotted bass.
68°F:	Growth or migration routes of salmonids and for egg development of perch and smallmouth bass
55°F:	Spawning and egg development of salmon and trout (other than lake trout)
48°F:	Spawning and egg development of lake trout, walleye, northern pike, sauger, and Atlantic salmon

chemical milieu. Having touched upon self-equilibrating homeostatic mechanisms, we would like to expand on selection-potential mechanisms which are indicators of environmental stability, and show how they simultaneously reflect and exert environmental stress. These factors are buffering potential, acidity, alkalinity, pH, solids, and hardness.

The ability to maintain equilibrium against changing conditions is referred to as the *buffering potential* of an organism or an environment. The natural water body exhibits a capacity to maintain self-equilibrium by keeping the hydrogen-ion concentration constant. Water molecules have a tendency to ionize, separating into H^+ and OH^-. In any given body of pure water, a small but constant number of water molecules are dissociated in this form. The number is constant because the tendency of water to dissociate is offset by the tendency of the ions to reunite; thus, even as some are ionizing, an equal number of others are forming covalent bonds. This is a state of dynamic equilibrium.

In pure water, the number of H^+ ions equals the number of OH^- ions. This is necessarily the case since neither ion can be formed without the other when only H_2O molecules are present. A solution acquires the properties of an acid when the number of H^+ ions exceeds the number of OH^- ions; conversely, a solution is basic (alkaline) when the OH^- ions exceed the H^+ ions. There is always an inverse relationship between the concentrations of H^+ and OH^- ions.

Chemists define degrees of acidity by means of the pH scale. The pH of the medium is as fundamental to organismic activity and development as is temperature. Although the pH range is subject to wide variation from one aquatic environment to another, each species and each niche within a given community can grow only within a certain pH range. Most rapid or luxuriant growth occurs in a narrow *optimum pH zone*: the constant level of hydrogen-ion concentration maintained by natural water bodies.

Organisms resist strong sudden changes in the pH of blood and their own intracellular and extracellular fluids by means of such buffer systems. Most life chemistry takes place at a pH of between 6 and 8. The blood, for example, maintains an almost constant pH of 7.4, despite the fact that it is the vehicle for a large number and variety of nutrients and other chemicals being delivered to the cells and for the removal of wastes, many of which are acids or bases. The capacity of a buffer system to resist changes in pH is greatest when the concentrations of proton donor and proton acceptor are equal. As the salt of a weak acid and a strong base, the buffer dissociates to provide bicarbonate ions which neutralize acidic or basic pollutants. When such pollutants change the ratio in either direction, the buffer becomes less effective, osmotic balances are altered, and the optimum pH zone can no longer be maintained. Waters buffered between pH's of 6 and 9 afford a stable environment for stream biota provided that contributory ions do not attain massive concentrations.

In evaluating the biological response to specific pollutants in a given body of water, one is immediately impressed with the weakness of the universal indicator concept. The search for a single species—or chemical—that accurately reflects the effects of toxic substances on the milieu as a whole has been frustrated by the wide range of responses of several species frequently chosen to act as indices of pollution.

For example, trout, which present the classic pollution-sensitive response, exhibit broad tolerances to pH extremes 3.3 to 10.7, whereas carp cannot survive indefinitely below a pH of 5.5. Algae are generally able to tolerate pH's down to 2.5, but they are surprisingly sensitive to the alkaline extremes of less than 8.5. Those algae which are sewage-tolerant can withstand concentrations of ammonia which kill most other forms, and neither variety produces widely usable results as a general index. The correlation between sewage-tolerant forms and ammonia (at acid pH's) is much higher, since ammonia is a normal constituent of domestic sewage; in clean water it forms less than 0.4 ppm; in mild pollution it forms 0.4 to 4.0 ppm; in gross pollution it forms 4 to 220 ppm.

Cyanides and sulfides, as respiratory enzyme poisons, select against predators because of their high oxygen demand. By lowering metabolic rates, they favor saprobic forms of life and thus have the same environmental effect as the thermal pollutants discussed earlier. Yet the problem of finding an organic index remains: Rainbow trout will die in 11 min when exposed to 0.2 ppm of cyanide, whereas carp will survive 90 min when exposed to 10.0 ppm of cyanide.

When heavy metals are at pH's that permit their solution, it should be noted that the mechanism of toxicity may again be through suffocation, this time through the denaturation of the mucosa covering the gill epithelium.

In general, chemical evaluations of environmental buffering potentials have produced more satisfactory and consistent results than the various not-so-universal indicators. These criteria of analysis are complexly interrelated. Following is a more detailed discussion of acidity, alkalinity, pH, solids, and hardness.

Acidity

Acidity is defined as the capacity to neutralize hydroxyl ions and is the combined effect of several substances and conditions. It is expressed as the calcium carbonate equivalent of the hydroxyl ions neutralized. In natural waters, it is a product of free carbon dioxide and weakly dissociated humic acids from swamps and peat deposits. In industrial wastes, acidity may originate primarily from acid mine drainage, sulfite wastes, and pickling liquors.

Acidities of healthy freshwater streams vary widely and are a function of the eutrophic features of the watershed. The toxicity of acids varies with generic types: mineral and organic. Mineral or inorganic acids exercise a toxic effect through their high dissociation constants and are additionally toxic when the sample has a low pH value. The consequent combination of hydrogen cations with the gill mucosa results in precipitation and coagulation of the proteinaceous material, causing asphyxiation.

Another consequence of artificial increases in acidity is the release of fatal concentrations of CO_2 which follows upon the interaction of the acids with the bicarbonate ion. The anion itself is of relatively little significance. The toxicity of weak organic acids, on the other hand, seems to correlate with the undissociated molecule which is absorbed as an internal poison. The penetrating molecule causes acidemia, inhibiting the exchange of respiratory gases. Because fish blood is so poorly buffered, it follows that mineral acids are toxic at much lower pH values than are organic acids.

Standards have not been specified for acidity in drinking water or for industry, except in such relatively isolated instances as concentrations of zinc solutions from galvanized pipe, or accelerations in iron corrision caused by excessive acidity.

The biological limits of this parameter have been defined in terms of pH and will be discussed in that section.

Alkalinity

The alkalinity of water is the capacity of a water to accept protons, i.e., to neutralize hydrogen ions. Alkalinity is usually imparted by the bicarbonate, carbonate, and hydroxide components of a natural or treated water supply and, to a lesser extent, by borates, silicates, phosphates, and organic substances. The cations most involved in alkalinity are Ca and Mg, because of their solubility in carbonic acid. These elements, together with iron, contribute to hardness, which refers to the soap-precipitating capacity of alkaline earths.

In itself, alkalinity is not considered harmful. However, the associated high pH values, hardness, and excessive dissolved solids may exercise a deleterious effect, particularly on the processing of food and beverages.

Although not particularly desirable for human consumption, such salts offer strong ecological advantages to the aquatic community through (1) the stabilization of hydrogen-ion concentration which reduces turbidity and increases primary productivity; (2) the flocculation of suspended solids; (3) increases in the supply of mineral nutrients; (4) reduction of cell permeability (and consequent vulnerability through intake of toxic agents); and (5) complexing and precipitation of otherwise toxic heavy metals.

The best waters for the support of a diversified aquatic life are those with pH values between 7 and 8, having an alkalinity of 100 to 120 mg/l (Huet, 1941). Alkalinity, then, acts as a buffer to help prevent any sudden change in pH which fish and other aquatic life forms cannot accommodate. To ensure the survival of the buffer, and environmental homeostasis, the total alkalinity should not be less than 20 mg/l.

pH

Basically, pH is an index of hydrogen-ion activity expressed as the negative log of the hydrogen-ion concentration. It should be understood that, although pH readings are used as indications of acidity and its reciprocal alkalinity, they are not a measure of either. Alkalinity and acidity values represent total capacities; pH is a measure of concentration.

RECOMMENDED LEVELS

Natural waters are generally alkaline (pH 6.5 to 8.5) and acidity is not a concern, particularly when the pH values are increased to 10. Although pH affects taste, corrosivity, efficiency of chlorination, and other treatment processes, there are no nationally set limits in domestic waters.

Recommended threshhold values for pH in various industrial processes vary widely: boiler feed water, 8.0 to 9.6; tanning, 6.2 to 8.0; pulp and paper, 6.7 to 7.3; brewing, 6.5 to 7.0; and carbonated beverages, 2.7 to 3.5. The corrosive action of water is increased at low pH values. This is not a function of pH alone, however, but it is also attributed to hardness, alkalinity, and carbon dioxide, the latter being particularly detrimental to concrete.

Since most inland waters containing fish have pH values which range from 6.7 to 8.6, with extremes of 6.3 to 9.0, it is recommended that no highly dissociated materials be added in amounts sufficient to lower the pH below 6.0 or to raise it above 9.0. Although propagation is possible beyond pH's of 4.0 and 9.4, conditions become marginal. The addition of weakly dissociated acids and alkalies should be regulated in terms of their own toxicity as established by bioassay procedures. pH remains a critical parameter for biological productivity for it determines the degree of dissociation of weak acids and bases which may be more toxic in the molecular than the ionic form. Hydrogen cyanide, hydrogen sulfide, and hypochlorous and tannic acids are all examples of compounds whose toxicity increases logarithmically with decreases in pH.

Similarly, low dissociated alkaline compounds are often toxic at pH values above 9.0. Therefore, the photosynthetic activity of the aquatic plants increases the toxicity of ammonium hydroxide by converting the carbonate to the hydroxide and increasing the pH. Since these high pH levels normally prevail for only a few hours, the harmful effects are minimal, unless pollutants continue to artificially raise them. In the same vein, increased sensitivity to ammonium hydroxide can result from decreased O_2 tension in the water, which in turn causes decreased CO_2 concentration and increased pH. A reduction of 1.5 pH units can cause a thousandfold increase in the toxicity of a cyanide complex. In addition to affecting dissociation rates, pH can affect toxicity through altering solubility. Heavy metals are highly soluble in acid waters. Also, the availability of many nutrient substances varies with acidity.

Solids

Solids in water are classified as either *dissolved* or *suspended*. Dissolved solids are capable of passing a fine mat of asbestos fiber in a Gooch crucible; suspended solids are retained on the asbestos mat. Both dissolved and suspended solids are further distinguished as fixed and volatile, i.e., those materials driven off by ignition at 600°C. The total suspended solids category is divided into floatable, settleable, and colloidal solids.

DISSOLVED SOLIDS

The major constituents of dissolved solids are carbonates, bicarbonates, sulfates, and the chlorides of sodium and calcium; minor components include nitrates of calcium, magnesium, sodium, and potassium, and traces of iron and manganese. These salts originate in soils and geological deposits but often appear in water as the result of the dumping of sewage and effluents by people.

Significant amounts of carbonates are found in sewage, industrial effluents, land runoff, water treatment sludges, and industrial wastes. Chlorides are leached from the soil and found in the discharge of human wastes, brines, industrial wastes, and street washings. Sulfates are of both natural and human origin, and their presence may or may not be attributable to pollution.

All such salts, aside from their individual physical-chemical properties, exert a profound influence on the aquatic biota through their contribution to hardness, turbidity, and possible synergistic and antagonistic interactions.

The 1962 Drinking Water Standards indicate that total dissolved solids should not exceed 500 mg/l where other more suitable supplies are available. Dissolved solids in industrial water supplies may cause foaming in boilers and interference with clearness, color, or taste. High concentrations of dissolved solids also tend to accelerate corrosion.

Dissolved solids may be of biological significance as poisonous or osmotically disruptive substances. The latter category comprises salts that are usually required in moderate concentrations for productive water. In this instance, total dissolved materials should not exceed 50 milliosmates in waters where diversified animal populations are to be protected. To maintain a state of dynamic equilibrium in a diversified aquatic community, total dissolved materials should not exceed base-line concentrations by more than one-third. When concentrations of dissolved materials are increased, bioassays and field studies should be instituted to determine the impact of these additives on the aquatic biota.

SUSPENDED SOLIDS

In natural water, suspended solids originate essentially from silt, organic detritus, and plankton.

The activities of human beings, however, have made a massive contribution to surface water solids of both inorganic (mining wastes and washings, and eroded soils) and organic (tars, grease, oil, wood fiber, plastics, and various settleable material from sewers) origin. Floating materials include sawdust, cannery wastes, hair and fatty material from tanneries, wood fibers, containers, scum, oil, garbage, tars, greases, and precipitated chemicals.

The 1962 Drinking Water Standards do not specifically set limits for suspended solids, but they are indirectly established by controls on turbidity (5 units) and total solids (500 mg/l).

Suspended solids in water interfere with many industrial processes by foaming in boilers, forming incrustations on equipment, and contributing to the deterioration of product quality.

In addition to being potentially toxic, suspended solids may reduce productivity to a marginal level by causing abrasive injuries and clogging the gills and respiratory passages of various aquatic fauna.

Indirectly, suspended solids are detrimental to all aquatic forms, and particularly the relatively sessile benthic macroinvertebrates. These solids blanket the stream bed, destroying food organisms and fish spawning grounds. Settling particulate matter traps bacteria and organic wastes, promoting the accumulation of such toxic anaerobic decomposition products as methane, hydrogen sulfide, and ammonium salts. Solids also contribute to turbidity, reducing light penetration and restricting photosynthesis and productivity. Suspended solids, due to their low specific heat, raise water temperature and decrease absorbed oxygen, while increasing metabolic demands. These factors conspire to produce a hostile environment for the predator, who is unable to satisfy the food and oxygen requirements of an increased metabolism. As in some of the cases cited earlier, saprobic conditions and, finally, eutrophication are the results.

Hardness

Originally, the hardness of a water was understood to be a measure of the capacity of the water for precipitating soap. Soap is precipitated chiefly by the calcium and magnesium ions commonly present in water but may also be precipitated by ions of other polyvalent metals, such as aluminum, iron, manganese, strontium, and zinc, and by hydrogen ions. In natural waters, all but the first two are present in insignificant concentrations. Hardness is thus defined as a characteristic of water which represents the total concentration of calcium and magnesium ions only, expressed as calcium carbonate. However, if present in significant amounts, other hardness-producing metallic ions should be included.

Since metals are not readily soluble in alkaline milieus, such waters will not be as toxic as soft (unbuffered) water, containing acid heavy-metal wastes. In addition to heavy metals, however, there are other wastes whose toxicities are pH-dependent: HCN, H_2S, tannic acid, and NH_4 compounds. In these instances, however, toxicity is correlated with the concentration of the undissociated molecule rather than solubility.

Hardness in excess of 100 mg/l is an inconvenience, since it causes soap waste and encrusts utensils (Phelps, 1948). Water of good quality should not exceed a hardness of 270 mg/l (Baughman, 1966), and levels above 500 mg/l are considered unsuitable for domestic uses. However, the harmful effects of hardness on the health of consumers has yet to be demonstrated (Thoesh and Bede, 1925).

Generally, biological productivity is directly correlated with hardness, because productivity depends on the specific combination of elements present. Calcium and magnesium contribute to hardness and to productivity, first as nutrient requirements and, second, by complexing with toxicants to form insoluble precipitates. However, most other elements which contribute to hardness reduce biological productivity. It is therefore recommended that the term "hardness" be avoided in dealing with water quality requirements for aquatic life.

These factors, then, acidity, alkalinity, pH, solids, and hardness, constitute the chemical indices of the buffering potential of an aquatic environment and allow us to measure the effects of pollutants as disruptive factors of homeostasis. It should be remembered that buffers are important to both the internal chemical balances of various organisms and to the aquatic community as a whole. The more closely equilibrated the acidic-basic ratios of ions, the greater is the ability of these buffers to resist the effects of pollution.

THE POLLUTION INDICATORS

THE FATE OF BACTERIA IN THE AQUATIC ENVIRONMENT

The vital conditions necessary for the maintenance of a homeostatic environment are reflected throughout all parts of the aquatic community, regardless of the requirements of any individual species. Microscopic organisms are particularly crucial because they occupy a prime place in the food chain, and by virtue of the fact that they are nearly as ubiquitous as water itself. In the case of bacteria, one encounters microorganisms which are often an essential element in the *internal* environments of various species, some of which may be members of the community, others of which may themselves be pollutors. As such, bacteria can be used as indicators of a multiplicity of conditions which cannot be explored simultaneously by any other means. For example, they are uniquely suitable for determining the presence or absence of pollution in general, and they indicate its probable particular source, if present.

Therefore, an understanding of the aquatic environment is impossible without some knowledge of the effects of bacterial densities, of the factors through which these densities are reduced or controlled, and of the particular species whose presence or absence may or may not be consistent with a healthy aquatic ecosystem.

In general, bacterial reduction may be effected by natural self-purification, predation, sunlight, adsorption and sedimentation, antibiotics and toxins, and salinity. Some of these have been extensively studied and much is known of their actions, but others remain obscure (Burm and Vaughan, 1966; Lawrence, 1958; Gameson and Saxon, 1967).

Natural purification may be defined as the combination of all the physical, chemical, and biological factors which tend to diminish the numbers of bacteria *not* naturally appearing within the given ecosystem for a specific period of time.

The *predators* of bacterial populations belong to the microphagic group of organisms that ingest microbes and can be considered one of the first links in the food-chain process of aquatic environments. Although the capacity of these predators to decimate bacterial populations has been confirmed by numerous studies, there is still controversy regarding their significance as determinants of population density.

Considerable information has been published on the bactericidal properties of *sunlight*, but relatively little is yet known of its regulatory effects on the aquatic microflora. However, it is generally agreed that sunlight is of secondary importance in the reduction of aquatic microbial populations. In water bodies of shallow depth, of course, it is of greater import than is the case for ocean depths.

Numerous studies have indicated that *adsorption* and *sedimentation* remove bacteria from the aquatic environment to the bottom stratum. Bacterial adsorption has been shown to have a direct relationship to the size and nature of the particles involved. For example, marine muds have been found to be highly adsorptive whereas sand particles are only slightly so. Thus, particle size and composition may partially explain the variation in bacterial populations of sediments.

Although *antibiotics* and *toxins* are generally considered constituents of the marine environment, they may be present in freshwater environments as well, since the microbes which produce these agents are ubiquitous. Numerous reports have demonstrated the presence of antibiotics or toxins in estuarine and oceanic environments. The rapid disappearance of *E. coli* and other sewage bacteria from receiving waters and, more recently, studies employing coliphages indicate the toxic properties of seawater for these microorganisms (Carlucci and Pramer, 1960a, 1960b; Carlucci et al., 1961; Orlob, 1956).

The *salinity* of seawater, which averages about 0.35 percent and which is considerably less at the freshwater interface, may adversely affect the survival of bacteria through either specific toxicity or osmotic effect. Although one would expect greater diversity here than in either milieu separately, the reverse appears to be the case.

As a general rule, bacteria survive longer at lower temperatures. This can be demonstrated for both freshwater and marine environments. Coliform bacteria in ocean waters survive in large numbers at 6 to 10°C and die at 20 to 25°C.

A multitude of other factors, such as pH, turbidity, chlorination, and industrial wastes must not be disregarded as determinants of bacterial survival, though their functions as control mechanisms remain obscure. In general, it may be concluded that since each factor that influences bacterial survival is a variable, affected by other factors, it is necessary to conduct special investigations to determine the overall effects of factors which lead to bacterial disappearance in a particular environment.

TOTAL BACTERIAL COUNTS

The first tentative standards for water quality were proposed by Koch and were based on total plate counts. Those waters showing counts more than 100 per milliliter on a gelatin medium incubated 3 days at 20°C were judged unsanitary. The 20° temperature was challenged by some workers, and later water samples were analyzed in duplicate with one set of plates incubated at 37°C, and the other at 20°C. The counts at both temperatures were compared, and waters showing a high count after incubation at 37°C were defined as polluted by animal sources, and those yielding excessive counts at 20°C were considered to contain large numbers of microorganisms of soil origin (Prescott et al., 1946).

It must be stressed, however, that the "total" bacterial count concept is quite nebulous in terms of laboratory determinations. The present plate count methods are limited in that they enumerate only those bacteria capable of growth on the culture medium provided, and under a specific incubation regimen. Unfortunately, there is no one substrate or one temperature that will stimulate and support the growth of every bacterial species and type.

It is extremely difficult to determine the total plate count of a freshwater sample because there is no single culture medium that is not selective for a particular metabolic pathway. Any bacterial inventory of natural waters purporting to be more than qualitative is immediately suspect. In this context, consequently, it is not inconsistent to find in the literature bacterial counts in rivers ranging from hundreds to several thousands per milliliter. The difficulty in obtaining a central tendency is compounded further by a host of environmental factors: the mineral and organic content of the watershed soils, extent of vegetative coverage, turbidity, flushing action of rainfall, and the quality (nutrient content or toxicity) of the effluent. All these factors act in concert and often synergistically either to enhance or to limit the indigenous microflora. The bacteria native to natural waters may be classified as the sheathed higher bacteria including the sulfur and iron forms, the stem or caulobacteria group, the spiral forms, the pigmented and nonpigmented rods, the coccus group, including the *Sarcina* and *Micrococcus* genera, the nitrogen-fixing bacteria (*Azotobacter*), and the nitrifying bacteria (*Nitrosomonas* and *Nitrobacter*).

Variability notwithstanding, however, the total bacterial counts of receiving waters can provide an extremely valuable statistic in monitoring the various industrial pollutants, provided enough base-line data has been accumulated to determine the normal amplitude of variation from the mean. Further it must be realized that the significance of these data is not as absolute measures but as relative values to be compared in space and time within the same river system.

BACTERIAL INDICATORS OF FECAL AND MUNICIPAL POLLUTION

A bacterial indicator of pollution is any organism whose presence or absence indicates deterioration of water quality. More precisely, bacterial indicators are associated primarily with denoting water contamination by excreta of warm-blooded animals, including human beings, domestic and wild animals, and birds.

Feces from warm-blooded animals regularly include a microflora comprising a great variety of bacterial genera and species. These include *Streptococcus*, *Lactobacillus*, *Escherichia*, *Enterobacter*, *Staphylococcus*, *Proteus*, *Pseudomonas*, *Clostridium perfringens*, and certain spore-forming bacteria. In addition, various pathogens may be isolated from fecal material released into the environment sporadically, depending on the state of community health, the nature and degree of waste treatment, and other factors. These enteric, pathogenic bacteria are *Salmonella*, *Shigella*, *Brucella*, *Mycobacterium*, *Vibrio*, and *Leptospira* and are implicated in a variety of diseases.

A number of viruses may also be classified as enteric, causing, for example, infectious hepatitis and poliomyelitis. In addition, members of the Coxsackie and ECHO groups have been postulated as causal agents of diarrheal and upper-respiratory diseases of unknown etiology, but these are apparently infective by the waterborne route. Protozoa, such as *Endamoeba histolytica*, the causal agent in amoebic dysentery, must also be included with these enteric waterborne agents (Geldreich et al., 1966).

The routine practice of water quality monitoring includes testing for the presence and numbers of bacteria that are known pollution indices. These indicator organisms may be grouped as the coliform organisms and certain related species, the fecal streptococci, and other miscellaneous bacteria. It is generally assumed that evidence of water contamination by intestinal wastes of warm-blooded animals indicates a health hazard.

Clostridium perfringens, formally called *C. welchii*, has been used as an indicator organism by the British for quite some time. This organism is a widely distributed species and is usually present in the intestinal tract of warm-blooded animals in large numbers. It is a gram-positive, anaerobic, spore-forming rod, whose spores cause a distinct swelling of the vegetative cell. The organism ferments carbohydrates very rapidly and produces the classic "stormy fermentation." The presence of this species indicates that pollution has occurred, but its isolation does not necessarily indicate unsafe water, because of the extreme resistance of the spores to environmental conditions. It is virtually impossible to base an estimate of the time of pollution solely on the presence of *C. perfringens*.

As mentioned earlier, efficient methods for the isolation of pathogenic bacteria of enteric origin have been developed and are presently employed. These procedures demonstrating the presence of pathogenic bacteria in water conclusively prove that the water is of unsanitary quality and hazardous to people drinking or coming into contact with it. The limitation of using pathogenic bacteria for this purpose is due to the lack of a routine procedure that will detect all enteric pathogens. Likewise, quantitative methods are not available for routine application. The intermittent release of these pathogens makes it impossible to regard a water as safe in their absence. Finally, because of the length of time necessary for isolation and identification of enteric pathogens, the public would have been exposed to these organisms before a positive identification could be verified.

The enteric pathogens find their application primarily in tracing the sources of infectious agents in epidemilogic investigation. They are also invaluable in water quality studies concerned with enforcement cases since there can be no question regarding the origin of these bacteria and the public health hazard of the water which harbors them.

It follows from this, as well as from our earlier discussion of the impracticality of the "universal indicator" concept, that it is impossible to select *a* bacterial group as *the* indicator of pollution. Obviously, the ideal group would give uniform information when used in all types of water. Although this ideal is unattainable, certain bacterial groups have been employed with great effectiveness. In selecting a bacterial indicator, the following criteria are important: It should be present in water when pathogenic bacterial constituents of fecal contamination are present; the density of the indicator organism should have some direct relationship to the degree of fecal pollution and should survive somewhat longer in water than enteric pathogens; similarly, the indicator must be absent in nonpolluted or bacterially safe water and lend itself to routine quantitative methodology; finally, such an indicator should be harmless to human beings and other animals. The merits and limitations of the main indicator groups will be discussed elsewhere in terms of their applications in evaluating water quality.

THE COLIFORM BACTERIA

In 1885, Escherich recovered certain bacteria from human feces in such consistently high numbers that he designated them "the characteristic organisms of human feces." The *Escherichia coli*, now the official type species of the group, have been substantiated along with much of Escherich's original concept (Escherich, 1885). However, what was first regarded as a single bacterial species is in fact a heterogeneous complex of bacterial species and variants. *Escherichia coli* is found not only in human feces but also in many environmental contexts such as sewage, all categories of surface waters, soil, on vegetables, etc.

As defined in "Standard Methods for the Examination of Water and Wastewater" (American Public Health Association, 1971): "The coliform group includes all of the aerobic and facultative anaerobic, Gram-negative, non-sporeforming rod-shaped bacteria which ferment lactose with gas formation within 48 hours at 35°C." The term "coliform" or "coliform group" is an inclusive one, comprising (1) *Escherichia coli, E. aurescens, E. freundii, E. intermedia*; (2) *Enterobacter aerogenes, E. cloacae*; and (3) biochemical intermediates between the *Escherichia* and *Enterobacter* genera. Although there is no provision in the definition for "atypical" or "aberrant" coliform strains, an individual strain of any of the above species may fail to meet one of the criteria of the group. Such an organism, by definition, *can not* be considered a member of the coliform group, though a taxonomist may correctly classify it as such.

Since coliform organisms are also an important component of the microflora of the soil, it was only natural for the early workers to attempt to subdivide this group into "fecal" and "nonfecal" categories. Single-test differentiations between coliforms of "fecal" origin and those of "nonfecal" origin are based on the assumption that typical *E. coli* and closely related strains are of direct fecal origin, whereas *Enterobacter aerogenes* and its close relatives are not. The latter assumption has not fully been borne out by investigations at the Environmental Protection Agency (Geldreich et al., 1962a, 1962b; Clark and Kabler, 1964).

A number of single differential tests have been proposed to distinguish between "fecal" and "nonfecal" coliforms and are cited here without discussing their relative merits.

1. *Determination of gas ratio*: Fermentation of glucose by *E. coli* results in gas production of hydrogen and carbon dioxide in equal amounts. Under identical conditions, *Enterobacter aerogenes* yields twice as much carbon dioxide as hydrogen. Further studies, based on the above, suggested an absolute correlation between H_2/CO_2 ratios and the terminal pH resulting from glucose fermentation. This led to the use of the following tests, particularly test 6.

2. *Methyl red test*: Glucose fermentation by *E. coli* usually results in a terminal reaction of the culture medium in the range of pH 4.2 to 4.6. This yields a red color or *positive* test with the addition of methyl red indicator. By contrast, the "aerogenes group" results in a terminal reaction greater than pH 5.6, thus yielding a yellow color or a *negative* methyl red test.

3. *Indole*: Typical *E. coli* strains are capable of producing indole from trytophane when this amino acid is incorporated in a nutrient broth. The *Enterobacter aerogenes* does not make this cleavage, and test results will be negative. Care must be taken with any method used for indole detection, as the results may be greatly influenced by the analytical procedure.

4. *Voges-Proskauer or acetylmethyl carbinol test*: This test is for acetylmethyl carbinol, which is a derivative of 2,3,-butyleneglycol and a result of glucose fermentation in the presence of peptone. *Enterobacter aerogenes* produces this chemical (positive test) whereas *E. coli* does not. It has been suggested by some that all coliform bacteria produce acetylmethyl carbinol in glucose metabolism. However, they assert, those cultures giving a negative test possess an added enzyme system which can further degrade this compound to other end products that do not give a positive test with the analytical procedure. Cultures giving a positive test for acetylmethyl carbinol would lack this enzyme system.

5. *Citrate utilization*: Cultures of *E. coli* cannot utilize citrates as the sole source of carbon (negative test), but members of the "aerogenes group" can (positive test). Very often an indicator, such as bromthymol blue, is incorporated into the citrate agar medium in order to demonstrate the typical alkaline reaction (pH 8.4 to 9.0) resulting from citrate utilization.

6. *Elevated temperature or Eijkman test*: This test is the basis of evidence that *E. coli* and other coliforms of fecal origin are capable of fermenting carbohydrates (glucose and lactose) at temperatures significantly higher than 37°C. Organisms of nonfecal origin are not able to grow at the elevated temperature (negative test). Although many media and techniques have been proposed, EC broth, developed by Perry and Hajna (1944), is used as a confirmatory

medium at 44.5 ± 0.2°C for 24 h. This is the current recommended medium and method of choice. Although the EC terminology of the medium suggests *E. coli*, this should not be regarded as a specific procedure for the isolation of this species (Eijkman, 1904).

A similar medium, boric acid lactose broth, developed by Levine and his associates (1955), gives results virtually identical with those obtained by EC broth but requires an incubation period of 48 h.

It is essential that all incubation for elevated temperature tests be done in a water bath. "Standard Methods" stipulates the temperature to be 44.5 ± 0.2°C (although temperatures ranging between 43.0 and 46.0°C with a tolerance of ±0.5°C have been suggested). Such a limited tolerance requires the use of a forced-circulation water bath to maintain the required temperature. In addition, the reliability of the elevated temperature tests is subject to further influence by the time required for the inoculated cultures to reach the designated incubation temperature. It is imperative that the cultures be placed in the water bath within 20 min after inoculation, and that a water level be maintained sufficient to completely submerge the entire EC broth.

In 1938, Parr made a literature survey of biochemical tests employed to differentiate between coliforms of fecal and nonfecal origin. Based on this summary and on his own studies, Parr recommended that a series of tests be utilized for differentiation that includes the indole, methyl red, Voges-Proskauer, and citrate tests. This series of reactions is designated by the mnemonic IMViC. Employing this scheme, any coliform culture can be described by an IMViC code according to the reaction to each test. Thus, a typical *E. coli* would have a code + + − −, and a typical *Enterobacter aerogenes* culture would have a code − − + +. Those cultures falling between these two specific IMViC codes are generally referred to as "intermediates" (Parr, 1938).

During the 1960s, scientists at EPA Laboratories were engaged in a survey of the natural distribution of coliform bacteria, in an evaluation of procedures for the differentiation of fecal and nonfecal origin. The results of this work are extremely significant because (1) rigid uniformity of laboratory methods was applied throughout; (2) massive numbers of cultures were examined; (3) a greater variety of environmental and biological sources was studied than in any previous project; and (4) samples were taken from a wide geographic range at all seasons of the year.

The results of these studies have indeed had important consequences for evaluating the efficacy of various testing procedures. It was indicated that there is a high correlation between known or probable fecal origin and the typical *E. coli* IMViC code (+ + − −). On the other hand, appreciable numbers of *Enterobacter aerogenes* and other IMViC types, which some regard as "nonfecal" segments of the coliform group, were found in human feces. The majority of coliforms attributable to fecal origin tend to be limited to a relatively small number of IMViC codes (+ + − −, − + − −, + − − −).

The distribution of coliform types from human sources should be regarded as a representative value for distributions taken from other sources. Investigations have indicated that there can be large differences in the distribution of IMViC types from person to person or even from an individual. Conversely, coliform bacteria recovered from such sources as undisturbed soil, vegetation, and insect life represent a wider range of IMViC codes without clear dominance of any one type. It was found that the most prominent IMViC code from nonfecal sources is the intermediate type, $-+-+$, constituting nearly half of the coliform cultures recovered from soils, and a high percentage of those from vegetation and insects. (Therefore, it appears that if any coliform segment could be termed a soil type it would be IMViC code $-+-+$.)

Since there is no known way to exclude the influence of limited fecal pollution from small animals and birds, it is not surprising that typical *E. coli* cultures are recovered in relatively small numbers from sources judged "unpolluted" by sanitary surveys. Thus, the presence of this small but demonstrable percentage of "fecal" coliforms in environmental sources is attributed to warm-blooded wildlife and not to gross pollution.

The results of these studies, based on a number of different criteria for differentiating between coliform of fecal origin and those from other sources, show that IMViC type $++--$ appears to be a reasonably accurate indicator of fecal cultures. The collective IMViC types, $++--$, $+---$, and $-+--$, not only signify cultures of probable fecal origin but also exclude most coliforms not present in large numbers in feces of warm-blooded animals.

Excellent correlation was found when the elevated temperature test was applied to isolates of human or highly probable fecal origin. The elevated temperature test yielded results equal to or better than those obtained from the IMViC code. Moreover, this test has the advantage in speed and ease of performance, yielding quantitative results for each water sample. It is, therefore, regarded as the method of choice for differentiation between coliforms of probable fecal origin and those of the normal microflora of soil and water.

The merits of the coliform group as indicators of pollution can be summarized: (1) The absence of these microorganisms is evidence of bacteriologically safe water; (2) the density of coliforms is generally proportional to the amount of fecal pollution present; (3) the presence of enteric pathogens in water is also accompanied by a greater density of coliforms; (4) coliforms are always present in the intestines of warm-blooded animals and are eliminated in large numbers in their feces; (5) coliforms survive longer than enteric pathogenic bacteria in the aquatic environments; and (6) coliforms are usually harmless to human beings and can be quantitated by routine laboratory procedures.

The limitations of this group are: (1) In addition to their occurrence in the intestines of warm-blooded animals, some of the members of the coliform bacteria are widely distributed in nature; (2) some strains of *Enterobacter* may multiply in polluted waters ("aftergrowth"), thereby compounding the difficulties of evaluating water quality and the age of the contaminant; (3) coliform determinations are subject to interference either by members of the *Pseudomonas* group or by synergistic conditions whereby gas is produced from lactose by two or more noncoliform organisms.

In discussing the merits of the fecal coliform component, it should be stated that (1) over 95 percent of the coliform bacteria isolated from intestines of warm-blooded animals ferment lactose, with the production of gas at the elevated temperature; (2) these organisms are found with relative infrequency in the environment, except in association with fecal pollution; (3) the survival time of fecal coliforms is shorter in the aquatic environment than for the coliform group as a whole and thus their presence is indicative of relatively recent pollution; (4) aftergrowth of fecal coliforms is not generally observed in the aquatic environment except in waters with concentrations of high carbohydrate wastes from sugar beet refineries and other food processing plants.

The necessary data for a comprehensive evaluation of fecal coliforms remain to be gathered, because they have been adopted as a standard test only within the past few years. No correlation has been established between ratios of total coliforms and fecal coliforms for determining sanitary quality of water. Likewise, little is known concerning the survival of fecal coliforms in polluted waters when compared with enteric pathogenic bacteria. It should be noted that approximately 3.6 percent of the coliforms isolated from feces of warm-blooded animals do not yield a positive fecal coliform test when the elevated temperature test is employed as the criterion of differentiation.

FECAL STREPTOCOCCI

Fecal streptococci are employed as bacterial indicators of fecal pollution (Mallmann and Seligman, 1950). In general, a fecal streptococcus may be defined as any streptococcus commonly found in high numbers in the feces of human beings and warm-blooded animals. The other terms used to designate these bacteria are "enterococci" and "group D streptococci." The former are characterized by specific biochemical reactions, and the latter by serological procedures. Although the fecal streptococcus, enterococcus, and group D streptococcus groups overlap, they are not synonymous, and since the emphasis here is on indicators of unsanitary environments, fecal streptococcus is deemed the more appropriate term.

A rigid definition of the fecal streptococcus group is not possible since present knowledge of this group is somewhat limited. It has been defined as gram-positive cocci; generally in pairs or short chains; growing in the presence of bile salts; usually capable of development at 45°C; producing acid but not gas in mannitol and lactose; failing to attack raffinose; failing to reduce nitrate to nitrite; producing acid in litmus milk and precipitating the casein in the form of a loose but solid curd; and exhibiting a greater resistance to heat, to alkaline conditions, and to high concentrations of salt than most vegetative bacteria. It must be underscored that streptococci departing from one or more criteria cannot be disregarded when isolated from water, even though some consider that growth both at 45° and in 40% bile is the most significant indication of fecal origin (Ministry of Health, 1956).

Streptococcus fecalis, S. fecalis var. *liquefaciens, S. fecalis* var. *zymogenes, S. durans, S. bovis, S. equinus, S. salivarious,* and *S. fecalis* biotypes, usually growing at 10°C, make up the fecal streptococcus group.

Although it is generally agreed that the presence of fecal streptococci, indicates pollution by warm-blooded animals, there is still some controversy over their distribution in nature (Medrek and Litsky, 1960; Litsky et al., 1955; Morris and Weaver, 1954; Mundt et al., 1962). Only 71 percent of samples from various mammalian species yielded enterococci, whereas reptiles showed an 83 percent incidence, and common rodents and wild birds yielded few isolations. Insects have been considered accidental mechanical carriers of fecal streptococcus, but Geldreich (1966) reported extremely high densities in some species.

The evidence to date indicates that pure water or virgin soil is free from fecal streptococci.[1] These organisms, unlike some of the coliforms, do not multiply in water. It is also believed that the fecal streptococci are of value in marine water analysis because of their high tolerance of salt.

The fecal streptococci are not considered pathogenic although some strains are known to be isolated from patients with endocarditis. A few outbreaks of foodborne infections have been alleged to be caused by fecal streptococci, but these conclusions were based on circumstantial evidence.

Great variation in the ratios of coliforms, fecal coliforms, and fecal streptococci has been reported. Since coliforms may originate in nonfecal sources, the fecal coliforms are a more reliable indication for comparison with the streptococcal densities. A comparison between the densities of fecal coliform and fecal streptococci in human and animal feces shows their ratios to be greater than 4:1 for man and less than 0.7 for animals.

Fecal coliform–fecal streptococcus ratios of greater than 4 are regarded as evidence of pollution from human origin, or pollution from mixed origin, the greater part of the latter being derived from human sources. A fecal coliform–fecal streptococcus ratio of less than 0.7 indicates that the pollution originated with livestock or poultry wastes. Such ratios are usually found near feed lots, stockyards, and stormwater runoffs. Ratios falling between 4.0 and 0.7 represent the "gray area" where interpretations must be made with caution. To be sure, a ratio of 3.5, for example, would be suggestive of human pollution, whereas a ratio of 0.9 would be more suggestive of animal origin.

Because of the differences in the rates of disappearance of members of the two bacterial groups, the greatest reliability of this test has been found in ratios taken for water samples not more than 24 h flow time from the pollution source.

In addition to the ratio of indicator organisms, it is often useful to determine the specific strain of fecal streptococci. The streptococcal population of human beings is made up of over 80 percent *Streptococcus fecalis* and the *S. salivarius-mitus* group. Cows and horses carry significant numbers of *S. bovis* and *S. equinus,* respectively, and the presence of these strains would *not* accompany human sewage. Fowl, however, seem to resemble people in their streptococcal flora. The *S. salivarius* group, a uniquely human group that dies out rapidly in water, is considered by some to be an index of recent pollution from human sources.

In the analysis of samples of untreated water, the presence of fecal streptococci may be construed as evidence of pollution by warm-blooded animals, even in those cases where the source and significance of the coliform group have been questioned. The presence of the streptococcus group should be interpreted as indicating that at least a part of the coliforms were of fecal origin. The absence of fecal streptococci, however, does not indicate that the water is bacteriologically safe until the survival rates have been determined.

---

[1] Recently, Mundt (1973) described a very simple method for differentiating fecal streptococci originating from human waste and those normally found on insects and various vegetation. His data indicate that isolates from fecal origin produced a hard, acid curd in litmus milk, while those from the latter group form a soft, rennet-like curd.

MICROBIAL INDICATORS OF INDUSTRIAL POLLUTANTS

At the species level, unlike the total count technique, quantitative data may easily be generated. As with fecal pollution in which the coliform and fecal streptococci groups are employed as indicators, specific industrial effluents may also be monitored by quantifying the response of the microbiota. *Lactobacillus* sp. and *Streptococcus lactis* are found in great numbers in dairy manufacturing effluents, as is *Klebsiella pneumoniae* in wood and pulp mill waste water. Similarly members of *Acetobacter* and *Saccharomyces* are predominant in the vinegar and brewing industries and may be used to monitor these effluents. Likewise, the many species of mold and Actinomycetes used in the production of antibiotics can be employed as indicators of pollution from this industry. In general, it may be said that any organism grown in great numbers in an industrial process can be used as the indicator for monitoring the effluent treatment prior to discharge.

There is still a third option, rapidly gaining favor among pollution biologists, which utilizes the community rather than the individual response. This ecological tack finds pertinence in the Federal Water Pollution Control Act Amendments of 1972 (PL 92-500) which now focuses on "the protection of aquatic life and wildlife by 1983." There is an extremely significant shift of emphasis here from acute to chronic toxicity. When the value of water extends beyond its utilization for agricultural, domestic, and industrial ends, to ecological concerns, criteria restricted to acute toxicity and standard bacteriological parameters no longer suffice. It is necessary to consider the response of the entire aquatic community when evaluating the more subtle, subclinical disruption exerted by chronic toxicity. The size and complexity of aquatic communities, however, restrict focus to those aspects of community structure and function amenable to measurement, yet sensitive to stress. Consequently many biologists infer environmental suitability from the species diversity of a particular sector of the aquatic community. One such segment has been the diatom populations of the periphyton microflora.

Species diversity consists of two partially independent facets: (1) number of species present (richness) and (2) the degree to which the organisms are apportioned among these species (evenness). Richness is a function of both the numbers and kinds of suitable habitats and the niche diversity. Evenness appears to be more immediately responsive to changes in the environment. Thus as habitats disappear or become altered, portions of the food web disintegrate, subdivide, or enlarge, and corresponding changes in species diversity are inevitable. Without sufficient time to adapt to shifts in environmental quality, a community will typically undergo a drop in diversity. Only those species preadapted to the altered media survive; fewer still are able to exploit the change. When unchecked by normal biological controls, these populations grow logarithmically.

The most widely used measure for diversity which reduces numbers and relative abundance to a single number running from about 5 to 0 was the Shannon-Wiener formula. Diversity measured by this index, however, has validity only if sampling is random, the population area is definable, and specimen numbers per species is high. Since microfloral monitoring is predicated on the use of suspended glass slides as artificial substrates, these preconditions cannot be satisfied. It is more correct to calculate diversity by Brillouin's probability formula

$$B = \frac{N!}{N_1 N_2! \cdots N_s!}$$

where B is the total diversity of a population of N organisms distributed among S species in the proportions $N_1, \ldots, N_s$. To compare the diversities of collections of different sizes, B is divided by N (collection size) to obtain H, the exact population value of diversity per individual. [Brillouin (1962)].

The reduction of so vast an amount of data to a single number has to result in a loss of information, but it does provide the engineer or management with a handle that would normally be obscured by a bewildering catalog of species. Biologically, it is at best an oversimplification, but it is a giant step for it utilizes an ecologically sound approach to interpret the community response to stress.

LITERATURE CITED

American Public Health Association (1971): "Standard Methods for the Examination of Water and Wastewater," 13th ed., Washington, D.C.

American Water Works Association (1950): "Water Quality and Treatment," 2d ed., New York.

Baughman, J. L. (1966): Oyster Production Decline Worries Maryland Oystermen, *South. Fisherman*, **6**:99.

Brillouin, L. (1962): "Science and Information Theory," 2d ed., Academic, New York.

Burm, R. J., and R. D. Vaughan (1966): Bacteriological Comparison Between Combined and Separate Sewer Discharges in Southeastern Michigan, *J. Water Pollut. Control Fed.*, **38**:400–409.

Carlucci, A. F., and D. Pramer (1960a): An Evaluation of Factors Affecting Survival of *Escherichia coli* in Sea Water. II. Salinity, pH, and Nutrients, *Appl. Microbiol.*, **8**:247–250.

—— and —— (1960b): An Evaluation of Factors Affecting Survival of *Escherichia coli* in Sea Water. IV. Bacteriophages, *Appl. Microbiol.*, **8**:254–256.

——, P. V. Scarpino, and D. Pramer (1961): Evaluation of Factors Affecting Survival of *Escherichia coli* in Sea Water. V. Studies With Heat- and Filter-sterilized Sea Water, *Appl. Microbiol.*, **9**:400–404.

Clark, H. F., and P. W. Kabler (1964): Reevaluation of the Significance of the Coliform Bacteria," *J. Am. Water Works Assoc.*, **56**:931–936.

Eijkman, C. (1904): Die Gärungsprobe bei 46° als Hilfsmittel bei der Trinkwasseruntersuchung, *Centralbl. Bakteriol., Abth. I, Orig.*, 37, 742.

Escherich, T. (1885): "Die Darmbakterien des Neugeborenen und Säuglings," *Fortschr. Med.*, **3**:515–547.

Gameson, A. L. H., and J. R. Saxon (1967): Field Studies on Effect of Daylight on Mortality of Coliform Bacteria, *Water Res.*, **1**:279–295.

Geldreich, E. E. (1966): Sanitary Significance of Fecal Coliforms in the Environment, *FWPCA Publ.* WP-20-3, U.S. Department of the Interior.

—— et al. (1962a): Type Distribution of Coliform Bacteria in the Feces of Warm-Blooded Animals, *J. Water Pollut. Control Fed.*, **34**:295–301.

—— et al. (1962b): The Fecal Coli-Aerogenes Flora of Soils from Various Geographic Areas, *J. Appl. Bacteriol.*, **25**:87–93.

—— et al. (1966): "Sanitary Significance of Fecal Coliforms in the Environment," U.S. Department of the Interior, *F.W.P.C.A. Publ.* WP-20-3, Washington, D.C.

Huet, M. (1941): "pH Values and Reserves of Alkalinity," *Commun. Sta. Rech. Groenendael*, no. 1; *Water Pollut. Abs.* 21 (November 1948).
Hutchinson, S. E. (1962): "A Treatise On Limnology," vol. I, Wiley, New York.
Lawrence, C. H. (1958): Sewage Effluent Dilution in Sea Water, *Water Sewage Works*, **105**:116–122.
LeClerc, J. and C. Dujanquier (1946): The Specific Action of Certain Ions in Water on the Quality of Beer, *Bull. Assoc. Ec. Brass. Univ. Louvain*, **42**:64.
Levine, M., R. H. Tanimoto, J. Arakaki, and G. Fernandes (1955): Simultaneous Determinations of Coliform and *Escherichia coli* Indices, *Appl. Microbiol.*, **3**:310.
Litsky, W., W. L. Mallmann, and C. W. Fifield (1955): Comparison of MPN of *Escherichia coli* and Enterococci in River Water, *Am. J. Public Health*, **45**:1949.
Mallmann, W. L., and E. B. Seligman, Jr. (1950): A Comparative Study of Media for Detection of Streptococci in Water and Sewage, *Am. J. Public Health*, **40**:286–89.
Medrek, T. F., and W. Litsky (1960): Comparative Incidence of Coliform Bacteria and Enterococci in Undisturbed Soil, *Appl. Microbiol.*, **8**:60–63.
Ministry of Health, London (1956): The Bacterial Examination of Water Supplies, *Rep. Public Health Med. Subjects*, **71**:34.
Morris, W., and R. H. Weaver (1954): Streptococci as Indices of Pollution in Well Water, *Appl. Microbiol.*, **2**:282–285.
Mundt, J. O. (1973): Litmus Milk as a Distinguishing Feature between *Streptococcus faecalis* of Human and Non-human Origins, *J. Milk Food Technol.*, **36**:363.
Mundt, J. O., J. H. Coggin, Jr., and L. F. Johnson (1962): Growth of *Streptococcus faecalis* var. *liquefaciens* on Plants, *Appl. Microbiol.*, **10**:552–555.
Orlob, Gerald T. (1956): Viability of Sewage Bacteria in Sea Water, *Sewage Ind. Wastes*, **28**:1147–1167.
Parr, L. W. (1938): Coliform Intermediates in Human Feces, *J. Bacteriol.*, **36**:1.
Perry, C. A., and A. A. Hajna (1944): Further Evaluation of EC Medium for the Isolation of Coliform Bacteria and *Escherichia coli*, *Am. J. Public Health*, **34**:735.
Phelps, S. B. (1948): "Public Health Engineering," Wiley, New York.
Prescott, S. C., C. E. A. Winslow, and M. McCrady (1946): "Water Bacteriology," Wiley, New York.
Thoesh, J. C., and J. F. Bede (1925): "The Examination of Water and Water Supplies," 3d ed., McGraw-Hill, New York.
U.S. Department of Health, Education, and Welfare (1962): "Public Health Service Drinking Water Standards," *U.S. Public Health Serv. Publ.* 956.
U.S. Department of the Interior (1968): "Report on the Committee on Water Quality Criteria," Washington, D.C.

13
PETROLEUM MICROBIOLOGY

Robert D. Schwartz
and
William W. Leathen

Petroleum microbiology has developed as a science within the last three-quarters of a century. It encompasses many disciplines, such as biology, chemistry, geology, physics, and engineering. Thus, a petroleum microbiologist should have an understanding of these sciences as they apply both to microbiology and to the petroleum industry. Petroleum chemists and engineers, generally, are concerned with the mechanics and chemical aspects of seeking and extracting crude oil from the earth and refining it into more usable chemical compounds. Although the petroleum microbiologists, chemists, and engineers have common interests in the same auxiliary sciences, they were, until recently, unacquainted with the others' field of endeavor. When they joined forces to solve some of the problems due to microorganisms which had arisen in the petroleum industry, it became apparent that the propagation of microorganisms on petroleum or petroleum fractions might also have certain advantages. The combined effort of several disciplines, coupled with environmental concerns, has accounted for the recent advances in petroleum microbiology.

It is the purpose of this chapter to (1) delineate, briefly, the history of petroleum microbiology; (2) describe environmental effects of microbes on petroleum and petroleum products; (3) describe, briefly, the biochemical aspects of microbial oxidation of hydrocarbons; and (4) present several beneficial aspects of petroleum microorganisms.

As the field of petroleum microbiology is large, the interested reader is referred to the numerous references listed, particularly the books by Davis (1967), Sharpley (1964), and Beerstecher (1954).

HISTORY

Historically, the development of petroleum microbiology may be divided into four eras: The first includes observations from earliest recorded history until 1920; the second is the period between World War I and World War II, approximately 1920 to 1940; the 20 years from 1940 to 1960 constitute the third; and the present period, beginning in 1960, continues at the writing of this chapter.

Early in recorded history the disappearance of oil from seepages emanating from the ground and spillages on water was observed. *Oil seeps*, or leakages from the earth, are present where petroleum-bearing geologic formations have been either ruptured or eroded. Crude oil, usually associated with gaseous hydrocarbons (methane, ethane, propane, and butane), exudes from the earth. Upon contact with topsoil, the oil soon disappears. Likewise, petroleum washed up on the shore from bodies of water disappears more rapidly than could be accounted for by chemical decomposition. After many years of observation, these phenomena came to be attributed to the microflora of the soil.

As observations of petroleum decomposition accumulated, it became clear that microbes were deeply involved. In 1895, Miyoshi provided the first report of a microorganism which grew on paraffin, a waxy constituent of crude oil. A mold was isolated and identified as *Botrytis cinera*. Ten years later, working independently, Kaserer and Sohngen reported that methane was oxidized by a bacterium. Sohngen identified the organism as *Bacillus methanicus*; currently the microorganism is called *Methanomonas methanica*. Although general interest in the degradation of petroleum continued into the early 1920s, little consideration was given to the chemical reactions brought about by the microbes; their characterization and identification received even less attention. It was in the next era that physiologic and taxonomic studies were carried out on these organisms.

The second period in the development of petroleum microbiology (1920–1940) is distinguished by a tremendous growth in the field of microbiology in general. Concurrently, advances were being made in fermentation technology. Although in use for centuries, fermentation procedures were being reviewed as commercially feasible processes. Citric acid, acetone-butanol, glycerol, and other fermentations were developed on an industrial scale. Except for citric acid, these processes succumbed to synthesis by the organic chemist when the petroleum industry entered the field of petrochemicals. Petroleum and petrochemical companies rapidly increased in size and number, as did their problems with microbes. Studies of metabolic mechanisms of hydrocarbon utilization were begun, and in some instances, the microorganisms were described and given specific epithets.

In this period, the petroleum industry first began to recognize the importance of microorganisms in some of their operations. Drilling muds were observed to "sour" or spoil. Microbes were involved, and the addition of inhibitors, or microbicides, effectively preserved the materials. Considerable attention was given to the microbiological analyses of injection water used in the secondary recovery of oil. Often, the bacterial content of the waters promoted plugging in oil-bearing geological formations. The bacteria produced slimes or gels, which clogged the pores of oil-bearing rock strata, preventing efficient recovery of oil. An unusually high incidence of dermatitis was observed among workers handling petroleum-based cutting oils in the machine metal industry. Some workers, it was found, developed a hypersensitivity to cutting fluids, causing skin irritations which were often followed by bacterial infections. The problem, traced to the microbial contamination of cutting oils, was alleviated by the addition of microbicides. Although little attention was given to the characterization and physiologic activities of the microbes concerned, the importance of microbiology in the petroleum industry was established.

Between 1940 and 1960, it became apparent to both interested academicians and industrialists that petroleum microbiology would be a fruitful field for investigation, from the viewpoints of pure and applied research. The educator-researcher was interested in the physiologic activities which permitted microbes to metabolize hydrocarbons; the petroleum technologist was concerned with methods to preserve or to protect petroleum products from microbial contamination. Accordingly, microbial metabolism was studied extensively and many metabolic pathways for the degradation of petroleum and petroleum products were delineated. (Foster, 1962; van der Linden and Thijsse, 1965). Many microorganisms, isolated from hydrocarbon habitats, were characterized and placed in already existing taxonomic systems. Indeed, the role of both bacteria and fungi in the petroleum field became well established. Concurrently, the petroleum industry was inquiring into the microbial contamination of certain products, especially fuel oils and fuels for jet aircraft. Bacteria and fungi were isolated from these hydrocarbon products and were found to contribute to the formation of sludges, which hindered efficient operation of domestic furnaces and of jet aircraft. This detrimental activity of microbes was a mammoth problem and was the cause of great concern in the petroleum industry. Needless to say, petroleum technologists were readily attracted to the field as the industry contributed financial support to microbiological research. Indeed many petroleum companies started their own microbiology laboratories to conduct both fundamental and applied research studies. Microbiology had become recognized as an integral part of petroleum science.

Research in petroleum microbiology, initiated to alleviate microbial problems which plagued the oil industry, and spurred by environmental concerns, soon extended to studies of the advantages of propagating microorganisms on hydrocarbons. Thus was launched the current period in the history of petroleum microbiology.

This includes environmental aspects such as controlling and dispersing oil spills, land reclamation and protection, and removing oil from ballast water; hydrocarbon fermentations for the production of useful products and cells (single-cell protein); oil recovery using biopolymers, microorganisms, and surfactants; energy production (methane, methanol); desulfurization; metals removal; and waste treatment.

ENVIRONMENTAL ASPECTS

MICROBIAL CORROSION

Microbial action may cause metallic corrosion by (1) producing corrosive metabolic products (inorganic and organic acids, sulfur, mercaptans, ammonia, oxygen, etc.); (2) producing differential aeration and concentration cells on the metal surface; (3) cathodic depolarization; (4) disruption of protective films; (5) breakdown of corrosion inhibitors (Iverson, 1968). The problem is of great practical importance to the petroleum industry as corrosion of fuel storage tanks, pipelines, drilling equipment, process machinery, etc., is costly.

Most work on microbial corrosion has been devoted to studying mechanisms of iron and steel deterioration in neutral, anaerobic environments. However, other materials, such as aluminum, copper, brass, and concrete are also susceptible to microbial corrosion, and the reactions may occur both aerobically and anaerobically. Under anaerobic conditions, for example, *Desulfovibrio* can reduce sulfate to sulfide as it oxidizes organic compounds to acetic acid. In the presence of nitrate, *Thiobacillus denitrificans* can oxidize the sulfide to sulfuric acid. If conditions become aerobic, other obligate chemolithotrophic thiobacilli can also oxidize the sulfide to sulfuric acid. Ferrobacilli, which are also chemolithotrophs, can oxidize ferrous to ferric iron. Note that corrosion can occur with or without deterioration of the product being stored and that the presence of water is necessary.

Storage tanks are often coated with protective polymers to prevent direct exposure of the metal or concrete to microorganisms, their products, and the material being stored. In addition, many compounds have been tested and used as both corrosion and microbial inhibitors, including inorganic chromates, nitrates, and nitrites, organoborons, phosphates, glycols, and numerous other organics. Unfortunately, many of these protective polymers and inhibitors are themselves subject to microbial deterioration. For example, nitrate, an inhibitor of aluminum (aircraft fuel tanks) corrosion, can be reduced to nitrite (ineffective) by hydrocarbon-oxidizing bacteria.

DETERIORATION OF PRODUCTS

The ability of some microorganisms to metabolize hydrocarbons often promotes deterioration of manufactured petroleum products, as evidenced by changes in viscosity and volatility, by discoloration, noxious odors, and changes in anticorrosive properties. Although many products are adversely affected by microbial action, only three examples will be discussed: asphalts, domestic fuel oils, and jet aircraft fuels.

Asphalts

Asphaltic materials are concentrated residues accumulated during the refining of petroleum. They comprise, principally, asphaltenes (compounds soluble in benzene but not in pentane) and resins, which contain large proportions of aromatic compounds. These materials are used in road building, pipeline coating, roofing material, and as agricultural mulch.

Soil associated with asphalt-base highways is a suitable habitat for the microorganisms, usually bacteria, which are capable of enhancing the deterioration of asphalt. Pseudomonads, flavobacteria, micrococci, and mycobacteria are all indicated as agents which accelerate the decomposition of asphalt roadways.

Pipelines are often coated with asphaltic material for protection against corrosion. Subsequently, bacteria capable of utilizing hydrocarbons develop in much greater numbers in soil adjacent to asphalt-coated pipe. In some areas where environmental conditions are especially favorable for bacterial growth, asphalt covering of pipe may deteriorate rapidly, exposing the metal of the pipe to moisture and subjecting it to both microbial and chemical corrosion. Actinomycetes and corynebacteria, as well as pseudomonads, flavobacteria, micrococci, and mycobacteria are involved in the degradation of asphaltic pipe coatings.

Recently, asphaltic mulches and films have been suggested for use in agriculture to retain moisture and to control weeds. Such asphaltic materials may be expected to decompose within a reasonable period of time. However, considerable research will be required to establish the useful life of such products and to assure the absence of residual material which might be deleterious to the soil.

Domestic Fuel Oils

Microbial contamination of domestic fuels, though recognized for many years, did not constitute a significant problem until the widespread use of furnace and diesel fuels in the mid-1950s. Expanded transportation and storage facilities to meet the demand for such fuels created many locations for accumulation of water, and ample habitats capable of supporting microbial growth were formed.

Water may be introduced into fuels from a variety of sources. At the refinery, fuels are washed with water, and though separation of the water and oil phases occurs, small quantities of water are transported with the fuel to storage tanks. In storage, separation continues by gravity, water accumulating in the bottom of the tank. Water may purposely be added to fuel to prevent loss of volatile hydrocarbons by seepage from the bottom of storage tanks. In some cases, fuel may be removed from storage tanks by displacement with water, rather than by directly pumping the product. Changes in temperature and weather cause condensation of atmospheric moisture in the headspace of storage tanks; droplets of water, thus formed, find their way into the fuel and either remain there or drop to the bottom of the storage tank.

Water in fuel lodges in dead spots in transport systems and in bottoms of storage tanks. Microorganisms proliferate at fuel-water interfaces, the degree of microbial activity depending in part upon the interfacial area. Although viable microorganisms predominate at the interface, often forming thick tough mats of cellular material, considerable numbers of microbes travel short distances into the fuel phase. In tanks recently disturbed by pumping or other mechanical action, microorganisms are distributed throughout the fuel; these gradually settle to the bottom of the tank. For the most part, mid-tank samples and well-settled fuels are relatively free of microorganisms. However, organisms which remain suspended in the fuel may enter the transportation system and clog filters, gauges, and small-bore tubing. A filter, removed from a domestic heating unit, is shown in Fig. 13-1 (Knecht and Watkins, 1963). A membrane,

Figure 13-1 Filter from a domestic heating unit. [From A. T. Knecht and F. M. Watkins (1963), *Dev. Ind. Microbiol.*, 4:15–23.]

approximately $\frac{1}{16}$ in thick, comprising microbes and microbial debris covered the screen. Many filamentous fungi and several species of bacteria were responsible for clogging the filter. Cultural studies indicated that the microorganisms were viable and capable of utilizing fuel components. A screen or filter of this sort provides an ideal surface for microbial activity. Nutrients, including oxygen, water, minerals, and organic matter, are carried into the screen by the unsettled fuel. Waste products, which normally would accumulate and inhibit microbial growth, are carried away as the fuel-water mixture passes through the filter. Thus, an ideal dynamic culture system is created, and large quantities of microbial cells and slime may be produced.

The principal detrimental effect of microbial sludge formation in domestic and diesel fuels is the inconvenience caused by shutdown of furnaces and diesel engines. Loss of petroleum products due to microbial utilization of constituents is probably negligible in a large storage facility.

Jet Aircraft Fuel ‡

Jet aircraft fuels are low-boiling kerosenes with boiling points in the range of 65 to 255°C and contain mostly aliphatic hydrocarbons with limited amounts of various cyclic compounds. Microbial activity in jet fuels causes the formation of sludge, deterioration of fuel tank linings, and corrosion of metals.

Microbiological sludge was first observed in jet aircraft fuel (JP-4) in 1956. Such deposits caused malfunctions of refueling equipment and were a potential hazard to efficient flight operations. It was found that both bacteria and fungi were present and grew prolifically in media containing mineral salts and jet fuel. The bacteria were identified as strains of *Pseudomonas, Enterobacter, Bacterium*, and *Bacillus*. Whereas the most prevalent fungi belonged in either the genus *Hormodendrum* or *Cladosporium*, other genera including *Penicillium, Aspergillus, Spicaria*, and *Helminthosporium* were represented.

Although many groups of bacteria and fungi may be isolated from jet fuels, some of the isolates are unable to remain viable; thus, the microbial inhabitants of jet fuels are considered to be composed of both indigenous and exogenous microbes. The indigenous group comprises microorganisms which inhabit and metabolize the fuel and may represent as much as 60 percent of the population (Leathen and Kinsel, 1963).

The microbial sludge, a gelatinous, filamentous deposit, originates at the fuel-water interface and is composed of microbial filaments and capsular material, often accompanied by debris which settles out of the fuel during storage. The capsular materials are usually high-molecular-weight polysaccharides, and the debris is often insoluble particles of dirt or metal.

The detrimental effects of microorganisms observed during processing, transportation, storage, and use of petroleum products are controlled, if not eliminated, by (1) use of protective linings, such as polysulfide polymers; (2) addition of microbial inhibitors to products; and (3) good housekeeping practices, such as protecting from undue contamination by water. It is essential that the protective liners and inhibitors are themselves not favorable substrates for microbial growth and are compatible with the petroleum product; i.e., the compounds must not interfere with the normal product properties such as combustibility, viscosity, or volatility (Bakanavskas, 1958; Hedrick et al. 1964; Walters and Elphick, 1968).

Note that the problem of controlling microbial activity is compounded by the presence of heterogeneous populations involving numerous synergistic interactions.

PETROLEUM AND PETROCHEMICAL SPILLS

In nature, wherever oil comes in contact with soil or water, the ingredients necessary for microbial metabolism usually are present. Although oil seepages must have existed for thousands of years, lakes of oil do not exist, partially oxidized petroleum derivatives are not found, and hydrocarbon concentrations in the atmosphere above natural oil seepages are low. Chemical oxidation and physical evaporation cannot alone explain the disappearance of the oil. Thus, failure of oil to accumulate can be explained only by constant attrition of the hydrocarbons by microbiological action. Similar phenomena occur in cases of man-made petroleum and petrochemical spills, be they from shipping disasters, oil-well "blowouts," deliberate release by tankers and cargo vessels, i.e., oil-contaminated ballast water, or process operations.

When oil is spilled in an aqueous environment a series of processes occur that alter its composition. Collectively, these processes are termed *weathering*. The first major weathering factor acting on the oil is evaporation. The rate and extent of evaporation are dependent on the type of oil, temperature, wind, solar radiation, and thickness of the oil. As evaporation proceeds, many of the lighter, more volatile components are lost and the remaining oil becomes more viscous. Depending on the oil and prevailing conditions, 20 to 60 percent can be lost by evaporation in less than 24 h. Some components of the oil undergo photochemical and oxidative reactions, and some components may dissolve (Regnier and Scott, 1975).

Although microbial oxidation is the main reason the oceans and beaches of the world are not covered with oil, the rate of degradation under natural conditions is usually slow. As outlined by Friede et al. (1972), organism availability, nutrient availability, and temperature are major factors limiting the rate of biodegradation. To overcome the low concentrations of hydrocarbon-oxidizing organisms naturally found in oil-free environments, seeding of oil slicks with suitable mixtures of organisms has been suggested. To be effective, major problems in producing, storing, and delivering large quantities of the active, mixed cultures required must be solved. In addition, many types of mixed cultures would be required to be effective on different petroleums and under different environmental conditions, i.e., temperature (arctic to tropical), salinity, (freshwater to brine), etc. Nutrient limitations, usually nitrogen and phosphorus, may also be overcome by adding these during the seeding. However, the nutrients would have to be in a form that adheres to the oil-water interface where the organisms are found.

As the organisms grow at the oil-water interface, they produce surfactants that emulsify and disperse the oil. The effect of dispersing the oil and the degradation products, as well as the organisms and their other metabolites, into the water column is of paramount importance. The dispersed material may more readily enter the food chain and be more dangerous than the slick itself. Major efforts are presently under way to determine the ecological fate and effect of oil. Thus use of hydrocarbon-oxidizing organisms and their products to protect shorelines in imminent danger of being damaged by an incoming oil slick, reclaiming oil-damaged land, removing oil from ballast water, and general petroleum-petrochemical pollution control are also being actively investigated (American Petroleum Institute, 1975).

OIL RECOVERY

When an oil well is drilled and oil is found, the initial recovery consists of that oil that comes to the surface because of pressure within the formation. As the rate of oil that is recovered decreases, additional wells are drilled and water (or steam) is injected to force more of the oil to the producing well and to the surface. These two stages are referred to as *primary* and *secondary recovery*, respectively, and may account for 40 to 60 percent of the oil in the formation. The remaining 40 to 60 percent is trapped in the porous rock structure of the oil-bearing formations.

It has been suggested that viable microorganisms be injected into the oil-bearing formation to aid in recovering this trapped oil. Presumably, the growing microorganisms would produce acids (to dissolve the rock) and/or surfactants (to decrease the oil's viscosity) as they metabolize part of the oil, thereby releasing the oil from the porous rock and increasing oil recovery.

Injecting nutrients into the well has also been suggested as a means of stimulating the growth of microorganisms indigenous to the oil-bearing formation. For these methods to be successful, problems of supplying nutrients, attaining the high cell concentrations and cell mobility needed in the formation, and preventing plugging of the porous rock and the oil well itself must be overcome.

A more recent method for recovering this trapped oil involves the injection of complex mixtures of surfactants and the use of viscosity enhancers to control the mobility of the oil through the porous rock and to the production well. These procedures are referred to as *tertiary recovery* procedures.

One of the more promising viscosity enhancers is a biopolymer produced by the bacterium *Xanthamonas campestris*. The polymer is a high-molecular-weight polyelectrolytic polysaccharide. It must retain its properties at temperatures up to 80°C (and in some cases as high as 150°C) and at salinities from 0 to 15 percent and must be protected from biodegradation. The polymer must be stable in underground formations for periods ranging from months to years.

Biopolymer production by microbial fermentation is one of the newer areas where microbiology is impacting on the petroleum industry. New and improved polymers are needed for tertiary recovery as well as for other applications such as drilling muds.

MICROBIAL OXIDATION OF HYDROCARBONS

Cultivation of microorganisms on petroleum substrates is not difficult, as evidenced by the hundreds (if not thousands) of organisms that have been isolated and grown on these substrates. Table 13-1 lists some of the genera of microorganisms reported to grow on various hydrocarbon substrates. The compounds oxidized range from the simpler aliphatic hydrocarbons (methane, n-alkanes, isoalkanes, olefins, cycloalkanes, phenylalkanes, etc.) to the simple as well as the more complex aromatic hydrocarbons (benzene, toluene and other substituted benzenes, naphthalene, anthracene, phenanthrene, etc.). Not all compounds are completely oxidized to CO_2 and H_2O.

Table 13-2 shows some examples of the metabolic pathways thought to operate in hydrocarbon-oxidizing microorganisms. Alkanes and alkenes can undergo monoterminal, diterminal, or subterminal oxidation, leading to a variety of products. Aromatic oxidation generally leads to a catechol intermediate before entering the major biochemical pathways. Details of the numerous pathways and the organisms that use them can be found in reviews by Markovetz (1972), Gibson (1972), Allen et al. (1971), Klug and Markovetz (1971), Kallio (1969), Nyns and Wiaux (1969), McKenna and Kallio (1965), Van der Linden and Thijsse (1965).

Many hydrocarbons that cannot be used for growth can be oxidized if present as a cosubstrate in a system in which a growth-supporting substrate (hydrocarbon or nonhydrocarbon) is being oxidized. The phenomenon is termed cooxidation (Raymond et al., 1971) and is useful in carrying out various biotransformations, including selective hydroxylations of drugs and hormones. Table 13-3 gives several examples. Note that, generally, the growth substrates are alkanes or the alkyl groups of substituted cyclic hydrocarbons, and the cooxidized compounds are complex, substituted, cyclic compounds. More will be

Table 13-1 Some genera of microorganisms reported to utilize petroleum or petroleum fractions for growth.

Bacteria	Filamentous fungi*
Corynebacteria	Absidia
Actinomyces	Acremonium
Micromonospora	Aspergillus
Streptomyces	Botrytis
Mycobacterium	Cephalosporium
Nocardia	Chaetomium
Methanomonas	Chloridium
Pseudomonas	Cladosporium
Acinetobacter	Colletotrichum
Micrococcus	Cunninghamella
Flavobacter	Dematium
Alkaligines	Epicoccum
Bacillus	Fusarium
Arthrobacter	Gliocladium
Achromobacter	Graphium
	Helicostylum
Yeast*	Helminthosporium
Candida	Monilia
Cryptococcus	Mucor
Endomyces	Oidiodendron
Hansenula	Paecilomyces
Mycotorula	Penicillium
Pichia	Rhizopus
Rhodotorula	Scolecobasidium
Torulopsis	Spicaria
Trichosporon	Syncephalastra
Saccharomyces	Trichoderma
Debaryomyces	
	Algae
	Prototheca
	Chlorella

*Yeast and filamentous fungi from M. J. Klug and A. J. Markovetz (1971): Utilization of Aliphatic Hydrocarbons by Microorganisms, *Adv. Microbiol. Physiol.*, 5:1–43.

said on this later, when products from petroleum and petrochemicals are discussed.

A common feature in these pathways is the initial incorporation of oxygen into the substrate. The source of the oxygen is molecular O_2, and the enzymes responsible are called *oxygenases* (Hayaishi, 1974). There are two major

Table 13-2 Some hydrocarbon oxidation pathways found in microorganisms

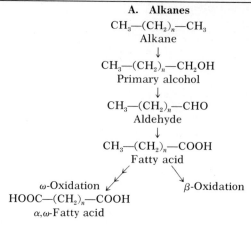

A. Alkanes

$CH_3-(CH_2)_n-CH_3$
Alkane
↓
$CH_3-(CH_2)_n-CH_2OH$
Primary alcohol
↓
$CH_3-(CH_2)_n-CHO$
Aldehyde
↓
$CH_3-(CH_2)_n-COOH$
Fatty acid

ω-Oxidation ↙   ↘ β-Oxidation
$HOOC-(CH_2)_n-COOH$
α,ω-Fatty acid

B. Alkenes

$CH_3-(CH_2)_n-CH=CH_2 \rightarrow CH_3-(CH_2)_n-CH\!-\!-\!CH_2$
$\backslash O /$

$HOOC-(CH_2)_n-CH=CH_2$ Alkene   1,2-Epoxide
 ↓   ↓
1-Alkeneoate $CH_3-(CH_2)_n-CH-CH_3$ $CH_3-(CH_2)_n-CH-CH_2$
| | |
OH OH OH
Secondary alcohol   1,2-diol
↓
$CH_3-(CH_2)_n-C-CH_3$
||
O
Methyl ketone

---

categories of oxygenases. Monooxygenases incorporate one atom of oxygen into the substrate and the other atom is reduced to water. Dioxygenases catalyze the incorporation of both oxygen atoms of O_2 into the substrate. Generally the monooxygenases are operative in the oxidation of paraffinic structures. The dioxygenases oxidize aromatic rings. The oxygenase systems found in hydrocarbon-oxidizing microorganisms are varied and complex, are multiprotein systems, usually require specific electron transport systems and cofactors, and are difficult to isolate. Nonetheless, several systems are yielding to intensive investigations, and the nature of oxyfunctionalization of hydrocarbons at the molecular level is beginning to emerge (Gunsalus et al., 1974; Coon et al., 1973; Jurtshuk and Cardini, 1972).

Success in this area may well lead to immobilized enzyme processes for carrying out useful and unique hydrocarbon transformations. These enzymes belong to the same group of enzymes responsible for drug and steroid hydroxylations in higher organisms, as well as in microorganisms.

Table 13-2 Some hydrocarbon oxidation pathways found in microorganisms (*Continued*)

C. Aromatics

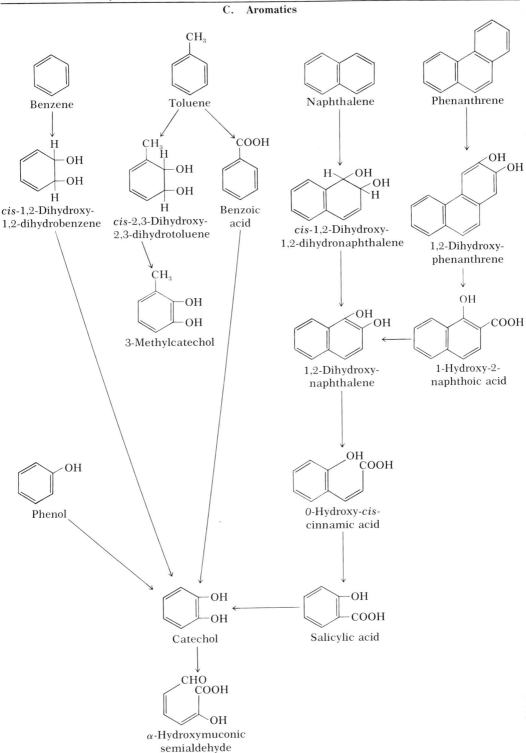

Table 13-3 Examples of biotransformations using the phenomenon of cooxidation

Microorganism	Growth substrate	Biotransformation substrate	Product
Pseudomonas methanica	Methane	CH_3-CH_3	Ethanol acetaldehyde acetic acid
Pseudomonas oleovorans	*n*-Octane	$CH_2=CH-(CH_2)_4-CH=CH_2$ 1,7-Octadiene	$CH_2=CH(CH_2)_4-CH-CH_2$ with epoxide O 7,8-Epoxy-1-octene $CH_2-CH(CH_2)_4-CH-CH_2$ with two epoxides 1,2-7,8-Diepoxyoctane
Arthrobacter sp.	Corn-steep liquor yeast extract	$CH_3-(CH_2)_{13}-CH_3$ $CH_3-(CH_2)_{14}-CH_3$	Pentadecanones Hexadecanols
Nocardia salmonicolor	*n*-Hexadecane	CH_2-CH_3 (on phenyl) Ethyl benzene	CH_2-COOH (on phenyl) Phenylacetic acid
Nocardia corollina	*n*-Hexadecane	2,6-Dimethyl naphthalene (with CH_3 groups)	6-Methyl-2-naphthoic acid (COOH and CH_3 on naphthalene)

Gram-negative rod (Job 5)	2-Methylbutane	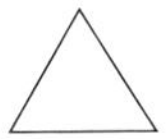 Cyclopropane	CH_3CH_2CHO Propionaldehyde
		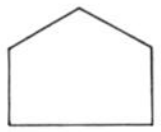 Cyclopentane	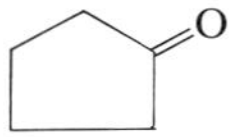 Cyclopentanone
Mycobacterium sp.	*n*-Decane	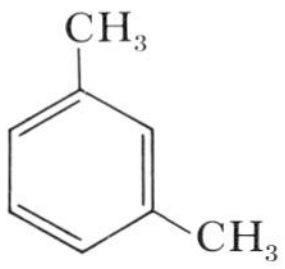 *m*-Xylene	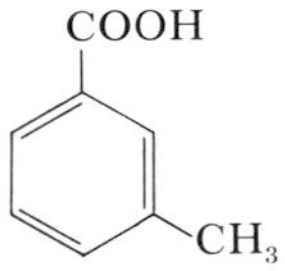 *m*-Toluic acid
Corynebacterium sp.	*n*-Hexadecane or glucose	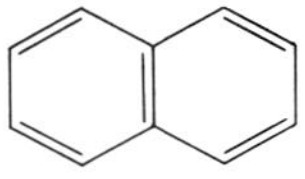 Naphthalene	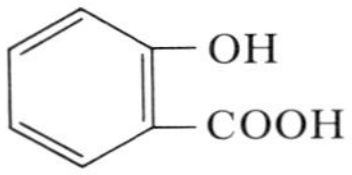 Salicylic acid

From time to time reports appear claiming the demonstration of anaerobic degradation of hydrocarbons. These claims have not been substantiated, and some have been retracted. The overwhelming evidence indicates that hydrocarbon oxidation is strictly an aerobic process.

MICROBIAL PRODUCTS FROM HYDROCARBON FERMENTATIONS

Abbott (1974a, 1974b) and Abbott and Gledhill (1971) compiled an exhaustive, comprehensive review, including the patent literature on the products from microbial hydrocarbon fermentations and biotransformations. They included the various organisms, product yields, and pathways used and restricted the use of the term "hydrocarbon" to mean those compounds containing only carbon and hydrogen. We will follow this same procedure but realize that many of the more than 350 known steroid hydroxylations are also accomplished by hydrocarbon-oxidizing microorganisms (Vezina and Rakhit, 1974).

First, let us look at some of the advantages and disadvantages of using hydrocarbons as substrates in microbial fermentations. Purity, availability, low density, high carbon content, and competitive cost make hydrocarbons attractive as fermentation substrates. Conventional sugar substrates, such as corn-steep liquor, molasses, and grains, often contain large amounts of extraneous material that can hinder the fermentation and product recovery and make for difficult-to-treat waste streams. The low solubility of many hydrocarbons facilitates separation of the aqueous and hydrocarbon phases following the fermentation and enhances product recovery by partitioning the water-soluble and hydrocarbon-soluble products. Hydrocarbons may also increase cell permeability and thereby increase cell and product yield. These advantages apply to situations in which the substrate is used for growth and the products are similar to those produced in conventional fermentations, i.e., antibiotics, amino acids, nucleic acids, sugars, polysaccharides, vitamins, organic acids, enzymes, etc.

Alternatively, the phenomenon of cooxidation and the availability of mutants blocked at specific points in hydrocarbon metabolic pathways, coupled with the wide array of hydrocarbons available, makes possible the quantitative synthesis of unique products by limited (one or two step) oxidations.

The major disadvantages of hydrocarbon fermentations are the large amount of heat evolved and the high oxygen demand. The cooling required due to excessive heat evolution may be minimized by using thermophilic microorganisms. New fermentation facilities, such as air-lift fermentors, designed to provide the oxygenation and mixing needed in hydrocarbon processes would also have to be constructed. Also, it is difficult and expensive to convert the inadequate conventional fermentation plants to facilitate hydrocarbon-based processes.

As presented by Abbott (1974a, 1974b) and Gledhill (1974), products from alkanes can be divided into two groups: (1) *biosynthetic products* synthesized via intermediates of alkane degradation, which are not structurally similar to the alkane substrate and are common cell metabolites similar to those produced in conventional fermentations; (2) *transformation products* in which most of the carbon skeleton of the product is derived intact from the alkane or alkene. Transformation products may be produced by cooxidation or by accumulation of early intermediates in the biodegradative pathway. Table 13-4 lists several of the

Table 13-4 Some biosynthetic and transformation products from alkanes*

Biosynthetic products

Amino acids	All	Nucleic acid-related
Organic acids	α-Ketoglutarate	Inosine
	Citric	Xanthosine
	Isocitric	Orotic acid
	Dipicolinic	Orotidine
	Fumaric	5′-Inosinic acid
	Anthronic	5′-Guanylic acid
	Corynomycolenic	5′-Adenylic acid
	Corynomycolic	Cyclic-AMP
	n-Docasanoic	Ornithine
	n-Ditriacontanoic	Citrulline
Enzymes		Lipase
		Alkaline protease
		Catalase
Vitamins and pigments	Biotin	
	Riboflavin	
	Vitamin B₁₂	Antibiotics: Chloramphenicol (precursor)
	Porphyrins	Carbohydrates: Glucose
	Carotenoids	Ribose
	Pyocyanine	Arabitol
	Vitamin B₆	Mannitol
	Gibberellin	Erythritol
	1-Phenazine-carboxylic acid	Trehalose
		Other: Polysaccharides
		L-Homoseryl-L-lysine

Transformation products

Acids	Alcohols	Ketones	Other
Acetic	Ethanol	Acetone	1,2-Epoxyoctane
Propionic	1-Decanol	2-Butanone	7,8-Epoxy-1-octene
Pentanoic	2-Decanol	2-Pentanone	1,2,7,8-Diepoxyoctane
Hexanoic	3-Decanol	2-Hexanone	1-Decene
Heptanoic	4-Decanol	2-Heptanone	cis-7-Hexadecene
Octanoic	5-Decanol	3-Decanone	cis-9-Octadecene
Nonanoic	2-Hexadecanol	4-Decanone	Acetaldehyde
Decanoic	3-Hexadecanol	5-Decanone	Trehalose glycolipid
4-Pentenoic	4-Hexadecanol	2-Pentadecanone	
10-Hydroxydecanoic	1,2-Hexadecanediol	3-Pentadecanone	
Adipic (hexanedioic)		4-Pentadecanone	
Suberic (octanedioic)			
Sebacic (decanedioic)			
Isovaleric			

*Some from B. J. Abbott (1974a, 1974b), Alkane and Alkene Transformation Products, and Biosynthetic Products from Alkanes, in A. I. Laskin and H. A. Lechevalier (eds.), "Handbook of Microbiology," vol. IV, pp.99–105, 107–115, CRC Press, Inc., Cleveland.

products produced in significant quantities. Details on organisms, specific alkane substrates, and product yields are found in the references cited above.

Of particular interest are the syntheses of glutamic acid and citric acid. Both products have long been produced by conventional carbohydrate fermentations. L-Glutamic acid is used as a flavor enhancer (monosodium glutamate). Citric acid is used as an acidulant in beverages, in effervescent salts, powders, and tablets, and may replace phosphates as sequesterants in detergents.

Recently, Liquichimica (Italy), using Japanese technology, commercialized a C_{11}–C_{18} alkane-based L-glutamic acid fermentation. Yields of 75 g/l are reported.

Alkane-based citric acid fermentations have been commercialized by Charles Pfizer Co. and Miles Laboratories (the latter company using Japanese technology). Yields as high as 112 g/l are reported.

The Japanese (Kyowa Hakko Kogyo) have also commercialized an alkane-based fermentation for producing a chloramphenicol precursor.

Biosynthetic products, as opposed to transformation products, from aromatic and cycloparaffinic hydrocarbons are few. These compounds are relatively toxic, and microorganisms generally grow on them slowly. Often, the alkyl chain of an alkylaromatic compound is used as growth substrate and the product is the remaining partially oxidized aromatic moiety. The major application of cyclic hydrocarbon fermentations is therefore biotransformation. Products may find use as medicinals, food additives, sequesterants, dye intermediates, perfume bases, biocides, organic polymer intermediates, and other specialty chemicals. Table 13-5 (also see Table 13-3) lists some major classes of products from aromatic and cyclic paraffinic hydrocarbons. Note that over 200 products have been reported from these substrates. None are yet commercialized using a microbial hydrocarbon-based process.

SINGLE-CELL PROTEIN[1] ‡

Single-cell protein, or SCP, is the name given to a variety of microbial products that are produced by fermentation. When properly produced, these materials make satisfactory proteinaceous ingredients for animal feed or human food. The production of protein from petroleum is by far the most advanced microbiological process in the petroleum industry. Most of the major petroleum companies, to various degrees, support or have supported research and development programs in this area, and two processes have been commercialized. Below we will (1) briefly review the history of SCP and its current status; (2) discuss some of the advantages and disadvantages of the major classes of microorganisms and the substrates used; (3) present a general process for SCP production; and (4) discuss the quality of the product. Our discussion will not be limited to hydrocarbons in the strict sense of the word or solely to petroleum company involvement.

The literature on SCP is huge. The interested reader is referred to recent texts by Davis (1974), Gounelle de Pontanel (1972), and Mateles and Tannenbaum (1968) for more details.

---

[1]The authors acknowledge and thank Dr. A. I. Laskin (Exxon Research and Engineering Co.) and D. D. MacLaren (Exxon Enterprises, Inc.) for providing much of the material used in the Single-Cell Protein section.

HISTORY AND CURRENT STATUS

The use of microorganisms for food has been known since ancient times. Yeast has been used in the baking and brewing industries for thousands of years. However, the first modern effort to manufacture microorganisms for food purposes was in Germany during World War I with the first commercial production of torula yeast. This production continued in the years between the wars and reached a level of over 15,000 tons a year for incorporation into human foods. Bacteria are commonly eaten in such foods as yogurt, sauerkraut, cheese, and sausage. Thus, the use of SCP for animal feed or human food is not such a major innovation as might appear.

The modern history of SCP began in the late 1950s when the petroleum industry began experiments on the use of microorganisms for removal of paraffin wax and sulfur-containing fractions from crude oil. In 1964, the British Petroleum Company announced success in cultivating a yeast, *Candida lipolytica*, on broad petroleum fractions, such as gas-oils and slack waxes, with the simultaneous removal of the paraffin component from the crude fraction. It was found that these organisms contained over 50 percent of high-quality protein, and so what started out as a by-product became the primary object of research to produce SCP.

As shown in Table 13-6, the technology for animal feed applications has progressed to the point where there are numerous demonstration and semicommercial plants in operation. There are also two full-scale commercial operations in Italy, and numerous announcements of other commercial plants have been made.

The technology for food-grade protein is not as far advanced (Table 13-7); only four companies have reached demonstration or semicommercial scale and several others, including Exxon-Nestlé, are working in pilot plants. Much more product testing and process and market development must be done, including clinical trials, before large-scale SCP operations for human food are undertaken.

Here it should be noted that it takes 6 lb of SCP to produce 1 lb of animal protein. Hence, direct human consumption is more efficient than first feeding and then eating the animal.

Why so much interest in SCP from petroleum, relative to other conventional feed and food sources? (1) Production is independent of the availability of arable land and the vagaries of weather; (2) SCP production requires relatively little water; (3) there is a broad, stable base of raw materials; (4) microorganism growth rates are much faster than plants or animals. For example, bacteria and yeasts can grow with mass doubling times of 20 to 120 min; grass and some plants require 1 to 2 weeks; cattle require 1 to 2 months. Put another way, 1000 lb of soybeans will produce about 100 lb of protein per day, whereas 1000 lb of bacteria could conceivably produce 1×10^{14} lb of protein per day; (5) continuous processing is possible.

MICROORGANISMS

As shown in Tables 13-6 and 13-7, bacteria, yeast, fungi, and algae are used in SCP processes. There are both advantages and disadvantages in using each of these microorganisms for animal or human consumption.

Table 13-5 Some substrates and products from aromatic and cycloparaffinic hydrocarbons

Substrate category		Example
Mononuclear:	Benzene	Benzene → cis-1,2-Dihydro-1,2-dihydroxy benzene
	Methylbenzenes	p-Xylene → p-Toluic acid
	Alkylbenzenes	n-Amylbenzene → Phenylacrylic acid (cinnamic acid)
Polynuclear:	Naphthalene	Naphthalene → 1,2-Naphthaquinone
		Naphthalene → Salicylic acid
	Anthracene	Anthracene → 2-Hydroxy-3-naphthoic acid
	Phenanthrene	Phenanthrene → 7,8-Benzocoumarin

Table 13-5 Some substrates and products from aromatic and cycloparaffinic hydrocarbons, (*Continued*)

Substrate category		Example
Mononuclear:	Benzene	
Cycloalkanes:	Monocyclic	Cyclohexane → Cyclohexanol
	Bicyclic	Decalin → Adipic acid
	Alkyl-substituted	Ethylcyclohexane → 4-Ethylcyclohexanol
Cycloalkenes:	Monocyclic	Cyclooctene → Cyclooctanone
	Bicyclic	Tetralin → α-Tetralol
Terpenes:		p-Cymene → Cumic acid
		1-p-Menthene → β-Isopropylpimelic acid

SOURCE: W. E. Gledhill (1974), Microbial Transformations of Cyclic Hydrocarbons, in A. I. Laskin and H. A. Lechevalier (eds.), "Handbook of Microbiology," vol. IV, pp. 45–98, CRC Press, Inc., Cleveland.

Table 13-6 Feed-grade SCP plants

Plant class	Company	Plant location	Substrate	Type of organism	Plant size, tons/year
Demonstration	British Petroleum	United Kingdom	n-Paraffin	Yeast	4000
	Chinese Petroleum	Taiwan	n-Paraffin	Yeast	1000
	Dianippon	Japan	n-Paraffin	Yeast	?
	ICI	United Kingdom	Methanol	Bacteria	1000
	Kanegafuchi	Japan	n-Paraffin	Yeast	5000
	Kohjin	Japan	?	Yeast	2400
	Kyowa Hakko	Japan	n-Paraffin	Yeast	1500
	Milbrew	United States	Whey	Yeast	5000
	Shell	Holland	Methane	Bacteria	1000
	Svenka-Socker	Sweden	Potato Starch	Yeast	2000
Semicommercial	British Petroleum	France	Gas oil	Yeast	20,000
	United Paper Mills	Finland	Sulfite Waste	Yeast	10,000
	USSR	USSR	?	Yeast	20,000
Commercial	British Petroleum	Italy	n-Paraffin	Yeast	100,000
	Liquichimica	Italy	n-Paraffin	Yeast	100,000
Other systems under consideration	LSU-Bechtel	......	Cellulose Waste	Bacteria	
	Tate & Lyle	......	Citric acid Waste	Fungi	
	ICAITI (Guatemala)	......	Coffee Waste	Fungi	
	IFP	......	CO_2-Sunlight	Algae	
	General Electric	......	Feed lot Waste	Bacteria	
	Mitsubishi	......	Methanol	Yeast	
	Finnish Pulp & Paper	......	Paper pulp Waste	Fungi	

Bacteria are usually high in protein (50 to 80 percent) and have rapid growth rates and an amino acid distribution superior to that of soy protein, and they are versatile. They can grow on a wide range of substrates, i.e., methane, methanol, ethanol, paraffins, etc. The principal disadvantages are small size and low density that makes harvesting from the fermentation broth difficult (more costly); high nucleic acid content relative to yeast and fungi; public acceptability.

All single-cell proteins contain varying amounts of nucleic acids. Although these materials are not harmful to animals, they can be detrimental to human beings, tending to increase the uric acid level in the blood. In extreme cases this can cause uric acid poisoning, or gout; therefore the nucleic acid level must be reduced to 1 to 3 percent if the SCP is to be used for human food. Obviously, this adds an additional processing step and increases the cost.

Table 13-7 Food-grade SCP plants

Demonstration plants				
Company	Location	Substrate	Organism	Plant size, tons/year
Amoco Foods	United States	Ethanol	Yeast	5000
Boise Cascade	United States	Sulfite waste	Yeast	6000
St. Regis Paper	United States	Sulfite waste	Yeast	5000
Slovnaft-Kojetin	Czechoslovakia	Ethanol	Yeast	1000

Other systems under consideration		
Company	Substrate	Type of organism
Exxon-Nestlé	Ethanol	Bacteria
Dianippon	Molasses	Yeast
RHM-DuPont	Starch and carbohydrates	Fungi
Kraftco	Whey	Yeast

The public, in general, is conditioned to thinking of all "bacteria" as harmful, producing disease. This is erroneous, but extensive education programs will be required before bacterial protein gains acceptance by the general public. A similar problem exists with public acceptance of "food" from petroleum.

Yeast have as advantages their larger size (easier to harvest), lower nucleic acid content, long history of use as a food, high lysine content, and ability to grow at acid pH. Disadvantages include lower growth rates, lower protein content (45 to 65 percent), and lower methionine content than bacteria. Lysine and methionine are essential amino acids.

Filamentous fungi have advantages in ease of harvesting and texture (functional properties) but have their limitations in lower growth rates, lower protein content, and acceptability. Major limitations of algae are their cellulostic cell walls (not digested by human beings) and their propensity for concentrating heavy metals.

SUBSTRATES

Although many substrates have been tried, we will look at four that have received the widest attention: n-paraffins, methane, methanol, and ethanol.

Normal paraffins were initially selected by many organizations for their particular SCP processes. Both bacteria and yeasts can be propagated on a wide range of n-paraffins, and the material is available and in high purity. Limitations include the very low solubility of shorter-chain paraffins (up to about eight carbon atoms in length) and insolubility of higher ones, high oxygen demand, and a large amount of heat evolution during the fermentation. The latter requires costly methods for cooling the fermentor. A high oxygen demand means that more energy is required to supply the oxygen needed; this too is costly. The same is true for low solubility, as energy is required to disperse the substrate in order to make it available to the microorganism.

Methane has advantages of low cost, availability (particularly in the Middle

East where billions of cubic feet are "flared" daily because of the inability to store, transport, and/or use the gas at the production site), purity, and restricted utilization by microorganisms (helps to control contamination). In addition, methane is readily separated from the fermentation broth. Disadvantages include its low solubility, explosion hazard, high oxygen demand, excessive heat evolution, lower growth rates achieved, and limitations in the type and desirability of the organisms that will utilize it.

Methanol has several advantages including low cost, availability, high purity, low oxygen demand, low heat evolution, and ease of storage and handling. In addition, methanol is readily miscible with water. Limitations include the limited number of organisms that can utilize methanol (can be an advantage), volatility, and toxicity to the organism at higher concentrations.

As a fermentation substrate, ethanol is not only available and in high purity but is soluble and accepted by the public. Its major disadvantage is its relatively high cost.

With a wide selection of organisms and substrates and a knowledge of their advantages and disadvantages, the choice of a particular substrate and microorganism will be determined by the monetary burden which the process can tolerate and still be economical.

PROCESS

The elements of a typical SCP process are shown schematically in Fig. 13-2. The heart of the process is the fermentor. There are several types, including conventional stirred, draft tube, air lift, and combinations of these. The inputs to the fermentor include water, the substrate on which the organism grows, oxygen, nutrients, and NH_3 to supply nitrogen and to control the pH. All these streams must be sterilized to prevent contamination of the culture by extraneous

Figure 13-2 Elements of a typical single-cell protein process.

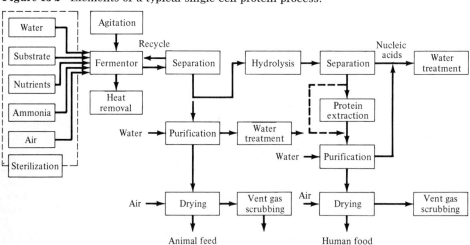

organisms. This can be done either by heating and/or by appropriate filtration techniques.

The fermentation reaction is exothermic, requiring the removal of heat in order to maintain the temperature at the optimum level for maximum growth rate. Since the fermentation is usually conducted at temperatures of less than 45°C, this heat removal is costly because of the small temperature differences possible with most cooling water sources. In fact, refrigeration may be required. The limiting step in the growth of the organism is the rate of oxygen transfer from the air entering the fermentor. This rate can be increased by breaking up the air bubbles to increase the surface for oxygen transfer. Thus, some form of agitation or power input is another important cost factor.

The fermentation is usually carried out continuously. A portion of the fermentor broth is continuously withdrawn to a separator where the supernatant is preferably recycled back to the fermentor in order to reduce sterilization costs and waste treatment requirements. The ease of this separation depends upon the type of organism and whether it has been treated to increase its coagulation ability. Centrifuges are usually used for this separation.

The concentrated biomass leaving the centrifuge may be recovered directly and, after washing, either spray dried or roller dried for use in animal feed. However, as previously mentioned, if the SCP is to be used for human food, it is necessary to remove most of the nucleic acids and, in addition, it is desirable to increase the functionality of the product for optimum food applications. This posttreatment involves a variety of techniques to solubilize the nucleic acids by hydrolysis and, subsequently, to solubilize various amounts of the protein by cell wall rupture. The treated biomass is then recovered by centrifugation and the supernatant is sent to waste treatment, another costly step because of low solids concentration. The concentrated food product is then washed and again centrifuged before being spray dried to prepare the final product.

PRODUCT

Single-cell protein basically comprises proteins, fats, carbohydrates, ash ingredients, water, and other elements such as phosphorus and potassium. The specific composition depends on the particular organism and the feedstock on which it grows. Some typical compositions are compared in Table 13-8 with soy and fish meal. What is normally termed "crude protein" is determined by measuring total nitrogen content, which includes true protein as well as nucleic acids. As shown, the true protein content can range from less than 40 percent in the case of fungi and algae to about 65 percent for bacteria.

If SCP is to be used successfully, there are five key criteria to be met: safety, nutritional value, acceptability, functionality, economics. Although these criteria are more stringent for human food use (and our discussion below will concentrate on this), they are also important for feed use.

Obviously, SCP must be safe to eat. Besides reducing nucleic acids, the product must be free of possible toxic substrates, endotoxins, etc. Safety can be established only by conducting extensive toxicological tests on animals (acute and chronic) followed by clinical trials.

Nutritional value must be high, is dependent on the chemical (amino acid)

Table 13-8 Compositions of representative single-cell proteins

Organism	Feedstock	Composition, wt %				
		Crude protein	True protein	Lysine	Methionine	Fats
Bacteria	Methanol	80	65	5.8	2.2	8
	Methane	60	50	4.3	3.0	10
Yeast	n-Paraffin	60	53	7.4	1.8	9
	Gas oil	69	60	7.8	1.6	2
	Ethanol	54	45	6.7	1.5	6
Fungi	Carbohydrate	35–50	30–40	6.5	1.5	5
Algae	CO_2	45–60	40–50	4.6	1.4	5
	Fish meal	60–65	50–60	7.0	2.6	5–10
	Soy meal	45–50	40–45	6.5	1.4	1.5

composition of the SCP, and is established by animal feeding tests. Generally, SCPs are higher in quality than their soy counterparts, having superior amino acid profiles.

Acceptability of SCP will be dependent on many factors. We have mentioned the psychological problems associated with eating "bacteria." There are also religious and sociological problems to be overcome, cost and convenience, as well as organoleptic characteristics to be met, i.e., flavor, appearance, mouth feel, aftertaste, chewiness, etc.

If SCP is to be introduced into such common staples as hamburger, hot dogs, bread, doughnuts, cakes, soups, spreads, or soft drinks, it must have the "functionality" to provide at least the same characteristics as the proteins that it replaces. As with soy protein, there is a whole range of functionalities that must be achieved with various forms of flour, concentrates, and high-protein isolates. These include coagulability, extrusability, spinnability, gelability, solubility, color, texture, water binding, fat binding, cohesiveness, and adhesiveness. No single form can provide the wide variety of properties desired by the food technologist.

Considerations such as political factors, competitive factors, organism, substrate, processing, functionality, and acceptability will affect the economic viability of SCP processes and are extraordinarily complex and changing daily. Commercialization of SCP is just beginning. The economic viability of these first operations in competition with soy and fish meals rests on limited volume, premium value markets such as milk replacer in Europe, government incentives, and/or low-cost feedstocks such as may be available in the oil-producing countries. Long-term viability will depend on improvements in technology, reducing costs, and worldwide producibility limits on soy and fish meals, raising their prices.

The use of SCP as a food ingredient is still in the final stages of research and development with only small demonstration plants currently operating. These operations produce first-generation whole cell products which, although wholesome and nutritious, have low functionality, limiting their use to relatively low-value food applications competing with soy flour and concentrate. For maximum utility and ultimate economic viability, technology must be developed to produce a full range of protein ingredients, including concentrates and isolates, analogous to what has been done with soy. Process and market development work in this direction is encouraging, particularly because of the very high-quality protein which SCPs contain. Good yields of functional and spinnable materials have been produced on a pilot scale, indicating the possibilities for a family of products, including meatlike analogs, which should be competitive with soy.

Although SCP has the potential to be a major source of highly functional protein, economic viability is yet to be demonstrated.

CONCLUDING REMARKS

We have tried to present the diversity of microbial involvement in the petroleum industry. A few specific areas have been discussed, and those only briefly. However, several areas have been omitted: refinery and chemical plant waste treating, including oily sludges and waste oils; microbial prospecting; microbial processing of coal, shale, and oil sands; microbial desulfurization; microbial hydrocarbon biosynthesis and petroleum biogenesis; microbial production of fuels (methane, methanol, ethanol, hydrogen); microbial or biochemical fuel cells; microbial effects on detergents (Bird and Lynch, 1974; Hammond et al., 1973; Eckenfelder and Ford, 1970; Davis, 1967; Beerstecher, 1954).

An immense task faces petroleum microbiologists who seek to explore the ability of microorganisms to carry out individual reactions in the nearly unlimited field of hydrocarbon chemistry. The task is one for team research. The microbiologist must seek the aid of and cooperate with experts in other disciplines. The expertise of the bioanalytical chemist, the bioengineer, the biochemist, the nutritionist, and many others must be relied upon to meet the challenge of the future. With a team effort, detrimental microbial reactions can be brought under control and the beneficial aspects of petroleum microbiology may be fully realized.

LITERATURE CITED

Abbott, B. J. (1974a): Alkane and Alkene Transformation Products, in A. I. Laskin and H. A. Lechevalier (eds.), "Handbook of Microbiology," vol. IV, pp. 99–105, CRC Press, Inc., Cleveland.

———(1974b): Biosynthetic Products from Alkanes, in A. I. Laskin and H. A. Lechevalier (eds.), "Handbook of Microbiology," vol. IV, pp. 107–115, CRC Press, Inc., Cleveland.

———and W. E. Gledhill (1971): The Extracellular Accumulation of Metabolic Products by Hydrocarbon Degrading Microorganisms, *Adv. Appl. Microbiol.*, **14**:249–388.

Allen, J. E., F. W. Forney, and A. J. Markovetz (1971): Microbial Subterminal Oxidation of Alkanes and Alk-l-enes, *Lipids*, **6**:448–452.

American Petroleum Institute (1975): Proceedings of Joint Conference on Prevention and Control of Oil Pollution, Washington, D.C.

Bakanavskas, S. (1958): Bacterial Activity in JP-4 Fuel, U.S. Air Force, *WADC Tech. Rep.* 58-32, Wright-Patterson Air Force Base, Ohio.

Beerstecher, E. (1954): "*Petroleum Microbiology*," Elsevier, New York.

Bird, C. W., and J. M. Lynch (1974): Formation of Hydrocarbons by Microorganisms, *Chem. Soc. Rev.*, **3**:309–328.

Coon, M. J., A. P. Autor, R. F. Boyer, E. T. Lode, and H. W. Strobel (1973): On the Mechanism of Fatty Acid, Hydrocarbon and Drug Hydroxylation in Liver Microsomal and Bacterial Enzyme Systems, in T. E. King, H. S. Mason, and M. Morrison (eds.), "Oxidases and Related Redox Systems," vol. 2, pp. 529–553, Wiley, New York.
Davis, J. B. (1967): "Petroleum Microbiology," Elsevier, New York.
Davis, P. (ed.) (1974): "Single Cell Protein," Academic, New York.
Eckenfelder, W. W., and D. L. Ford (1970): "Water Pollution Control," Jenkins Publishing Co., New York.
Foster, J. W. (1962): Bacterial Oxidation of Hydrocarbons, in O. Hayaishi (ed.), "Oxygenases," pp. 241–271, Academic, New York.
Friede, J., P. Guire, R. K. Gholson, E. Gaudy, and A. F. Gaudy (1972): Assessment of Biodegradation Potential for Controlling Oil Spills on the High Seas, United States Coast Guard Department of Transportation, Project DAT 204-72.
Gibson, D. T. (1972): The Microbial Oxidation of Aromatic Hydrocarbons, *CRC Crit. Rev. Microbiol.*, **1**:199–223.
Gledhill, W. E. (1974): Microbial Transformations of Cyclic Hydrocarbons, in A. I. Laskin and H. A. Lechevalier (eds.), "Handbook of Microbiology," vol. IV, pp. 45–98, CRC Press, Inc., Cleveland.
Gounelle de Pontanel, H. (ed.) (1972): "Proteins from Hydrocarbons," Academic, New York.
Gunsalus, I. C., J. R. Meeks, J. D. Lipscomb, P. Debrunner, and E. Munck (1974): Bacterial Monooxygenases—the P_{450} Cytochrome System, in O. Hayaishi (ed.), "Molecular Mechanisms of Oxygen Activation," pp. 559–613, Academic, New York.
Hammond, A. I., W. D. Metz, and T. H. Maugh II (1973): "Energy and the Future," American Association for the Advancement of Science, Washington, D.C.
Hayaishi, O. (ed.) (1974): "Molecular Mechanisms of Oxygen Activation," Academic, New York.
Hedrick, H. G., C. E. Miller, J. E. Halkias, and J. E. Hildebrand (1964): Selection of a Microbiological Corrosion System for Studying Effects on Structural Aluminum Alloys, *Appl. Microbiol.*, **12**:197–200.
Iverson, W. P. (1968): Mechanisms of Microbial Corrosion, in A. H. Walters and J. J. Elphick (eds.), "Biodeterioration of Materials," pp. 28–43, Elsevier, New York.
Jurtshuk, P., and G. E. Cardini (1972): The Mechanism of Hydrocarbon Oxidation by a *Corynebacterium* Species, *CRC Crit. Rev. Microbiol.*, **1**:239–289.
Kallio, R. E. (1969): Microbial Transformations of Alkanes, in D. Perlman (ed.), "Fermentation Advances," pp. 635–648, Academic, New York.
Klug, M. J., and A. J. Markovetz (1971): Utilization of Aliphatic Hydrocarbons by Micro-organisms, *Adv. Microbiol. Physiol.*, **5**:1–43.
Knecht, A. T., and F. M. Watkins (1963): The Influence of Microorganisms on the Quality of Petroleum Products, *Dev. Ind. Microbiol.*, **4**:17–23.
Leathen, W. W., and N. A. Kinsel (1963): The Identification of Microorganisms That Utilize Jet Fuel, *Dev. Ind. Microbiol.*, **4**:9–16.
Markovetz, A. J. (1972): Subterminal Oxidation of Aliphatic Hydrocarbons by Microorganisms, *CRC Crit. Rev. Microbiol.*, **1**:225–237.
Mateles, R. I., and S. R. Tannenbaum (eds.) (1968): "Single-cell Protein," M.I.T., Cambridge, Mass.
McKenna, E. J., and R. E. Kallio (1965): The Biology of Hydrocarbons, *Ann. Rev. Microbiol.*, **19**:183–208.
Nyns, E. J., and A. L. Wiaux (1969): Biology of Hydrocarbons, *Agricultura (Louvain)*, **17**:3–56.
Raymond, R. L., V. W. Jamison, and J. O. Hudson (1971): Hydrocarbon Cooxidation in Microbial Systems, *Lipids*, **6**:453–457.
Regnier, Z. R., and B. F. Scott (1975): Evaporation Rates of Oil Components, *Environ. Sci. Tech.*, **9**:469–472.

Sharpley, J. M. (1964): *Tech. Doc. Rep.* ASD-TDR-63-752. Wright-Patterson Air Force Base, Ohio.
Van der Linden, A. C., and G. J. E. Thijsse (1965): The Mechanisms of Microbial Oxidations of Hydrocarbons, *Adv. Enzymol.,* **27**:469–546.
Vezina, C., and S. Rakhit (1974): Microbial Transformation of Steroids, in A. I. Laskin and H. A. Lechevalier (eds.), "Handbook of Microbiology," vol. IV, pp. 117–441, CRC Press, Inc., Cleveland.
Walters, A. H., and J. J. Elphick (eds.) (1968): "Biodeterioration of Materials," Elsevier, New York.

GENERAL REFERENCES

Gunsalus, I. C., and V. P. Marshall (1972): Monoterpene Dissimilation: Chemical and Genetic Models, *CRC Crit. Rev. Microbiol.,* **1**:239–289.
Jones, A. (1974): "World Protein Resources," Wiley, New York.
Jones, J. G. W. (ed.) (1973): "The Biological Efficiency of Protein Production," Cambridge University Press, New York.

14
ELEMENTS OF HEAT AND GASEOUS STERILIZATION
Charles R. Stumbo

Sterilization as discussed here refers to the means employed in the use of certain lethal agents to free materials of microorganisms that might reproduce in or on such materials or in or on materials to which the microorganisms might be transferred. It will not necessarily refer to freeing a material of all viable microorganisms; that is, a material may be sterilized with respect to one or more species but not with respect to all, pasteurization of milk being a good example. Loss of the power of cell reproduction will be considered microbial death, and dead cells will be considered destroyed. From the practical standpoint, this connotation is quite satisfactory, because it is how a microbial cell affects its environment that is of primary concern.

Presently, there are several agents employed to sterilize materials used in, or produced by, industry. It is the purpose of this treatment to discuss considerations involved in determination and evaluation of processes employed in the use of the more prominent of these lethal agents; namely, moist heat, dry heat, and ethylene oxide in the vapor phase. It is believed that such considerations will at least touch upon elements involved in sterilization by almost any other lethal agent. To further confine the discussion, primary consideration will be given to those microorganisms that are generally most resistant to lethal agents, that is, bacteria; a material freed of viable bacteria by the application of a lethal agent is usually freed of other microorganisms.

ORDER OF DEATH OF BACTERIA

Death of bacteria subjected to moist heat has been discussed in detail by Rahn (1945a, 1945b) and Stumbo (1973). It was pointed out by Stumbo that death of bacteria subjected to moist heat is generally logarithmic, but that certain factors may cause apparent deviations from this order. These same things are true with respect to death of microorganisms subjected to the other lethal agents considered here. Available evidence overwhelmingly indicates that death of cells in a pure culture when exposed to any of these agents is logarithmic, that is, decrease in number of viable cells is an exponential function of exposure time (see Fig. 14-1). Moreover, available evidence indicates that observed deviations of semilogarithmic survivor curves from straight lines are generally due to influences other than resistance or susceptibility of a particular microbial species to lethal effects of the sterilants. Therefore, in the treatments to follow, death of a single species upon exposure to the lethal agents under consideration will be assumed to be strictly logarithmic.

Microbial death being logarithmic allows it to be described mathematically in the same manner as a unimolecular or a first-order bimolecular chemical reaction. Only one substance reacts in a unimolecular reaction, and its rate of decomposition is directly proportional to its concentration; decomposition of phosphorous pentoxide is a good example. In a first-order bimolecular reaction one reactant is in such great excess that variation in its concentration is negligible, and rate of decomposition of the second reactant is directly proportional to its concentration; hydrolysis of sucrose, when water is present in great excess, is a good example.

As pointed out by Stumbo (1973), this may be expressed mathematically as follows:

$$-\frac{dC}{dt} = kC \tag{14-1}$$

or

$$-\frac{dC}{C} = k\, dt \tag{14-2}$$

where C = concentration of decomposing reactant
k = proportionality factor
$-dC/dt$ = rate at which concentration decreases

Integrating Eq. (14-2) between limits of concentration C_1 at time t_1 and concentration C_2 at time t_2, we obtain

$$-\int_{C_1}^{C_2} \frac{dC}{C} = k \int_{t_1}^{t_2} dt \tag{14-3}$$

and $-\ln C_2 - (-\ln C_1) = k(t_2 - t_1)$ \hfill (14-4)

or

$$k = \frac{\ln C_1 - \ln C_2}{t_2 - t_1} = \frac{2.303}{t_2 - t_1} \log \frac{C_1}{C_2} \tag{14-5}$$

If we consider $t_2 - t_1$ as the time increment, Eq. (14-5) may be written as follows:

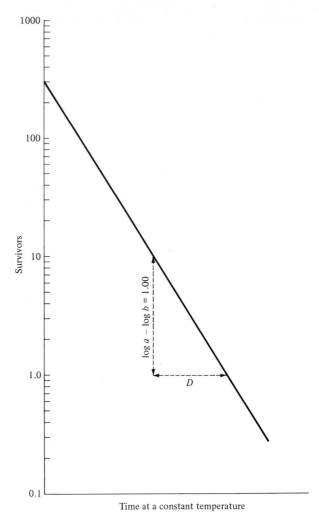

Figure 14-1 Logarithmic survivor curve expressed by the equation
$$t = D(\log a - \log b)$$

$$k = \frac{2.303}{t} \log \frac{C_0}{C} \tag{14-6}$$

or $\quad t = \dfrac{2.303}{k} \log \dfrac{C_0}{C} \tag{14-7}$

where C_0 = initial concentration of reactant
C = concentration after reaction time t

The microbial survivor curve is depicted in Fig. 14-1. If the initial number of cells is represented by a [comparable to C_0 in Eq. (14-7)] and the number of surviving cells after exposure time t is represented by b [comparable to C in Eq. (14-7)],

$$t = \frac{2.303}{k} \log \frac{a}{b} \tag{14-8}$$

It will be noted from Fig. 14-1 that the time required to destroy 90 percent of the cells is the time required for the curve to traverse one log cycle. This decimal reduction time is represented by the symbol D (Katzin et al., 1943). It is obvious from Fig. 14-1 that the slope of the survivor curve may be expressed as

$$\frac{\log a - \log b}{D} = \frac{1}{D} \tag{14-9}$$

Substituting in the equation of a straight line,

$$y = mx \tag{14-10}$$

we obtain

$$\log a - \log b = \frac{1}{D} t \tag{14-11}$$

or $\quad t = D (\log a - \log b) = D \log \dfrac{a}{b} \tag{14-12}$

where t = time of exposure to lethal agent
$\quad\quad D$ = time required to destroy 90 percent of cells
$\quad\quad a$ = initial number of viable cells
$\quad\quad b$ = number of viable cells after exposure time t

In comparing Eqs. (14-8) and (14-12), it becomes obvious that

$$D = \frac{2.303}{k} \tag{14-13}$$

and that both k and D represent the slope of the survivor curve. As we proceed it will become obvious that D is the more convenient term to use in comparing resistance of different microorganisms.

Although D is a convenient parameter for describing resistance of a given microbial species to some lethal agent, or for comparing resistance of different species, another resistance parameter is necessary for calculating and evaluating temperature-dependent sterilization processes, if different temperatures are involved in the total process. Different species and strains of microorganisms vary in their relative resistance to different temperatures and also in their relative resistance to a chemical in the vapor phase applied at different temperatures. The parameter employed to account for this difference in relative resistance is z, which will be discussed in the next section.

DESTRUCTION CURVES

Thermal destruction (TD) curves describe the relative resistance of microorganisms to different lethal temperatures. Thermochemical destruction (TCD) curves describe the relative resistance of microorganisms to a given concentration of a chemical (in this case, in the vapor phase) applied at different temperatures. In either case, the destruction curve is constructed by plotting the logarithm of D in the direction of ordinates against exposure temperature in the direction of abscissae (see Figs. 14-2 and 14-3). Over the range of concern in any given sterilization process, TD and TCD curves are essentially straight lines.

The term z, employed in the calculation and evaluation of sterilization processes to account for relative resistance, is generally taken as equal, numerically, to the number of degrees Fahrenheit required for the TD or TCD curve to traverse one log cycle. If the Celsius temperature scale is preferred, the value of

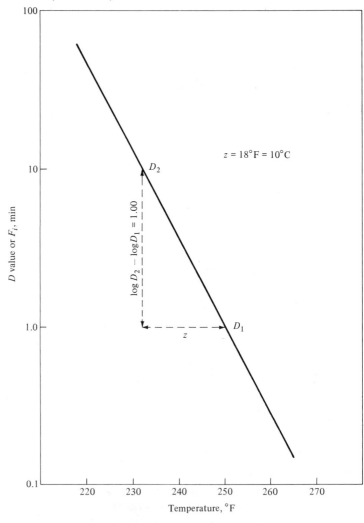

Figure 14-2 Thermal destruction curve passing through 1 min at 250°F (moist heat).

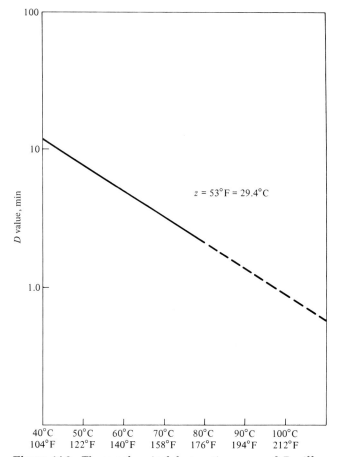

Figure 14-3 Thermochemical destruction curve of *Bacillus subtilis* spores exposed to a mixture of ethylene oxide and dichlorodifluoromethane. [ETO (ethylene oxide) concentration 700 mg/l; relative humidity, 33 percent.] [From Tien szu Liu et al. (1968), Dichlorodifluoromethane–Ethylene Oxide Mixture as a Sterilant, *Food Technol.*, **22**:86.]

z may be expressed in degrees Celsius required for the TD or TCD curve to traverse one log cycle. In either case, z represents the slope of the curve because it is mathematically equal in value to the reciprocal of the slope.

Referring to Fig. 14-2 or Fig. 14-3 and considering a portion of either that traverses one log cycle, it is obvious that the slope may be expressed as follows:

$$\frac{\log D_2 - \log D_1}{z} = \frac{1}{z} \tag{14-14}$$

The general equation of the TD or TCD curve may be conveniently written

$$\log D_2 - \log D_1 = \frac{1}{z}(T_1 - T_2) \tag{14-15}$$

where $D_2 = D$ value corresponding to temperature T_2 and time to destroy 90 percent of cell population at temperature T_2

$D_1 = D$ value corresponding to temperature T_1 and time required to destroy 90 percent of the cell population at temperature T_1

Q_{10} AND z

In the literature it will be found that both z (in degrees Fahrenheit) and Q_{10} (in degrees Celsius) are employed to represent the slope of the straight line obtained by plotting logarithms of death times against temperatures. As in comparing D with k, z appears to be the more convenient term to use in calculating and evaluating sterilization processes. However, for the sake of clarity, the relationship between z and Q_{10} should be understood. Q is defined as a quotient describing how much more rapidly death proceeds at a Celsius temperature T_2 than it does at a lower Celsius temperature T_1. For comparative purposes, it is common to give the coefficient for an increase of 10°C, that is, for $T_2 - T_1 = 10°C$.

The relationship between z and Q_{10} was pointed out by Rahn (1945a) and Stumbo (1973). It can best be demonstrated by starting with Eq. (14-7), that is,

$$k = \frac{2.303}{t} \log \frac{C_0}{C}$$

Now let k_2 be the death rate constant at temperature T_2, and k_1 the death rate constant at temperature T_1, which is lower than T_2, and t_2 and t_1 the corresponding reaction times. Then,

$$k_2 = \frac{2.303}{t_2} \log \frac{C_0}{C} \tag{14-16}$$

and $$k_1 = \frac{2.303}{t_1} \log \frac{C_0}{C} \tag{14-17}$$

Dividing Eq. (14-16) by (14-17), we obtain

$$\frac{k_2}{k_1} = \frac{t_1}{t_2}$$

This ratio is the temperature coefficient for n degrees equal to $T_2 - T_1$. Then

$$Q_n = Q_1{}^n = \frac{t_1}{t_2}$$

in which

Q_1 = coefficient for 1°C

Then

$$Q_{10} = Q_1{}^{10}$$

By definition,

$$n = 0.555z°C \quad \text{and} \quad \frac{t_1}{t_2} = 10$$

Therefore,

$$Q_1{}^n = Q_1{}^{0.555z} = \frac{t_1}{t_2} = 10$$

and $0.555z \log Q_1 = \log 10 = 1$

or $\quad \log Q_1 = \dfrac{1}{0.555z} \quad$ (14-18)

Since $Q_{10} = Q_1{}^{10}$

$$\log Q_{10} = 10 \log Q_1 = \frac{10}{0.555z} = \frac{18}{z}$$

and $\quad z = \dfrac{18}{\log Q_{10}}$

when z is expressed in degrees Fahrenheit. To convert z in degrees Fahrenheit to z in degrees Celsius, simply multiply by the factor 0.555.

BASIC CONSIDERATIONS IN DETERMINING STERILIZATION REQUIREMENTS

Sterilization requirements are usually based on reduction in the number of viable cells of the most resistant microbial species likely to be encountered. Spores of many bacterial species are far more resistant to most sterilants than are vegetative forms and spores of other microorganisms. Therefore, in most applications, sterilization requirements are based on some given reduction in the number of the more resistant bacterial spores. (This is especially true for dry-heat and vapor-phase sterilization, wherein the objective is usually to eliminate all microbial forms.) However, if the objective is to sterilize a material or object with respect to pathogenic microorganisms, with no concern for the reduction in nonpathogens, then the resistance of the more resistant pathogen, or pathogens, would logically be the basis for establishing sterilization specifications.

Regardless of whether sterilization requirements are based on the resistance of the more resistant microorganisms present or on the resistance of less resistant forms, the basic principles of establishing sterilization specifications are identical. The resistance parameters D and z, if sterilization is temperature-dependent, characterizing resistance of the key microorganism to be eliminated must be known or estimated. The initial number of cells of the key microorganism contaminating the unit to be treated must be known or estimated. Also, the chance of survival which will be acceptable must be decided. These things are essential to the calculation and evaluation of sterilization processes because death of microorganisms exposed to the sterilants under consideration is logarithmic and the time for any degree of reduction of viable cells is described by Eq. (14-12), that is,

$$t = D(\log a - \log b)$$

It should be pointed out that, in accordance with this equation, when the value of b is less than 1, it represents a probability of survival per unit volume of material considered. Any quantity of material may be considered as unit volume, that is, 1 ml, 100 ml, 10 gal, or whatever, as long as volumes containing a and b are equal. It follows that, whatever is considered as unit volume, there is no treatment that will sterilize an infinite number of such volumes; the logarithmic curve never reaches zero. For example, let us assume that there is initially one spore of *Clostridium botulinum* per container of 10^{12} containers of a food material to be heat-treated and that unit volume is considered the volume of the 10^{12} containers. Let us assume further that the resistance of these spores in the material being considered is characterized by a D_{250} of 0.21. The heat treatment given all containers is equivalent in sterilizing capacity to 2.52 min at 250°F. What is the most probable number of spores that would survive in the 10^{12} containers?

Since

$$t = D(\log a - \log b)$$
$$2.52 = 0.21(\log 10^{12} - \log b)$$
$$12 = 12 - \log b$$
$$\log b = 0$$
$$b = 1$$

This tells us that one spore would most probably survive in the 10^{12} containers or that most probably one container would not be sterilized. It should be realized, however, since we are dealing with probability, that there is a certain chance that all 10^{12} containers would be sterilized and, conversely, there is a certain chance that two, three, or more spores would survive and that these might occur in one, two, three, or more containers. There would never, of course, be more nonsterile containers than surviving spores. The above computation describes only the most probable event.

Now suppose we had taken as unit volume the volume of one container. Then

$$2.52 = 0.21(\log 1 - \log b)$$
$$\log b = -12$$
$$b = 10^{-12}$$

This tells us that there is 1 chance in 10^{12} that any particular container of the 10^{12} would not be sterilized; again, that in all 10^{12} containers there would most probably be one survivor or one nonsterile container.

Let us now consider the probability of survivors in the 10^{12} containers if they are divided into four equal groups, that is, $10^{11.398}$ containers per group. The probability of survival in the 10^{12} containers would equal the sum of the probabilities of survival in the four different groups. To test this, let us first compute the probability of survival in each of the four groups if they are all given the same process, that is, a heat process equivalent to 2.52 min at 250°F.

For each group

$$2.52 = 0.21(\log 10^{11.398} - \log b)$$
$$12 = 11.398 - \log b$$
$$\log b = -0.602$$
$$b = 0.25 = \text{most probable number of survivors in one group}$$

Then in the four groups

$$b = 0.25 \times 4 = 1.00 = \text{sum of probable survivors in four groups}$$
$$= \text{probable survivors in all } 10^{12} \text{ containers}$$

Now let us consider that each of the four groups is given a different severity process than that given any other, say the equivalent of 2.52, 2.31, 2.10, and 1.89 min at 250°F, respectively.

Probable survivors in first group:

$$2.52 = 0.21(\log 10^{11.398} - \log b)$$
$$b = 0.25$$

In second group,

$$2.31 = 0.21(10^{11.398} - \log b)$$
$$b = 2.50$$

In third group,

$$2.10 = 0.21(10^{11.398} - \log b)$$
$$b = 25.0$$

In fourth group,

$$1.89 = 0.21(10^{11.398} - \log b)$$
$$b = 250$$

Now in the four groups we would expect

$$0.25 + 2.50 + 25.0 + 250 = 277.75 \text{ survivors} = \text{number of survivors to be expected in all } 10^{12} \text{ containers}$$

It is all too often the tendency, when process severity variation occurs, say from operation to operation, to average the lethal values of processes actually given in order to determine survival probability. That this is an erroneous and dangerous procedure may be readily shown. Taking the above process, for example, we find the average of the four lethal values to be 2.205. Considering this to be the representative process given all 10^{12} containers, we find

$2.205 = 0.21(\log 10^{12} - \log b)$
$b = 31.8$ as the number of survivors predicted

that is, only about 11.5 percent of the 277.75 which would actually survive the four different processes.

If the effective mean sterilizing value of a number of processes is desired, the number of survivors from each process should be determined and the sum of survivors from all taken as the value of b and substituted into the equation of the survivor curve. For example, for the four processes considered above,

$b = 277.75$
and $t = 0.21(\log 10^{12} - \log 277.75)$
$t = 2.010 =$ effective mean

This is in contrast to the arithmetic mean of 2.205.

The equation of the survivor curve is applicable for determining sterilization time under any prescribed set of conditions. Accounting for temperature dependence of lethal effectiveness of chemical vapors will be discussed later. In application of the above equation, the resistance parameter D may be determined experimentally, or estimated through knowledge of what might be expected based on previously determined resistance of the organism of concern. However, it is often difficult to determine what is the organism of chief concern because the most resistant organism may be present in very low numbers and a less resistant one may be present in very high numbers. Because sterilization time depends on a as well as D it often happens, with microbial species with resistances that are not greatly different, that the organism of chief concern is the species with the lower resistance. By way of example, assume that we are dealing with spores of two species, and that the D_{300} (dry heat) of the more resistant one has been determined to be 2.36 and that of the less resistant one to be 2.14. Further, assume that the number of spores of the more resistant one, per unit to be sterilized, is 10, and the number of spores of the less resistant is 10,000. Which one is the organism of chief concern? Suppose that the probability of survival sought is 10^{-5}, the value of b in either case. Then for the organism of higher resistance,

$t = 2.36(\log 10 - \log 10^{-5})$
$t = 2.36 \times 6 = 14.16$

Whereas, for the organism of lower resistance,

$t = 2.14(\log 10,000 - \log 10^{-5})$
$t = 2.14 \times 9 = 19.26$

In heat sterilization practice, it is usual procedure to take the value of a as the total number of the more resistant organisms; generally the total number of spores that can resist a "pasteurization" treatment (e.g., 20 min at 80°C) is employed. Then if the D value characterizing the most resistant organism likely to be encountered is used in the calculation, the sterilization time computed will be such as to reduce the probability survival to, or beyond, that sought. This is the safest and most practical procedure. Sometimes the total microbial cell count is employed as the a value; however, going to this extreme is totally unnecessary because, generally speaking, the large majority of the total microbial cell population will be of low resistance compared with that of the minority.

As pointed out before, in temperature-dependent sterilization reactions, the equation

$$t = D(\log a - \log b)$$

is applicable only for computing sterilization time for an article exposed to some one given constant temperature. If the sterilization process involves a number of different influencing temperatures, which is usually the case, the effects of the different temperatures may be integrated to obtain the total effect. The procedure of choice for accomplishing this will be explained fully later.

RESISTANCE OF MICROORGANISMS TO MOIST AND DRY HEAT

There appears to be little or no correlation between resistance of different microbial species to dry and moist heat. However, there is one general similarity: Bacterial spores are more resistant to both dry and moist heat than are vegetative cells. Generally speaking, bacterial spores are several hundred times more resistant to dry heat than they are to moist heat. Also, z values characterizing relative resistance to different dry-heat temperatures are much higher than those characterizing relative resistance to moist-heat temperatures. Comparative values observed for some of the most resistant bacterial species are given in Table 14-1.

It has been postulated by some that the large difference in z may indicate a difference in mechanism of action of dry and moist heat; no one knows the mechanism of action of either. If death by moist heat should be in accordance with the theory of Rahn (1929, 1945b), the large difference in z would not necessarily indicate a difference in mechanism of action. Rahn proposed that loss of reproductive power of a bacterial cell when subjected to moist heat is due to the denaturation of one gene essential to reproduction. He reasoned that, since the death of bacteria resembles a unimolecular or first-order bimolecular reaction, death of a single cell must be due to the denaturation of a single molecule, and, since the size of a gene (Fricke and Demerec, 1937) is that of a small protein molecule, a gene would consist of only one or two molecules.

If this is the correct explanation for death by moist heat and if denaturation of the molecular gene is due to a "fatal blow" by one or more surrounding vibrating molecules, it is feasible that the same explanation might apply to death by dry heat. With moist heat the moisture content of the cell should allow for great freedom for molecular vibration, the intensity of which should depend for the most part on temperature alone; as the temperature increases the intensity of vibration increases, accounting for increase in lethality with increase in temperature.

With dry heat, the cell is dehydrated; the higher the temperature, the greater is the dehydration. Dehydration would restrict freedom for molecular vibration; the greater the dehydration, the greater is the restriction. (This is in contrast to a moist-heat environment in which the content of cell water should remain virtually constant.) This greater dehydration with increase in temperature might well account for the larger z values characterizing relative resistance to dry-heat temperatures.

Regardless of what the mechanism of action is, and whether it is the same for moist and dry heat, the fact remains that the order of death of bacteria subjected to either moist or dry heat is essentially logarithmic. This fact, in the words of Rahn (1945a), " . . . permits us to compute *death rates* and to draw conclusions from them which are independent of any explanation. Death rates make it possible to compare the heat resistance of different species at the same temperature, or the heat resistance of one species at different temperatures." Such comparisons are now expressed through the resistance parameters D and z; see Table 14-1.

Knowing the D value at one temperature and the z value, comparative sterilization times (if instantaneous heating and cooling are assumed) for any other temperatures may be computed. By way of example, taking the D_{350} value and the z value given in Table 14-1 as characterizing the resistance of *Bacillus subtilis* spores to superheated steam, the D_{300} value may be computed as follows:

$$\log D_2 - \log D_1 = \frac{1}{z}(T_1 - T_2) \qquad (14\text{-}15)$$

$$\log 0.57 - \log D_1 = \frac{1}{42}(300 - 350)$$

$$D_1 = 0.95 = D_{300}$$

Assume that sterilization has been defined as a $12D$ reduction in population. Then the sterilization time at 350°F may be computed as follows:

$$t = D_{350}(\log 10^{12})$$
$$t = 0.57 \times 12 = 6.85 \text{ min}$$

And the sterilization time at 300°F may be computed as follows:

$$t = 0.95 \times 12 = 11.4 \text{ min}$$

The procedure, though it is useful in showing the comparative effectiveness of different lethal temperatures, is not adequate for calculating and evaluating sterilization processes wherein an infinite number of temperatures are existent during the process, as is always the case when time is required in heating and cooling the article being sterilized. The procedure here involves integration of the lethal effects of all temperatures existent during the process. Since the same procedure applies in calculating and evaluating vapor-phase sterilization processes, it will be described in detail later.

RESISTANCE OF MICROORGANISMS TO ETHYLENE OXIDE

As in comparing resistance of microorganisms to moist and dry heat, there is apparently little correlation between the resistance of microorganisms to ethyl-

Table 14-1 Heat resistance characteristics of a number of bacterial species (spores only)

Species	Moist heat*		Dry heat		
	D_{250}	z, in °F	D_{350}[†]		z, °F
Bacillus subtilis	...	...	(SHS)	0.57	42[‡]
B. subtilis	...	...	(N + He)	0.17	31[‡]
B. subtilis	...	...	(A, O_2CO_2)	0.13	31[‡]
B. stearothermophilus	4.0	18	(SHS)	0.14	26[§]
B. polymyxa	0.005	14	(SHS)	0.13	28[§]
Clostridium botulinum	0.20	15	(A)	0.20	61[¶]
P.A. 3679	1.50	18	(SHS)	0.13	60[§]
P.A. 3679	...	...	(He)	0.44	39[‡]
P.A. 3679	...	...	(A)	0.30	39[‡]

*Sources of data: Charles P. Collier and Charles T. Townsend (1956), The Resistance of Bacterial Spores to Superheated Steam, *Food Technol.*, **10**:477; H. D. Michener et al. (1959), Search for Substances Which Will Reduce the Heat Resistance of Bacterial Spores, *Appl. Microbiol.*, **7**:166; C. R. Stumbo et al. (1950), Nature of Thermal Death Time Curves for P.A. 3679 and *Clostridium botulinum*, *Food Technol.*, **4**:321; C. R. Stumbo et al. (1964), A Procedure for Assaying Residual Nisin in Foods, *J. Food Sci.*, **29**:859.

[†]Heating media:
    SHS: superheated steam
    N: nitrogen
    He: helium
    A: air
    O_2: oxygen
    CO_2: carbon dioxide

[‡]From data by I. J. Pflug (1960, 1963), Thermal Resistance of Microorganisms to Dry Heat: Design of Apparatus, Operational Problems and Preliminary Results, *Food Technol.*, **14**:483; personal communication.

[§]From data by Charles P. Collier and Charles T. Townsend (1956), The Resistance of Bacterial Spores to Superheated Steam, *Food Technol.*, **10**:477.

[¶]From data by F. W. Tanner and G. M. Dack (1922), *Clostridium botulinum*, *J. Infect. Dis.*, **31**:92.

ene oxide and to either dry or moist heat. This is not surprising since death by ethylene oxide is due to chemical activity, probably the alkylation of protein cellular components (see Fraenkel-Conrat, 1944; Phillips, 1949). Evidence (Brookes and Lawley, 1961; Shull, 1963; Michael and Stumbo, 1969) is rapidly accumulating to indicate that microbial death is primarily due to the alkylation of DNA, treatment with ethylene oxide impairing or destroying the microbial cell's ability to replicate DNA. When DNA replication ceases, cell multiplication of course ceases, at least beyond one generation (Lawley and Brookes, 1968; Michael and Stumbo, 1969). Such a mechanism of action would explain the logarithmic nature of death of microorganisms subjected to the action of ethylene oxide.

A number of factors influence the efficacy of ethylene oxide as a sterilant. Most influential are ethylene oxide concentration in the sterilant atmosphere, relative humidity of the sterilant atmosphere, temperature, and possibly the nature of "inert" diluent gases employed. In regard to the latter factor, most work with pure ethylene oxide indicates that the optimum relative humidity for

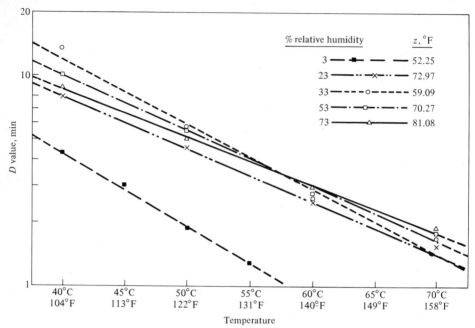

Figure 14-4 Thermochemical destruction time curves for *Clostridium botulinum* spores suspended in distilled water, deposited on paper discs, preconditioned at specified relative humidities, and exposed to Oxyfume Sterilant 12 at the same relative humidities. [From L. N. Kuzminski et al. (1969), Factors Influencing the Death Kinetics of *Clostridium botulinum* 62A on Exposure to a Dichlorofluoromethane–Ethylene Oxide Mixture at Elevated Temperatures, *Food Technol.*, **34**:561–567.]

lethal action is between 30 and 35 percent (Kaye and Phillips, 1949; Phillips and Kaye, 1949; Bruch 1961; Gilbert et al., 1964) whereas, for a mixture of ethylene oxide (12 percent) and dichlorodifluoromethane (88 percent) the optimum relative humidity for lethal action appears to be of the order of 5 percent (Kuzminski et al., 1969). The reason for this is not yet known. The influence of relative humidity, in the case of an ethylene oxide–dichlorodifluoromethane mixture (Oxyfume Sterilant 12), is illustrated in Fig. 14-4. For the same mixture, the influence of temperature is illustrated in Fig. 14-5. Table 14-2 gives D and z values characterizing the death of spores of three different bacterial species exposed to an ethylene oxide–dichlorodifluoromethane mixture. It will be noted that spores of *Bacillus subtilis* are slightly more resistant than spores of *Clostridium botulinum* and, at the lower temperatures, considerably more resistant than spores of *B. coagulans*.

RESISTANCE MEASUREMENT

Numerous methods have been developed for measuring the resistance of microorganisms subjected to the lethal agents under discussion.

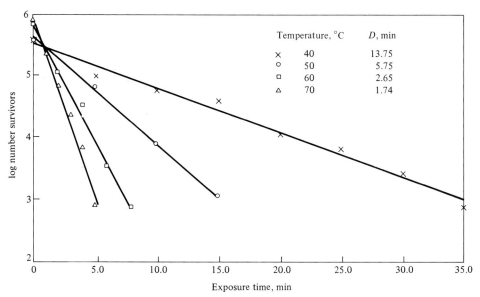

Figure 14-5 Survivor curves for *Clostridium botulinum* spores suspended in distilled water, deposited on paper discs, preconditioned at 33 percent relative humidity, and exposed to Oxyfume Sterilant 12 at specified temperatures (ETO concentration, 700 mg/l; relative humidity, 33 percent.) [From L. N. Kuzminski et al. (1969), Factors Influencing the Death Kinetics of *Clostridium botulinum* 62A on Exposure to a Dichlorofluoromethane–Ethylene Oxide Mixture at Elevated Temperatures, *Food Technol.*, **34**:561–567.]

Table 14-2 Resistance parameters for spores of three species of bacteria exposed to an ethylene oxide–dichlorodifluoromethane mixture. Ethylene oxide concentration, 700 mg/l; relative humidity 33 percent

Species	D at 30°C 86°F	40°C 104°F	50°C 122°F	60°C 140°F	70°C 158°F	80°C 176°F	z
Bacillus subtilis, (ATCC 9524)	...	17.0	7.70	3.55	1.62	0.74	53°F 29.4°C
Clostriduim botulinum (Type A-Str. 62)	...	12.2	6.14	3.00	1.48	...	59°F 32.8°C
Bacillus coagulans	10.9	7.00	4.51	2.89	...	...	94°F 53.3°C

SOURCE: Tien szu Liu et al. (1968), Dichlorodifluoromethane-Ethylene Oxide Mixture as a Sterilant, *Food Technol.*, **22**:86; L. N. Kuzminski et al. (1969), Factors Influencing the Death Kinetics of *Clostridium botulinum* 62A on Exposure to a Dichlorodifluoromethane-Ethylene Oxide Mixture at Elevated Temperatures, *Food Technol.*, **34**:561–567; and D. F. Blake and C. R. Stumbo (1970), Ethylene Oxide Resistance of Microorganisms Important in the Spoilage of Acid and High-Acid Foods, *Food Technol.*, **35**:26–29.

MOIST HEAT

The most common methods now in use for measuring resistance of microorganisms to moist heat may be roughly classified as follows:

1. Thermal death time (TDT) tube method (Bigelow and Esty, 1920)
2. Flask method (Levine et al., 1927)
3. Thermal death time (TDT) can method (American Can Company, 1943)
4. Thermoresistometer method (Stumbo, 1948)
5. Unsealed TDT tube method (Schmidt, 1950)
6. Capillary tube method (Sterns and Proctor, 1954)

These and the advantages and disadvantages of each have been discussed in detail by Stumbo (1973).

DRY HEAT

Hot gases studied as dry-heat sterilants include air, superheated steam, nitrogen, helium, oxygen, and carbon dioxide. Of these, air and superheated steam are employed most commonly. There is still a great paucity of information regarding the relative efficacy of the different gases, but what evidence there is indicates that the different gases, as germicides per se, are essentially inert, the germicidal effects, for the most part, being due to temperature alone (see Table 14-1).

As with moist heat, various methods have been employed for measuring the resistance of microorganisms to dry heat. Few of these have been described in the literature but each consists primarily of some sort of chamber in which the hot dry "inert" gases may be contained while samples of material, inoculated with test microorganisms, are introduced and exposed for predetermined times. The most sophisticated method thus far described is that of Pflug (1960).

ETHYLENE OXIDE

Ethylene oxide is germicidal when applied at ordinary temperatures; however, as shown above, its effectiveness is influenced by temperature and relative humidity of the exposure environment. Therefore, methods employed in measuring death kinetics of microorganisms exposed to ethylene oxide generally include means for controlling both temperature and relative humidity. As with hot gases, any method requires an exposure chamber in which the gas may be contained while samples of material inoculated with test microorganisms are introduced and exposed for predetermined times. Because, in the presence of moisture, hydrolysis of ethylene oxide to ethylene glycol is catalyzed by a number of metallic ions, all equipment parts in contact with the gas should be of relatively "nonreactive" materials such as glass, "stainless" metals, rubber, and Teflon. For details of experimental methods which have been used successfully in kinetic studies, see Phillips (1949), Kaye and Phillips (1949), Kaye et al. (1952), Ernst and Shull (1962), Vondell (1962), and Liu et al. (1968).

TREATMENT OF RESISTANCE DATA

There are two procedures commonly employed for treating resistance data, depending on how the data are obtained.

1. Construction of survivor curves by plotting logarithms of numbers of survivors against corresponding times at some prescribed set of exposure conditions. D values may be taken directly from such plots.
2. Assumption of the logarithmic order of death, and calculation of a D value from the "initial" number of viable cells and number of surviving cells after some given time of treatment. Here, when possible, the "initial" number of cells is taken as that number observed after cell activation or equilibration lag has occurred and the time reckoned from this point to the end of treatment. (This is sometimes referred to as the "end-point" technique.)

The first procedure is quite satisfactory as long as reasonably straight lines are obtained. If straight lines are not obtained, it is often difficult to judge which portion of the curve is describing only microbial death, or to draw a straight line that best represents death alone. A number of factors totally unrelated to microbial death may cause deviations of survivor curves from straight lines or produce survivor curves of different slopes than would be the case if death alone were being described. Among such factors are:

1. Activation, by the sterilant, of early and rapid cell multiplication or spore germination, the latter being the more common. In either case, activation causes an initial rise, or no change in observed count, before the recession of survivors becomes orderly.
2. Starting with a flocculated (clumped) cell suspension. Here, because survivor counts are estimated by numbers of colonies produced by survivors, the count may remain virtually constant until the action of the sterilant reduces the number of viable cells per clump to one; only then will the straight-line orderly reduction be observed.
3. Flocculation or deflocculation of cells during the early part of sterilant treatment. Here, a fall or rise, respectively, may be expected which cannot be attributed to lethal action of the sterilant.
4. Using a test suspension which contains two or more species or strains of different sterilant resistance. In this case, one or more "breaks" would occur in the survivor curve, each straight-line portion of which would describe, predominantly, death of one species or strain.
5. Nature or adequacy of the media used for survivor subculture and incubation time to allow repair of injured cells. Deviations of survivor curves because of cell injury short of death are known to occur; that is, the use of more favorable subculture media and extended incubation periods with some given media are known to produce survivor curves different from those that would be obtained otherwise. The most common observation is change in slope of the survivor curve, owing to the extent of damage being greater, the more severe the treatment which cells have survived. Cells suffering the greater damage would be expected to be more fastidious in their nutrient requirements for recovery and require longer under any given set of conditions to

initiate cell proliferation. It is interesting to speculate, though there is some sparse evidence, that the chief damage caused by the sterilants is to the DNA replication system, which if not too severe may be repaired when given adequate time in a nutrient environment superior to that which would be minimal for reproduction of uninjured cells.

For a more detailed discussion of causes of deviations of survivor curves from straight semilogarithmic lines see Stumbo (1973).

Some workers (Stumbo et al., 1950; Schmidt, 1957; and others) feel that the second procedure given above for treating data is the more satisfactory and practical in handling sterilant resistance data and in calculation of sterilizing processes. In this method the "initial" number is usually the viable cell count taken after a short period to allow for cell activation and for the treatment system to reach equilibrium; this may be a few seconds or several minutes, depending on the cells under study and the nature of the sterilant treatment system. To obtain the initial count, the spores may be activated in a medium in which clumping does not occur, the suspension then being serially diluted for estimate of numbers by colony count. Or, the suspension may be diluted subsequent to activation so that, on the average, less than one activated cell will be transferred to each separate volume of subculture medium. The latter allows the application of "most probable" techniques to arrive at counts by considering the ratio of subcultures positive for growth to those negative for growth.

The second point for use in calculation of the death rate parameter D is most readily arrived at in the following manner. A number of times of exposure to some given set of treatment conditions are chosen that will reduce the numbers of survivors to very low levels compared with the number of cells initially present in the population. These times should be such that subcultures, from replicate samples treated for the shortest time are all positive for growth, for the next longer time (or few times) are neither all positive nor all negative for growth, and for the next longer time (or times) are all negative for growth. When the number of viable cells in the population is so reduced by the treatment, most colonies developing in subculture will arise from single cells; or most subcultures that are positive for growth will be so by virtue of growth arising from one or very few surviving cells.

In this "end-point" technique, if a clear solid medium is used for subculture, colonies developing may be counted directly to arrive at the number of survivors. If a liquid medium, or a turbid solid medium, is used for subculture the number of survivors may be estimated by considering only those treatment times that resulted in some, but not all, of the subcultured replicate samples showing growth. A number of statistical procedures are available for estimating the most probable number of survivors according to data of this type. These procedures are all based on the premise that, though growth is obtained from only part of the replicate treated samples, each positive subculture is not necessarily positive as a result of growth from a single surviving cell; some may be positive because of growth from two, three, or more surviving cells.

Stumbo et al. (1950) applied the equation of Halvorson and Ziegler (1932) to estimate the most probable number of survivors represented by growth from some but not all treated samples. The equation follows:

$$\bar{x} = \frac{2.303}{e} \log \frac{n}{q} \tag{14-19}$$

where $\bar{x}$ = most probable number of surviving cells per unit volume of treated material
  e = volume, in chosen units, of treated sample
  n = total number of replicate samples exposed for given time to one set of treatment conditions
  q = number of replicate samples exposed, for given time to one set of conditions, that were negative for growth in subculture

When this technique is used to estimate the number after activation and the number surviving treatment time t, D may be calculated as follows:

$$D = \frac{t}{\log a - \log b} \tag{14-20}$$

where a and b are the most probable numbers of initial and surviving cells, respectively, in the population. When a clear solid subculture medium is employed, in which colony counts may be made, values of a and b to satisfy Eq. (14-20) are computed from dilution factors in the usual manner.

It is obvious from Eq. (14-19) that, when sample size is taken as unity, the value of e is 1 and $\bar{x}$ would equal the most probable number of survivors per sample. Also, if it is desired to compute the most probable number of cells in some given number of replicate samples, the right side of the equation is simply multiplied by the number of replicate samples. For example, suppose we desire to compute the most probable number per sample in 12 replicate samples,

$$\bar{x} = 2.303 \log \frac{12}{q}$$

Now, further suppose we desire to obtain the most probable number surviving in all 12 of the replicate samples,

$$\bar{x} = 2.303 \log \frac{n}{q} \, 12$$

Schmidt (1954) presented a method for treating resistance data based on a procedure for bioassay suggested by Reed (1936). As in the method just described, only the initial number a and the "end-point" number b of cells are employed to determine D values. The difference in this method and that of Stumbo et al. (1950) is in the procedure used to determine the most probable number c of cells surviving. Schmidt's method employs probability paper to estimate the time of sample exposure resulting in 50 percent of the subcultures being sterile. This is taken as the LD_{50} point. It corresponds to 0.69 surviving cells per sample. For example, if 12 replicate samples are used,

$$b = \bar{x} = 2.303 \log \frac{12}{6} = 0.69$$

Values of D are calculated by the equation

$$D = \frac{LD_{50}}{\log a - \log 0.69}$$

As shown by Schmidt (1957), D values obtained in this manner are generally in good agreement with those obtained by the method of Stumbo et al. Considering the similarity of the two procedures, this would be expected.

Other methods have been suggested for calculating D values for resistance data (see Lewis, 1956). These methods may be statistically somewhat superior to those just discussed, but in comparison, they are extremely laborious and time-consuming. It is doubtful, in view of the precision presently possible in resistance determinations, that the use of the more complicated statistical procedures is warranted. This is well indicated by the close correlation between D values calculated by the method of Stumbo et al. (1950) and other probability methods (see Reynolds et al., 1952; Schmidt, 1954; Esselen and Pflug, 1956; Lewis, 1956).

As was pointed out earlier, the "end-point" technique is preferred by some over the multiple-point technique because it tends to eliminate the description by destruction curves of occurrences other than microbial death, such as flocculation, deflocculation, etc. Moreover, it requires much less time and effort to obtain resistance parameters.

CALCULATION AND EVALUATION OF TEMPERATURE-DEPENDENT STERILIZATION PROCESSES

As stated above, in the calculation or evaluation of temperature-dependent sterilization processes, two resistance parameters are employed to describe the influence of temperature. These are D and z as defined above. D represents the rate of reaction and therefore the slope of the curve describing reduction in number of viable cells with time of exposure in a given system at constant temperature. It follows that the value of D decreases as the system temperature increases. Because most sterilization systems cannot be heated to some given temperature instantaneously, held at this temperature for some given time, then cooled instantaneously, relative resistance of microorganisms exposed to the various temperatures (usually an infinite number) must be accounted for. The parameter z, as explained earlier, is employed for this purpose. It represents the slope of the destruction curve obtained by plotting logarithms of D values against corresponding temperatures. Therefore, two equations are basic to the calculation and evaluation of sterilization processes; the equation of the survivor curve,

$$t = D (\log a - \log b) \tag{14-12}$$

and the equation of the destruction or thermochemical destruction curve,

$$\log D_2 - \log D_1 = \frac{1}{z}(T_1 - T_2) \tag{14-15}$$

UNIT OF LETHALITY AND LETHAL RATE

It is not only convenient but virtually essential to conception to express the sterilizing capacity of any temperature-dependent process in terms of its equivalent in time at some given temperature, e.g., in time at 250°F, at 300°F, or at some other designated temperature. A system operating at one temperature for a given time will cause some given degree of destruction; the longer the time, the greater will be the degree of destruction. Therefore, a unit of lethality may be chosen, such as 1 min at 250°F or at some other temperature. Taking, for example, 1 min at 250°F as the unit of lethality, the total lethal value of a sterilization process may be expressed in terms of its equivalent in minutes at 250°F. By so doing, the relative lethality of different processes may be compared directly; e.g., one process might be equivalent to 3 min at 250°F and another equivalent to 6 min at 250°F, the second process being twice as severe as the first.

When a number of different temperatures are operative in a process, in order to evaluate the effects of these different temperatures, it is necessary to express the time, at each, which is equivalent in lethal effect to 1 min at the reference temperature. The time at some other temperature equivalent to 1 min at a reference temperature is represented by the symbol Fi.

Since Fi represents time, as does D, Eq. (14-15),

$$\log D_2 - \log D_1 = \frac{1}{z}(T_1 - T_2)$$

may be rewritten to give

$$\log Fi_2 - \log Fi_1 = \frac{1}{z}(T_1 - T_2) \qquad (14\text{-}21)$$

where Fi_1 = time at some given reference temperature T_1
Fi_2 = time to cause same degree of destruction at another temperature T_2

Passing the destruction curve through 1 min at 250°F, as in Fig. 14-2, and taking Fi_1 equal to D, Eq. (14-21) becomes

$$\log Fi_2 = \frac{250 - T_2}{z} \qquad (14\text{-}22)$$

Then taking Fi_2 as time at any other temperature T, as simply Fi,

$$\log Fi = \frac{250 - T}{z}$$

or

$$Fi = \log^{-1} \frac{250 - T}{z} \qquad (14\text{-}23)$$

By Eq. (14-23) the time at any temperature occurring during the process to cause the same amount of destruction as 1 min at 250°F may be computed. If any temperature other than 250° is chosen as the reference temperature, all that is necessary is to substitute this temperature for 250 in Eq. (14-23), allowing processes evaluated to be expressed in terms of their equivalents at the chosen reference temperature. Also, if Celsuis is preferred to Fahrenheit, substitution of equivalents is all that is required. Equation (14-23) may be computer-programmed and Fi tables readily established for convenience.

From the above, it is obvious that for any given temperature there is a lethal rate involved, that is, the rate at which destruction proceeds at this temperature. It is also obvious that total lethality conferred by some given temperature operative for some given time would be the product of the lethal rate and time. For example if the lethal rate for microorganisms exposed to 240°F were 0.2 compared with 1 at 250°F, the total lethality of a 10-min process at 240°F would be equivalent to 0.2 × 10 or 2 min at 250°F.

Lethal rate represented by the symbol L is equal in value to the reciprocal of Fi. That is,

$$L = \log^{-1} \frac{T - 250}{z} \tag{14-24}$$

As in the case with Eq. (14-23), Eq. (14-24) may be computer-programmed and lethal rate tables readily established for convenience.

INTEGRATION OF LETHALITY

As mentioned earlier, it is not possible to raise the temperature of a material instantaneously to some designated temperature nor to cool it instantaneously from this temperature. Consequently, during the heating of a material to a more influential temperature and cooling from this temperature, an infinite number of less influential temperatures occur. Although less influential than the temperature reached, and where a steady state may be maintained for a period of time, the influences of these lower temperatures are usually quite appreciable and their overall contribution to the sterilizing capacity of a process must be accounted for. Accounting for these influences, in terms of lethality, requires integration. To integrate them it is most convenient to express them in their equivalents in minutes at some reference temperature, that is, in terms of L multiplied by time.

It is obvious that to account for the influences of all temperatures existent during a temperature-dependent process a curve must be established which designates all the different temperatures existent during heating, holding, and cooling of the material being sterilized. An adequate number of temperatures to establish such a curve may be obtained by pyrometer measurement employing thermocouples and a galvanometer. Very sensitive and convenient potentiometers are now available for this purpose.

Having a curve describing temperatures existent during a process and having the resistance parameters characterizing the microorganisms to be eliminated, it is a relatively simple job to determine the sterilizing capacity of a given process or to determine another with greater or lesser sterilizing capacity.

If the time-temperature curve is subject to rather simple mathematical description, as is the case for canned foods which heat either "solely" by convection or by conduction, integration to obtain total lethality may be conveniently accomplished by established mathematical procedures (see Stumbo, 1973). If, however, the time-temperature curve is not subject to simple mathematical description, which is often the case, graphical integration of lethal effects represented may be readily performed. (Graphical integration, of course, may be performed as well if the time-temperature curve is subject to mathematical description.) The graphical procedure of choice will be described here; it is a slight modification of published procedures (Bigelow et al., 1920; Schultz and Olson, 1940).

Because of its versatility the graphical integration procedure is often referred to as the "General Method." It is based on the premise that each point on the time-temperature curve represents not only a time and a temperature but a lethal rate as well. The original General Method (Bigelow et al., 1920) consisted of establishing a curve by plotting lethal rates represented, by the points on the time-temperature curve, against times at which the corresponding temperatures occurred. The area under such a lethality curve is proportional to the total lethality of the process because, in taking the area, all lethal rates represented are mutiplied by corresponding times during which they were operative, and the products are summed, or total lethality is taken as the product of lethal rate and time. Although this method is accurate, it is time-consuming because lethal rates to be used must be computed for a large number of temperatures on the time-temperature curve, in order to construct the lethality curve. This inconvenience was eliminated in the Improved General Method (Schultz and Olson, 1940) which employs specially ruled coordinate paper on which each horizontal line is positioned to represent the lethal rate corresponding to some temperature existent during the process. These lines are placed in the direction of ordinates and time lines are placed in the direction of abscissae (Fig. 14-6).

To construct lethal-rate paper, two horizontal parallel lines are drawn unit distance apart on a plain sheet of paper. It is usually convenient to draw these 10 in apart, although any distance will suffice. The bottom line is labeled zero and the top line with some temperature as high or somewhat higher than that which can occur in the sterilization process to be employed. The total distance between these two lines is taken as unity, which designates the top line as representing unit lethal rate ($L = 1$). For the purpose of demonstration, let the top line be labeled 250°F, as in Fig. 14-6, thereby designating the lethal rate at 250°F as 1 and the unit of lethality as 1 min at 250°F. Then the lethal rate of 1, if the lines are 10 in apart, represents 10 in on the vertical scale. Next locate parallel lines in between, representing lethal rates between 0 and 1. To do this, lethal rates are computed for a number of temperatures below 250°F which will likely occur during a sterilization process. First, the lethal rate for 249°F may be computed.

$$L = \log^{-1} \frac{T - 250}{z} \qquad (14\text{-}24)$$

Taking z as 18°F,

$$L = \log^{-1} \frac{249 - 250}{18} = 0.88$$

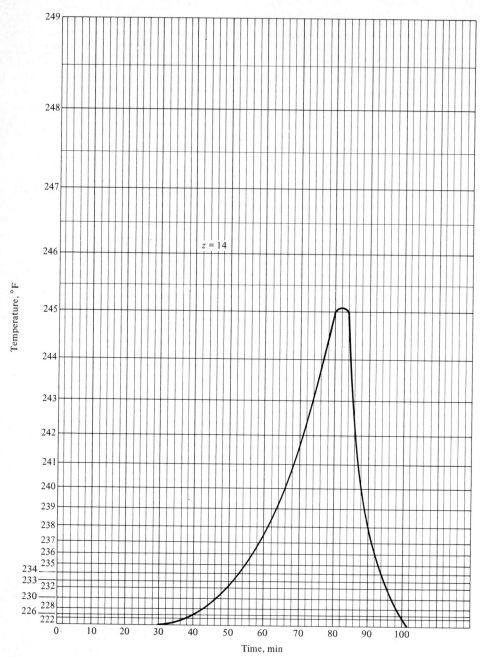

Figure 14-6 Lethality curve on specially ruled lethal-rate paper.

Since on the vertical scale of the paper a lethal rate of 1 represents 10 in, a lethal rate of 0.88 would represent 8.8 in. Accordingly, a parallel line is drawn 8.8 in above the zero line and labeled 249°F. This procedure is repeated for temperatures of 248, 247, 246 and so on until the lethal rate represented is comparatively very small. Lethal-rate paper so constructed, except for a z of 14, appears as Fig. 14-6. It is obvious, since lethal rate is a function of z, that lethal-rate paper must be constructed for the z value characterizing the resistance of the organism to be eliminated in the sterilization process.

Now, having a series of temperatures existent during the course of heating and cooling in a sterilization process, these temperatures may be plotted directly against time (on the x axis) to obtain a lethality curve; see Fig 14-6. Because of the way in which the paper was constructed, this curve actually represents a plot of lethal rates versus times, and the area beneath the curve is proportional to total lethality (lethal rate multiplied by time). The area in square inches beneath the curve is most conveniently determined by use of a planimeter, although it may be determined by excising the portion under the curve, weighing it, and dividing its weight by the weight of 1 in² of the paper.

Although for construction of lethal-rate paper it is convenient to take the top line as representing lethal rate at 250°F, once the paper is constructed the top line may be labeled with any temperature desired above that which may occur in the process, and the other lines labeled correspondingly; that is, the top line may be labeled 260°F, the next down 259°F, and so on. This is true because, for a given value of z, the ratio of lethal rates at two different temperatures is the same so long as the temperature interval is the same. For example, for $z = 18$,

$L_{250} = 1$ and $L_{249} = 0.88$
$L_{260} = 3.60$ and $L_{259} = 3.16$

and

$$\frac{1}{0.88} = 1.14 = \frac{3.60}{3.16}$$

The next step is to determine the total lethal value, represented by the lethality curve, in terms of its equivalent in minutes at some given reference temperature. This may be done with use of the following simple equation.

$$F = \frac{mA}{Fi\ d} \tag{14-25}$$

where F = total lethality in terms of its equivalent in minutes at designated reference temperature
m = number of minutes represented by 1 in on time scale
A = area under lethality curve, in²
Fi = number of minutes at the temperature used to label top line on lethal-rate paper, equivalent in lethal value to 1 min at designated reference temperature
d = number of inches from top to bottom line on lethal-rate paper

By way of example, suppose the top line of the lethal-rate paper is labeled 260°F and it is desired to determine the lethal value of a sterilization process in terms of its equivalent in minutes at a reference temperature of 250°F. The temperature curve is plotted on the lethal-rate paper and the area beneath it is found to be 6 in². The distance between the top and bottom lines on the lethal-rate paper is 10 in, the z value characterizing the resistance of the organism to be eliminated is 18°F, and the number of minutes represented by 1 in on the time scale of the lethal-rate paper is 10.

Since

$$Fi = \log^{-1} \frac{250 - T}{z}$$

$$Fi_{260} = \log^{-1} \frac{250 - 260}{18} = 0.278$$

then, $F_{250} = \dfrac{10 \times 6}{0.278 \times 10} = 21.4$

That is to say, the total lethal value of the process applied should have accomplished the same degree of sterilization as would have been accomplished if the material to be sterilized in the system had been heated instantaneously to 250°F, held at this temperature for 21.4 min, and cooled instantaneously to a nonlethal temperature, or, in the case of a gas, the lethality of which was temperature-dependent, was removed coincident with instantaneously cooling the system to a noninfluential temperature.

OUTER SURFACE STERILIZATION

This is the simplest among the many sterilization problems. We first must establish or assume some value to represent the contamination load, that is, the number of most resistant microbial cells present on the article prior to the sterilization treatment. Next we must decide upon the probability of survival we are willing to accept, that is, 1×10^{-4}, 1×10^{-12}, or whatever. It is not uncommon in routine practice, where it is impractical to determine the contamination load each time an article is to be sterilized, to use some degree of reduction, such as $5D$, $12D$, or $20D$ as the sterilization specification regardless of what the initial contamination load may be; for example, in the sterilization of low-acid canned foods a $12D$ reduction in the number of the most resistant spores of *C. botulinum* is accepted as the minimum sterilization specification.

In addition to the above, we must know or assume values of D and z characterizing the resistance of organisms to be eliminated. Again in routine practice, it is not uncommon to assume that organisms of greatest known resistance will be present, that is, to expect the worse, and to base sterilization specifications on established resistance parameters for these organisms.

In temperature-dependent sterilization of surfaces it is not uncommon to ignore temperature effects during heating of an article to and cooling it from some designated treatment temperature, considering these effects as added safety factors. In such cases, timing is from the instant the treatment temperature is reached to the instant cooling is begun. When the time required for heating and cooling is not appreciable, this is a safe and acceptable procedure.

However, when such time is appreciable, as it might well be in the sterilization of such an item as a spacecraft vehicle, it may be desirable to determine heating and cooling curves and account for lethal effects contributed during heating and cooling through integration by the General Method discussed above. Note should be made here of the heat sensitivity of materials being sterilized. If the materials are relatively insensitive, that is, not appreciably injured by heat, time becomes the factor of prime consideration. If time consumed during heating and cooling is of little importance, lethal effects conferred during heating and cooling may well be considered as added safety. If time is important or the material is injured by excessive heating, lethal effects during heating and cooling should be accounted for so as to reduce the total exposure time.

Now let us consider some of the simplest sterilization problems. First is the sterilization by moist heat of such items as glassware in the laboratory and surgical instruments in the hospital. Assume they are to be sterilized with respect to some of the more resistant bacterial spores known, e.g., spores of *Bacillus stearothermophilus*. From previous studies it is known that resistance of these spores to moist heat in a neutral pH medium is characterized by a D_{250} value of about 4.00 and a z value of about 18°F (10°C). Assuming that these values are representative, that a 4D reduction in spore population would be satisfactory, and that lethal effects contributed during heating and cooling would be negligible, sterilization time at 250°F (121.1°C) would be computed as follows:

$t = 4 \, (\log 10^4)$
$t = 16 \text{ min}$

If greater than a 4D reduction were desired, say 12D, then

$t = 4 \, (\log 10^{12})$
$t = 48 \text{ min}$

Suppose 240°F (115.6°C) were chosen as the sterilization temperature. Knowing the value of z, D_{240} may be computed, since

$$\log D_2 - \log D_1 = \frac{1}{z}(T_1 - T_2) \qquad (14\text{-}15)$$

Taking T_1 as 250°F and T_2 as 240°F,

$\log D_2 - \log 4 = \frac{1}{18}(250 - 240)$
$\log D_2 = \frac{10}{18} + 0.602$
$\log D_2 = 1.158$
$D_2 = 14.4 = D_{240}$

Then for a 4D reduction,

$t_{240} = 14.4 \, (\log 10^4) = 57.6 \text{ min}$

If it is desired to use Celsius temperatures, simply make the appropriate substitutions in Eq. (14-15). That is,

$$\log D_2 - \log 4 = \frac{1}{10}(121.1 - 115.6)$$

Now suppose the time to heat to, and cool from, the so-called sterilization temperature were for one reason or another very important or that the article to be surface sterilized were of such mass that sterilization might be accomplished without the surface temperature ever reaching what has been designated as the sterilization temperature. In this case, if accuracy were to be realized, it would be essential to obtain a curve describing temperatures from start of heating to completion of cooling. This would be done as discussed earlier. Using these temperatures, a lethality curve would be constructed on lethal-rate paper as previously described and all lethal effects summed up by taking the area under the curve. The sum of these lethal effects would then be converted by Eq. (14-25) to its equivalent in minutes at some reference temperature. In establishing a sterilization process the temperature-time curve obtained in trial may show that the process represented would give a greater or lesser degree of sterilization than desired. The process giving the desired degree of sterilization may be approximated closely by shifting the cooling portion of the lethality curve (on lethal-rate paper) to the left or right, respectively, until the desired area under the lethality curve is obtained. After this adjustment, the process, or exposure time, may be read on the x axis of the lethal-rate graph. By way of example, suppose the sterilization process desired were equivalent to 16 min at 250°F (121.1°C) and in trial it had been found that the temperature curve obtained represented the equivalent of 20 min at 250°F (121.1°C). To approximate the desired process, the portion of the lethality curve representing cooling would be shifted to the left (keeping it parallel to the original) until the area under the lethality curve, after adjustment, represented a total lethality equivalent of 16 min at 250°F (121.1°C). If such an adjustment were reasonably small, the approximation should be very close. If the adjustment were considerable it might be desirable to redetermine a temperature curve and make further appropriate adjustment.

The procedures followed in establishing surface sterilization processes employing dry heat or ethylene oxide as the sterilant are identical; only the organisms of chief concern and the resistance parameters employed would be different. A few examples will serve to illustrate. Suppose we have a number of articles to sterilize by the application of ethylene oxide. We want to accomplish a 12D reduction of *B. subtilis* spores on these articles. We found a D_{122} value of 7.7 min and a z value of 53°F as characterizing resistance at 50°C (122°F) of *B. subtilis* spores exposed to an atmosphere (88% dichlorodifluoromethane and 12% ethylene oxide) containing 700 mg of ethylene oxide per liter and conditioned to 33 percent relative humidity.

The time to accomplish a 12D reduction, assuming instantaneous attainment of equilibrium in the sterilization system and instantaneous removal of sterilant at the end of the exposure period, would be computed as follows:

$$t = D(\log a - \log b)$$

in which

$$\log a - \log b = 12$$

Then $t = 7.7 \times 12 = 92.4$ min

If a shorter time would be more convenient and a higher temperature were not deleterious to the materials, an exposure temperature of 110°C (230°F) might be employed. Since $z = 53°F$,

$$\log D_{122} - \log D_{230} = \frac{1}{53}(230 - 122)$$

$$D_{230} = 0.067$$

and $t = 0.067 \times 12 = 0.8$ min $= 48$s

It should be noted that virtually instantaneous attainment of equilibrium in a sterilization system and virtually instantaneous removal of sterilant at the end of the exposure period may be accomplished in the following manner.

1. Provide an exposure chamber which may be vacuumized or pressurized.
2. Procedure.
    a. With the materials to be sterilized in the chamber, draw a vacuum, of about 30 in mercury, on the chamber.
    b. Introduce air, nitrogen, or some other relatively inert gas which has been preconditioned to the desired temperature and relative humidity.
    c. Maintain these conditions of temperature and relative humidity until equilibrium has been reached. (Keeping the chamber atmosphere under a slight controlled pressure will aid greatly in maintaining the conditions.) The time required to attain equilibrium will depend on the temperature desired, initial temperature and heat capacity of the chamber and its contents, and on the initial moisture conditions of the chamber contents. Temperature and moisture may be checked by pyrometer and hygrometer measurements.
    d. When equilibrium has been established, again vacuumize the chamber.
    e. Immediately introduce sterilant gas, preconditioned to the desired temperature and relative humidity, and maintain at a pressure to give the desired concentration of sterilant.
    f. At the end of the prescribed exposure period, aspirate the sterilant gas.

If it were desired to use the sterilant gas also as the heating medium, and account for sterilization effects during temperature rise, the above preconditioning with an inert gas would be eliminated. In this case one would need to establish the temperature-rise curve and evaluate the sterilizing capacity of the total process by the General Method discussed earlier. (Here, the relative humidity conditioning might be assumed to parallel the temperature rise. This should be a reasonably safe assumption when only surface sterilization is considered.) Knowing the temperature-rise rate, the following procedure may be followed in establishing a process equivalent in sterilizing capacity to the 92.4 min at 50°C (122°F) determined above as required. (The same gaseous mixture will be considered and the Fahrenheit temperature scale will be employed.)

1. For a process during which the maximum temperature reached should be no more than about 131°F, take the unit of lethality as 1 min of exposure at 122°F.
2. Calculate lethal-rate values for a number of temperatures between 86 and 122°F, using the equation

$$L = \log^{-1} \frac{T - 122}{z}$$

For example,

$$L_{122} = \log^{-1} \frac{122 - 122}{53} = 1$$

$$L_{121} = \log^{-1} \frac{121 - 122}{53} = 0.957$$

and

$$L_{120} = \log^{-1} \frac{120 - 122}{53} = 0.917$$

and so on down.

3. Construct lethal-rate paper.
   a. On a plain sheet of paper rule two horizontal parallel lines 10 in apart.
   b. Consider the distance between the top and bottom lines as representing unit lethality; then, in distance, an L of 0.957 equals 9.57 in and an L of 0.917 equals 9.17 in. Accordingly, draw horizontal lines, parallel to the top and bottom lines, 9.57 and 9.17 in, respectively, above the bottom line. Continue this procedure with lethal-rate values for lower temperatures until a satisfactory horizontal grid has been prepared.
   c. Next draw vertical parallel lines intersecting the horizontal lines and unit distance apart (1 in or smaller increments may be employed). These lines describe the process time scale and should be sufficient in number to allow total process time to be represented.
4. For the process under consideration, label the top line of the lethal-rate paper 135 and the bottom line zero. Next, starting from the top, label in-between lines 134, 133, 132, and so on until all horizontal lines are labeled. (It will be noted in constructing the paper that lethal rates of the lower temperatures considered are very small and lines must be drawn so that the distance between any two of them will represent more than a 1° temperature interval, perhaps 2, 4, 8°, or more).
5. Next label the vertical lines from left to right so that 1 in represents some given time interval, in this case say 20 min.

6. Plot temperature-time process data on the lethal-rate paper to obtain a lethality curve.
7. Determine the area under the curve, in square inches.
8. Calculate the lethal value of the process in terms of its equivalent in minutes at 122°F by

$$F_{122} = \frac{mA}{Fi_{135}\,d}$$

Compute Fi_{135} as follows:

$$Fi_{135} = \log^{-1}\frac{122 - 135}{53}$$

$$Fi_{135} = 0.57$$

Then,

$$F_{122} = \frac{20 \times A}{0.57 \times 10} = \frac{2A}{0.57}$$

Suppose the area found were 30 in². Then

$$F_{122} = \frac{60}{0.57} = 105 \text{ min}$$

This is a more severe process than the equivalent of 92.4 min at 122°F desired.

9. Determine the exact exposure time to obtain the lethality desired as follows:
   a. Determine the required area under the lethality curve.

$$92.4 = \frac{2A}{0.57}$$

$$A = 26.3 \text{ in}$$

   b. Shift the vertical line on the lethal-rate paper, representing the time at which the sterilant gas is aspirated from the exposure chamber, to the left until the area under the temperature-rise curve is exactly 26.3 in.
   c. Read the proper exposure time from the intersection of this adjusted vertical line and the time scale.

If the first evaluation should show that the process applied is less severe than the equivalent of 92.4 min at 122°F desired, extrapolate the temperature-rise curve and shift the vertical line representing the time of sterilant aspiration to the right until the desired area under the curve is obtained.

MULTIPLE HARD-SURFACE ITEMS

The same procedure as employed to determine the sterilization process for a single item may be employed if the total number of resistant microbial cells on all items is taken as the a value in the equation

$$t = D(\log a - \log b)$$

that is, the *sum* of the microbial populations of all the different items is taken as a and *not* the arithmetic average population per unit area or per item. A simple example will illustrate this point and the proper procedure to follow.

Suppose we consider a number of items, or one item with a number of different components, to be sterilized by an ethylene oxide treatment. Suppose, further, that the desired ethylene oxide concentration in and the desired exposure temperature of the system may be attained virtually instantaneously and discontinued virtually instantaneously as described above. For purpose of illustration, the number of items or number of components (such as in a spacecraft vehicle) will be limited to four; the same procedure would apply if there were more than four. In this case, the Celsius temperature scale will be used, but the Fahrenheit scale could be used as well.

We have determined that all items are contaminated with a number of bacterial spores which will resist a moist heat pasteurization treatment of 80°C applied for 20 min. We have also determined or assumed that these spores exposed to an atmosphere of 12% ethylene oxide in dichlorodifluoromethane, at a pressure to give 700 mg of ethylene oxide per liter, are characterized by a D_{50} of 10 min. In addition, we have determined the following:

Item or component	Surface area and contamination	Spore load
1	600 in² 10 spores/in²	6,000
2	10,000 in² 6 spores/in²	60,000
3	300 in² 30 spores/in²	9,000
4	6000 in² 3 spores/in²	18,000
Total spore load (a) =		93,000

What would be the exposure time at 50°C under the above prescribed conditions that would result in a 10^{-4} probability of survival on all four items or components (not on any one). Designating this time as F,

$F = D_{50} (\log a - \log b)$
$F = 10 (\log 93{,}000 - \log 10^{-4})$
$F = 89.68$ min

A few further computations will show that it is the *total* number of spores on the four items and *not* the average number per item that must be taken as the value of a. (This point is emphasized because the arithmetic average is so often incorrectly taken as the total spore load.)

First let us compute the probability (p) of survival on each item or component (all exposed for 89.68 min). Then, for item 1,

$89.68 = 10(\log 6000 - \log b_1)$
$\log b_1 = -5.190$
$b_1 = 0.0646 \times 10^{-4} = p_1$

For 2,

$89.68 = 10(\log 60{,}000 - \log b_2)$
$\log b_2 = -4.190$
$\quad b_2 = 0.6460 \times 10^{-4} = p_2$

For 3,

$89.68 = 10(\log 9000 - \log b_3)$
$\log b_3 = -5.014$
$\quad b_3 = 0.0968 \times 10^{-4} = p_3$

For 4,

$89.68 = 10(\log 18{,}000 - \log b_4)$
$\log b_4 = -4.712$
$\quad b_4 = 0.194 \times 10^{-4} = p_4$

Now, since the total probability (P) of survival on all four items or components is the *sum* of probabilities on all,

$P = p_1 + p_2 + p_3 + p_4$

Entering the above numerical values, we find that $P = 1$, as it should.

The following will emphasize the error that would have incurred if the arithmetic average of the four populations had been taken as a. In this case, a would have been taken as 23,250 and

$89.68 = 10(\log 23{,}250 - \log b)$
$\quad b = 0.293 \times 10^{-4} = P =$ incorrect predicted probability (less than one-third true value)

In the above example, had the items or components not been preconditioned with respect to temperature prior to the introduction of the sterilant gas, temperature-rise data for the surface of each item or component during treatment would have been required. With these data, lethality conferred to the surface of each item or component during treatment could have been integrated by the General Method, as explained earlier, and the exposure time to obtain a 10^{-4} probability of survival on all items or components determined. For such an evaluation the z, as well as the D_{50}, value characterizing spore resistance would have been needed. (This type of evaluation is more complex but requires no procedural techniques not already discussed.)

INTERNAL SURFACE STERILIZATION

Many articles are made up of a multitude of smaller elements the surfaces of which may be contaminated, e.g., batting-filled mattresses and pillows, packets of cloth and paper items, etc. The concern is still with surface sterilization but with surfaces of the internal elements as well as the external surface of the whole article. In the sterilization of such articles, penetration of the sterilant is a major factor; the sterilant must contact all internal surfaces if it is to be effective.

Normally, the interstitial voids of such articles will be filled with air at ambient temperatures. For effective sterilization this air must be replaced by the sterilant, whether the sterilant is steam, hot air, or a chemical in the vapor phase. Interchange by diffusion at atmospheric pressure is extremely slow and unreliable. Therefore, such articles should be placed in a chamber and the air removed by vacuumization. The vacuum should then be dissipated by the introduction of the sterilant gas; to obtain the sterilant concentration desired, pressurization may be employed. For temperature-dependent sterilization, the heat capacity of the materials may be satisfied by proper venting and flushing of the chamber or by supplying an external source of heat. If the system cannot be vacuumized, extensive flushing will be required to accomplish the necessary exchange. In humidity-dependent sterilization, as with certain chemicals in the vapor phase, the water capacity of absorbent materials must be satisfied. Again, this can be accomplished by proper venting and flushing. Temperature and humidity conditions should be monitored by pyrometer and hygrometer measurements. In any event, the surface of every counterpart of a martrix-type material must be exposed to the designed sterilization conditions if satisfactory results are to be expected; the sterilant must contact the contaminant in order to be effective.

When the desired sterilization conditions are provided, sterilization efficiency may be predicted or evaluated in the same manner as dicussed above for outer surface sterilization. The contamination load (a in the equation) is the total number of contaminants located throughout the structure of the material to be sterilized. If the total contaminant load is unknown or impracticable to obtain, a given load reduction (say $10D$, $12D$, or some other) may be taken as the sterilization specification. In this case, if a fixed set of sterilization conditions can be attained virtually instantaneously and "stilled" virtually instantaneously, the sterilization exposure time may be computed as follows:

$$t = D \text{ (log of load reduction)}$$

If attainment and cessation are not virtually instantaneous, but during their course appreciable lethality is conferred, total exposure time to obtain a prescribed contaminant reduction may be determined by the General Method discussed above. In such a case, temperature rise and fall in the slowest heating portion of the pack should be the basis for evaluation, if integration of lethal effects throughout the pack is impossible or impractical. Integration of effects throughout an object will be discussed in the following section.

INTERNAL STERILIZATION OF GAS-IMPERMEABLE OBJECTS

Many objects which are impermeable to any gas may be contaminated internally, e.g., foods or other materials in hermetically sealed containers and objects formed of smaller entities by compression at temperatures inadequate to free the smaller entities of contamination. Such objects must be sterilized by the introduction of lethal energy. The necessary energy may be in the form of heat or some type of radiation. Since heat is by far the most commonly used sterilizing agent, attention here will be confined to considerations involved in internal sterilization by heat, although it should be noted that the basic principles, relating to determination and evaluation of sterilizing processes, are virtually the same in consideration of heat and radiation energies. Death of bacteria exposed to either heat or radiation energy is logarithmic.

In the case of gas-impermeable objects, heat is transmitted from the heating medium to the object at the surface of the object; it is distributed in the product by either convection or conduction. In convection heating, the product enclosed in a gas-impermeable barrier, such as a liquid in a glass or metal container, circulates during heating and cooling so that temperature throughout the material mass at any given instant during heating or cooling is virtually uniform. In practice, temperature rise and fall at the geometrical center of the material mass is taken as representative of temperature rise and fall throughout the mass; this is considered a safe and adequate approximation. Employing such a temperature rise and fall curve, all lethal effects conferred during heating, holding, and cooling of an object may be integrated by the General Method discussed above or mathematically as described by Stumbo (1973). The mathematical procedure is perhaps as accurate as the graphical and much simpler in application.

For products heated by conduction there is insufficient product movement within the material mass to aid significantly in heat distribution. Heat accepted at the object surface is transferred from particle to particle within the mass. This results in decreasing severity of heating from outer surface to geometrical center of the material mass. This in turn results in an infinite number of lethal intensities in the total mass. Then, to assess the overall lethal effect of a given heat process, an infinite number of lethal effects must be integrated. This can be done if the temperature distribution pattern in the entire mass at any given instant during heating and cooling can be determined.

Patterns, at different times during heating and cooling, can be described for conventionally shaped objects such as cylinders or bricks. They have been described for cylindrical-shaped objects (see Stumbo, 1973). Knowing these patterns, temperature rise and fall curves may be constructed describing heating and cooling at any point in a material mass. Therefore, lethality of heat at any point in the mass may be determined and expressed in terms of bacterial population reduction at that point. In turn, knowing the initial population load, the number of survivors, or better the probability of survival, at each point may be determined. The probability of survival in the entire mass will equal the sum of probabilities of survival at all points in the mass.

Perhaps this concept can best be explained by starting with the equation of the survivor curve,

$$t = (\log a - \log b)$$

(Assume that contamination is evenly distributed in the object mass.) Let us express the lethality of all heat received by the material mass of any object in terms of its equivalent in minutes at some reference temperature. Designate this by the symbol F_s. Then for population reduction in the entire mass,

$$F_s = D_r(\log a - \log b) \tag{14-26}$$

where $D_r =$ decimal reduction time at reference temperature
$a =$ initial number of microbial cells making up population of total mass
$b =$ number of surviving cells in total mass

Similarly, let us express the lethality of all heat received by any point in the mass in terms of its equivalent in minutes at the same reference temperature. Then, designating this by the symbol F_λ,

$$F_\lambda = D_r(\log a - \log b) \tag{14-27}$$

where $a =$ initial number of cells in population at any point
$b =$ number of surviving cells at any point

Since the volume of any point will be infinitesimal, it is convenient to think in terms of the fractions of the initial population which may survive rather than absolute numbers of cells, that is, b/a. Rearranging Eq. (14-26), we find that for the entire object

$$\frac{b}{a} = 10^{-F_s/D_r}$$

and, by rearranging Eq. (14-27), that for any point in the mass

$$\frac{b}{a} = 10^{-F_\lambda/D_r}$$

Obviously the fraction of the population surviving in the entire object multiplied by the volume of the object must be equal to the sum of the fractions surviving at all points multiplied by their respective volumes. If the volume of the entire object is taken as unity, then

$$10^{-F_s/D_r} \times 1 = \int_0^1 10^{-F_\lambda/D_r} dv$$

This integral has been solved to produce an equation which is readily applicable for determination and evaluation of sterilizing processes for conduction-heating foods in cylindrical containers (see Stumbo, 1953, 1973). The use equation is extremely simple and easy to apply. It should be applicable for determining and evaluating sterilizing processes for any cylindrical objects which heat by conduction.

If the above integral is insolvable because of object shape, nonuniformity of material composition regarding heat transfer, or for some other reason, the only recourse is to consider that the object, at all points, heats and cools as does the slowest heating and cooling point in it. If, in sterilization computations, a is still taken as the number of resistant microbial cells in the entire object mass, the probability of survival after the sterilization treatment will be less than that used as b in the computations. In other words, the sterilization treatment determined will be more severe than that which would be determined as necessary if integration of survival probabilities throughout the mass were possible. It should be noted that direct microbiological examination for sterility is, if not impossible, certainly impractical when remote probabilities of survival such as 10^{-4} to 10^{-12} are sought.

LITERATURE CITED

American Can Company (1943): "The Canned Food Reference Manual," p. 248, New York.
Bigelow, W. D., and J. R. Esty, (1920): Thermal Death Point in Relation to Time of Typical Thermophilic Organisms. *J. Infect. Diseases*, **27**:602
——G. S. Bohart, A. C. Richardson, and C. O. Ball (1920): Heat Penetration in Processing Canned Foods, *Natl. Canners' Assoc. Bull.* 16 L.
Blake, D. F., and C. R. Stumbo (1970): Ethylene Oxide Resistance of Microorganisms Important in the Spoilage of Acid and High-acid Foods, *Food Technol.*, **35**:26-35.
Brookes, P., and P. D. Lawley (1961): The Reaction of Mono- and Difunctional Alkylating Agents with Nucleic Acids, *Biochem. J.*, **80**:496.
Bruch, C. W. (1961): Gaseous Sterilization, *Ann. Rev. Microbil.*, **15**:245.
Collier, Charles P., and Charles T. Townsend (1956): The Resistance of Bacterial Spores to Superheated Steam, *Food Technol.*, **10**:477.
Ernst, R. R., and J. J. Shull (1962): Ethylene Oxide Gaseous Sterilization. I. Concentration and Temperature Effects, *Appl. Microbiol.*, **10**:337.
Esselen, W. B., and I. J. Pflug (1956): Thermal Resistance of Putrefactive Anaerobe No. 3679 in Vegetables in the Temperature Range of 250-290°F, *Food Technol.*, **10**:557.
Fraenkel-Conrat, H. L. (1944): Action of 1,2-Epoxides on Proteins, *J. Biol. Chem.*, **154**:227.
Fricke, H., and M. Demerec (1937): The Influence of Wave-length on Genetic Effects of X-rays, *Proc. Natl. Acad. Sci. U.S.*, **23**:230.
Gilbert, G. L., V. M. Gambill, D. R. Spiner, R. K. Hoffman, and C. R. Phillips (1964): Effect of Moisture on Ethylene Oxide Sterilization, *Appl. Microbiol.*, **12**:496.
Halvorson, H. O., and N. R. Zeigler (1932): Application of Statistics in Bacteriology, *J. Bacteriol.*, **25**:101.
Katzin, L. I., L. A. Sandholzer, and M. E. Strong (1943): Application of the Decimal Reduction Time Principle to a Study of the Resistance of Coliform Bacteria to Pasteurization, *J. Bacteriol.*, **45**:265.
Kaye, S., and C. R. Phillips (1949): The Sterilizing Action of Gaseous Ethylene Oxide. IV. The Effect of Moisture, *Am. J. Hyg.*, **50**(3):296.
——, H. F. Irminger, and C. R. Phillips (1952): The Sterilization of Penicillin and Streptomycin by Ethylene Oxide, *J. Lab. Clin. Med.*, **40**:67.
Kuzminski, L. N., G. L. Howard, and C. R. Stumbo (1969): Factors Influencing the Death Kinetics of *Clostridium botulinum* 62A on Exposure to a Dichlorodifluoromethane-Ethylene Oxide Mixture at Elevated Temperatures, *Food Technol.*, **34**:561-567.
Lawley, P. D., and P. Brookes (1968): Cytotoxicity of Alkylating Agents towards Sensitive and Resistant Strains of *Escherichia coli* in Relation to Extent and Mode of Alkylation of Cellular Macromolecules and Repair of Alkylation Lesions in Deoxyribonucleic Acids, *Biochem. J.*, **109**:433.
Levine, M., J. H. Buchanan, and G. Lease (1927): Effect of Concentration and Temperature on Germicidal Efficiency of Sodium Hydroxide, *Iowa State Coll. J. Sci.*, **1**:379.
Lewis, J. C. (1956): The Estimation of Decimal Reduction Times, *Appl. Microbiol.*, **4**:211.
Liu Tien szu, Gloria L. Howard, and C. R. Stumbo (1968): Dichlorodifluoromethane-Ethylene Oxide Mixture as a Sterilant, *Food Technol.*, **22**:86.
Michael, G. T., and C. R. Stumbo (1969): Unpublished data.

Michener, H. D., P. A. Thompson, and J. C. Lewis (1959): Search for Substance Which Will Reduce the Heat Resistance of Bacterial Spores, *Appl. Microbiol.,* **7**:166.

Pflug, I. J. (1960): Thermal Resistance of Microorganisms to Dry Heat: Design of Apparatus, Operational Problems and Preliminary Results, *Food Technol.,* **14**:483.

——(1963): Personal communication.

Phillips, C. R. (1949): The Sterilizing Action of Gaseous Ethylene Oxide. II. Sterilization of Contaminated Objects with Ethylene Oxide and Related Compounds: Time, Concentration, and Temperature Relationships, *Am. J. Hyg.,* **50**(3):280.

——and S. Kaye (1949): The Sterilizing Action of Gaseous Ethylene Oxide, I. Review, *Am. J. Hyg.,* **50**:270.

Rahn, O. (1929): The Size of Bacteria as the Cause of the Logarithmic Order of Death, *J. Gen. Physiol.,* **13**:179.

——(1945a): Physical Methods of Sterilization of Microorganisms, *Bacteriol. Rev.,* **9**:1.

——(1945b): Injury and Death of Bacteria, *Biodynamica Monogr.* 3.

Reed, L. J. (1936): Statistical Treatment of Biological Effects of Radiation, in B. M. Duggar (ed.), "Biological Effects of Radiation," vol. 1, p. 227, McGraw-Hill, New York.

Reynolds, H., A. M. Kaplan, F. B. Spencer, and H. Lichtenstein (1952): Thermal Destruction of Cameron's Putrefactive Anaerobe 3679 in Food Substrates, *Food Res.,* **17**:153.

Schmidt, C. F. (1950): A Method for Determination of Thermal Resistance of Bacterial Spores, *J. Bacteriol.,* **59**:433.

——(1954): Thermal Resistance of Microorganisms, in G. F. Reddish (ed.), "Antiseptics, Disinfectants, Fungicides and Sterilization," 1st ed., pp. 720–759, Lea & Febiger, Philadelphia.

——(1957): Thermal Resistance of Microorganisms, in G. F. Reddish (ed.), "Antiseptics, Disinfectants, Fungicides and Sterilization," 2d ed., pp. 831–884, Lea & Febiger, Philadelphia.

Schultz, O. T., and F. C. W. Olson (1940): Thermal Processing of Foods in Tin Containers. III. Recent Improvements in the General Method of Thermal Process Calculations—a Special Coordinate Paper and Methods of Converting Initial and Retort Temperatures, *Food Res.,* **5**:399.

Shull, J. J. (1963): Microbiological Aspects of Ethylene Oxide Sterilization, Paper presented at the annual meeting of the Parenteral Drug Association in New York City, Oct. 31, 1963.

Sterns, J. A., and B. E. Proctor (1954): A Micro-method: An Apparatus for Multiple Determination of Rates of Destruction of Bacteria and Bacterial Spores Subjected to Heat, *Food Technol.,* **8**:130.

Stumbo, C. R. (1948): A Technique for Studying Resistance of Bacterial Spores to Temperatures in the Higher Range, *Food Technol.,* **2**:228.

——(1953): New Procedures for Evaluating Thermal Processes for Foods in Cylindrical Containers, *Food Technol.,* **7**:309.

——(1973): "Thermobacteriology in Food Processing," 2d ed., Academic, New York.

——, J. R. Murphy, and Jeanne Cochran (1950): Nature of Thermal Death Time Curves for P.A.3679 and *Clostridium botulinum, Food Technol.,* **4**:321.

——, Larry Voris, B. G. Skags, and B. Heinemann (1964): A Procedure for Assaying Residual Nisin in Foods, *J. Food Sci.,* **29**:859.

Tanner, F. W., and G. M. Dack (1922): *Clostridium botulinum, J. Infect. Dis.,* **31**:92.

Vondell, R. M. (1962): "Studies on the Kinetics of the Bactericidal Action of Ethylene Oxide in the Vapor phase," Ph. D. dissertation, University of Massachusetts, Amherst, Mass.

SPECIES INDEX

SPECIES INDEX

The editors recognize the importance of the bacteriological, botanical, and zoological codes of nomenclature. However, because taxonomic descriptions and microbial systematics are not a primary purpose of this book, we have accepted the names of organisms as presented by each author.

Acetobacter acetigenum, 183
Acetobacter orleanse, 183
Acetobacter xylinoides, 183
Achromabacter perolens, 270
Actinomucor elegans, 190
Actinomyces olivaceus, 231, 232
Actinomyces streptomycini, 239
Aerobacillus colistinus, 64
Aerobacter aerogenes, 176
Agaricus campestris var. *bisporus*, 191
Alcaligenes recti, 342
Alcaligenes viscolactis, 269
Alcaligenes viscous var. *dissimilis*, 330
Alternaria dianthicola, 342
Alternaria humicola, 318
Artemisia absinthium, 182
Aschochyta linicola, 95
Aspergillus chevalieri, 275
Aspergillus flavus, 100, 165, 299, 342
Aspergillus glaucus, 265
Aspergillus nidulans, 223–225
Aspergillus niger, 19, 86, 149, 150, 157, 231, 238, 265, 268, 285
Aspergillus ochraceum, 91
Aspergillus oryzae, 149, 150, 157, 158, 166, 190, 238
Aspergillus sclerotiorum, 103
Aspergillus sogae, 238
Aspergillus tamarii, 100
Aspergillus terreus, 339

Bacillus brevis, 66
Bacillus caucasi, 166
Bacillus cereus, 165, 262, 271, 292, 298, 299, 342
Bacillus circulans, 230, 263
Bacillus coagulans, 262, 274, 426, 427
Bacillus coagulans var. *thermoacidurans*, 264, 270
Bacillus macerans, 264
Bacillus megaterium, 259, 269
Bacillus methanicus, 385
Bacillus mycoides, 342

Bacillus polymyxa, 65, 264, 268, 294, 425, 427
Bacillus sphaericus, 97, 99, 342
Bacillus stearothermophilus, 262, 264, 270, 274, 425, 439
Bacillus subtilis, 13, 14, 40–43, 64, 112, 147, 149–151, 161, 262, 264, 265, 268, 290, 417, 424–426, 440
Biscula pinocola, 318
Bispora effusca, 318
Bordetella bronchiseptica, 13
Botryodiplodia malorum, 342
Botrytis cinera, 342, 385
Brucella abortus, 286
Brucella melitensis, 286
Brucella suis, 286
Byssochlamys fulva, 271, 275

Calonectria decora, 95, 96, 101
Camarosporium ambiens, 318
Candida lipolytica, 401
Candida tropicalis, 20
Candida utilis, 265
Cephalosporium acremonium, 234, 235
Cephalosporium carpogenum, 342
Ceratocystis pilifera, 317
Chaetomium elatum, 317
Chaetomium fumicola, 317
Chaetomium globusum, 317, 339
Chlamydomonas eugametos, 237
Cladosporium sphaerospermum, 342
Claviceps purpurea, 299
Clostridium botulinum, 258, 259, 262, 265–267, 275, 276, 284–289, 420, 425–427, 438, 440
Clostridium butyricum, 262, 266, 268, 269
Clostridium histolyticus, 150, 269
Clostridium nigrificans, 266, 270
Clostridium pasteurianum, 266, 271
Clostridium perfringens, 257, 262, 265–267, 274, 285, 286, 289, 290, 374
Clostridium putrefaciens, 269

Clostridium sporogenes, 266, 269, 270, 274, 276, 285
Clostridium thermosaccharolyticum, 262, 266, 270
Clostridium welchii, 374
Coniophora cerebella, 316, 317, 323
Coniophora puteana, 316
Coniothyrium fuckelii, 318
Corticium evolvans, 316, 322
Corticium galactinum, 316
Corynebacterium diphtheriae, 241
Coxiella burnetii, 286, 300, 301
Curvularia lunata, 90–92, 95
Cylindrocarpon radicicola, 100

Echinodontium tinctorium, 216, 217
Emericellopsis glabra, 237
Endomycopsis fibuliger, 268
Entamoeba histolytica, 286, 374
Enterobacter aerogenes, 330, 375–377
Enterobacter cloacae, 268, 375
Escherichia aurescens, 375
Escherichia coli, 13, 23, 26, 125, 176, 195, 203, 204, 209, 211–215, 218, 220, 235, 237, 243, 263, 265, 267, 271, 285, 292, 294, 296, 297, 372, 375–377
Escherichia freundii, 375
Escherichia intermedia, 375
Euglena gracilis, 17

Flammula connisans, 316
Flammula spumosa, 322
Flavobacterium dehydrogens, 102
Flavobacterium invisibile, 342
Flavobacterium marinum, 342, 344, 345
Flavobacterium proteus, 176
Fomes annosus, 316, 322, 324
Fomes ignarius, 316
Fomes laricis, 316
Fomes nigricans, 316
Fomes nigrolimitatus, 316
Fomes pini, 316, 317
Fomes roseus, 316
Fursarium culmorum, 101
Fursarium flocciferum, 342
Fursarium javanicum, 100
Fursarium oxysporum, 238
Fursarium sporotrichiella, 299

Ganoderma appalanatum, 316

Gonyaulax catenella, 299
Gonyaulax tamarensis, 299
Gymnodinium brevis, 299

Halobacterium cutirubrum, 269
Helicosporium aureum, 318
Helminthosporium speciferum, 342
Hymenochaete corrugata, 322
Hypomyces haematococeus, 100
Hypoxylon cohaerans, 316

Irpex fusco-violaceus, 316

Klebsiella aerogenes, 263, 265, 268, 271, 292
Klebsiella pneumoniae, 13, 25, 36, 281

Lactobacillus acidophilus, 51, 167, 187
Lactobacillus brevis, 271
Lactobacillus bulgaricus, 166, 167, 187, 262, 268
Lactobacillus casei, 16–19, 271
Lactobacillus citrovorum, 167
Lactobacillus delbruckii, 13, 185
Lactobacillus lactis, 7–9
Lactobacillus leischmannia, 17, 51
Lactobacillus pastorianus, 176
Lactobacillus plantarum, 13, 17, 54, 56, 58, 167
Lactobacillus thermophilus, 269
Lactobacillus viridescens, 17, 265
Lentinus lepideus, 316, 322, 324
Lenzites saepiaria, 316, 317, 322, 324
Lenzites trabea, 316, 317
Leuconostoc citrovorum, 167, 176, 187
Leuconostoc mesenteroides, 111, 113, 167

Margarinomyces fasciculatis, 324
Merulius himantioides, 316
Methanomonas methanica, 385
Microbacterium lacticum, 269
Micrococcus albus, 342
Micrococcus candidus, 342
Micrococcus glutamicus, 125
Micrococcus lysodeikticus, 150
Micrococcus radiodurans, 275, 284, 285
Micrococcus roseus, 265, 268
Micrococcus ureae, 342

SPECIES INDEX / **455**

Mucor griseocyamas, 95
Mucor mucedo, 269
Mucor pusillus, 157
Mucor racemosus, 269
Mycobacterium tuberculosis, 65, 257, 286

Neurospora crassa, 17, 221–224, 234
Neurospora sitophila, 268
Nocardia corallina, 103, 396
Nocardia mediterranea, 65
Nocardia salmonicolor, 396

Ochromonas malhamensis, 17, 51
Odontia bicolor, 316
Oidium aurantiacum, 268
Omphalia campenella, 316
Ophiostoma caerulescens, 317
Ophiostoma picae, 317
Ophiostoma pini, 317

Paecilomyces varioti, 342
Paxillus panuoides, 322
Pediococcus acidilactici, 189
Pediococcus cerevisiae, 18, 19, 189
Penicillium chrysogenum, 73, 100, 197, 225, 229, 238, 242–244, 246, 249
Penicillium expansum, 268
Penicillium lilacinum, 100
Penicillium notatum, 2, 234, 285
Penicillium oxalicum, 342
Penicillium roqueforti, 324, 334
Penicillium rubrum, 299, 324
Penicillium stoloniferum, 197
Peniophora abietinus, 316
Peniophora gigantea, 316, 322
Peniophora luma, 316
Peniophora septrionalis, 316
Peniophora velutina, 316
Phialophora richardsiae, 318
Pholiota adiposa, 316
Phoma glomerata, 342
Photobacterium fisheri, 21
Pleurotus ostreatus, 316
Polyporus abietinus, 316, 317, 322, 323
Polyporus ancepts, 317
Polyporus asseus, 316
Polyporus balsameus, 316, 317
Polyporus betulina, 316
Polyporus chioneus, 316
Polyporus circinatus, 316, 317

Polyporus hirsutus, 316
Polyporus paragamenus, 316, 317
Polyporus schweinitzii, 316, 317
Polyporus zonatus, 316
Polystictus versicolor, 317
Poria asiatica, 316
Poria cocos, 316
Poria monticola, 316
Poria subacida, 316
Poria tsugina, 316
Poria vaporia, 316
Proteus melanovogenes, 270
Proteus vulgaris, 112, 271, 292
Pseudomonas aeruginosa, 341
Pseudomonas fluorescens, 230, 236, 237, 265, 267, 270
Pseudomonas fragi, 269
Pseudomonas graveolens, 270
Pseudomonas methanica, 396
Pseudomonas mucidolens, 270
Pseudomonas multivorans, 236
Pseudomonas oleovorans, 396
Pseudomonas testosteroni, 99, 100
Pullularia pullulans, 324, 342, 343, 345, 350

Redicoccus cerevisiae, 176
Rhizopus arrhizus, 86, 88, 90
Rhizopus nigricans, 79, 86–88, 90, 91, 95, 96, 101, 268
Rhizopus oligosporous, 167, 190
Rhizopus oryzae, 167

Saccharomyces carlsbergensis, 17, 166, 175, 177
Saccharomyces cerevisiae, 17, 51, 150, 163, 166, 175, 177, 182, 183, 187, 222, 223, 285
Saccharomyces cerevisiae var. *ellipsoideus*, 166, 182
Saccharomyces fragilis, 150
Saccharomyces microsporus, 237
Saccharomyces rouxii, 166, 265, 277
Saccharomyces saké, 166, 183
Salmonella binza, 285
Salmonella enteritidis, 285, 294
Salmonella gallinarum, 25
Salmonella heidelberg, 294
Salmonella manhattan, 275
Salmonella meleagridis, 275
Salmonella montevidea, 275

Salmonella newport, 265
Salmonella oranienburg, 265, 275, 285
Salmonella pullorum, 275
Salmonella senftenberg, 275, 285
Salmonella thompson, 285
Salmonella typhi, 293
Salmonella typhimurium, 218, 237, 263, 267, 275, 285, 294, 295
Salmonella worthington, 275
Sarcina flava, 342
Sarcina lutea, 13
Schizophyllum commune, 316
Sepedonium chrysospermum, 96
Septomyxa affinis, 97, 98, 100
Serratia marcescens, 268, 271
Shigella boydii, 296
Shigella dysenteriae, 296
Shigella flexneri, 296
Shigella sonnei, 296
Sphaerotilus natans, 330, 331
Sporedonema sebi, 265
Sporotrichum sulfurescents, 96
Stemphylium consortiale, 342
Stereum abierinum, 316
Stereum chailletii, 316
Stereum frustulosum, 317
Stereum hirsutum, 316
Stereum purpureum, 316
Stereum rugosusculum, 316
Stereum sanguinolentum, 316, 322, 324
Stereum subpileatum, 317
Stereum sulcatum, 316
Staphylococcus aureus, 13, 14, 23–28, 36, 40, 43, 71, 165, 237, 261–265, 267, 271, 275, 285, 286, 291–293
Streptococcus bovis, 379, 380
Streptococcus cremoris, 167, 187, 268
Streptococcus durans, 379
Streptococcus equinus, 379, 380
Streptococcus faecalis, 13, 15–19, 265–268, 275, 297, 379, 380
Streptococcus faecalis var. *liquefaciens*, 379
Streptococcus faecalis var. *zymogenes*, 379
Streptococcus faecium, 265, 285
Streptococcus hemolyticum, 150
Streptococcus lactis, 166, 167, 187, 188, 268, 292, 381

Streptococcus salivarious, 372, 380
Streptococcus thermophilus, 167, 187, 262, 269
Streptomyces antibioticus, 240
Streptomyces aureofaciens, 70, 72–74, 221, 229, 239–241
Streptomyces coelicolor, 219, 220, 240, 241
Streptomyces erythreus, 64, 239, 240
Streptomyces fradiae, 229, 239, 240
Streptomyces griseoflaveus, 240
Streptomyces griseus, 9, 221, 228, 239, 240
Streptomyces nodosus, 64
Streptomyces olivaceus, 221
Streptomyces orientalis, 66
Streptomyces rimosus, 72, 73, 75, 239–241
Streptomyces roseochromogenes, 85, 93, 94
Streptomyces scabies, 240
Streptomyces venezuelae, 64
Streptomyces violaceoruber, 220, 240, 241
Streptomyces viridifaciens, 10
Streptomyces viridochromogenes, 240

Thamnidium elegans, 269
Thermobacterium acidophilus, 51
Thiobacilus dentrificans, 387
Torula nigra, 342
Trametes sirialis, 316, 322
Trechispora brinkmannii, 316
Trechispora raduloides, 316
Trichoderma viridae, 150, 157, 324
Trichosporium heteromorphum, 318, 324
Trichosporon pullalans, 265
Trichurus terrophilus, 318

Ustilago maydis, 238

Vibrio cholerae, 286
Vibrio metschnikovi, 265
Vibrio parahaemolyticus, 298

Xanthamonas campestris, 392
Xeromyces bisporus, 265, 275

SUBJECT INDEX

SUBJECT INDEX

Page numbers in **boldface** type indicate reference is within a table or illustration.

Acid measuring methods:
  pH, 20, 22
  titrimetric, 20, 22
Agar diffusion, 62
Alcoholic beverages:
  classes, 184
  flavor substances, 186
  microorganisms, 185
  substrates, 185, 186
Amino acid assay:
  automated method, 114, 115
  broth culture method, 112, 113
  inhibition reversal method, **112**
  lactic acid bacteria, 15, 16
  manometric method, 114
  microbial selection for, 107–109
  plate method, 110, 111
  Streptococcus faecalis, **15**
  titrimetric method, 113, 114
  turbidimetric method, 113
Amino acids:
  alanine, 107
  arginine, 15, 114, 119
  aspartic acid, 114, 115, 125
  auxotrophs, 115, 116
  culture mutation, 115
  culture stability, 126
  cysteine, 119
  cystine, 119
  essential, 1, 15
  glutamic acid, 15, 106, 109, 114, 116, 118, 119, 122
  histidine, 15, 114
  homoserine, 115, 116, 125, 126
  isoleucine, 15, 115, 125
  leusine, 15, 125
  lysine, 15, 106, 107, 111, 113–116, 119, 122, 125, 126
  metabolic control mechanisms, 125, 126
  methionine, 15, 104, 109, 115, 125, 126
  microbial synthesis, 106–**110**
  nutritional requirements: carbon sources, 117, **118**
    nitrogen sources, 118, 119
  ornithine, 114, 116
  phenylalanine, 112, 116

Amino acids:
  production: aeration, 122–125
    pH factor, 119, 120
    temperature, 122
  proline, 116
  serine, 119
  threonine, 15, 107, 109, 115, 116, 119, 125, 126
  tryptophan, 15, 107, 109, 119
  tyrosine, 114
  valine, 15, 116, 125
Amylases, 130, 135
Analytical microbiology:
  introduction to, 13–15
  test organism selection, 13, 14
Anemia, pernicious, 6, **7**, **8**
Antibiotics:
  agricultural uses of, 68
  amphotericin B, 20, 67, 69
  assays: buffers in, 29, 30
    concentration and turbidity of, 26, 27
    diffusion method, 31–35, 40–43
    diffusion and turbidimetric methods, comparison of, 34, 35
    dose-response lines, 25, 26, **27**, **28**
    by leakage of salts, 21, 22
    media, 44, 45
    pH, 29, 30
    mode of action, 20
    penicillin G, standard curve, **28**
    petri dish method, 43, 44
    photometric method, 21, 22, **36**, 37–40
    serial dilution method, 35
    standard line, 27, 38
    turbidimetric method, 34, 35
      automation, 35
      impurities, 34, 35
  bacitracin, 67–69
  broad spectrum, 61, 67
  capriomycin, 29
  cephalosporins, 14, 26, 29, 36, 61, 69, 230, 234, 235
  chloramphenicol, 20, 26, 36, 61, 67, 69
  chlortetracycline, 61, 68, 70, 72, 74, 75, 77, 229, 239
  clindamycin, 67

Antibiotics:
  clinically effective, **64**, 65, 66
  colistin, 67, 69
  concentration and turbidity equations, 26, 27, **28**
  definition, 60, 61
  demeclocycline, 70, 72
  dose response lines, 25–27
  doxycycline, 70
  effectiveness and limitations, 63, 67, 68
  erythromycin, 14, 20, 26, 29, 36, 67, 69, 76, 239
  gentamycin, 67, 69
  griseofulvin, 67, 69
  historical development, 61
  hygromycin B, 29
  industrial importance, 70
  industrial production, 71–74, 76, 77
  intermediate-spectrum, 67
  in vitro test systems, 62
  kanamycin, 67, 69
  lincomycin, 67–69
  methacycline, 70
  microbiological assays for, 20
  minocycline, 70, 71
  mode of action, 20, 68, 69
  moenomycin, 68
  monensin, 68
  narrow-spectrum, 67, 68
  neomycin, 20, 36, 229
  new: clinical development, 77
    preclinical development, 77
    screening methods, 61–63
  nonmedical uses, 68
  novobiocin, 69
  nystatin, 67, 69, 246
  oxytetracycline, 68, 70, 72, 73, 75, 239, 229
  patulin, 22
  penicillin, 2, 4, 14, 20, 22, 26, 28, 29, 36, 38, 60, 61, 63, 67–69, 71, 73, 75, 198, 227–230, 233, 234, 238, 243, 246, 249, 251, 252
  penicillinase, 36
  polymyxin, 67, 69
  rifamycin, 69
  semisynthetic, 61
  spectinomycin, 69
  streptomycin, 20, 22, 29, 36, 61, 67–69, 76, 228, 231
  tetracycline, 20, 26, 36, 61, 67, 69, 70, 72, 73, 75–77, 228, 229, 236
    physicochemical properties, **71**

Antibiotics:
  tylosin, 14, 20, 26, 29, 36, 68
  tyrothricin, 22, 61, 67, 69
  vancomycin, 14, 20, 67, 69
Anticodon, 197, 202
Antimetabolites, 112
Antimutagenic agents, 209
Ascorbic acid, 53
Asexuality, 222
Auxotrophy, 115, 116, 246, 247

Bacteriophage, 211, 214, 216
Baker's yeast, 21
Barley, 168
Basic research:
  application, 4, 6–8
  definition, 1, 2
  discipline, 5
  efficiency, 9, 10
  experimentation, 4–7
  hypothesis, 3
  objectives, 3
  organization, 6
  physical parameters, 6, 7, 11
  principles, 3, 6, 9, 11
Beer:
  aging, 172
  components, 167
  fusel oils, 173
  pasteurization, 172
  types, 177
  of yeast, 177
Botulism, 287–289
Brewing process:
  beer and lager, 170–173
  biochemistry, 172–174
  historical development, 166, 167
  liquid adjuncts, 177
  microbiology, 173–177
  recent trends, 177
  yeast, 172–177
Broth dilution, 62

Calibration lines:
  antibiotic, 28, 29
  computer calculation, 28, 29
  inverted logarithmic, 28, 29
Casein hydrolysates, 52
Cell mass, 23
Cheese:
  classification, 189

Cheese:
  microorganisms, 188
  production, 188, 189
Chemotherapy, historical development of, 60, 61
Chromosome, 214
Cistron, 193
Codon, 197, 202, 203
Coenzyme A:
  pantothenic acid assay, 54, **55**, 56
  structural form, **55**
Coenzymes, 146, 147
  vitamin assays, 48, 49, **50**, 51, 52
Conductivity bridge, measuring salt leakage, 22
Conjugation, 211–213
Crossing over, 219
Cyanocobalamin, 6, 7, 53

Deoxyribonucleic acid (DNA), 193, 195, 198, 219, 250
  functions, 197
  helix, 194
  polymerase, 200
  replication transcription, 199, 200
  tautomers, 205
Deoxyribosides, 51
Derepression, 204
Diffusion assays, 40–43
  antibiotic, 31–35
  computation of potency, 33
  cup plate method, **31, 32**
  dose-response line, 32–34
  equations, 31, 32
  measuring response, 33
  petri-dish method, 43, 44
  plate method, mechanics, 40–43
  validity, 33, 34
Dikaryons, 221
Diploids, 221, 223
Distillation, alcoholic beverages, 184–186

Ecosystem:
  buffering, 365–371
  carbondioxide, 361
  metabolism, 359–361
  nuisance organisms, 362, 363
  nutrients, 362, 363
  oxygen, 359–361
  temperature, 363–365
  toxicity studies, 357, 358

Ecosystem:
  turbidity, 361
Enzyme action:
  equations, 141
  factors, 143–148
Enzyme activators, 146, 147
Enzyme assays, 148, 149
Enzyme commission, 140
Enzyme inhibitors, 147, 148
Enzyme purification, 149
Enzymes:
  chemical nature, 137–140
  classes, 140–142
  functions, 141, 143
  industrial applications, 129–137
  intracellular, 163, 164
  microbial production: cultivation methods, 151, 152
    industrial methods, 156–162
    laboratory methods, 152–156
  mutants, 149–151
  reaction specificity, 138
  sources, 128, 129, 131–134
  substrate specificity, 138
  uses, 131–134
  in vitamin assays, 48, 49, **50**
Eosomes, 212
Episomes, 212, 215
Ethylene oxide:
  resistance to measuring methods, 428
  resistance of microorganisms, 424–428
Eukaryons, 221
Experimental design, 1–3, 6, 9

Fact discovery, 2
Feedback mechanism, 204, 234
Fertility factor, 214
Folic acid, 18, 20, **21**
Folic acid complex, response of test bacteria, 18, **19**, **21**
Food poisoning, 286–291, 295–301
  algal, 299, 300
  Bacillus cereus, 298, 299
  Clostridium perfringens, 289, 290
  Escherichia coli, 296, 297
  food composition, 266, 267
  mycotoxins, 299
  planktonic, 299, 300
  rickettsial, 300, 301
  salmonellosis, 293–295
  shigellae, 295, 296
  staphylococci, 291–293

Food poisoning:
  streptococci, 297, 298
  Vibrio parahaemolyticus, 298
  viral, 300, 301
Food preservation:
  aseptic maintenance, 258
  chemical inhibitors, 276–281
  dehydration, water activity equations, 263–265
  drying, 258, 263
  filtration, 258, 286
  heat methods, 258, 272–275
  radiation, 282–286
Food spoilage:
  aerobic, 264, 266, 271
  anaerobic, 264–266, 271
  microaerophilic, 264–266, 271
  microorganisms in selected foods, 268–271
  oxygen as factor, 264–266
Foods:
  fermentation: defined, 165
    microorganisms, **166**, 187–191
    nonyeasts, **167**
  fermentation in: bread, 187
    cheese, 188, 189
    cocoa, 189
    dairy products, 187, 188
  microbial growth: equations, 259–261
    temperature coefficient equations, 262, 263
    temperature factor, 261–263

Gene, 205
Glucose oxidase, 136, 137
Growth substance assays, 16, **17**, **18**

Haploids, 221
Heterokaryons, 223, 225–238, 248, 249
Homokaryons, 229
Hybridization, 220, 239
Hydrocarbons:
  fermentation products, 398–406
  microbial oxidation, 392–398
Hypothesis, 6

Induction, 204, 233
Inhibition end product, 203
Isoenzymes, 139

Lactic acid microbes, 13, 15, 18, 20, **21**, 22, 52, 53
Lactobacilli in vitamin assays, 49, 51–54, 56, **58**
Leakage of salts, 14, 21, 22
Lysogeny, 211, 216, 220, 241, 243

Malt, 168, 169
Malting process, 167, 168, **169**, 170
Meiosis, 222
Michaelis constant, 141, 142
Microbiological assay:
  acid as response, 20
  for amino acids, 14, 15
  for antibiotics, 14
  by antiluminescence, 22
  birth of, 48
  carbon dioxide as response, 20
  cell multiplication, 25, 26, 35
  cell sizes, 23, 24, 26
  errors, 24
  organisms used, 13, 14
  pH as response, 20, **21**, 22
  photometer calibration, 25, 26
  responses, 14, 15
  trace elements, **19**
  turbidity measuring, 22–24
Mitosis, 210, 222, 223
Mutagenic agents, 208, 227, 233, 239, 244, 245, 249
  acridines, 209
  aminopurines, 209
  anthracenes, 245
  butadiene, 245
  ethyl methane sulfonate, 245
  hydroxylamine, 209
  N-methyl-*N'*-nitro-*N*-nitrosoguanidine (NG), 209, 229, 245
  nitrogen mustard, 245
  nitroso methylmethane, 245
  nitrous acid, 245
  triethylene melamine, 245
  UV, 208, 238
Mutation, 205, 207, 208, 221, 225–231, 233, 236, 244–246, 252
  mutant types, 232, 236, 247, 248, 250
  rate, 210, 248

Neosomes, 203

Oleic acid, 52, 57, 58
Oxygen tension, 7

Paint:
  biodeterioration, 340–344
  microbicides, 350–353
  microorganism control, 346–350
Paint binders biodeterioration, 344–346
Paint films microflora, 341–344
Pantothenic acid assay, 54, 56, **57**, 58
Paper, biodeterioration, 325–329
    additives, 337–340
    microbial slime control, 330–337
Papermaking:
  paper mill, 313–315
  pulp mill, 310, 312, 313
  schematic flow diagram, 311
  wood handling, 310
Parasexuality, 223–226, 237, 238, 246
Patternization, 198, 200, 203
Peptide chains, 139
Petroleum:
  microbial corrosion, 387
  microbial deterioration: asphalt, 387, 388
    domestic fuel oils, 388, 389
    jet aircraft fuels, 389, 390
    oil recovery, 391, 392
    oil spills, 390, 391
    petrochemical spills, 390, 391
Petroleum microbiology, history, 385, 386
Phagicin, 243
Photobacteria, 15
Photometers, calibration, 25, 26
Photometric assay, 21, 22, **36**, 37–40
  inoculum, 37
  mechanics, 38, 39
  penicillin G, 36–40
  response and answers, 39, 40
  sample preparation, 37, 38
  standard solutions, 37
  test organism, 37
Photoreactivation, 208
Polyploidy, 223
Polysomes, 198, 202
Prontosil, 60
Proteases, 135, 136
Prototrophic, 248
Pulp, biodeterioration, 320–324
    cellulose loss, 320–322
    discoloration, 323, 324
    production loss, 324, 325

Pulp, biodeterioration:
    strength loss, 323
Pulpwood:
    biodeterioration, 315–320
    wood rots, 306, 317, 318, 320
Pyrrolnitrins, 230, 236

Recombination, 210, 214, 220, 221, 236, 237, 239–241, 248, 250, 252
  parasexual, 238
  sexual, 210
  somatic, 24
  syncytic, 220, 238
Repression, 204, 232–234, 236
  catabolite, 235
Resistance selection, 248
  transfer factor, 216
Respirometric method, 20
Reversion, 236
Ribonucleic acid (RNA), 1, 193, 195–197
  messenger (mRNA), 195, 196, 198, 199, 202, 203
  polymerase, 200
  ribosomal (rRNA), 195, 197, 199
  tautomers, 205
  transfer (tRNA), 195, 196, 199, 202, 203
  viral (vRNA), 194
Ribosomes, 198, 202

Salvarsan, 60
Serial dilution assay, 35
Sexduction, 211, 215, 218
Single-cell protein:
  composition, 407–409
  microorganisms, 400–405
  petroleum substrates, 405, 406
  process, 406, 407
Statalon, 197
Sterilization:
  evaluation of temperature-dependent processes, 432–438
  gas, 446–449
  heat: microbial death equations, 413–415
    requirements, 419–423
    resistance of microorganisms, 423–425, 428
    thermal destruction curves, 415–419
  internal surface treatments, 445, 446
  outer surface treatments, 438–445
  treating resistance data, 429–432

Steriod processes, analytical methods, 83, 84
Steroid production, economics of, 84, 85
Steroids:
  C_{18} and C_{19} production, 100
  cortisone, 79
  dehydrogenation: enzymes, 99, 100
    microorganisms, 96–99
  9 α-fluorocortisol 16-hydroxylation, 93, 94
  hydroxylation, enzymes, 94–96
  Kendall's compound, 79
  microbial production, 100, 101
  progesterone, 89–93
    11 α-hydroxylation, 85–93
  structures, 80–83
Strain selection, 226
  development, 227, 228, 230, 231, 233, 248, 251, 252
  variants, 227, 228, 233
Streptomycete isolates, 61
Sulfanilamide, 60
Syntrophisms, 226

Titration assay, 20
Trace metal assays, **19**
Transcription, 200
Transduction, 211, 217, 218, 221, 239
Transformation, 211, 218, 220, 221
Transition, 206, 208
Translation, 199
Transversion, 206
Turbidimetric assay:
  automated, 35
  flow birefringence error, 24, 25
Typhus, transmission of, 4, 5

Vinegar:
  acetic acid in, 184
  microorganisms in, 183, 184
  types of, 183
Vitamin assay:
  development, 48
  dose-response line, 30, 31
  enzyme factor, 48, 49, **50**
  errors, 30
  media, 52, 53
  as microchemical technique, 54
  microorganisms, 49, 51–54, 56, **58**
  organisms used, 16, **17**, 18, **19**
  of pantothenic acid, 54, 56, **57**, 58

Vitamin assay:
  procedure selection, 53, 54
  response specificity, 51, 52
  standard curve, 30, 31
  turbidimetric, 30
  vitamin B_{12}, 30
Vitamins:
  B complex, 47, 48
  B_{12}, 6–9, 49, 51
  biotin, 53, 57, 108, 109, 115
  calcium pantothenate, 56
  chemistry and occurrence, 53, 54
  deficiency, 49
  definition, 47
  discovery, 47, 48
  fat soluble, 49
  nicotinic acid, 48, 49
  pantothenic acid, 125
  pyridoxine, 48
  riboflavin, 14, 17, 48
  thiamine, 48, 51
  water soluble, 49

Water:
  acidity, 367, 368
  alkalinity, 368
  bacteria, 371–373
  bacterial counts, 373
  bµffering, 365–367
  carbon dioxide content, 361
  coliform bacteria, 375–379
  fecal streptococci, 379, 380
  nutrients, 363
  oxygen content, 359–361
  pH of, 368, 369
  solids, 369–371
  temperature, 363–365
  turbidity, 361
Water pollution:
  biological aspects, 356, 357
  indicators for, 371–382
  sociological aspects, 355, 356
  toxic substances, 357, 358
Wine:
  champagne, 182
  nongrape sources, 183
  production process, **179**, 180, 181
  sweet table (dessert), 181, 182
Wine production:
  must composition, 180
  yeast, 178, 182, 183

Wort, 167

Yeast:
   in beer fermentation, **166**, 167–169, 172–177
   in distillation of alcoholic beverages, 184, 185
   in food fermentation, **166**, 187, 188, 190

Yeast:
   in wine fermentation, **166**, 178, 180, 182, 183
Yeast autolysate, 109
Yeast enzyme system, 48
Yeast hydrolysate, 109

Zygote, 214